Legumes Research - Volume 2

Edited by Jose C. Jimenez-Lopez and Alfonso Clemente

Published in London, United Kingdom

IntechOpen

Supporting open minds since 2005

Legumes Research - Volume 2
http://dx.doi.org/10.5772/intechopen.104821
Edited by Jose C. Jimenez-Lopez and Alfonso Clemente

Contributors
Durai Pandian, K.E. Poorani, H. Abirami, M.B. Anusha, Silvina Mónica Alvarez, Gabriel Giezi Boldrini, Glenda Daniela Martin Molinero, María Verónica Pérez Chaca, Nidia Noemí Gómez, Motunrayo Ganiyat Akande, Toshihiro Nakamori, Mitsutaka Kohno, Peter Joseph Shaba, Dickson A. Musa, Evans C. Egwim, A. Uthman, Mahmut Camlica, Gulsum Yaldiz, Mayra Nicolás-García, Cristian Jiménez-Martínez, Madeleine Perucini-Avendaño, Brenda Hildeliza Camacho-Díaz, Antonio Ruperto Jiménez-Aparicio, Gloria Davila-Ortiz, Krishna Kumar Rai, Nagendra Rai, Shashi Pandey-Rai, Retno Indrati, Fateme Hekmatpour, Mansour Torfi Mozanzadeh, Elisha Onyango, Saima Parveen, Imran Pasha, Amina Jamil, Farah Ahmad, Carlo Giuseppe Rizzello, Michela Verni, Erica Pontonio, Marco Montemurro, Hai Ha Pham Thi, Thanh Luan Nguyen, Ndenum Suzzy Shitta, Alex Chukwudi Edemodu, Abush Abebe Tesfaye, Wosene Gebreselassie Abtew, Zainab Muhammad Bello, Sanusi Muhammad, Adamu Aliyu Aliero, Ibrahim Aliyu Dabai

First published in London, United Kingdom, 2022 by IntechOpen
IntechOpen is the global imprint of INTECHOPEN LIMITED, registered in England and Wales, registration number: 11086078, 5 Princes Gate Court, London, SW7 2QJ, United Kingdom
Printed in Croatia

British Library Cataloguing-in-Publication Data
A catalogue record for this book is available from the British Library

Additional hard and PDF copies can be obtained from orders@intechopen.com

Legumes Research - Volume 2
Edited by Jose C. Jimenez-Lopez and Alfonso Clemente
p. cm.
Print ISBN 978-1-80356-914-7
Online ISBN 978-1-80356-915-4
eBook (PDF) ISBN 978-1-80356-916-1

For EU product safety concerns:
IN TECH d.o.o., Prolaz Marije Krucifikse Kozulić 3, 51000 Rijeka, Croatia, info@intechopen.com or visit our website at intechopen.com.

Meet the editors

Dr. Jose C. Jimenez-Lopez obtained a BS in Biochemistry in 1998, a BS in Biological Sciences in 2001, and an MS in Agricultural Sciences in 2004, all from the University of Granada, Spain. He obtained a Ph.D. in Plant Cell Biology from the Spanish National Research Council (CSIC) in 2008. He was a postdoctoral research associate at Purdue University, USA (2008–2011), and a Marie Curie research fellow (2012–2015) at the University of Western Australia and CSIC. Currently, he is a tenured scientist at CSIC, working in the functionality, health benefits, and molecular allergy aspects of seed proteins from crop species of agro-industrial interest (mainly legumes). He is an author of more than 70 journal articles, 30 book chapters, and 2 patents. He has contributed to more than 130 international congresses. Dr. Jimenez-Lopez is an active member of different scientific societies, an editorial board member for international journals, and an editor of multiple books.

Dr. Alfonso Clemente is a senior scientist and the director of the Zaidín Experimental Station (EEZ), Spanish National Research Council (CSIC), Spain. He has carried out postdoctoral stays at the Institute of Food Research (1999–2000), John Innes Centre (2001–2002), and Sainsbury Laboratory (2003–2004) in the United Kingdom. An expert in gastrointestinal health, Dr. Clemente investigates the beneficial influence of dietary components resistant to the digestive process, using several in vitro and in vivo approaches. He has led numerous research projects with collaborations at national and international levels. During 2015–2021, Dr. Clemente was president of the Spanish Legume Association (AEL), coordinating numerous national and international activities and scientific meetings. He has been a board member of the Spanish Society of Microbiota, Probiotics, and Prebiotics (SEMiPyP) since 2016.

Contents

Preface

Legumes, with their outstanding nutritional and nutraceutical properties, are an important and cost-effective source of high-quality proteins (20%–50% of seed content) for the human diet. They encompass the necessary genetic diversity to cope with different environmental stresses that threaten agriculture and food security. Soon, humanity will face many global changes, many of them exacerbated by climate change, such as global food security, the production of waste, and the release of greenhouse gases into the atmosphere. Strategies are needed to increase sustainable sources of protein to feed the growing population while promoting recycling and a circular economy.

In this context, legumes may be key crops to meet the needs for sources of plant-based proteins for humans and livestock at an affordable cost. Legume proteins play an important role in nutrition as well as overall health. Nutraceutical aspects of a variety of legume seed compounds are being investigated, particularly their anti-inflammatory effects, which are helpful for ameliorating and treating diseases such as cardiovascular disease, type II diabetes, obesity, metabolic syndrome, cancer, and so on. Legume proteins are storage proteins belonging to different families, including Vicilin (7S globulin), Legumin (11S globulin), 2S albumin, glutelins, enzymes, enzyme inhibitors, and lectins, which integrate part of the defensive mechanism of the seed. These positive properties of legume proteins may be a consequence of their unique structure and functional features, raised properties of derived peptides from their hydrolysis, or even modification of legume proteins for improved functionality and digestibility.

Legume proteins from pea, lentil, lupine, chickpea, and other types of beans are valuable for functional food production, that is, producing foods with improved nutritional and technological properties. They show a wide range of techno-functional characteristics such as emulsification and stability activity, foam formation and stabilization, gel formation, and water holding capacity. These technological characteristics are fundamental for the final effect that food containing legume proteins will have on human health. Knowing the advantages and potential disadvantages of legume seed proteins for food making will allow industrial applications of these proteins to be more easily achieved.

This book provides an overview of the health benefits, functional properties, and industrial applications of legume seed compounds.

Jose C. Jimenez-Lopez
Department of Biochemistry, Cell and Molecular Biology of Plants,
Estacion Experimental del Zaidin,
Spanish National Research Council (CSIC),
Granada, Spain

Alfonso Clemente
Department of Physiology and Biochemistry of the Animal Nutrition,
Estacion Experimental del Zaidin,
Spanish National Research Council (CSIC),
Granada, Spain

Chapter 1

Legumes, Sustainable Alternative Protein Sources for Aquafeeds

Fateme Hekmatpour and Mansour Torfi Mozanzadeh

Abstract

Aquaculture produce a great portion of aquatic derived proteins for human in the world. It has the highest and the fastest growth rate among the protein producing industries. Fish meal (FM) is the main and the most expensive ingredient for aquafeeds production. It provides protein, essential amino acids, energy, minerals and vitamins in aquafeeds. Given the current rapid development of aquaculture industry the competition for limited global supplies of FM may reduce its availability and elevate its price. Thus, finding high quality, economic and environmentally friendly alternative protein sources (APS) for aquafeeds production is vital for sustainability of the aquaculture industry. Among various APS, legumes have been proved to be promising APS because they have medium protein content with suitable amino acid profile, high digestible protein and energy levels, and appropriate minerals and vitamins for the most cultured aquatic species. They also are cost-effective and highly accessible. However, they contain various anti-nutritional factors that may reduce feed palatability and may negatively affect growth and health of cultured aquatic animal species. This chapter provide information regarding legumes and their derivatives as APS, their nutritional quality and their potential drawbacks. In addition, strategies for increasing the efficiency of legumes in aquafeeds are reviewed and discussed.

Keywords: additives, anti-nutritional factors, aquaculture, nutrients digestibility, essential amino acids

1. Introduction

The aquafeed market is estimated to account for USD 50.6 billion in 2020 and with compound annual growth rate of 7.2%, it is projected to reach USD 71.6 billion by 2025 [1]. Two main factors amplify such a lucrative revenue in aquafeed market including increase in global seafood consumption (122% from 1990 to 2018) and fast grow rate in aquaculture production (527% from 1990 to 2018) [2]. In fact, the annual growth rate of aquaculture industry was about 10% during the 1990s and about 5.8% annually from 2000 to 2018 indicating aquaculture is the fastest growing food production sector in the world [3]. According to Food and Agriculture Organization of the United Nations [2] the aquaculture accounted for 52% (54.3 million tonnes (MT)) of global fish production that means this industry supplied 17% of total animal proteins for the global population [2]. About 70% of aquaculture production relied on the aquafeeds, which is the main expenditure (~ 40%–70% of the total expenses) and the largest input in this industry [4, 5]. Thus, the development of the aquafeed industry along with the improvement of feed efficiency are prerequisite to achieve the projected aquaculture production. On the other hand, aquafeed production industries mainly depend on marine-derived ingredients namely fish meal (FM) and fish oil (FO), but increasing demands for these marine-derived feedstuffs along with overexploitation and/or static tendency in capture fisheries of small pelagic fish resulted in uncertain supply and the inflation of their prices that adversely affect the profitability margins of aquaculture [6, 7]. It has been reported that about 70% to 80% of all produced FM is used in aquafeed industry [3]. Thus, seeking out environmentally friendly and economic alternative ingredients for substitution of FM and FO in aquafeeds formulation is a fundamental goal for aquaculture sustainability [7].

An alternative feedstuff for FM should possess some properties such as high availability, commercial competitive cost as well as ease of shipping and storage [8]. In addition, by considering nutritional aspects, an alternative protein source (APS) should have low quantities of fiber (< 6%), carbohydrates including starch (< 20%) and non-starch polysaccharides (NSP)(< 8%), anti-nutritional factors (ANF), high digestible protein level (≥ 48%), appropriate essential amino acid (EAA) profile (arginine > 3, lysine > 3.5, methionine > 1.5, threonine > 2.2% of total AA profile) and suitable palatability [6, 8]. On the other hand, it has been predicted that the application of FM in aquafeeds will be dropped to under 10% in some most popular aquaculture species such as omnivorous (*e.g.* carp, catfish and tilapia, 1-2% FM) and carnivorous fish species (*e.g.* marine fish, salmon and trout, 5-10% FM) [9]. Such tremendous shift from FM to APS in aquafeeds has resulted in new challenges in performance, feed utilization, welfare, and final product quality of cultured aquatic animal species [6].

Among alternative plant protein sources (APPS), legumes are the most abundant protein rich ingredients for applying in aquafeeds [10]. However, these APPS have several drawbacks such as a wide range of protein contents, EAA imbalances or inadequacies (*e.g.* sulfur amino acids including cysteine, methionine and taurine), low bioavailability of some minerals or microelements as they bounded with phytic acid (*e.g.* phosphorous, iodine, calcium and selenium) and the presence of high amounts of various ANF (*e.g.* antitrypsin factors, phytates, saponins, and polyphenols), which consequently impose some challenges in their use in aquafeeds formulations [8, 11]. Although some processed protein derivatives of legumes such as soy protein concentrate, soy protein isolate or pea protein isolate contain high protein levels (65-90%) and have low ANF concentrations, but they are too expensive to be used in most aquafeeds. In addition, the agriculture industry also restricted to develop production of these APPS without putting extra stress on land, water, and phosphorous resources. In this chapter it was aimed to highlight the opportunities and challenges in application of legumes as APPS and providing some strategies for enhancing their efficiency in aquafeeds.

2. Legumes, alternative protein sources for aquafeeds

The family Fabaceae (Leguminosae) or commonly named legumes are the third largest family of flowering plants with almost 770 genera and over 19,500 species [12, 13]. The largest and the most important economically subfamily of the legumes, are the Faboideae, which are the source of primary crops including dry seeds (*e.g.* lentils, broad beans, beans and peas), flavoring plants (*e.g.* carob and lupins), fodder plants (*e.g.* alfalfa) and oilseeds (*e.g.* peanut or groundnut and soybeans). The seeds are the most important part of the legumes that used as ingredients for aquafeeds manufacturing. In this section, it has been tried to introduce the most important legumes used in the aquaculture industry. In 2020, the global production of some legumes such as peanut, dry pea, chick pea, cow pea, lentil and lupins were 48.757, 48.75, 14.184, 14.246, 8.786, 5.734 and 1.01 MT, respectively [14].

2.1 Soybean (*Glycine max*)

Soybean (*Glycine max*) is the leading oilseed crop in the world and occupies the first place as a global APS for FM due to its large amounts of protein (~ 40%) and reasonable EAA profile as well as its most availability in the global market [8]. Moreover, owing to genetic engineering technologies and plant biotechnology available for the crop, the output traits of soybean also improved for aquaculture purposes [15]. The global production of soybean was estimated to be 336.7 MT in 2020 and projected to reach to 371.3 MT in 2030 with annual growth rate of 1.8% from 2020 to 2030. A huge amount of this crop used for oil extraction that yield a cake with high levels of protein that will be processed to a wide range of soy products including soy flour, soybean meal (SBM), full-fat SBM, soy protein concentrate (SPC), soy protein isolate (SPI), soy protein hydrolysate (SPH) and fermented SBM (FSBM).

2.2 Soybean meal

Unprocessed SBM contains 30-50% protein, about 30% carbohydrates including oligosaccharides (*e.g.* stachyose and raffinose), starch and NSP (*e.g.* cellulose, hemicellulose, and pectins; **Table 1**) [16–18]. Except for cystine, the amounts of the EAA (mainly lysine, methionine and threonine; **Table 2**), taurine and tyrosine in SBM are generally lower than FM [8]. The protein level and EAA content can be enhanced by chemical processing of soy flakes to SPC and SPI or through hydrolysation and fermentation of SBM. In addition, high levels of ANF in SBM can be reduced by conventional processing (*e.g.* heat, autoclaving, and use of solvents).

2.3 Soy protein concentrate

Soy protein concentrate is commonly obtained by fractionation of SBM through aqueous alcohol that not only enhance protein content (~ 70%) [19–21] by extracting carbohydrates but also remove ANF (*e.g.* saponin), soluble oligosaccharides and fiber [21–23] that eventually increase its palatability and protein digestibility [24]. Soy protein concentrate has been getting more attention in aquafeed industry because of its well-balanced AA profile compared to other APPS [23, 25].

2.4 Soy protein isolate

Soy protein isolate is produced by refinement of SBM through a series of aqueous extractions with different pH levels that increase its protein content over 78% [26]. As, the processing of SPI does not have the aqueous alcohol extraction step, the saponin content of SPI (~0.8%) is higher than that in SPC (~0%) [27, 28]. Because of high protein content in SPI, the inclusion levels of this ingredient in aquafeeds could be lower than SBM and SPC that eventually reduce the total ANF input in aquafeed. On the other hand, it has been confirmed that the nutrients digestibility of SPI was higher than that of SPC, when it was replaced by 40% of FM in feed for rainbow trout (*Oncorhynchus mykiss*) [29].

Protein sources	Bioactive component																		
	Oxidized and polymerized lipids	Histamine and putrescine	Nitosamines	Protease/trypsin inhibitors	Amylase inhibitors	Saponins	Phytate	Glucosinolates	Sterols	Oxalate	Anti-vitamins	Polyphenols/Tannins	Lectins	Phytoestrogens	Alkaloids	Cyanogens	NSP	Oligosaccharide	Antigenic proteins
Fish meal	●	●	●																
Soybean	●			●		●	●	●	●		●		●	●			●	●	●
Faba bean				●		●	●		●			●	●						●
Lupin				●		●	●		●			●		●	●	●	●	●	
Peas				●	●	●	●		●	●	●	●	●			●		●	
Pea nut				●	●		●		●			●			●				
Alfalfa						●			●		●	●		●	●				
carob seed germ meal				●		●			●			●							
Soy/Lupin/Pea/Faba/Alfalfa PC						●	●		●			●							
Soy/Lupin/Pea PI						●													
Novel varieties				●	●	●	●		●		●						●	●	●

PC: protein concentrate; PI: Protein Isolate.

Table 1.
Bioactive component present in legumes and means of alleviation.

DOI: http://dx.doi.org/10.5772/intechopen.99778

Protein sources	Protein content(%)	Lys(CS)	Met(CS)	Cys	Arg(CS)	Trp	Leu(CS)	Ile(CS)	Phe	His	Thr(CS)	Tyr	Val	Arg/Lys (CS)	Met/Lys (CS)
							g.100 g^{-1} protein								
Fish meal	70	7.5	2.7	0.8	6.2	0.6	7.2	4.2	3.8	2.4	4.1	2.7	4.9	0.83	0.36
Faba bean	29.0	6.3 (16↓)	0.8 (70.4↓)	1.2	9.0 (45.2↑)	0.3	7.1 (1.4↓)	4.1 (2.4↓)	4.0	2.6	3.5 (14.6↓)	1.2	4.6	1.43 (72.3↑)	0.13 (64↓)
Faba bean PC	63	6.63 (11.6↓)	0.71 (73.7↓)	1.1	8.68 (40↑)	1	7.57(	4.1 (2.4↓)	4.35	2.5	3.49 (14.9↓)	3.35	4.48	1.31 (57.8↑)	0.11 (69.4↓)
Lupin (*Lupinnus angustifolius*)	33.8	4.7 (37.3↓)	0.7 (74.1↓)	1.5	11.0 (77.4↑)		6.9 (4.2↓)	4.2(=)	4.0	2.7	3.4 (17.1↓)	3.6	3.9	2.34 (182↑)	0.15 (58.3↓)
dehulled lupin (*Lupinus albus*) meal	42	5(33.3↓)	0.8 (70.4↓)	1.6	11.3 (82.3↑)		7.3 (1.4↑)	4.2(=)	3.9	2.3	3.8 (7.3↓)	4.8	4	2.26 (172.3↑)	0.16 (55.6↓)
Lupin PI	61	3.44 (54.1↓)	0.6 (77.8↓)	0.33	9.02 (45.5↑)		5.25 (27.1↓)	2.46 (41.4↓)	2.95	1.97	2.62 (36.1↓)	3.11	2.29	2.62 (215.7↑)	0.17 (52.8↓)
Fermented Lupin	40	5.57 (25.7↓)	0.82 (69.6↓)	1.4	9.75 (57.3↑)	0.83	7.3 (1.4↑)	4.6 (9.5↑)	4.25	3.3	4.1(=)	4.25	4.75	1.75 (110.8↑)	0.15 (58.3↓)
Pea seed	23.9	7.2(4↓)	1.0(63↓)	1.4	8.4 (35.5↑)		7.1 (1.4↓)	4.2(=)	4.7	2.5	3.8 (7.3↓)	3.1	4.8	1.17(41↑)	0.14 (61.1↓)
Filed pea PC	46.5	7.61 (1.5↑)	0.9 (66.7↓)	1.25	8.32 (34.2↑)	0.99	7.31 (1.5↓)	4.26 (1.4↑)	4.95	2.39	3.51 (14.4↓)	3.31	4.73	1.1 (32.5↑)	0.12 (66.7↓)
Pea PI	80	6(20↓)	0.78 (71.1↓)	0.25	7.4 (19.4↑)		7.13(1↓)	3(28.6↓)	4.75	2	3.13 (23.7↓)	3.25	3.5	1.23 (48↑)	0.78 (116.7↑)
Bambara nut		8(6.7↑)	0.64 (76.3↓)	2.41	7.48 (20.6↑)	0.6	10.2 (41.7↑)	5.45 (29.6↑)	7.69	3.86	4.43(8↑)	3.13	6.24	0.94 (13↑)	0.08 (77.8↓)
Full fat soy bean meal	36	6.3(16↓)	1.29 (52.2↓)		7.44 (20↑)	1.44	7.08 (1.7↓)	5.31 (26.4↑)	5.2	2.58	4.18(2↑)		4.97	1.18(42↑)	0.20 (44.4↓)
Soy bean (expeller)	43.5–49.3	6.3(16↓)	1.4 (48.2↓)	1.6	7.5(21↑)	1.2	7.7 (6.9↑)	4.6 (9.5↑)	5.1	2.7	3.7 (9.8↓)	3.5	4.5	1.19 (43.4↑)	0.22 (38.9↓)

Protein sources	Protein content(%)	Lys(CS)	Met(CS)	Cys	Arg(CS)	Trp	Leu(CS)	Ile(CS)	Phe	His	Thr(CS)	Tyr	Val	Arg/Lys (CS)	Met/Lys (CS)
	g.100 g^{-1} protein														
Soybean (dehulled)	53.5	6.3(16↓)	1.4 (48.2↓)	1.6	7.3 (17.7↑)	1.4	7.7 (6.9↑)	4.6 (9.5↑)	5.1	2.7	3.8 (7.3↓)	3.5	4.8	1.16 (39.8↑)	0.22 (39.8↓)
Soy PC	67–72	6.3(16↓)	1.3 (51.9↓)	1.25	7.3 (17.7↑)	1.5	7.9 (9.7↑)	4.6 (9.5↑)	5.1	2.6	4.3 (4.9↑)	3.5	4.8	1.16 (39.8↑)	0.21 (41.7↓)
Soy PI	90–92	6.0 (20↓)	1.0(63↓)		7.2 (16.1↑)	1.2	7.8 (8.3↑)	4.5 (7.1↑)	5.2	2.5	3.5 (14.6↓)		4.6	1.2 (44.6↑)	0.17 (52.8↓)
Fermented soy bean	48.9	6.2 (17.3↓)	1.6 (40.74↓)		7.8 (25.8↑)	1.23	8.2 (13.9↑)	4.91 (16.9↑)	5.5	2.66	4.1(=)		4.1	1.3 (56.6↑)	0.26 (37.8↓)
Pea nut meal (corticated-decorticated)	32–46.5	2.29 (69.5↓)	0.53 (80.4↓)		5.28 (14.8↓)	0.48	3.18 (55.8↓)	1.80 (57↓)	2.37	1.26	1.40 (65.9↓)		2.08	2.3(177↑)	0.23 (36.1↓)
Pea nut meal expeller (corticated-decorticated)	34.1–46.5	2.46 (67.2↓)	0.43 (84.1↓)		5.01 (19.2↓)	0.37	3.03 (57.9↓)	1.58 (62.4↓)	2.32	1.05	1.27 (69↓)		2.06	3.43 (313↑)	0.29 (19.4↓)
(Chloroplastic) alfalfa leaf PC	53.22	4.57 (39.1↓)	0.3 (88.9↓)		5(19.3↓)	4	8(11.1↑)	4.57 (8.8↑)	5.2	2.4	4.5 (9.8↑)		5.2	1.1 (32.5↑)	0.07 (80.6↓)
Cytoplasmic alfalfa leaf PC	69.24	5.3 (29.3↓)	0.35 (87↓)		5.8 (6.4↓)	4	9(25↑)	5.25 (25↑)	5.88	2.97	5.2 (26.8↑)		5.8	1.1 (32.5↑)	0.07 (80.6↓)
carob seed germ meal	34.8	6.49 (13.5↓)	1.41 (47.8↓)		14.1 (127↑)		7.1	3.88 (7.6↓)	3.64	3.5	4.14(1↑)		4.31	2.17 (161.4↑)	0.22 (38.9↓)

PC: protein concentrate; PI: Protein Isolate; Lys: lysine; Met: methionine; Cys: cysteine; Arg: arginine; Trp: tryptophan; Leu: leucine; Ile: Isoleucine; Phe: phenylalanine; His: histidine; Thr: threonine; Tyr: tyrosine; Val: valine.

Table 2.
Amino acid content of legumes products in comparison to fish meal.

2.5 Soy protein hydrolysates

Soy protein hydrolysate is produced by restricted enzymatic hydrolysis of soy products that consequently improves their nutritional and practical characteristics [30]. The enzymes generally use for hydrolysis of soy protein are mainly endo and exopeptidases (*e.g.* leucine aminopeptidase) that derived from fermentation of selected strains of bacteria (*e.g. Bacillus licheniformis*) [30]. The most important characteristics of SPH are the maximized digestibility of protein, excellent protein solubility and minimized ANF that ultimately enhance its efficiency. Significant solubility of the SPH is mainly due to the formation of short chain hydrophilic polypeptides and the elimination of insoluble fractions through a sedimentation step by a centrifugation process [31]. In addition, SPH has numerous bioactive low molecular weights peptides with health improving properties such as antioxidative and immunostimulatory compounds [32–34].

2.6 Fermented soybean meal

The fermentation process in aquaculture usually uses for enhancing the nutritional value and reducing ANF in alternative protein sources for incorporating into aquafeeds [35]. Fermentation of SBM by yeast (*e.g. Saccharomyces cerevisiae*), fungus (*e.g. Aspergillus niger*) or bacterial strains (*e.g. Bacillus spp., L. plantarum P8, Pediococcus acidilacticstrains*) can improve its nutritional quality and digestibility by providing low molecular weight peptides, increasing bioavailability of minerals and reducing its ANF (*e.g. trypsin inhibitors*) [35–37]. Moreover, fermented SBM has more protein content (~10%) than SBM with negligible change of its EAA profile [37]. In addition, fermented SBM provide probiotic characteristics and can increase efficiency of aquafeeds by elevating trypsin and fibrinolytic enzymes activities [35].

2.7 Transgenic soybean

"Genetically modified plants are typically created by the addition or deletion of existing innate genes in the plant's own genome or transferring external non-host genome through DNA splicing" [38]. Genetic approaches were applied for creating a prototype soybean that synthetize and accumulate a n-3 long chain polyunsaturated fatty acid (*i.e.* eicosapentaenoic acid, EPA) and a carotenoid (*i.e.* astaxanthin) in the seed [15]. Soybean contains very low levels of lutein (10 μg/g seed), but the expression of the phytoene synthase gene in transgenic soybean increases the accumulation of ß-carotene up to 800 μg/g seed and significantly reduces lutein content (~29%). The expression of fatty acid elongases and Δ5 desaturase in transgenic soybean increase the synthesis of EPA up to 5%, but EPA content need to be improve in transgenic soybean to better reflect FO fatty acid profile [15]. It should be mentioned that the annual sales of astaxanthin reaches to over USD 200 million and inclusion of such trait in transgenic soybean can impressively enhance the attractiveness of such product especially for incorporating in aquafeeds for salmon, trout and shrimp [39, 40].

2.8 Pea protein

Pea (*Pisum sativum*) is another promising APPS for aquaculture species with highly digestible protein and energy levels and it is a good source of digestible starch (~40-50%) [41, 42]. This legume contains low levels of ANF (*e.g.* tannins) and does not have trypsin inhibitors, but contain high levels of saponins. The protein (~21-25%) and methionine contents in peas also lower compared to soybean. Pea protein concentrate (PPC) is a pea derived product that produced by fine grinding dehulled peas and air processing to remove fiber and carbohydrates. The PPC contains higher protein and lower ANF compared to unprocessed pea meal, thus it is more suitable APPS for aquafeeds. Like other plant protein derived products, extrusion and micronizing processes improve protein and energy digestibility of pea meals [41].

2.9 Peanut

Peanut (*Arachis hypogaea*) is the fourth largest oilseed crop in the world and peanut pulp that remain after the oil extraction can be used as an APPS in aquafeeds [43, 44]. Peanut meal (PNM) is a residue after solvent extraction of whole shelled peanuts and considered as a great APPS due to its higher protein content (~47.8%) than SBM, higher palatability and the same cost as SBM [45, 46]. However, the protein quality of PNM is inferior compared to SBM and it contains lower levels of lysine and methionine than SBM, but a higher level of arginine [43, 45]. Lysine and methionine deficiencies in PNM can be met by using crystalline amino acids.

2.10 Lupines

Lupines include many legumes species with a considerable protein (~ 35%) but low lipid (~8–10%) levels [21, 47]. About 80% of lupines species the can be used as feed ingredients, particularly *Lupinus angustifolius,* is produced in Australia. Among different lupines, Andean lupin (*L. mutabilis*) seed contain ~50% protein (dry matter) and its derivatives such as dehulled, deoiled and lupine protein concentrate contain higher protein content (~61%) [48]. Although lupines have low levels of lysine and sulfur amino acids, they contain more arginine content than soybean [49]. Four species of lupines including *L. angustifolius, L. albus, L. luteus and L. mutabilis* named as "sweet lupins" as they contains low levels of alkaloids and because of their high protein contents they have great potential as APPS [50]. However, using lupins in aquafeeds are still limited because of their low protein digestibility and the presence of various ANF [51].

2.11 Faba bean

Faba bean, (*Vicia faba* L.) is a legume with high amount of protein (~ 20 to 41%), carbohydrate (~51% to 68%), B-vitamins and minerals depending on its variety [52, 53]. Its protein composition is mainly consisted of albumins (20%) and globulins (80%) and rich in glutamic and aspartic acids. But, the levels of sulfur amino acids and tryptophan residues are low [54]. The main carbohydrates in faba been are starch (~41–53%), low molecular weight carbohydrates (*e.g.* raffinose , stachyose, and verbascose), and fiber mainly hemicellulose [52, 55]. Faba bean contains some ANF such as trypsin inhibitors and lectins, condensed tannin, phytic acid, vicine and convicine [56]. Processing of faba bean protein does provide ingredients with higher protein contents such as faba bean protein concentrate (~55% crude protein) and faba bean isolate (~80% crude protein) that contain lower levels of ANF [57–59].

2.12 Other protein sources

Carob seed (*Ceratonia siliqua*) germ does have a high protein content (~45–50% crude protein) and it is cheaper than SBM [60]. Carob seed germ meal is produced from the germ of the carob seed after the separation of the gums and the fibrous [61–63]. However, it contains high levels of tannins [64].

Alfalfa (*Medicago sativa*) protein concentrate is another APPS that produced by pressing fresh alfalfa foliage (mainly leaves and stems) to make a protein-rich juice which is centrifuged and heated to fractionate proteins from the juice [65]. This byproduct contains reasonable protein level (~52% crude protein) with high amounts of lysine, threonine, and methionine. It also contains high levels of vitamins and antioxidants such as carotenoids, but low content of fiber and ANF (*e.g.* phytic acid or lectins) [65, 66].

3. Anti-nutritional factors in legumes

The ANF are defined as compounds that disturb feed utilization and can affect the health condition and production of livestock [67]. Legumes contain various ANF such as saponins, tannins, phytic acid, gossypol, lectins, protease inhibitors, amylase inhibitors, antivitamin factors, metal binding ingredients, goitrogens, etc. (**Table 1**) that combine with nutrients and reduce bioavailability of them in aquafeeds [8]. Some ANF such as protease inhibitors and phytates abate digestibility of proteins and energy as well as reduce mineral absorption that consequently results in malnutrition and microelements deficiencies.

The ANF can be divided into four classes [66]:

I. Substances that affect dietary protein utilization (*e.g.* protease inhibitors, tannins and lectins).

II. Substances that influence dietary mineral utilization (*e.g.* phytates, gossypol, oxalates and glucosinolates)

III. Antivitamins

IV. Miscellaneous (*e.g.* non-starch polysaccharides, mycotoxins, alkaloids, pyrimidine glycosides, phytoestrogens and saponins).

4. Improvement of legumes efficiency in aquafeeds

Digestibility of an ingredient is a pivotal parameter for determining its potential for use in the aquafeeds [68]. In order to validate the nutritional quality of a feedstuff, determination of apparent digestibility coefficient (ADC) of its dry matter and nutrients is necessary. As previously mentioned, generally legumes contain high amounts of starch and NSP and ANF [66] that negatively affect ADC in most fish and shrimp species. Carnivorous fish species are more susceptible to legumes. The ADC of crude protein of legumes are generally over 0.80, indicating high quality of protein provided by these APS. However, the ADC of gross energy in these research showed great fluctuations from 0.5 to 0.7 [69]. It has been reported that the ADC of legumes in diet mainly depends on fish species. Thus, ADC of legumes in omnivorous species such as Nile tilapia is higher than carnivorous fish such as rainbow trout [69].

Several strategies were applied for improving nutrients digestibility in legumes such as processing techniques (*e.g.* dehulling, soaking, extrusion cooking, fermentation etc.), using novel and new variety of plant protein sources (*e.g.* transgenic legumes), nutritional programming and selective breeding of fish to be more adapted to legumes in aquafeeds, modulation of gut microbiota (*e.g.* probiotics and short chain fatty acids), and inclusion of additives (*e.g.* acidifiers, CAA, phospholipids etc.) in aquafeeds [53, 70]. Here the most efficient strategies for reducing ANF in APPS were described:

4.1 Conventional strategies

Several physical processing strategies applied for removing, inactivating or reducing ANF (*e.g.* trypsin inhibitors, glucosinolates, tannins and saponins) contents in APPS including heat and/or soaking in water, dehulling and germination, roasting or autoclaving as well as extrusion and micronizing (infrared heat) [71] (**Table 3**). These conventional methods positively improve digestibility of legumes; however, these strategies are not conclusive in eradicating the adverse influences of ANF in legumes [72]. Moreover, some strategies such as heat damage lead to loss of some amino acids and adversely affect quality of proteins and carbohydrates through Malliard reactions [73, 74]. Furthermore, soaking in water may result in leaching water-soluble nutrients by this process.

4.2 Exogenous enzymes and phytase

It has been confirmed that inclusion of carbohydrase exogenous enzymes such as xylanae, ß-glucanase and cellulase as well as phytase in aquafeeds can reduce the negative effects of NSP and phytate on digestion [75, 76]. Exogenous carbohydrases by facilitating carbohydrate digestion and reducing feed polymerization degree is going to decrease its viscosity and liberate carbohydrate oligomers [77]. In addition, carbohydrases by neutralizing NSP can increase the digestibility of energy, macronutrients and bioavailability of minerals because NSP reduce accessibility of enzymes to substrates and there is a relationship between phytate and NSP in PPS [75, 78]. In addition, carbohydrases may improve host's gut health by supporting the propagation of beneficial microbiota in the gut that can facilitate fermentation of NSP and consequently increase the amounts of organic acids and especially short chain fatty acids production [78, 79].

4.3 Acidifiers

A plethora of studies confirmed that high amounts of dietary FM could be substituted with APPS by supplementing diet with short-chain fatty acids and acidifiers [80, 81]. In fact, acidification of plant protein based aquafeeds with acidifiers increase the bioavailability of minerals and trace elements and they neutralize or alleviate the negative impacts of ANF on nutrients digestibility [60, 82–84]. In addition, acidifiers by reducing the chyme pH through the gut can induce the pepsin activity [85]. Moreover, reduction of the chyme pH triggers the release of gastrointestinal hormones (*e.g.* secretin and cholecystokinin) that stimulate secretion of pancreatic digestive enzymes, which in turn elevates the digestibility of protein and minerals. Furthermore, it has been reported that acidifiers by controlling the appetite through the parasympathetic nerve system including orexigenic neurotransmitters that increase feed efficiency [86].

4.4 Gut microbiota as ANF biodegrading agent in fish

The application of the gut "indigenous" microbiota as probiotics can improve feed digestibility by supplying exogenous enzymes (*e.g.* cellulase, phytase, tannase and xylanases) and by eradicating and/or reducing ANF of the plant protein

Reduction strategies	Bioactive component															
	Protease/trypsin inhibitors	**Amylase inhibitors**	**Saponins**	**Phytate**	**Glucosinolates**	**Oxalate**	**Anti-vitamins**	**Polyphenols/Tannins**	**Sterols**	**Lectins**	**Phytoestrogens**	**Alkaloids**	**Cyanogens**	**NSP**	**Oligosacharide**	**Antigenic proteins**
Milling				●				●		●						
De-hulling								●				●				
Soaking	●			●				●			●	●				●
Heat/Autoclaved	●	●	●	●	●	●	●	●		●	●					●
Solvent extraction			●	●			●		●		●	●			●	●
Germination	●			●				●								●
Acid addition	●		●	●			●							●	●	
Alkalin addition							●	●				●		●	●	●
enzymes addition	●	●		●										●		●
Fermentation	●		●	●		●		●			●	●	●	●	●	●
Micronutrient addition	●			●					●	●			●			
New Variates				●	●							●				

Table 3.
Legume bioactive compounds neutralizing strategies.

ingredients in the fish gut [87]. A plethora of studies have recognized cellulase-producing bacteria such as *Citrobacter sp. C. freundii, Enterobacter* sp. *Bacillus coagulans*, *B. cereus*, *B. subtilis* P6, *B. velesensis* P11, *B. pumilus*, *B. tequilensis* (KF640219), *B. megaterium* (KF640220) and *B. altitudinis* from the gut of various cultured fish species including Chinese carps [88, 89], Indian major carps [90, 91], tilapia (*Oreochromis mossambica*) [92], bata fish [93], murrels (*Channa punctatus*) [94], pacu (*Piaractus mesopotamicus*) and piaucom-pinta (*Leporinus friderici*) [95]. Using the above mentioned microorganisms as potential probiotics in aquafeeds or applying these microorganisms for fermentation of plant protein ingredients can provide great potential for eradicating ANF and improving their nutritional quality by boosting up EAA, minerals and vitamins bioavailability and increasing digestibility of protein and energy.

4.5 Other functional feed additives

Supplementing PP-based aquafeeds with additives can improve the digestibility of feed's nutrients. In this context, it has been reported that supplementation of a diet contained high levels of legumes including SPC and pea protein (32%) with phosphatidylcholine pronouncedly improved lipid digestibility in Atlantic salmon [96].

On the other hand, it has been confirmed that the replacement of FM with PP sources could reduce cholesterol content in aquafeeds that may disturb bile acids synthesis in fish and result in low digestibility of lipid [97]. In this regard, it has been reported that supplementing SBM-based diets with cholesterol remarkably improved growth in channel catfish [98], turbot (*Scophthalmus maximus*) [85] and rainbow trout [99].

As mentioned earlier legumes are deficient in taurine or its precursors (*i.e.* cysteine and methionine) and some aquatic animal species especially marine fish unable, or have low ability to synthesize taurine [100]. Taurine is the main component of bile acids and increase the bile-salt dependent lipase activity in fish [101]. It has been proved that supplementation of soy protein-based aquafeeds with taurine improved growth performance, lipid metabolism, palatability, digestibility and overall nutritional quality of feeds in marine fish species such as common dentex (*Dentex dentex*) [101] and European sea bass larvae [102] and juveniles [103, 104].

Moreover, it has been confirmed that replacement of FM with plant protein sources with high levels of ANF (*i.e.* saponins, oligosaccharides, fibers and high molecular weight proteins) disturb bile metabolism in fish and may adversely affect fish productivity [105]. Bile acids as an emulsifier enhance digestion and absorption of lipid and lipid soluble nutrients through emulsification of lipids and activation of bile salt dependent lipase [105]. It also facilitates the excretion of cholesterol and toxic metabolites. The ANF in PP sources may induce gut inflammation that reduce resorption of bile acids or they may bind with bile salts and trigger extra excretion of bile acids into gut [105]. Thus, supplementing legume protein-based diet with bile acids may improve their efficiency and alleviate their negative effects on fish performance. For example, supplementing SBM-based diet with 1.5% bovine bile salts significantly improved growth rate in rainbow trout [106].

4.6 Nutritional programming and selective breeding

In recent years some studies were carried out on early nutritional programming of fish for increasing the acceptance of their offspring to new ingredients in aquafeeds. For instance, it has been reported that substitution of dietary FM and FO with vegetal feedstuffs through nutritional programming in brooders elevated the acceptance of vegetal ingredients and PP-based diets in rainbow trout [107] and gilthead seabream [108] offspring. In this regard, it has been reported that early nutritional programming in Atlantic salmon with a plant-based aquafeed enhanced growth rate and feed efficiency for 24% and 23%, respectively compared to those fed a diet contained FM and FO and then challenged with a plant-based aquafeed [109].

Recently, a new strain of rainbow trout (ARS-KO) was created by the US department of Agriculture by selective breeding over the course of two decades and this strain can grow better when fed with soy protein-based diets and does not develop enteritis [110, 111]. More research are required to be carried out in these genetic engineering to these novel techniques be advantageous and applicable at commercial stage.

5. Conclusions

Over the course of the past four decades, a great amount of knowledge has been gained in application of legumes as APS in aquafeeds, leading to a better comprehension regarding the impacts of these APPS on overall performance of different cultured species. Herbivorous and omnivorous fish and crustacean species have a great potential in utilization of legumes in their diets. Moreover, carnivorous species have mostly adapted to legumes-protein rich aquafeeds. However, in order to enhance the efficiency of legumes-protein based aquafeeds for carnivorous fish, further innovations and development is required by considering the cultured animal species and feed ingredients for increasing adaptability of cultured aquatic species to legumes. These innovations can be carried out in different aspects such as use of novel feedstuffs, eradication and/or reducing ANF, application of feed additives and use of precise feed formulations. The application of genetic engineering in legumes could result in the production of strains with low levels of ANF and make them appropriate for legumes-protein based aquafeeds. In addition, supplementing aquafeeds with functional feed additives can improve the efficiency of legumes for aquaculture nutrition. Using nutritional programming and applying genetic engineering also other novel strategies to provide new fish and crustacean strains with high capacity in acceptance and utilization of legumes in aquafeeds. Further studies are required to increase the efficiency of legumes in aquafeeds to support growth, health and welfare of cultured species.

Author details

Fateme Hekmatpour and Mansour Torfi Mozanzadeh*
South Iran Aquaculture Research Centre, Iranian Fisheries Science Institute (IFSRI), Agricultural Research Education and Extension Organization (AREEO), Ahwaz, Iran

*Address all correspondence to: m.torfi@areeo.ac.ir; mansour.torfi@gmail.com

References

[1] Markets, Markets . https://wwwma rketsandmarketscom/new-reportshtml. 2021.

[2] FAO. Sustainability in action. State of World Fisheries and Aquaculture Rome. 2020;200.

[3] FAO. The State of World Fisheries and Aquaculture 2018—Meeting the sustainable development goals. Rome. Licence: CC BY-NC-SA 3.0 IGO. 2018.

[4] Henry M, Gasco L, Piccolo G, Fountoulaki E. Review on the use of insects in the diet of farmed fish: past and future. Animal Feed Science and Technology. 2015;203:1-22. DOI: 10.1016/j.anifeedsci.2015.03.001.

[5] Chakraborty P, Mallik A, Sarang N, Lingam SS. A review on alternative plant protein sources available for future sustainable aqua feed production. Int J Chem Stud. 2019;7:1399-404.

[6] Hardy RW. Utilization of plant proteins in fish diets: effects of global demand and supplies of fishmeal. Aquaculture Research. 2010;41(5): 770-6. DOI: 10.1111/j.1365-2109.2009. 02349.x.

[7] Tacon AG, Metian M. Feed matters: satisfying the feed demand of aquaculture. Reviews in Fisheries Science & Aquaculture. 2015;23(1):1-10. 10.1080/23308249.2014.987209.

[8] Gatlin III DM, Barrows FT, Brown P, Dabrowski K, Gaylord TG, Hardy RW, et al. Expanding the utilization of sustainable plant products in aquafeeds: a review. Aquaculture research. 2007;38 (6):551-79. 10.1111/j.1365-2109.2007. 01704.x.

[9] Tacon AG, Metian M. Global overview on the use of fish meal and fish oil in industrially compounded aquafeeds: Trends and future prospects. Aquaculture. 2008;285(1-4):146-58. DOI: 10.1016/j.aquaculture.2008.08.015.

[10] Tacon AG, Hasan MR, Metian M. Demand and supply of feed ingredients for farmed fish and crustaceans: trends and prospects. FAO Fisheries and Aquaculture technical paper. 2011(564):I.

[11] Oliva-Teles A, Enes P, Peres H. Replacing fishmeal and fish oil in industrial aquafeeds for carnivorous fish. Feed and feeding practices in aquaculture. 2015:203-33. 10.1016/ B978-0-08-100506-4.00008-8.

[12] McNeill J, Barrie F, Buck W, Demoulin V, Greuter W, Hawksworth D, et al. International Code of Nomenclature for algae, fungi and plants (Melbourne Code): Koeltz Scientific Books Königstein; 2012.

[13] Turland NJ, Wiersema JH, Barrie FR, Greuter W, Hawksworth DL, Herendeen PS, et al. International Code of Nomenclature for algae, fungi, and plants (Shenzhen Code) adopted by the Nineteenth International Botanical Congress Shenzhen, China, July 2017: Koeltz Botanical Books; 2018.

[14] Faostat F. Crops. Food and Agriculture Organization of the United Nations Available online at. 2020.

[15] Park H, Weier S, Razvi F, Peña PA, Sims NA, Lowell J, et al. Towards the development of a sustainable soya bean-based feedstock for aquaculture. Plant biotechnology journal. 2017;15(2): 227-36. 10.1111/pbi.12608.

[16] Liu K. Chemistry and nutritional value of soybean components. Soybeans: Springer; 1997. p. 25-113.

[17] Refstie S, Svihus B, Shearer KD, Storebakken T. Nutrient digestibility in Atlantic salmon and broiler chickens related to viscosity and non-starch

polysaccharide content in different soyabean products. Animal Feed Science and Technology. 1999;79(4):331-45. DOI: 10.1016/S0377-8401(99)00026-7.

[18] Elangovan A, Shim K. The influence of replacing fish meal partially in the diet with soybean meal on growth and body composition of juvenile tin foil barb (Barbodes altus). Aquaculture. 2000;189(1-2):133-44. 10.1016/S0044-8486(00)00365-3.

[19] Salze G, McLean E, Battle PR, Schwarz MH, Craig SR. Use of soy protein concentrate and novel ingredients in the total elimination of fish meal and fish oil in diets for juvenile cobia, Rachycentron canadum. Aquaculture. 2010;298(3-4):294-9. DOI: 10.1016/j.aquaculture.2009.11.003.

[20] Mamauag R, Koshio S, Ishikawa M, Yokoyama S, Gao J, Nguyen B, et al. Soy peptide inclusion levels influence the growth performance, proteolytic enzyme activities, blood biochemical parameters and body composition of Japanese flounder, Paralichthys olivaceus. Aquaculture. 2011;321(3-4): 252-8. 10.1016/j.aquaculture.2011.09.022.

[21] Hartviksen MAB. Replacement of fishmeal with alternative proteins in diets for Atlantic salmon (Salmo salar L.): A study on the microbiota, morphology and function of the intestine 2015.

[22] Bureau DP, Harris AM, Cho CY. The effects of purified alcohol extracts from soy products on feed intake and growth of chinook salmon (Oncorhynchus tshawytscha) and rainbow trout (Oncorhynchus mykiss). Aquaculture. 1998;161(1-4):27-43. DOI: 10.1016/S0044-8486(97)00254-8.

[23] Storebakken T, Shearer K, Roem A. Growth, uptake and retention of nitrogen and phosphorus, and absorption of other minerals in Atlantic salmon Salmo salar fed diets with fish meal and soy-protein concentrate as the main sources of protein. Aquaculture Nutrition. 2000;6(2):103-8.

[24] Storebakken T, Shearer K, Roem A. Availability of protein, phosphorus and other elements in fish meal, soy-protein concentrate and phytase-treated soy-protein-concentrate-based diets to Atlantic salmon, Salmo salar. Aquaculture. 1998;161(1-4):365-79. DOI: 10.1016/S0044-8486(97)00284-6.

[25] Gamboa-Delgado J, Rojas-Casas MG, Nieto-López MG, Cruz-Suárez LE. Simultaneous estimation of the nutritional contribution of fish meal, soy protein isolate and corn gluten to the growth of Pacific white shrimp (Litopenaeus vannamei) using dual stable isotope analysis. Aquaculture. 2013;380:33-40. DOI: 10.1016/j.aquaculture.2012.11.028.

[26] Alam MS, Teshima S-i, Koshio S, Ishikawa M, Uyan O, Hernandez LHH, et al. Supplemental effects of coated methionine and/or lysine to soy protein isolate diet for juvenile kuruma shrimp, Marsupenaeus japonicus. Aquaculture. 2005;248(1-4):13-9. DOI: 10.1016/j.aquaculture.2005.04.015.

[27] Ireland PA, Dziedzic SZ. Effect of hydrolysis on sapogenin release in soy. Journal of Agricultural and Food Chemistry. 1986;34(6):1037-41. DOI: 10.1021/jf00072a027.

[28] Anderson RL, Wolf WJ. Compositional changes in trypsin inhibitors, phytic acid, saponins and isoflavones related to soybean processing. The Journal of nutrition. 1995;125(suppl_3):581S-8S. DOI: 10.1093/jn/125.suppl_3.581S.

[29] Glencross B, Evans D, Dods K, McCafferty P, Hawkins W, Maas R, et al. Evaluation of the digestible value of lupin and soybean protein concentrates and isolates when fed to

rainbow trout, Oncorhynchus mykiss, using either stripping or settlement faecal collection methods. Aquaculture. 2005;245(1-4):211-20. 10.1016/j.aquaculture.2004.11.033.

[30] Hrckova M, Rusnakova M, Zemanovic J. Enzymatic hydrolysis of defatted soy flour by three different proteases and their effect on the functional properties of resulting protein hydrolysates. Czech journal of food sciences. 2002;20(1):7-14.

[31] Kristinsson HG, Rasco BA. Biochemical and functional properties of Atlantic salmon (Salmo salar) muscle proteins hydrolyzed with various alkaline proteases. Journal of agricultural and food chemistry. 2000; 48(3):657-66. DOI: 10.1021/jf990447v.

[32] Chen H-M, Muramoto K, Yamauchi F, Fujimoto K, Nokihara K. Antioxidative properties of histidine-containing peptides designed from peptide fragments found in the digests of a soybean protein. Journal of agricultural and food chemistry. 1998; 46(1):49-53. 10.1021/jf970649w.

[33] Peñta-Ramos E, Xiong Y. Antioxidant activity of soy protein hydrolysates in a liposomal system. Journal of food science. 2002;67(8):2952-6.

[34] Maruyama N, Fukuda T, Saka S, Inui N, Kotoh J, Miyagawa M, et al. Molecular and structural analysis of electrophoretic variants of soybean seed storage proteins. Phytochemistry. 2003; 64(3):701-8. DOI: 10.1016/S0031-9422(03)00385-6.

[35] Dawood MA, Koshio S. Application of fermentation strategy in aquafeed for sustainable aquaculture. Reviews in Aquaculture. 2020;12(2):987-1002. 10.1111/raq.12368.

[36] Kim B. Physiological functions of chongkukjang. Food Industry and Nutrition. 1999;4:40-6.

[37] Hong K-J, Lee C-H, Kim SW. Aspergillus oryzae GB-107 fermentation improves nutritional quality of food soybeans and feed soybean meals. Journal of medicinal food. 2004;7(4): 430-5. DOI: 10.1089/jmf.2004.7.430.

[38] Osmond AT, Colombo SM. The future of genetic engineering to provide essential dietary nutrients and improve growth performance in aquaculture: Advantages and challenges. Journal of the World Aquaculture Society. 2019;50(3):490-509. 10.1111/jwas.12595.

[39] Higuera-Ciapara I, Felix-Valenzuela L, Goycoolea F. Astaxanthin: a review of its chemistry and applications. Critical reviews in food science and nutrition. 2006;46(2):185-96. 10.1080/10408690590957188.

[40] Hussein G, Sankawa U, Goto H, Matsumoto K, Watanabe H. Astaxanthin, a carotenoid with potential in human health and nutrition. Journal of natural products. 2006;69(3):443-9. 10.1021/np050354+.

[41] Cruz-Suarez LE, Ricque-Marie D, Tapia-Salazar M, McCallum IM, Hickling D. Assessment of differently processed feed pea (Pisum sativum) meals and canola meal (Brassica sp.) in diets for blue shrimp (Litopenaeus stylirostris). Aquaculture. 2001;196(1-2):87-104. 10.1016/S0044-8486(00)00572-X.

[42] Hickling D. Canadian feed peas industry guide. Pulse Canada, Winnipeg, Manitoba. 2003:1-36.

[43] Liu Xh, Ye Jd, Wang K, Kong Jh, Yang W, Zhou L. Partial replacement of fish meal with peanut meal in practical diets for the Pacific white shrimp, Litopenaeus vannamei. Aquaculture Research. 2012;43(5):745-55. 10.1111/j.1365-2109.2011.02883.x.

[44] Yıldırım Ö, Acar Ü, Türker A, Sunar MC, Kesbic OS. Effects of

Replacing Fish Meal with Peanut Meal (Arachis hypogaea) on Growth, Feed Utilization and Body Composition of Mozambique Tilapia Fries (Oreochromis mossambicus). Pakistan Journal of Zoology. 2014;46(2):497-502.

[45] Batal A, Dale N, Café M. Nutrient composition of peanut meal. Journal of applied poultry research. 2005;14(2): 254-7. DOI: 10.1093/japr/14.2.254.

[46] Jiang H, Qi P, Wang T, Chi X, Wang M, Chen M, et al. Role of halotolerant phosphate-solubilising bacteria on growth promotion of peanut (Arachis hypogaea) under saline soil. Annals of Applied Biology. 2019;174(1): 20-30. DOI: 10.1111/aab.12473.

[47] Draganovic V. Towards sustainable fish feed production using novel protein sources. Wageningen: Wageningen University; 2013.

[48] Molina-Poveda C, Lucas M, Jover M. Evaluation of the potential of Andean lupin meal (Lupinus mutabilis Sweet) as an alternative to fish meal in juvenile Litopenaeus vannamei diets. Aquaculture. 2013;410:148-56. 10.1016/ j.aquaculture.2013.06.007.

[49] Glencross BD. Feeding lupins to fish: A review of the nutritional and biological value of lupins in aquaculture feeds: Department of Fisheries– Research Division, Government of Western Australia; 2001.

[50] Drew M, Borgeson T, Thiessen D. A review of processing of feed ingredients to enhance diet digestibility in finfish. Animal Feed Science and Technology. 2007;138(2):118-36. DOI: 10.1016/j. anifeedsci.2007.06.019.

[51] Estensoro I, Ballester-Lozano G, Benedito-Palos L, Grammes F, Martos-Sitcha JA, Mydland L-T, et al. Dietary butyrate helps to restore the intestinal status of a marine teleost (Sparus aurata) fed extreme diets low in fish meal and fish oil. PLoS One. 2016;11 (11):e0166564. DOI: 10.1371/journal. pone.0166564.

[52] Revilla I. Impact of thermal processing on faba bean (Vicia faba) composition. Processing and impact on active components in food: Elsevier; 2015. p. 337-43.

[53] Alinasabhematabadi L. Evaluation of apparent nutrient digestibility in Atlantic salmon fed diets with differently processed faba bean: Norwegian University of Life Sciences, Ås; 2019.

[54] Multari S, Stewart D, Russell WR. Potential of fava bean as future protein supply to partially replace meat intake in the human diet. Comprehensive Reviews in Food Science and Food Safety. 2015;14(5):511-22. DOI: 10.1111/ 1541-4337.12146.

[55] Vidal-Valverde C, Frias J, Sotomayor C, Diaz-Pollan C, Fernandez M, Urbano G. Nutrients and antinutritional factors in faba beans as affected by processing. Zeitschrift für Lebensmitteluntersuchung und-Forschung A. 1998;207(2):140-5. DOI: 10.1007/s002170050308.

[56] Helsper J, Van Norel A. Antinutritional Factors in Faba Bean and Resistance to Soil-Borne Diseases. Developments in Agricultural and Managed Forest Ecology. 23: Elsevier; 1991. p. 387-91.

[57] De Santis C, Crampton VO, Bicskei B, Tocher DR. Replacement of dietary soy-with air classified faba bean protein concentrate alters the hepatic transcriptome in Atlantic salmon (Salmo salar) parr. Comparative Biochemistry and Physiology Part D: Genomics and Proteomics. 2015;16: 48-58. 10.1016/j.cbd.2015.07.005.

[58] De Santis C, Ruohonen K, Tocher DR, Martin SA, Krol E,

Secombes CJ, et al. Atlantic salmon (Salmo salar) parr as a model to predict the optimum inclusion of air classified faba bean protein concentrate in feeds for seawater salmon. Aquaculture. 2015;444: 70-8. 10.1016/j.aquaculture.2015.03.024.

[59] Król E, Douglas A, Tocher DR, Crampton VO, Speakman JR, Secombes CJ, et al. Differential responses of the gut transcriptome to plant protein diets in farmed Atlantic salmon. BMC genomics. 2016;17(1): 1-16. DOI: 10.1186/s12864-016-2473-0.

[60] Hossain MA, Pandey A, Satoh S. Effects of organic acids on growth and phosphorus utilization in red sea bream Pagrus major. Fisheries Science. 2007;73 (6):1309-17. DOI: 10.1111/ j.1444-2906.2007.01469.x.

[61] Alexis MN, Papaparaskeva-Papoutsoglou E, Theochari V. Formulation of practical diets for rainbow trout (Salmo gairdneri) made by partial or complete substitution of fish meal by poultry by-products and certain plant by-products. Aquaculture. 1985;50(1-2):61-73. DOI: 10.1016/ 0044-8486(85)90153-X.

[62] Martínez-Llorens S, Baeza-Ariño R, Nogales-Mérida S, Jover-Cerdá M, Tomás-Vidal A. Carob seed germ meal as a partial substitute in gilthead sea bream (Sparus aurata) diets: amino acid retention, digestibility, gut and liver histology. Aquaculture2012. p. 124-33.

[63] Couto A, Barroso C, Guerreiro I, Pousão-Ferreira P, Matos E, Peres H, et al. Carob seed germ meal in diets for meagre (Argyrosomus regius) juveniles: Growth, digestive enzymes, intermediary metabolism, liver and gut histology. Aquaculture. 2016;451: 396-404. DOI: 10.1016/j. aquaculture.2015.10.007.

[64] Avallone R, Plessi M, Baraldi M, Monzani A. Determination of chemical composition of carob (Ceratonia siliqua): protein, fat, carbohydrates, and tannins. Journal of food composition and analysis. 1997;10(2):166-72. DOI: 10.1006/jfca.1997.0528.

[65] Koegel R, Straub R. Fractionation of alfalfa for food, feed, biomass, and enzymes. Transactions of the ASAE. 1996;39(3):769-74. 10.13031/2013.27559.

[66] Francis G, Makkar HP, Becker K. Antinutritional factors present in plant-derived alternate fish feed ingredients and their effects in fish. Aquaculture. 2001;199(3-4):197-227. 10.1016/ S0044-8486(01)00526-9.

[67] Makkar H. Antinutritional factors in foods for livestock. BSAP Occasional Publication. 1993;16:69-85. DOI: 10.1017/S0263967X00031086.

[68] Allan GL, Parkinson S, Booth MA, Stone DA, Rowland SJ, Frances J, et al. Replacement of fish meal in diets for Australian silver perch, Bidyanus bidyanus: I. Digestibility of alternative ingredients. Aquaculture. 2000;186 (3-4):293-310. 10.1016/S0044-8486(99) 00380-4.

[69] Magalhaes S, Cabrita A, Valentao P, Andrade P, Rema P, Maia M, et al. Apparent digestibility coefficients of European grain legumes in rainbow trout (Oncorhynchus mykiss) and Nile tilapia (Oreochromis niloticus). Aquaculture nutrition. 2018;24(1): 332-40. DOI: 10.1111/anu.12564.

[70] Colombo SM. Physiological considerations in shifting carnivorous fishes to plant-based diets. In: Benfey TJ, Farrell AP, Brauner CJ, editors. Fish Physiol. 38: Academic Press; 2020. p. 53-82.

[71] Samtiya M, Aluko RE, Dhewa T. Plant food anti-nutritional factors and their reduction strategies: An overview. Food Production, Processing and Nutrition. 2020;2(1):1-14. DOI: 10.1186/ s43014-020-0020-5.

[72] Mohanta K. Plant feed resources of India. In: Sogbesan OA, Mohanta KN, Sahoo PK, Mitra G, Jayasankar P, editors. Invited Papers on Application of Solid State Fermentation Technology in Aquaculture. Central Institute of Freshwater Aquaculture, Bhubaneswar, India2012. p. 100-13.

[73] Glencross B, Hawkins W, Curnow J. Nutritional assessment of Australian canola meals. I. Evaluation of canola oil extraction method and meal processing conditions on the digestible value of canola meals fed to the red seabream (Pagrus auratus, Paulin). Aquaculture Research. 2004;35(1):15-24. DOI: 10.1111/j.1365-2109.2004.00974.x.

[74] Glencross B, Hawkins W, Curnow J. Nutritional assessment of Australian canola meals. II. Evaluation of the influence of the canola oil extraction method on the protein value of canola meals fed to the red seabream (Pagrus auratus, Paulin). Aquaculture Research. 2004;35(1):25-34. DOI: 10.1111/j.1365-2109.2004.00975.x.

[75] Castillo S, Gatlin III DM. Dietary supplementation of exogenous carbohydrase enzymes in fish nutrition: a review. Aquaculture. 2015;435:286-92. DOI: 10.1016/j.aquaculture.2014.10.011.

[76] Maas RM, Verdegem MC, Dersjant-Li Y, Schrama JW. The effect of phytase, xylanase and their combination on growth performance and nutrient utilization in Nile tilapia. Aquaculture. 2018;487:7-14. DOI: 10.1016/j.aquaculture.2017.12.040.

[77] Vahjen W, Osswald T, Schäfer K, Simon O. Comparison of a xylanase and a complex of non starch polysaccharide-degrading enzymes with regard to performance and bacterial metabolism in weaned piglets. Archives of animal nutrition. 2007;61(2):90-102. DOI: 10.1080/17450390701203881.

[78] Adeola O, Cowieson A. Board-invited review: opportunities and challenges in using exogenous enzymes to improve nonruminant animal production. Journal of animal science. 2011;89(10):3189-218. DOI: 10.2527/jas.2010-3715.

[79] Kiarie E, Nyachoti C, Slominski B, Blank G. Growth performance, gastrointestinal microbial activity, and nutrient digestibility in early-weaned pigs fed diets containing flaxseed and carbohydrase enzyme. Journal of Animal Science. 2007;85(11):2982-93. DOI: 10.2527/jas.2006-481.

[80] Ng WK, Koh CB. The utilization and mode of action of organic acids in the feeds of cultured aquatic animals. Reviews in Aquaculture. 2017;9(4):342-68. DOI: 10.1111/raq.12141.

[81] Sarker M, Satoh S, Kamata K, Haga Y, Yamamoto Y. Partial replacement of fish meal with plant protein sources using organic acids to practical diets for juvenile yellowtail, Seriola quinqueradiata. Aquaculture Nutrition. 2012;18(1):81-9. DOI: 10.1111/j.1365-2095.2011.00880.x.

[82] Khajepour F, Hosseini SA. Citric acid improves growth performance and phosphorus digestibility in Beluga (Huso huso) fed diets where soybean meal partly replaced fish meal. Animal feed science and technology. 2012;171(1):68-73. DOI: 10.1016/j.anifeedsci.2011.10.001.

[83] Lin Y-H, Cheng M-Y. Effects of dietary organic acid supplementation on the growth, nutrient digestibility and intestinal histology of the giant grouper Epinephelus lanceolatus fed a diet with soybean meal. Aquaculture. 2017;469:106-11. DOI: 10.1016/j.aquaculture.2016.11.032.

[84] Sarker MSA, Satoh S, Kamata K, Haga Y, Yamamoto Y. Supplementation effect (s) of organic acids and/or lipid to plant protein-based diets on juvenile yellowtail, Seriola quinqueradiata

Temminck et Schlegel 1845, growth and, nitrogen and phosphorus excretion. Aquaculture Research. 2012;43(4): 538-45. DOI: 10.1111/j.1365-2109.2011.02859.x.

[85] Castillo S, Rosales M, Pohlenz C, Gatlin III DM. Effects of organic acids on growth performance and digestive enzyme activities of juvenile red drum Sciaenops ocellatus. Aquaculture. 2014; 433:6-12. DOI: 10.1016/j.aquaculture.2014.05.038.

[86] Zhang J, Zhong L, Chi S, Chu W, Liu Y, Hu Y. Sodium butyrate supplementation in high-soybean meal diets for juvenile rice field eel (Monopterus albus): Effects on growth, immune response and intestinal health. Aquaculture. 2020;520:734952. 10.1016/j.aquaculture.2020.734952.

[87] Ghosh K, Ray AK. Aquafeed formulation using plant feedstuffs: Prospective application of fish-gut microorganisms and microbial biotechnology. Soft Chemistry and Food Fermentation: Elsevier; 2017. p. 109-44.

[88] Bairagi A, Ghosh KS, Sen SK, Ray AK. Enzyme producing bacterial flora isolated from fish digestive tracts. Aquaculture International. 2002; 10(2):109-21. DOI: 10.1023/A:1021355406412.

[89] Li J, Tan B, Mai K. Dietary probiotic Bacillus OJ and isomaltooligosaccharides influence the intestine microbial populations, immune responses and resistance to white spot syndrome virus in shrimp (Litopenaeus vannamei). Aquaculture. 2009;291(1-2):35-40. DOI: 10.1016/j.aquaculture.2009.03.005.

[90] Ray AK, Roy T, Mondal S, Ringø E. Identification of gut-associated amylase, cellulase and protease-producing bacteria in three species of Indian major carps. Aquaculture Research. 2010;41 (10):1462-9. DOI: 10.1111/j.1365-2109.2009.02437.x.

[91] Banerjee S, Mukherjee A, Dutta D, Ghosh K. Non-Starch Polysaccharide Degrading Gut Bacteria in Indian Major Carps and Exotic Carps. Jordan Journal of Biological Sciences. 2016;9(1). DOI: 10.12816/0027010.

[92] Saha S, Roy RN, Sen SK, Ray AK. Characterization of cellulase-producing bacteria from the digestive tract of tilapia, Oreochromis mossambica (Peters) and grass carp, Ctenopharyngodon idella (Valenciennes). Aquaculture Research. 2006;37(4):380-8. DOI: 10.1111/j.1365-2109.2006.01442.x.

[93] Mondal S, Roy T, Ray AK. Characterization and identification of enzyme-producing bacteria isolated from the digestive tract of bata, Labeo bata. Journal of the World Aquaculture Society. 2010;41(3):369-77.

[94] Kar N, Ghosh K. Enzyme producing bacteria in the gastrointestinal tracts of Labeo rohita (Hamilton) and Channa punctatus (Bloch). Turkish Journal of Fisheries and Aquatic Sciences. 2008; 8(1):115-20.

[95] Peixoto SB, Cladera-Olivera F, Daroit DJ, Brandelli A. Cellulase-producing Bacillus strains isolated from the intestine of Amazon basin fish. . Aquaculture Research 2011;42:887–91.

[96] Krogdahl Å, Hansen AKG, Kortner TM, Björkhem I, Krasnov A, Berge GM, et al. Choline and phosphatidylcholine, but not methionine, cysteine, taurine and taurocholate, eliminate excessive gut mucosal lipid accumulation in Atlantic salmon (Salmo salar L). Aquaculture. 2020;528:735552. 10.1016/j.aquaculture.2020.735552.

[97] Kortner TM, Björkhem I, Krasnov A, Timmerhaus G, Krogdahl Å. Dietary cholesterol supplementation to a plant-based diet suppresses the complete pathway of cholesterol

synthesis and induces bile acid production in Atlantic salmon (Salmo salar L.). British journal of nutrition. 2014;111(12):2089-103. DOI: 10.1017/S0007114514000373.

[98] Twibell RG, Wilson RP. Preliminary evidence that cholesterol improves growth and feed intake of soybean meal-based diets in aquaria studies with juvenile channel catfish, Ictalurus punctatus. Aquaculture. 2004;236(1-4):539-46. 10.1016/j.aquaculture.2003.10.028.

[99] Chen Z, Zhao S, Liu Y, Yang P, Ai Q, Zhang W, et al. Dietary citric acid supplementation alleviates soybean meal-induced intestinal oxidative damage and micro-ecological imbalance in juvenile turbot, Scophthalmus maximus L. Aquaculture Research. 2018;49(12): 3804-16. DOI: 10.1111/are.13847.

[100] Reyshari A, Mohammadiazarm H, Mohammadian T, Torfi Mozanzadeh M. Effects of sodium diformate on growth performance, gut microflora, digestive enzymes and innate immunological parameters of Asian sea bass (Lates calcarifer) juveniles. Aquaculture Nutrition. 2019;25(5):1135-44. DOI: 10.1111/anu.12929.

[101] Sarker SA, Satoh S, Kiron V. Supplementation of citric acid and amino acid-chelated trace element to develop environment-friendly feed for red sea bream, Pagrus major. Aquaculture. 2005;248(1-4):3-11. DOI: 10.1016/j.aquaculture.2005.04.012.

[102] Liu Y, Chen Z, Dai J, Yang P, Xu W, Ai Q, et al. Sodium butyrate supplementation in high-soybean meal diets for turbot (Scophthalmus maximus L.): effects on inflammatory status, mucosal barriers and microbiota in the intestine. Fish & shellfish immunology. 2019;88:65-75. DOI: 10.1016/j.fsi.2019.02.064.

[103] Krome C, Schuele F, Jauncey K, Focken U. Influence of a sodium formate/formic acid mixture on growth of juvenile common carp (Cyprinus carpio) fed different fishmeal replacement levels of detoxified Jatropha curcas kernel meal in practical, mixed diets. Journal of Applied Aquaculture. 2018;30(2):137-56. DOI: 10.1080/10454438.2017.1412845.

[104] Gao Y, Storebakken T, Shearer KD, Penn M, Øverland M. Supplementation of fishmeal and plant protein-based diets for rainbow trout with a mixture of sodium formate and butyrate. Aquaculture. 2011;311(1-4):233-40. DOI: 10.1016/j.aquaculture.2010.11.048.

[105] Kortner TM, Penn MH, Björkhem I, Måsøval K, Krogdahl Å. Bile components and lecithin supplemented to plant based diets do not diminish diet related intestinal inflammation in Atlantic salmon. BMC veterinary research. 2016;12(1):1-12. DOI: 10.1186/s12917-016-0819-0.

[106] Mukhopadhyay Na, Ray A. Effect of fermentation on the nutritive value of sesame seed meal in the diets for rohu, Labeo rohita (Hamilton), fingerlings. 1999. DOI: 10.1046/j.1365-2095.1999.00101.x.

[107] Lazzarotto V, Corraze G, Leprevost A, Quillet E, Dupont-Nivet M, Médale F. Three-year breeding cycle of rainbow trout (Oncorhynchus mykiss) fed a plant-based diet, totally free of marine resources: consequences for reproduction, fatty acid composition and progeny survival. PloS one. 2015; 10(2):e0117609. DOI: 10.1371/journal.pone.0117609.

[108] Izquierdo M, Turkmen S, Montero D, Zamorano M, Afonso J, Karalazos V, et al. Nutritional programming through broodstock diets to improve utilization of very low fishmeal and fish oil diets in gilthead sea bream. Aquaculture. 2015;449: 18-26. DOI: 10.1016/j.aquaculture.2015.03.032.

[109] Clarkson M, Migaud H, Metochis C, Vera LM, Leeming D, Tocher DR, et al. Early nutritional intervention can improve utilisation of vegetable-based diets in diploid and triploid Atlantic salmon (Salmo salar L.). British Journal of Nutrition. 2017; 118(1):17-29. DOI: 10.1017/S0007114517001842.

[110] Overturf K, Barrows FT, Hardy RW. Effect and interaction of rainbow trout strain (Oncorhynchus mykiss) and diet type on growth and nutrient retention. Aquaculture Research. 2013;44(4):604-11. DOI: 10.1111/j.1365-2109.2011.03065.x.

[111] Abernathy J, Overturf K. Expression of antisense long noncoding RNAs as potential regulators in rainbow trout with different tolerance to plant-based diets. Animal biotechnology. 2019;30(1):87-94. DOI: 10.1080/10495398.2017.1401546.

Chapter 2

Pulses: A Potential Source of Valuable Protein for Human Diet

Saima Parveen, Amina Jamil, Imran Pasha and Farah Ahmad

Abstract

Nutritional profile of pulses has significant importance in human diet with respect to protein and mineral quality and bioavailability. Protein energy malnutrition is widespread throughout the world especially among the developing countries. Pulses being rich in macronutrients such as protein from 20 to 26% and low in calories are most suitable for product development for target-oriented population. During last decade, the demand for pulse-based products with high protein and fiber, low glycemic index, and gluten free with more antioxidant showed increasing trend by the consumers. Drift of end-use application of pulses generated interest for research in all disciplines such as breeding, agronomy, food, and nutrition, etc. A great share of plant protein in human diet may be a critical step for reducing dependence on animal origin protein source. This chapter will review contribution or choice of plant-based protein from legumes or pulses with good-quality protein based on amino acid composition. Additionally, this overview can give insight into the development of new product with balanced nutritional quality and high protein contents as a potential protein supply for malnourished population.

Keywords: malnutrition, pulse proteins, amino acid, bioavailability, digestibility, value addition

1. Introduction

Pulses are dry seeds of legume family grown in pods in varying shape and size. These are member of the *Leguminosae*, family *Phaseoleae*, subfamily *Papilionoideae* [1]. Chronological, archeological evidences showed that legumes and pulses were domesticated and originated from America [2]. Now consumed in every part of this globe especially by people in the developing countries as well as developed countries [3, 4]. In some areas such as Mexico, south and central American, and African countries, these are being consumed as staple foods where per capita intake may extend up to 40 kg per year [5].

Oilseeds are excluded from this category, which are solely grown or harvested for oil extraction purpose. A variety of pulses are grown with various shapes and size throughout the world. Most commonly consumed pulses include chickpea (*Cicer arietinum*), field peas (*Pisum sativum*), lentils (*Lens culinaris*), mung bean (*Phaseolus mungo*), dry broad bean (*Vicia faba*), moth beans (*Phaseolus aconitifolius*), lupins (*Lupinus albus*) etc. In addition, there are a large number of minor pulses that are grown and consumed in different parts of the world [6, 7].

Pulses are most commonly consumed food in the Asian countries as a culinary staple since ancient times. However, its cultivation is not as increased as other staple crops such as wheat, corn, barley, etc., and mass production of pulses is restricted in underdeveloped countries where per capita consumption is increased up to 125–140 kcal as compared with western world such as the United States where per capita consumption is minimal, i.e., only up to 27 kcal [8].

Plant-based protein could be the best substitute for animal-based protein to overcome protein energy malnutrition. Legumes are considered as a good source of protein having 12–40% protein on an average. Although 60% share of global protein consumption is occupied by plant-based protein, and remaining 40% is fulfilled by animal-based proteins [8], however, pulse consumption has been increased to certain regions of the world. Currently there is a great concern for the sustainable, clean label product, pulses and legumes are best suited as these are environment-friendly with no carbon footprints or CO_2 emission. Ultimately substituting animal-based protein with plant-based protein would be beneficial from both environment and consumers' health perspective. Similarly the consumers' demand for plant-based protein can be met by advanced research and innovative processing technique with efficient availability of good-quality protein providing key amino acids that play a vital role in the development, reproduction, and support of the human body. Pulses in combination with cereals provide one of the best solutions to protein energy malnutrition being complementary to each other having lysine and methionine, respectively [11], and combination of both cereals and pulses is complimentary for product development with balanced nutritional quality and high protein contents. These products can be claimed as a source of potential protein supply for malnourished population.

This chapter covers the importance of nutrients of pulses (legume grains) especially protein and their importance and strategies for industrial application and processing industry for developing target-oriented protein-enriched products.

2. Nutritional importance of pulses

Whole and split pulses are a good source of complex carbohydrates, protein, and dietary fiber, having significant amounts of vitamins and minerals. Protein content in legume grains range from 17 to 40%, contrasting with 7–13% of cereals, and being equal to the protein contents of meats (18–25%) [9].

2.1 Micronutrients in pulses

Pulses are a good source of fat-soluble vitamins especially folate, riboflavin, and thiamin. Folate is an important micronutrient to reduce the risk of neural tube defects in newly born during pregnancy. Pulses are rich in vitamin A also but poor source of vitamin C. Pulses are also a good natural source of essential minerals including iron, zinc, selenium, phosphorus, calcium, magnesium, potassium, and chromium. These minerals play an important in human body during different physiological processes such as iron is required for hemoglobin synthesis and antioxidative activity, copper and zinc for lipids and carbohydrates metabolism, calcium is essential nutrient for bone health, copper for enzyme activity and iron metabolism. Pulses are lower in sodium content and helpful for decreasing the trends of different diseases especially hypertension. Pulses are high in iron content, but their bioavailability is low. However, legumes can be a good source of ion if consumed with vitamin-C-rich foods. Iron absorption increased in this way plays an important role in prevention and treating anemia especially in women because during menstruation, there is a high risk of anemia [10].

2.2 Carbohydrates in relation to pulses

Carbohydrates are energy-giving macronutrients, present in pulses up to 60% (dry weight). Leguminous starch is digested slower as compared with starch from cereals and tubers and considered as low glycemic index food for blood glucose control making them suitable for consumption by diabetic patients and those with an elevated risk of developing diabetes. Pulses are gluten-free, a very suitable option for patients suffering from celiac disease and persons who are sensitive to gliadin and glutenin proteins. Pulses are a valuable source of dietary fiber 5–37% including both soluble and insoluble dietary fiber. The monomers of dietary fiber present in legumes are glucose, galactose, rhamnose, arabinose, fucose, xylose, and mannose. Pulses also contain significant amounts of resistant starch and oligosaccharides, mainly raffinose, which have been reported to possess prebiotic properties. These are fermented to short-chain fatty acids by probiotics, helpful for improving colonic health and reducing the risk of colon cancer [11].

2.3 Dietary fiber in pulses

Pulses are an excellent source of dietary fiber and other complex carbohydrates. A wide variation is present in the amount of dietary fiber with a significant ratio of soluble and insoluble fiber. Depending on the specie, total dietary fiber ranges from 14 to 32% of dry weight [12]. Various types of dietary fiber present in pulses that include galacto oligosaccharides, long- and short-chain soluble and insoluble polysaccharides, and resistant starch. Insoluble dietary fiber is helpful in laxation while soluble dietary fiber is related to reducing the cholesterol levels and maintaining the post-prandial glucose level. Both types of fiber can act as prebiotics and are helpful in supplying nutrients to gut microorganisms. Fiber-rich fractions of pulses can be added to processed foods to increase their fiber content. Despite the nutritional and health-promoting effects, pulse fiber can also utilized to improve the textural properties by binding and retaining fat and moisture in food items [13].

Pulse fibers are important for individuals seeking a healthy, disease-free lifestyle. High-fiber and low-glycemic diets are important for preventing and treating many diseases/conditions including diabetes, constipation, heart complications, piles, and also some cancers. Furthermore, dietary fiber especially soluble dietary fiber has the ability to improve glucose tolerance and helps to lower the cholesterol by forming a gel lining along the intestinal wall that acts as a protective layer, thus decreasing the glucose and cholesterol assimilation into the blood stream while insoluble dietary fiber helps in increasing fecal bulk and stimulating normal laxation because it has low densities [14]. Pulses are an invaluable part of the human diet. Dietary fiber fractions of pules have found use in the bakery, meat, extruded products, and beverage industries as stabilizers, texturing agents, bulking agents, fat replacers, and emulsion stabilizers. Legume starch isolates have been employed as thickening agents in soups and gravies in the food industry [15].

2.4 Fatty acids composition of pulses

Pulses are generally low in fat, free from saturated fatty acids. The fat in pulses constitutes significant amounts of mono- and polyunsaturated fatty acids (PUFAs). The highest amount of poly unsaturated fatty acids of 71.1% in kidney beans and mono-unsaturated fatty acids of 34% in chickpeas are reported. The polyunsaturated fatty acids are present in legumes that include essential omega 6 linoleic acid and omega 3 alpha linolenic acid. These fatty acids must be included in diet because these are essential and cannot be synthesized in human body [16].

3. Antioxidant capacity of pulses

Pulses contain non-nutrient bioactive compounds such as phytochemicals and antioxidants include isoflavones, lignans, protease inhibitors, trypsin and chymotrypsin inhibitors, saponins, alkaloids, phytoestrogens, and phytates. Most of these chemicals are termed "anti-nutrients" and although they are nontoxic. Most of these anti-nutrients are heat-labile, and since pulses are consumed after cooking, they do not pose a health hazard. These anti-nutritional substances can be removed by different procedures such as boiling, soaking, de-hulling, steaming, roasting, fermentation, and sprouting before consumption.

Different studies have shown that many of non-nutrient components are phytochemicals that exhibit antioxidant characteristics, which play an important role in human body to protect from different diseases such as cancers, osteoporosis, heart diseases, and other degenerative diseases. The antioxidant capacity of pulses is helpful to prevent or stop the oxidative process that leads to many degenerative diseases. As such, the incorporation of pulses into human diets all over the world could offer protection against chronic diseases. Therefore, legumes, especially pulses, should be explored for the development of innovative, value-added products [17].

4. Pulse proteins

Pulses are well thought to be a good source having protein ranging from 20 to 40% d.m [18]. Proteins from different pulses vary in composition and structure and have different functional properties. The major proteins found in most pulses comprise globulins and albumins.

4.1 Globulins

Globulins are soluble in salt solutions, and albumins are soluble in water. Globulins accounts for 70–80% of seed protein. These are primarily storage proteins. These proteins are further divided into two types, i.e., Legumins and Vicilins, also called 7S and 11-12S globulins on the basis of their sedimentation coefficient. Molecular weight of legumins vary from 300 to 400 kDa. Legumins have higher amount of sulfur-containing amino acids (methionine and cysteine) as compared with Vicilins [19] Molecular weight of viciline is 145–190 kDa. These proteins are trimers of monomers either identical or nonidentical. This globulin does not have cysteine residue and thus lacks disulfide bond.

4.2 Albumins

Albumins are most nutritive proteins in pulse seeds in terms of amino acid composition. Albumins are primarily composed of metabolic proteins including enzymatic and non-enzymatic proteins. Only 10–20% seed proteins are made up of albumins. These proteins are generally low in molecular weight (MW; 5–80 kDa) and higher in cysteine and methionine content than pulse globulins [19] Albumins may also contain some anti-nutritional components such as trypsin or chymotrypsin inhibitors, amylase inhibitors, hemagglutinins, lectins, etc.

4.3 Prolamins and glutelins

Prolamins and glutelins are present in minor quantity. Prolamins are soluble in alcohol, and glutelins are soluble in dilute acid/base. Protamine has high concentration of proline and glutamine, and glutelins have high concentration of methionine and cystine. Globulin is the major fraction of embryo and cotyledons of pulses [20].

5. Amino acids in pulses

All proteins are created from 20 different amino acid building blocks. Out of these 20, nine amino acids are those that cannot be produced by the body and are called "essential," and they must be obtained from food source. Each amino acid within the body is associated with specific function. Lysine and arginine are found to be associated with the release of growth hormone in young children. In the early years of age, protein intake is positively associated with height and weight. Hence children with lower serum level of essential amino acids particularly arginine, glycine, and glutamine lead to stunting and wasting. Pulses can provide potential ingredient for the intervention in such types of ailments. Branch chain amino acids (BCAA), e.g., leucine, are reported to play significant role in regulating signaling pathways of muscle protein, valine for repair, and isoleucine in muscle growth. Likewise each amino acid performs a specific function during different life stages, from infancy to elders [21].

Legumes containing relatively low quantity of methionine, essential amino acid, compared with egg, red meat, or poultry meat, are suitable as complimentary source with low lysine-containing cereals such as wheat. It can be source of good-quality protein. Protein quality is defined by Food and Agriculture Organization (FAO) as the capacity of a food protein source and diet to meet the protein and essential amino-nitrogen requirements [22]. Quality can be evaluated in terms of amino acid composition and protein digestibility. It is essential to have a good balance of amino acids in order to synthesize enough protein in the cells to keep the body healthy. Dietary intake of proteinaceous meal comprising sufficient quantity of balanced is essential for adults as well as growing children. If the dietary intake of amino acids is unbalanced, the amino acid that is most limiting becomes the bottleneck for the amount of protein synthesized.

Pulse consumption not only fulfills essential amino acid requirement within the body besides that lower methionine intake is responsible for reduction in oxidative stress by decreasing mitochondrial ROS generation and damage of the liver, which would ultimately increase longevity by this dietary manipulation. Amino acid composition of some common pulses is given in the table comprising essential and non-essential amino acids (**Table 1**).

6. Application of pulses in processing industry

Pulses and legumes have been recognized as valuable since hundreds of years ago, as a low-cost source of high-quality protein products such as flour, concentrates, and isolates. However, the pulse flour application on an industrial scale is only limited to soybean proteins and to lesser extent pea protein products, owing to insufficient information regarding functional properties of other pulse flours. Being rich in protein, with essential amino acid composition along with dietary fiber and other micronutrients such as minerals, vitamins, and folates, pulses are best suited for the formulating and enrichment of food products [31].

Pulse flour as a whole or pulse flour fractions can be utilized in combination with staple cereals such as wheat, rice, barley, etc., to overcome the amino acid lysine deficiency, which is deficient in wheat (a most commonly consumed staple crop)

Essential amino acids										
Amino acids/ type of pulse	**Arginine**	**Histidine**	**Isoleucine**	**Leucine**	**Lysine**	**Methionine**	**phenylalanine**	**Threonine**	**Valine**	**References**
White lupin	0.945	0.808	0.864	0.877	0.891	0.803	0.903	0.821	0.879	[23]
Lentil	0.831	0.36	1.75	0.180	0.723	1.18	0.88	0.990	0.971	[24–26]
Mung bean	0.945	0.808	0.864	0.877	0.905	0.803	0.908	0.829	0.841	[27]
Cowpea	0.750	0.311	0.453	0.778	0.756	0.224	0.756	0.385	0.506	[25, 28]
Green pea	0.724	0.245	0.456	0.745	0.817	0.113	0.524	0.385	0.5.9	[25]
Chickpea	0.488	0.296	0.425	0.759	0.600	0.290	0.789	0.389	0.554	[29]
Feba bean	1.048	0.306	0.373	0.646	0.634	0.060	0.242	0.346	0.454	[28, 30]
Non-essential amino acids										
Non-essential amino acida/ pulses	**Alanine**	**Aspartic acid**	**Cystine**	**Glycine**	**Glutamic acid**	**Proline**	**Serine**	**Tyrosine**	**Tryptophane**	
White lupin	0.954	0.849	0.808	0.851	0.904	0.831	0.852	0.882	—	[23]
Lentil	0.423	1.180	0.094	0.365	2.157	0.353	0.520	0.326	0.650	[24–26]
Mung bean	0.297	0.979	0.140	0.286	2.13	0. 429	0. 532	0.266	—	[27]
Cowpea	0.423	1.088	0.053	1.726	0.282	0.401	0.456	0.305	0.702	[25]
Green pea	0.524	1.106	0.183	0.451	1.750	0.383	0.564	0.373	0.505	[25]
Chickpea	0.497	1.104	0.606	1.730	0.373	0.385	0.546	0.280	0.0902	[29]
Feba bean	0.358	1.032	0.776	0.381	1.625	0.424	0.487	0.305	0.691	[28, 30]

Table 1.
Amino acid composition of some common pulses.

and methionine deficiency in pulses, making complimentary to each other to overcome essential amino acid dearth. Due to low cost and comparative functionality, pulse and pulse proteins find their way in numerous industrial processing applications in cereal-based foods as well as in dairy and meat replacers' food products having improved texture and finishing by increasing water absorption in dough and better.

Modification of protein through various methods of processing may reduce the protein denaturation and further value addition such as preparation of high-protein food supplements using defatted sesame seeds or flour, concentrates of mung bean, lentil, lupins, yellow pea utilized in baking as well as in dairy products. Application of heat, roasting, autoclaving, fermentation, frying, etc., brings more chances for further value addition [30, 32].

The enrichment of bread and other cereal-based confections with legume flours particularly in regions where protein utilization is inadequate has long been recognized [33]. Soybeans are most often modified into a paste, curd, or milk. Soy milk is suitable for lactose-intolerant consumers and emerges as a nondairy substitute for both milk and baby formula, who are unable to digest the lactose that naturally occurs in cows' milk. Tofu, or bean curd, is prepared from curds of soy milk. A variety of other products such as soy cheeses, yogurt, sour cream, and other dairy spreads are prepared from this raw material. Chickpea, mung bean, yellow pea flours have great potential in dairy industry for preparing Imitation cream, ice cream, yogurt owing to their emulsifying abilities, and a host of other varieties. An ice cream–like desert called Tofutti is another well-known tofu product. These are especially welcome products for lactose-intolerant individuals as well as for those wishing to avoid the saturated fat in dairy products [34].

The growing interest in gluten-free, vegan, and vegetarian diets has resulted in an increase in pulse consumption. Bread, a traditional and economical product commonly consumable food throughout the word as a main component of breakfast, is a source of calories and of complex carbohydrates, with a modest amount of essential amino acids such as lysine and threonine. Pulses flour must be included in combination to wheat flour for cereal-based commonly consumed products such as flat bread, leavened bread, pasta, croissants, crackers, chips, cookies, etc.

In many countries, mung bean is used to make mini sweet desserts of different shapes such as vegetables and fruits. Mung bean noodles and breads are also common. Mung beans are prescribed for patients in the hospitals and served with bread. Green gram has good nutritive value, and on germination, it is free of flatulence-causing agents.

Dietary diversification strategy involves combination of more than one type of food source especially diverse plant-based food to improve nutritional health of people who are suffering from malnutrition such as protein energy malnutrition [35]. Pulses and legumes can therefore complement each other when blended at optimum ratio in providing good-quality protein [36].

7. Role of protein in human body

Protein is involved in almost all the body functions taking place such as development of muscle, bone, skin, hair, etc. It makes up the enzymes that trigger many chemical reactions within the body, e.g., the hemoglobin is responsible for carrying oxygen in the blood. Protein plays an important role in growth and muscle building. Protein requirements increase during illness, pregnancy, breastfeeding, and after surgery or an injury. Enzymes are also protein in nature that control metabolic processes within and outside the cell. These enzymes' functions are critical during the process of digestion, blood clotting, muscle contraction, and energy production [37].

Some hormones are also made up of proteins; these are chemical messengers that conduct communication between cells, tissues, and organs. Some hormones are insulin, glucagon, human growth hormone, antidiuretic hormone, adrenocorticotropic hormone. Protein is helpful in maintaining the acid and base balance in body fluids including blood, gastric juice, etc. Proteins required to regulate body fluids as albumin and globulin are present in blood and are helpful to maintain body fluid by attracting and retaining the water. Protein is helpful in the formation of immunoglobulins or antibodies; these are necessary to fight against infections and play an important role in immune system. It also helpful in transporting nutrients, i.e., vitamins, minerals, blood sugars, oxygen, and cholesterol into cells, out of cells, and within cells. Proteins also play an important role in storing nutrients as ferritin is a form of protein that stores iron. Protein provides energy to body in the same amount as carbohydrates 4 g/kcal [38].

8. Outcomes of protein deficiency

Protein becomes deficit when intake of protein does not meet body requirement. According to an estimation, one billion people suffer from inadequate protein intake worldwide, protein deficiency is especially severe in Central Africa and South Asia, where up to 30% of children could not get sufficient amount of protein from their diet. Certain people in the developed countries are also at risk, who follow an imbalanced diet, as well as older people and hospitalized patients. Low protein consumption may result in compositional changes within the body that develop over a long period of time, such as muscle wasting. Kwashiorkor is the most severe form of protein deficiency. It often occurs in children in developing nations because of famine and lack of balanced diets. Protein deficiency can affect almost all aspects of body function [15].

Edema, characterized by swollen and puffy skin, is a classic symptom of kwashiorkor. Scientists believe it is caused by low amounts of human serum albumin and globulin, which is the most abundant protein in the liquid part of blood or blood plasma. When levels of albumin and globulin decrease in body, they are no longer able to regulate blood in blood vessels, and then fluid starts to build in spaces of cells, edema and swelling occur specially in stomach region [39].

Another common symptom of kwashiorkor is a fatty liver or fat accumulation in liver cells. Main cause of this is unknown, but some studies suggest that an impaired synthesis of fat-transporting proteins, known as lipoproteins, may contribute to the condition. Protein deficiency often leaves its mark on the skin, hair, and nails, which are largely made of protein as keratin protein present in hair. For instance, kwashiorkor in children is distinguished by flaky or splitting skin, redness, and patches of depigmented skin. Hair thinning, faded hair color, hair loss (alopecia), and brittle nails are also common symptoms [40].

Muscles are said to be body's largest reservoir of protein. When dietary protein is in short supply, the body tends to take protein from skeletal muscles to preserve more important tissues and perform body functions. Over the time this lack of protein leads to muscle wasting. Even moderate protein insufficiency may cause muscle wasting, especially in elderly people. Bones gives support and shape to body, are also at risk when there is protein deficiency, and risk of fractures increases. One study in postmenopausal women found that a higher protein intake was associated with a lower risk of hip fractures. The highest intake of protein is linked to a 69% reduced risk of fractures [41].

Besides maintaining muscle and bone mass, protein is essential for body growth. Thus, growing age deficiency or insufficiency is especially harmful to children who require a steady supply of protein. In fact, stunting is the most common sign of childhood malnutrition. In 2013, an estimated 161 million children suffered from stunted growth that reaches up to 38.9 million during 2020. Stunted growth is also one of the main characteristics of kwashiorkor in children. Similarly the rate of wasting rises from 149.2 million to 203.6 million during this decade. A protein deficiency can also affect the immune system. Impaired immune functionality may increase the severity of infections. For instance, one study in mice showed that following a diet consisting of only 2% protein was associated with a more severe influenza infection, compared with a diet that provides 18% protein [42].

9. Role of pulses/legumes protein in human health

Pulses can play an important role in food systems to provide global food security and fulfill the nutritional needs in future. Food systems fail to provide safe, sufficient, nutritious food to all due to urbanization, climate change, and increase in population [43]. Leguminous family is highly appreciated as it is a cheap and safe source of nutrients especially protein. Compared with different maize, that is a staple in different regions of world, pulses are a better and effective source of protein and are rich in micronutrients such as iron, zinc, thiamin, folate, and niacin. Nutrient concentration varies in different varieties, locations, and between grains. Human nutrient consumption and status clearly depend on the bioavailability of nutrients. Furthermore, pulses are a good source of essential amino acids especially in lysine (~64 mg/g of protein) and threonine (~38 mg/g of protein), which are complementary to most staple foods, helpful to improve the quality of protein of diet. Pulses offer potential health benefits where future demand of nutritious and cheap food commodities is increasing because of lack of resources and undernutrition [7].

Protein requirements is not same for everyone; it depends on many factors including body weight, muscle mass, age, and physical activity. The recommended daily allowance of protein is 0.4 g per pound of body weight and 0.8 g per kg of body weight. Athletes required a greater amount of protein ranging from 0.5 to 0.6 g per pound of body weight (1.2–1.4 g per kg), required for muscle maintenance and training recovery. Pulses contain 21–25% protein and provide double amount as compared with cereals [44].

Protein is a most satiating macronutrient than fat and carbohydrates, and it is present in a high amount in pulses. Protein elicits the secretion of satiety-related hormones in the small intestine such as peptide YY, Glucagon such as peptide-1. Fiber and protein both are helpful to control satiety; pulses are rich in both fiber and protein and ideal to reduce caloric intake and managing obesity [45].

Pulses are recommended by Canadian and American government agencies as part of a healthy diet. Both Canada's Food Guide (CFG) and the USDA MyPlate nutrition guides place pulses in the meat and alternative group. Animal protein is expensive and not acceptable for many people due to their beliefs and lifestyle. Pulse protein could be a great choice for vegetarian people and meet their essential requirements of amino acids for physiological processes and growth of children. Pulses supply a good amount of protein when consumed with cereals [46].

10. Protein digestibility of pulse-based diet

Protein digestibility can be defined as the percentage of protein or AA intake, absorbed by the digestive tract. It can also predict the estimated individual AA bioavailability [47, 48]. When protein-containing food is consumed, digestion begins from the stomach, and it triggers the release of the hydrochloric acid in the stomach by the partial cells of gastrointestinal mucosa of the gastrointestinal tract. Released acid activates the pepsinogen and converts it into active form, i.e., pepsin. This pepsin can break down the polypeptides into di- and tripeptides, which are ultimately broken down into amino acids; within the duodenum amino acids travel to the liver through hepatic portal vein and undergo de-amination. Amine groups are cleaved to form urea. The amino acids are simultaneously converted into non-essential amino acids or carbohydrate and fats or catabolized directly to energy. Since protein cannot be stored within the body, metabolism of amino acids is completed within few hours. If neither of these actions occurs, then it is released from the body in the form of urine or urea nitrogen contents (**Figure 1**) [49, 50].

When we come to digestibility of pulse-based diet, it is evident that some type of anti-nutritional factors inhibit the digestibility and availability of protein to the body. These may be some types of protein inhibitor components such as trypsin and chymotrypsin inhibitors that arrest the functionality of proteolysis. Trypsin is a digestive enzyme, and the presence of this inhibitor interferes with normal protein digestion in humans. Presence of less digestible protein fractions, high levels of insoluble fiber, and high concentrations of anti-nutritional factors lowers the digestibility of protein. However, processing, cooking, and germination improve the digestibility of pulses. The preparation involves soaking, autoclaving, roasting, fermentation, and germination to reduce anti-nutritional factors (phytic acid, tannins, and polyphenols), which inhibit mineral absorption and protein bioavailability [50–52].

Starch present in pulse grains provides certain health benefits due to its high amylose content. It promotes the formation of resistant starch that cannot be hydrolyzed during digestion. Dietary fiber remains undigested in the small intestine; afterward, it is fermented by the microbiota in the colon that is helpful in controlling weight management, diabetes, and has cholesterol-lowering effect. Fruits and vegetables rich in promoter substances (ascorbate and beta-carotene) for mineral absorption should be taken to enhance micronutrient content and bioavailability. When processed, cooked, or heat-treated, the process of protein digestion by gastrointestinal enzymes as the inhibitory effect on proteolytic enzymes is inactivated [30]. Cooking or heat treatment increases the enzyme activity 2–3 times, chelating activity was also found to be increased in different legumes by cooking. Overall, the

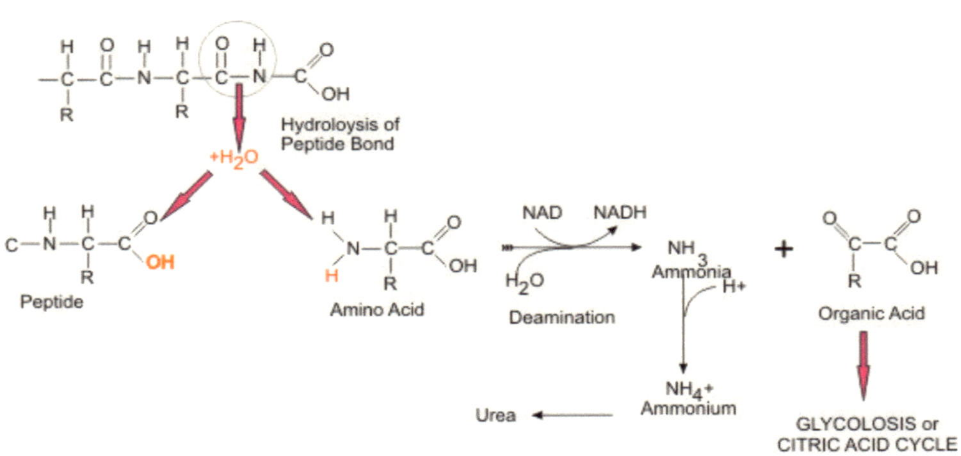

Figure 1.
Catabolism of protein.

total antioxidant capacity values denoted the increased electron donating capacity of the legume seed proteins after digestion with GI enzymes, which could thus act as better radical chain terminators or free-radical stabilizers, when legumes and pulses are treated with heat application [53]. Overall the nutritional value and bioavailability of nutrients, proteins, minerals, phenolic or antioxidant capacity, are increased or improved by processing methods applied to grains as compared when consumed in raw form.

11. Conclusions

Pulses are becoming the corner stone of food and agriculture industry by the time as the awareness of plant-based protein over animal-based protein is revealed and global food security needs to provide balanced diet are publicized. The role of lentils, chickpea, beans, and other pulses becomes even more significant. Future projections suggest 23% global increase in consumption of high protein, high-fiber legumes. To satisfy this need, multidisciplinary approach is required toward research and development sector; efforts in increasing pulse production and consumption lead toward the sustainable food security goal in regional and national food system. Research is needed to identify pulse varieties and innovative processing approaches to develop complimentary food products with balanced available nutrients. Diet diversity and effective processing conditions can improve the nutrient availability and ultimately help to overcome malnutrition and protein deficiency.

Acknowledgements

No funding is provided by any institute or university for this publication.

Conflict of interest

The authors declare no conflict of interest.

Author details

Saima Parveen[1*], Amina Jamil[2], Imran Pasha[2] and Farah Ahmad[2]

1 Post Harvest Research Center, Ayub Agricultural Research Institute, Faisalabad, Pakistan

2 University of Agriculture, Faisalabad, Pakistan

*Address all correspondence to: saimasehar3@gmail.com

References

[1] Graham PH, Ranalli P. Common bean (Phaseolus vulgaris L.). Field Crops Research. 1997;**53**:131-146 https://org/doi:10.1016/s0378-4290(97)00112-3

[2] Papa R, Acosta J, Delgado-Salinas A, Gepts P. A genome-wide analysis of differentiation between wild and domesticated Phaseolus vulgaris from Mesoamerica. Theoretical and Applied Genetics. 2005;**111**:1147-1158 https://org/doi:10.1007/s00122-005-0045-9

[3] Shimelis EA, Rakshit SK. Proximate composition and physicochemical properties of improved dry bean (Phaseolus vulgaris L.) varieties grown in Ethiopia. LWT- Food Science and Technology. 2005;**38**:331-338. DOI: 10.1016/j.lwt.2004.07.002

[4] Hayat I, Ahmad A, Masud T, Ahmed A, Bashir S. Nutritional and health perspectives of beans (Phaseolus vulgaris L.): An overview. Critical Reviews in Food Science and Nutrition. 2014;**54**:580-592. DOI: 10.1080/10408398.2011.596639

[5] FAO. Legumes Can Help Fight Climate Change, Hunger and Obesity in Latin America and the Caribbean. Santiago de Chile: Food and Agriculture Organization of the United Nations and World Health Organization; 2016 FAO. http://www.fao.org/americas/noticias/ver/en/c/409536/

[6] Chandran MDS. The pulses of life. In: Proceedings of Conference on Conversation and Sustainable Management of Ecologically Sensitive Regions in Western Ghats. Karnataka, India: Environmental Information System, CES, Indian Institute of Science; 2016. Available from: http://ces.iisc.ernet.in/energy

[7] Jager DI, Borgonjen Berg JK, Giller KE, Brouwer ID. Current and potential role of grain legumes on protein and micronutrient adequacy of the diet of rural Ghanaian infants and young children: Using linear programming. Nutrition Journal. 2019;**18**:12-28 doi.org/10.1186/s12937-019-0435-5

[8] FAOSTAT. Food and Agriculture Organization of the United Nations, Statistics Division. Food and Agriculture Data. Rome, Italy: FAOSTAT; 2018 Available online at http://faostat.fao.org

[9] Tharanathan RN, Mahadevamma S. Legumes – A boon to human nutrition. Trends in Food Science and Technology. 2003;**14**:507-518. DOI: 10.1186/s12937-019-0435-5

[10] Maphosa Y, Jideani VA. The role of legumes in human nutrition in functional food - Improve health through adequate food. In: María CH, editor. Book of Functional Food-Improve Health Through Adequate Food. United Kingdom: Intechopen; 2017. pp. 104-121. DOI: 10.5772/intechopen.69127

[11] Singh N. Pulses: An overview. Journal of Food Science and Technology. 2017;**54**:853-857 doi.org/10.1007/s13197-017-2537-4

[12] Brummer Y, Kaviani M, Tosh SM. Structural and functional characteristics of dietary fibre in beans, lentils, peas and chickpeas. Foodservice Research International. 2015;**67**:117-125. DOI: 10.1016/j.foodres.2014.11.009

[13] Tosh S, Yada S. Dietary fibres in pulse seeds and fractions: Characterization, functional attributes, and applications. Foodservice Research International. 2010;**43**:450-460. DOI: 10.1016/j.foodres.2009.09.005

[14] Anderson JW, Baird P, Davis RH, Ferreri S, Knudtson M, Koraym A, et al.

Health benefits of dietary fiber. Nutrition Reviews. 2009;**67**:188-205. DOI: 10.1111/j.1753-4887.2009.00189.x

[15] Temba MC, Njobeh PB, Adebo OA, Olugbile AO, Kayitesi E. The role of compositing cereals with legumes to alleviate protein energy malnutrition in Africa. International Journal of Food Science and Technology. 2016;**51**:543-554. DOI: 10.1111/ijfs.13035

[16] Ofuya ZM, Akhidue V. The role of pulses in human nutrition: A review. Journal of Applied Sciences and Environmental Management. 2005;**9**: 99-104. DOI: 10.1111/ijfs.13035

[17] Jamdar SN, Rajalakshmi D, Marathe SA. Effect of processing conditions and in vitro protein digestion on bioactive potentials of commonly consumed legumes. Food Bioscience. 2017;**20**:1-41. DOI: 10.1016/j.fbio.2017.07.007

[18] Patto VMC, Amarowicz R, Aryee ANA. Achievements and challenges in improving the nutritional quality of food legumes. Critical Reviews in Plant Sciences. 2015;**34**:105-143. DOI: 10.1080/07352689.2014.897907

[19] Boulter D, Croy RRD. Legume seed storage proteins. Advances in Botanical Research. 1997;**27**:2-70. DOI: 10.1016/S0065-2296(08)60280-3

[20] Gupta R, Dhillon S. Characterization of seed storage proteins of lentil (Lens culinaris M.). Annals of Biology. 1993;**9**:71-78

[21] Agarwai A. Proteins in pulses. Journal of Nutritional Disorders and Therapy. 2017;7:100-129. DOI: 10.4172/2161-0509.1000e129

[22] Hurrell RF. Influence of vegetable protein sources on trace element and mineral bioavailability. Journal of Nutrition. 2003;**133**:2973s-2977s. DOI: 10.1093/jn/133.9.2973s

[23] Nalle CL, Ravindran V, Ravindran G. Nutritional value of white lupins (Lupinus albus) for broilers: Apparent metabolisable energy, apparent ileal amino acid digestibility and production performance. Animal. 2012;**6**:579-585. DOI: 10.1017/S1751731111001686

[24] Kahraman A. Nutritional components and amino acids in lentil varieties. Selcuk Journal of Agriculture and Food Science. 2016;**30**:34-38

[25] Iqbal A, Khalid NA, Khan MS. Nutritional quality of important food legumes. Food Chemistry. 2006;**97**:331-335. DOI: 10.1016/j.foodchem.2005.05.011

[26] Khazaei H, Subedi M, Nickerson M, Villaluenga CM, Frias J, Vandenberg A. Seed protein of lentils: Current status, progress, and food applications. Food. 2019;**8**:391-414. DOI: 10.3390/foods8090391

[27] Brishti FH, Zarei M, Mohammad K, Fitry MRI. Evaluation of the functional properties of mung bean protein isolate for development of textured vegetable protein. International Food Research Journal. 2017;**24**:1595-1605

[28] Kerr P. World Pulse Day. Rome: Food and Agriculture Organization; 2021. Available from: https://pulses.org/world-pulses-day-guidelines-2020

[29] Arab E, Helmy I, Bareh G. Nutritional evaluation and functional properties of chickpea (*Cicer arietinum* L.) flour and the improvement of spaghetti produced from its. Journal of American Science Journal of American Science. 2010;**66**:1055-1072 http://www.jofamericanscience.org/journals/amsci/am0610/124_3751am0610_1055_1072.pdf

[30] Vogelsang OM, Petersen IL, Joehnke MS, Sørensen JC, Bez J, Detzel A, et al. Comparison of faba bean

protein ingredients produced using dry fractionation and isoelectric precipitation: Techno-functional, nutritional and environmental performance. Food. 2020;**9**:322. DOI: 10.3390/foods9030322

[31] Bresciani A, Alessandra M. Using pulses in baked products: Lights, shadows, and potential solutions. Food. 2019;**8**:451-470. DOI: 10.3390/foods 8100451

[32] Ogabaei M, Parkash J. Effect of primary processing of cereals and legumes on its nutritional quality: A comprehensive review. Journal of Cogent Food and Agriculture. 2016;**2**:1-15. DOI: 10.1080/23311932.2015.1136015

[33] Singh B, Singh JP, Shivkani K, Singh N, Kaur A. Bioactive constituents in pulses and their health benefits. Journal of Food Science and Technology. 2017;**54**:858-870. DOI: 10.1007/s13197-016-2391-9

[34] Anil B, Vijayakumar TP. Properties of industrial fractions of sesame seed (*Sesamum indicum* L.). International Journal of Agricultural and Food Science. 2013;**3**:86-89

[35] Singhal A, Karaca AC, Tyler R, Nickerso M. Pulse proteins: From processing to structure-function relationships. In: Goyal AK, editor. Grain Legumes. United Kingdom: IntechOpen; 2016. DOI: 10.5772/64020. Available from: https://www.intechopen.com/chapters/5095

[36] Okoye JI, Okaka JC. Production and evaluation of protein quality of bread from wheat cowpea flour blends. Contental Journal of Food Science and Technnology. 2009;**3**:1-7

[37] Hatzimanikatis V, Li C, Ionita JA, Broadbelt IJ. Metabolic networks: Enzyme function and metabolite structure. Current Opinion in Structural Biology. 2004;**14**:300-306. DOI: 10.1016/j.sbi.2004.04.004

[38] Hashimoto A, Kambe T. Mg, Zn and Cu transport proteins: A brief overview from physiological and molecular perspectives. Journal of Nutritional Science and Vitaminology. 2015;**61**: S116-S118. DOI: 10.3177/jnsv.61.S116

[39] Coulthard MG. Oedema in kwashiorkor is caused by hypo-albuminaemia. International Journal of Pediatrics. 2015;**35**:83-89. DOI: 10.1179/2046905514y.0000000154

[40] Badaloo A, Reid M, Soares D, Forrester T, Jahoor F. Relation between liver fat content and the rate of VLDL apolipoprotein B-100 synthesis in children with protein-energy malnutrition. The American Journal of Clinical Nutrition. 2005;**81**:1126-1132. DOI: 10.1093/ajcn/81.5.1126

[41] Munger RG, Cerhan JR, Chiu BC. Prospective study of dietary protein intake and risk of hip fracture in postmenopausal women. The American Journal of Clinical Nutrition. 1999;**69**:147-152. DOI: 10.1093/ajcn/69.1.147

[42] Hughes S, Kelly P. Interactions of malnutrition and immune impairment, with specific reference to immunity against parasites. Parasite Immunology. 2006;**28**:577-588

[43] FAO. Climate change and food security: Risks and responses. Rome: Food and Agriculture Organization of United Nations; 2016. Available from: http://www.fao.org/3/i5188e/i5188e.pdf

[44] Danielle K, Thomas C, Riviere T, Tieges Z, Greig CA. Dietary protein in older adults: Adequate daily intake but potential for improved distribution. Nutrients. 2017;**9**:184-194. DOI: 10.3390/nu9030184

[45] Mudryj NA, Yu N, Aukema MH. Nutritional and health benefits of pulses. Applied Physiology, Nutrition, and Metabolism. 2014;**39**(11):1197-1204. DOI: 10.1139/apnm-2013-0557

[46] Havemeier S, Erickson J, Slavin J. Dietary guidance for pulses: The challenge and opportunity to be part of both the vegetable and protein food groups. Analysis of the New York Academy of Sciences. 2017;**1392**:58-66. DOI: 10.1111/nyas.13308

[47] Boye J, Zare F, Pletch A. Pulse proteins: Processing, characterization, functional properties and applications in food and feed. Foodservice Research International. 2010;**43**:414-431. DOI: 10.1016/j.foodres.2009.09.003

[48] FAOSTAT. Food and Agriculture Organization of United Nations, Statistic Division. The State of Food Security and Nutrition in the World. Rome: Food and Agriculture Organization of United Nations; 2019 Available online at http://fao.org/state-of-food-security-nutrition/en/

[49] Lundquist P, Artursson P. Oral absorption of peptides and nanoparticles across the human intestine: Opportunities, limitations and studies in human tissues. Advances in Drug Delivery Review. 2016;**106**:256-276. DOI: 10.1016/j.addr.2016.07.007

[50] McArdle WD, Katch FI, Katch VL. Exercise Physiology: Nutrition, Energy and Human Performance. 7th ed. Philadelphia: Lippincott Williams & Wilkins; 2010 https://lib.ugent.be/catalog/rug01:001906078

[51] Morgan PT, Breen L. The role of protein hydrolysates for exercise-induced skeletal muscle recovery and adaptation: A current perspective. Nutrition and Metabolism. 2021;**18**:1-18. DOI: 10.1186/s12986-021-00574-z

[52] Gibala, Martin J. The Role of Protein in Exercise Recovery. Barrigton: Gatorade Sports Science Institute; [cited: 13 November 2012]. Available from: https://www.active.com/nutrition/articles/the-role-of-protein-in-exercise-recovery

[53] Gilani GS, Cockell AKA, Sepehr E. Effects of Ant nutritional Factors on Protein Digestibility and Amino Acid Availability in Foods. Journal of AOAC International. 2005;**3**:967-987. DOI: 10.1093/jaoac/88.3.967

Chapter 3

Bioactive Peptides from Legumes and Their Bioavailability

Retno Indrati

Abstract

Bioactive peptides (BPs) isolated from legumes have functional properties as healthy foods. These functional effects depend on their stability and bioavailability in the gastrointestinal tract before reaching the target organs. Therefore, it is necessary to disclose the factors that influence it and discuss the technical processing to develop its utilisation. This chapter discusses and summarises the bioactive activities of BPs from various legumes, factors and mechanisms related to the bio-assessability, stability, bio-availability and bioactivity of BPs. Furthermore, the development of BPs bioseparation was also discussed. The results show that the nature of BPs varies greatly depending on the legume source and the production method. Factors that influenced the bio-availability of BPs include molecular weight, charge, amino acid sequence, the presence of specific residues and hydrophobic amino acids, and resistance to the action of peptidase while in the digestive tract. However, some BPs showed increased bio-accessibility and bio-availability after being hydrolyzed by digestive enzymes. Processing technologies such as encapsulation allowing BPs to enter the body and undergo release and degradation by enzymes digestion. Further studies are required to understand the increase in the bioavailability of BPs, the safety of the food components produced, and their use in producing functional foods.

Keywords: Legumes, bioactive peptides, fermentation, germination, health benefits, bioavailability

1. Introduction

Legumes are a cheap and healthy source of nutrition because of their high protein content and complete constituent components, such as fats, essential amino acids, complex carbohydrates, vitamins, and minerals or dietary fibre. This high protein content plays an essential role in producing functional compounds such as bioactive peptides (BPs) that benefit the health and treatment of chronic diseases. For example, BP from legumes is used as an antioxidant compound to prevent degenerative diseases such as atherosclerosis, coronary heart disease, diabetes mellitus, and cancer [1, 2]. The number of deaths due to NCDs (non-communicable diseases), especially cardiovascular disease, cancer, chronic respiratory diseases, and diabetes, increases globally, both in low-income and rich countries. As a result, NCDs are still the cause of most global deaths each year [3]. One way to reduce the risk of NCD is controlling hypertension, regulating diet and obesity.

Protease enzymes have an essential role in producing BP as a result of protein hydrolysis. Food processing, microbial fermentation, germination, or other process involving protease enzymes are examples of proteolytic processes. The involvement of protease enzymes in producing BP is significant because BP is mainly composed of 2–20 amino acids [4]. Some countries have healthy food products from legumes. Examples of fermented foods such as natto [5], douchi [6], tempe [7], and others, have antihypertensive activity. Fermented foods represent, on average, one-third of total food consumption [8]. Fermented food has a delicious taste, easy to digest, nutritious and has beneficial properties. Such as antidiabetic, hypocholesterolemic, and anti-inflammatory activities [8, 9].

Many researchers have proven (both in vitro and in vivo using experimental animals) that BP from legumes has functional properties as a healthy food. However, this functional effect depends on the stability of BP to withstand the action of digestive enzymes while in the digestive tract on its way to reach the target organs [10]. In this target organ, BP will act to provide health effects for the body. One of the determining factors to be bioavailable is the number of amino acids, hydrophobic amino acid content, and resistance to digestive enzyme activity [11]. This chapter aims to describe the quality of various legumes, the BP of legumes and their effects on health. Also, an explanation of the factors that affect the stability, absorption, bioavailability and bio-activity of BP and food technology to develop functional food.

2. Legume, as a source of bioactive peptides

One of the excellent sources of essential amino acids and protein is legumes. As a source of bioactive peptides (BPs), an ingredient must have a high enough protein content. In addition, legumes also contain many components needed for body health, such as antioxidant compounds, resistant starch, dietary fibre and others [12]. However, it is a fact that the nutritional content and phytochemical composition among legumes vary widely, as shown in **Table 1**. The differences in their genetics, varieties, geographical location and climatic conditions may cause the nutritional content variation [18].

In general, the protein content of legumes ranged from 17.0 to 39.8% (w/w). Soybean is the legumes that have the highest protein content. Soybean is also the most studied legume regarding its function on health. According to FAO [22] world soybean production in 2019 was 333,671,692 tonnes (the highest among the types of legumes produced), of which Brazil produced 34.25% as the world's No. 1 producer country. Apart from soybean, some legumes also have a high protein content as a source of BPs, such as jack beans, velvet beans, lima beans, mung beans, and kidney beans (**Table 1**).

In addition to the protein content, it is also necessary to pay attention to the amino acid composition in choosing ingredients. Peptides with hydrophobic amino acids (Tyr, Phe, Trp, Ala, Ile, Val, and Met), positively charged amino acids (Arg and Lys), or contain Pro at the C end will have higher biological activity. For example, inhibition of ACE enzymes [23], Diabetes mellitus type 2 (T2DM) inhibitory activity [24], or other biological functions. Angiotensin-Converting Enzyme Inhibitor (ACEI), is a BP that affects lowering blood pressure. Meanwhile, DPP-IV inhibitors are compounds that can inhibit dipeptidyl peptidase-IV, an enzyme associated with T2DM disease [25]. So the presence of hydrophobic amino acids in short-chain peptides (between 2 and 20 amino acids in length) [4] is related to biological activities beneficial to health. Soybean, jack bean, velvet bean and mung bean are legumes that have high hydrophobic amino acid content (**Table 1**). Enzymatic breakdown through food processing produces short-chain peptides. For example, fermentation (to produce tempeh, soy sauce, natto, miso, douche, or other legume fermented products), germination (mung bean sprouts, soybean sprouts, or other sprouts), or other processes can break the polypeptide chain.

No	Legumes sources	Protein		Amino acid hydrophobic	
		(db, % w/w)	Reference	(% w/w protein)	Reference
1	Soybean, *Glycine max*	35.35–39.80	[13]	37.70	[14]
2	Jack bean, white, *Canavalia ensiformis* L. DC	22.80–35.30	[14]	8.60–43.50	[14]
3	Sword bean, red, *Canavalia gladiata* jacq. DC	32.4–35.0	[15]	26.62–29.47	[15]
4	Cowpea, *Vigna unguiculata* L. Walp	20.90–24.70	[16]	8.46	[16]
5	Bambara groundnut, *Vigna subterranea* L. Verde	17.00–17.30	[16]	6.45	[16]
6	Velvet bean, White, *Mucuna pruriens var. utilis*	28.82 ± 0.14	[17]	34.14	[17]
7	Velvet bean, Black, *Mucuna pruriens var. utilis*	26.26 ± 0.07	[17]	32.79	[17]
8	Kidney beans, *Phaseolus vulgaris* L.	21.80–29.20	[18]	7.96	[16]
9	Lima beans, brown, Phaseolus luna tus	28.06	[7]	4.68	[7]
10	Mung bean, Phaseolus radiata	26.80	[19]	39.75	[20]
11	Chick pea, Cicer Arieti num	19.68–22.75	[16]	35.3	[21]

The level of hydrophobic amino acids is the sum of the data available in the reference sources used.

Table 1.
Protein and hydrophobic amino acid content of several legumes.

Apart from these benefits, legumes contain substances that are considered anti-nutritional compounds [26]. Despite having a high protein content, some toxic anti-nutritional substances limit the use of legumes. Food processing, such as soaking (hydration), cooking, autoclaving, germination, and combination, could reduce or eliminate anti-nutritional compounds [27, 28]; these processes can increase the digestibility value of protein ingredients. **Table 2** shows some of the anti-nutritional compounds present in some legumes. Kalpanadevi and Mohan [26] said that the soaking process and continued germination was less effective in removing anti-nutrients. However, the process would be effective if the germination process was extended (96 hours) or continued with the heating process (or autoclaving process). With this combination process, the effect of anti-nutritional compounds can be eliminated, such as phenolics, tannins, hydrogen cyanide, phytic acid, trypsin inhibitors, oligosaccharides and Phyto-hemagglutination activity.

Other researchers stated that the fermentation process is also very effective for reducing/removing anti-nutritional compounds because the fermentation process is a combination of several processes, including soaking, heating, and proteolytic hydrolysis by starter microbes [31]. For example, the process of fermenting koro bean tempe (*Canavalia ensiformis*) for 48 hours can eliminate 100% of concanavalin-A (Con-A) and reduce almost 99% of its HCN content [31]. Moreover, the fermentation process causes the percentage of peptides with MW <1 kDa to increase to the highest [33]. Short-chain peptides (between 2 and 20 amino acids) are associated with biological activities beneficial to health [4]. As shown by peptides from *Phaseolus lunatus* and *Mucuna pruriens*, peptides with MW < 1 kDa have higher ACE inhibitory activity [7, 34].

No	Legume sources	Antinutritional compound								References
		Phenol (%)	Tannin (%)	L-DOPA (%)	Phytic acid (%)	Trypsin inhibitor (TIU/mg protein)	HCN (ppm)	Oxalate (%)	Total oligosaccharide (%)	
1	Cowpea, *Vigna unguiculata* L. Walp.	1.21	0.38	2.46	0.4	26.48	2.2	NA	NA	[26]
2	Velvet bean, Black, *Mucuna pruriens var. utilis*	2.84	0.26	3.64	0.45	43.1	3.1	NA	NA	[17]
3	Velvet bean, White, *Mucuna pruriens var. utilis*	3.13	0.34	3.24	0.42	43.5	2.1	NA	4.49–6.08	[17, 29]
4	Sword Bean, red, Canavalia gladiate jacq DC.	1.21–1.41	0.16–0.20	2.32–2.64	NA	30.43–34.34	NA	NA	NA	[16]
5	Kidney bean, *Phaseolus vulgaris* L.	NA	0.54–2.88	NA	1.68–2.41	3.65	NA	NA	12.51	[19, 30]
6	Chick pea, *Cicer arietinum*	NA	0.01–0.05	NA	NA	NA	13.5–40.5	0.18–0.45	NA	[17]
7	Jack Bean, white, *Canavalia ensiformis* L. DC.	3, 83	0.083	1.7	0.90	NA	59.95	NA	8.17	[28, 31]
8	Soybean, *Glycine max*	1.40–3.62	1.11–1.88	NA	0.51–2.45	23.67	NA	NA	8.30–10.11	[27, 32]

NA, data not available in the references used.
Data is prepared and recalculated so that it has the same units.

Table 2.
Anti-nutritional compounds in legumes.

3. Bioactive peptides

Bioactive peptides (BPs) are tiny fragments of dietary protein, consisting of 2–20 amino acids, have a molecular weight of less than three kDa, and promote health benefits. After entering the body, BPs can be absorbed in the intestine, carry out various metabolic pathways, and perform various physiological functions [4, 35, 36]. Several researchers have reported that legumes have various biological activities good for body health (**Table 3**).

The hydrolysis of legumes protein can produce these BPs. The enzymatic hydrolysis process occurs in the food processing process, for example, in the fruit ripening process, the fermentation process (producing soy sauce, tempe, natto, and other fermented products), or the germination process (producing soybean sprouts, green bean sprouts, and sprouts products others). In addition, the protein breakdown process can also be carried out by in vitro enzymatic hydrolysis, for example, using the alcalase in legume protein. Some examples of enzymatic hydrolysis are soybean hydrolysis (*Glycine max*) or mung bean hydrolysis (*Vigna radiata*), which produces BP hydrolysate as an ACE inhibitor [40, 45]. The in vitro enzymatic hydrolysis process will produce peptides with enormous structural diversity. Bioinformatics techniques using in silico studies can help select suitable peptide sources. Simulations of biological processes, such as enzyme hydrolysis, can use these in silico studies and further characterise processes and products using software/computers [61].

BPs can perform their activities and roles based on their structural properties, composition and amino acid sequence [62]. Biologically, the active peptides have similar structural properties, including the length of the amino acids, containing hydrophobic amino acids, and resistance to proteolysis [11]. For example, BPs with antioxidant activity have a length of 5–16 amino acids [63]. The structure of ACE inhibiting BPs contains arginine or lysine residues at the C-terminal will affect their activity [11]. Therefore, selecting the protease enzyme to form BPs is essential to produce biologically active peptides. For example, the Carlsberg enzyme subtilysin will hydrolyse peptide bonds with broad specificity to produce peptides with C terminal in the form of hydrophobic amino acids such as Phe, Tyr, Trp, Leu, Ile, Val and Met [64]. In addition, because of their relatively small size and high specificity, BPs can inhibit protein–protein interactions [65]. Some examples of the functional properties possessed by BPs are anti-hypertensive [66], antioxidants [67], hypocholesterolemia [68], antimicrobials [69], anti-inflammatory [70, 71], anti-cancer [59], and other functional properties. One type of BPs can have more than one functional property [9, 65]. To date, researchers are still developing comprehensive studies and reviews to confirm the therapeutic effect of BPs. This chapter will discuss the BPs of legumes and their functions.

3.1 Antidiabetic

Increased blood sugar levels are signs of diabetes caused by decreased insulin secretion, impaired insulin function, or both. In patients with T2DM, the body does not respond adequately to insulin action and the blood glucose level increases, a condition known as hyperglycemia [72]. Changing diet is one way of treating diabetes, besides losing weight, exercising, or taking drugs to increase glucose homeostasis [25]. Side effects from synthetic antidiabetic drugs are gastrointestinal disorders [73]. Other side effects are hypoglycemia and weight gain [74].

Meanwhile, some patients are intolerant of the drug [75]. Therefore, research to find BPs from food as a safe antidiabetic has recently increased to overcome these side effects [76]. Measuring the inhibitory activity of DPP-IV is one way to

	Legumes	Amino acid sequence	Reference
a. Antidiabetic activity			
	Kidney bean, *Phaseolus vulgaris* (L), Fermented	INEGSLLLPH	[9]
	Soybean, germinated	NNDDRDS, LSSTEAQQS, NAENNQRN, QQQQQGGSQSQ, EEPQQPQQ, IKSQSES	[37]
	Cowpea, *Vigna unguiculata*, germinated and enzymatic hydrolysis	TTAGLLE	[38]
	Black bean, *Phaseolus vulgaris*, hydrolysate	AKSPLF, ATNPLF, FEELN, LSVSVL	[39]
b. Antihypertensive activity			
	Kidney bean, *Phaseolus vulgaris* (L), Fermented	FVVAEQAGNEEGFE	[9]
	Mung Bean, *Vigna radiata*, Hydrolysate	KDYRL, VTPALR, KLPAGTLF	[40]
	Soybean, *Glycine max*, Hydrolysis	VLIVP, LAIPVNKP, LPHF, NVVGPLV, YLAGNQ, IPPGVPYWT, DQTPRVF, ASYDTKF, DTKF, PNNKPFQ, RPSYT	[41–45]
	Soybean protein, *Glycine max*, Hydrolysis	IVF, LLF, LNF, LSW, LEF	[46]
	Soybean germinated	RNLQGENEEEDSGA	[37]
	Fermented soybean, natto	VAHINVGK, YVWK	[5]
	Fermented soybean, soy sauce	GY, SY	[47]
	Garden pea, *Pisum sativum*, in silico	LRW	[48]
	Pigeon pea, in silico	VVSLSIPR	[49]
c. Hypocholestcrolcmic activity			
	Kidney bean, *Phaseolus vulgaris* L, Fermented	SGGGGGGVAGAATASR	[9]

	Legumes	Amino acid sequence	Reference
	Soybean, *Glycine max*, Hydrolysate	LPYP, IAVPGEVA, IAVPTGVA, YVVNPDNDEN, YVVNPDNNEN, LPYPR	[50–52]
	Soybean protein, *Glycine max*, in silico	SFGYVAE	[53]
	Black bean and cowpea, in silico	YAAAT	[54]
d. Antioxidant activity			
	Soybean, *Glycine max*, Hydrolysis	LLPHHADADY, LLPHH, LVNPH, DHQN, TTYY, LQSGDALRVPSGTTYY	[42]
	Soybean, *Glycine max*, Hydrolysis	RPSYT	[43]
e. Antimicrobial activity			
	Soybean, *Glycine max*, Hydrolysis	IIVVQGKGAIGF, ASRGIRVNGVAPGPVWTPIQPA, IIIAQGKGALGV,	[42]
	Soybean, *Glycine max*	IKAFKEATKVDKVVVLWTA, PGTAVFK	[55]
f. Anti-inflammatory activity			
	Black bean (*Phaseolus vulgaris* L), hydrolysis	IAISISGLL, CNKY, YETN, QAEEEF, MSAMSNAAA, DLPYSCR, ATL, NLG, EDAY, GYDHPMGGL, PVNF, EEAK, LGAL, DLK, LVVL, VPTK, TGVI, TTW, MEL, FNL, GFTPL, KYGDKSVY, IPVL, KTCENL, GGSSDKR	[56]
	Soybean (Glycine mox)	Lunasin	[57]
g. Anti-cancer activity			
	Soybean (*Glycine max*)	Lunasin	[58]
	Common beans (*Phaseolus vulgaris*), Extra long Autumn Purple Bean cultivar	ANEIYFSFQRFNETNLILQR	[59]
	Chickpea (Cicer anetinum)	ARQSHFANAQP	[60]

Table 3.
The amino acid sequence of several bioactive peptides from legumes.

determine whether BPs have an antidiabetic activity or not. The role of the DPP-IV enzyme is to inactivate incretins, especially GLP-1 and GIP. GLP-1 is a glucagon-like peptide, while GIP is a glucose-dependent autotrophic insulin peptide. Incretin is a hormone that vitalising insulin secretion. So the mechanism commonly used to control T2DM is to measure how much DPP-IV inhibition is [77].

Several low molecular weight BPs can induce insulin stimulation in blood intake, for example, the peptides present in fermented soybeans [78] or fermented kidney beans [9]. Some BPs that are isolated from black bean (*Phaseolus vulgaris* L) protein hydrolyzate effectively inhibits glucose transporter 2 (GLUT2) and glucose transporter, which depends on sodium 1 (SGLT1), which functions to lower blood glucose levels [39]. The BPs found in legumes (such as kidney beans, *Phaseolus vulgaris* L with ten amino acids) have antidiabetic properties [9] (**Table 3**). The table states that the process can produce antidiabetic peptides, including fermentation, germination, or enzymatic hydrolysis.

3.2 Anti-hypertensives

Controlling hypertension is essential to reduce the risk of cardiovascular complications such as coronary heart disease (which causes heart disease) and stroke, congestive heart failure, irregular heart rhythm, and renal failure [3, 79]. A healthy diet is a way to control hypertension. Eating foods high in BPs is very healthy. Several studies have shown that food ingredients derived from legumes have an anti-hypertensive function. The preparation of BPs uses three ways: fermentation of materials into fermented products, germination, and enzymatic hydrolysis. **Table 3** shows some of the research results.

The anti-hypertensive activity was measured by measuring the inhibitory activity of the ACE (Angiotensin I-converting Enzyme). The ACE will cut angiotensin I to produce angiotensin II (vasoactive peptide). This angiotensin II compound will bind to receptors on the walls of blood vessels causing contraction of blood vessels so that blood pressure rises [80]. The presence of BPs will bind to the ACE enzyme, thus inhibiting the action of ACE, and as a result, blood pressure can drop. Some legumes that are recognised to contain BPs that lower blood pressure include garden beans (*Pisum sativum*) [48], green beans [20, 40], soy (*Glycine max*) glycinin [41], kidney bean [9], and pigeon pea [49]. Fermented products also have anti-hypertensive activity, such as douchi, a traditional Chinese food fermented soybean [8]. Other fermented products are Korean soybean paste fermented with mixed cultures of bacteria [81, 82], and many other products from legumes.

Research on anti-hypertensive BPs from food is still being studied [65]. Anti-hypertensive BPs (isolated from food) have a higher tissue binding affinity than synthetic drugs, resulting in slower tissue loss [83]. For vigorous anti-hypertensive activity, the position of specific amino acid residues is critical. For example, valine and isoleucine are essential for ACE inhibition [84]. Increased ACE inhibitory activity occurs when the C-terminal is Proline [84]. Therefore, the strategy to produce peptides with high anti-hypertensive activity is to hydrolyse protein to produce proline containing peptides.

3.3 Hypo-cholesterolemic

Many researchers have studied and reviewed the ability of BPs as cholesterol-lowering agents [65]. The human body needs healthy cholesterol levels to produce vitamin D and steroid hormones, and bile acids. However, arteriosclerosis can occur when cholesterol in the blood forms plaque in the arteries. As a result, it can reduce oxygen supply to the heart, which leads to cardiovascular disease. While chemicals that lower blood cholesterol can cause liver damage or failure, myopathy [85] and diabetes [86, 87], or some people are sensitive to statins (cholesterol-lowering drugs) [88]. Therefore, the research for BPs that can lower cholesterol has increased over the years [65]. **Table 3** shows BPs in legumes (such as red beans and soybeans with 4–16 amino acids) with hypocholesterolemic activity.

Cholesterol reduction by peptides can occur due to inhibition of cholesterol micelle formation, inhibition of lipase activity and strong bile acid-binding [89]. Peptides from fermented soy milk show the ability to bind bile acids [90]. The solubility of cholesterol in lipid micelles will be reduced due to BPs [91], resulting in inhibition of cholesterol absorption in Caco-2 cells with one layer. For example, peptides from cowpeas can inhibit HMGCoA reductase and reduce the dissolution of cholesterol micelles in vitro [92]. A 36% reduction in plasma cholesterol levels could occur in the livers of rats consuming the a'-subunit. The tight binding of BPs with taurocholate, deoxytaurocholate, and glycodeoxycholate can also lead to decreased cholesterol absorption in the intestine [93]. Soybean peptides (LPYP, IAVPGEVA and IAVPTGVA) can activate the LDLR-SREBP 2 pathway to increase LDL uptake effectively. For moderate hypercholesterolemia, 30 g/ml lupine protein consumption effectively reduced the Proprotein Convertase Subtilisin/Kexin type 9 enzyme (PCSK9). Inhibiting HMGCoA reductase activity on HepG2 cells may explain the hypocholesterolemic effect of lupine protein hydrolysate [94]. In addition, peptides cause the regulation of lipoprotein b-VLDL cholesterol receptors to increase in rat liver [95].

3.4 Antioxidants

The antioxidant properties of peptides have more to do with their composition, structure, and hydrophobicity [62]. The amino acid sequence of these peptides can determine different biological activities. Amino acids Tyr, Trp, Met, Lys, Cys, and His are examples of amino acids that cause antioxidant activity [96]. BPs from some legumes have antioxidant properties, for example, soy peptides with 4–16 amino acids [42, 43] (**Table 3**). This table also shows that BP of Leu-Leu-Pro-His-His from soybean β-conglycinin hydrolysate has antioxidant properties. The amino acid leucine or proline at the N-end can increase its antioxidative activity [35]. Amino acids with aromatic residues can donate protons to electron-deficient radicals. This property enhances the radical scavenging character of amino acid residues. Amino acids in the C-terminal region can increase the antioxidant activity higher than in the N-terminal region. This increase in antioxidative activity relates to the nature of the electronic, hydrophobic, steric, and hydrogen bonding amino acids in the area [39]. Soy milk has significant antimutagenic and antioxidant activity. So fermented soy milk probably can prevent mutagenic and oxidative damage [97]. Consumption of douchi (fermented soy food) extract will increase the activity of SOD (Superoxide dismutase) in the liver and kidneys of mice. This consumption also reduces the serum TBARS (Thio Barbituric Acid Reactive Substance), which will increase catalase activity (CAT). These results may indicate the involvement of BPs and free amino acid components from douchi extract as antioxidants [98].

3.5 Antimicrobial

The ability of BPs as antimicrobial peptides (AMP) has also been widely researched and studied [65]. For example, Pina-Pérez and Ferrús-Pérez [55] studied AMP from several legumes against bacterial pathogens that cause foodborne diseases. AMP is generally active against a broad spectrum of microorganisms, including bacteria (Gram+ and Gram-), fungi, and viruses [99]. Some AMPs also show additional activity, such as antioxidant activity [100], immunomodulation [101] and wound healing activity [102]. Therefore, this AMP may be a better choice of antibiotics for pathogenic bacteria resistant to conventional antibiotics.

AMP has various characteristics, including amino acid length (between 12 and 50), amino acid composition, charge and position of disulfide bonds [103]. AMP isolated from soybeans showed that long-chain peptides had higher AMP activity than short peptides [55]. AMP interacts with microbes due to positive charges or hydrophilic and hydrophobic (amphipathic) terminal amino acids, recognised as a prominent structural motif. The charge, hydrophobicity and length of cationic AMP are directly related to their potential as antimicrobials [103]. AMP will cause changes in permeability and osmotic disturbances in bacterial cell membranes [104]. AMP can directly kill bacteria by creating pores through the bacterial cell membrane [101] or interacting with macromolecules in microbial cells [105]. The structure and sequence of peptide amino acids are the main factors for whether or not it is effective as an antimicrobial [104]. Some AMPs are rich in positively charged amino acids (arginine and lysine). Such AMP can enter microbial cells by inducing energy-dependent endocytic pathways such as micropinocytosis [106]. **Table 3** shows some of the AMP amino acid sequences from soybeans.

3.6 Anti-inflammatory

Inflammation is a natural immune system reaction to fight disease. Inflammation is generally associated with cancer because it involves the interaction of various immune cells that can lead to signals of proliferation, growth and invasion of tumour cells [107, 108]. There are two pathways of inflammatory-cancer interaction, namely the extrinsic pathway (inflammation facilitates cancer development) and the intrinsic pathway (genetic changes causing cancer to stimulate the inflammatory process to support tumour development) [109]. Bastiaannet and co-workers [110] and Crawford [111] reported that anti-inflammatory therapy could reduce or prevent cancer risk. This report shows the interaction between inflammatory-cancer. So far, lunasin, VPY and -glutamyl peptides have anti-inflammatory activity [70]. Lunasin exerts an anti-inflammatory effect by inhibiting the Akt-mediated NF-kB pathway [57]. BPs from legumes, particularly soybeans, can regulate several inflammatory markers, which include prostaglandin E2 (PGE2), nitric oxide (NO), induced nitric oxide synthase (iNOS), cyclooxygenase 2 (COX2), cytokines, and chemokines [70]. BPs from *Phaseolus vulgaris* L (the result of hydrolysis of alcalase and digestive enzymes pepsin and pancreatin) also showed similar results, inhibiting important markers and mediators of the inflammatory process [71]. Therefore BPs from *Phaseolus vulgaris* L can assist in managing diseases associated with chronic inflammatory processes such as T2DM and cancer. Several peptides from *Phaseolus vulgaris* L with anti-inflammatory activity are low molecular weight (MW) and contain 3–11 amino acids (**Table 3**). A lunasin-like peptide with low MW (5 kDa) inhibited the most potent pro-inflammatory markers than peptides with MW 8 and 14 KDa [112]. Biopsy of the small intestine mucosa showed repair in intestinal inflammation after supplementation of tempe in the diet [8]. Tempe contains many easily digestible compounds such as peptides and free amino acids that are affecting intestinal growth.

3.7 Anti-cancer

Many researchers are experimenting with cancer therapy using BPs from various legumes. The results are more promising, cheaper, and safer than cancer treatment using surgery, chemotherapy, or radiotherapy, which have adverse side effects due to the emergence of drug resistance and radio-resistance [113]. Extensive exploration has shown that a high intake of legumes can significantly reduce the risk of colorectal adenoma [114, 115] BPs with anti-cancer activity have a relatively low molecular weight (**Table 3**), as isolated from black soybean by-products have the sequence Leu/Ile-Val-Pro-Lys [116]. In comparison, lunasin from soybeans contains 43 amino acids [58]. The hallmark of lunasin is the Arg-Gly-Asp sequence which functions for adhesion to the extracellular matrix, and the 8 Asp sequence to bind chromatin [117]. Kim and co-workers [118] said that hydrophobic BPs isolated from soybeans could act as anti-cancer. Some legumes also have higher hydrophobic amino acids that are similar in levels to soybeans, such as mung bean, chickpea, and velvet bean (**Table 1**), thus potentially producing anti-cancer peptides.

The mechanism of inhibition of tumour growth by BPs varies depending on the variety of legume sources, namely by induction of extrinsic apoptosis [119], induction of chromatin condensation [59] or inhibition of inflammatory processes [120]. BP isolated from chickpea (*Cicer arietinum* L.) inhibits the proliferation of breast cancer cells effectively [60]; this BP has the sequence ARQSHFANAQP. Meanwhile, BP from *Phaseolus vulgaris* (cultivar extra-long autumn purple) can also inhibit the proliferation of human tumour cells by inducing apoptotic bodies and nitric oxide [59]. Apoptosis (programmed cell death) is a complex process coordinated by specific target proteins and, in many cases, possibly responsible for the potential anti-cancer effects [121]. Lunasin can reduce the incidence of skin tumours by 70% [58]. Also, it inhibits gastrointestinal cancer cells [122] and cardiovascular and immunological disorders [123]. The researcher reported that consumption of legumes could reduce the risk of 10 kinds of chronic diseases, including breast cancer, lung cancer and colon cancer [124]. Consumption of legumes in higher amounts will lead to a lower risk of death from cancer [125].

4. Bioacessability, stability, and bioavailability of bioactive peptides

Bioaccessibility, stability, and bioavailability are the main concerns in utilising bioactive peptides (BPs) from food ingredients to remain active in maintaining a healthy body. Bioaccessibility is the first step in the digestive system so that nutrients/BPs out of the food tissue and transported across the intestinal epithelial barrier into the blood circulation system. BP transport processes may involve passive transport (paracellular or passive diffusion) or active routes [126]. During the nutrient transport process, the stability of the material must be kept high, so the bioavailability of nutrients is maintained to be utilised by target cells or tissues. In the digestive tract, nutrients are released from the food matrix and converted into chemical forms that can bind to and enter intestinal cells or pass between them. Dietary factors can also affect the bioavailability of the BPs contained. Interactions between peptides and components of the food matrix can modulate their digestibility and alter the absorption route of the peptide [10]. The release of nutrients in the small intestine starts from chewing, which involves digestive enzymes in the mouth and then in the stomach mixed with acids and enzymes in gastric juices. This whole process is a process for making nutrients biologically accessible [127].

Although the number of active components in the food consumed is abundant, it cannot necessarily prevent disease because it depends on the amount available to function in target organs or tissues [128]. Bioavailability is the number of bioactive compounds that organisms can use effectively [129]. For example, when food contact with the mouth or gastrointestinal tract, various interactions can affect the bioavailability of food nutrients (e.g., the presence of fat can increase the bioavailability of quercetin in food) [130]. In studying the role of bioactive compounds in human health, several factors can inhibit the bioavailability of the active components for use in target organs or tissues [131]. For example, fruit antioxidants mixed with macromolecules form a food matrix such as carbohydrates, fats, and proteins [132].

From a nutritional point of view, bioavailability refers to several nutrient fractions or bioactive compounds that are ingested and can reach the systemic circulation and can finally be utilised [133]. Besides that, bioavailability is the fraction of a nutrient stored or available for a particular physiological function [134]. Another definition, bioavailability, is the amount of active metabolite from the oral dose fraction reaching systemic circulation [135]. The bioavailability of oral BPs is limited because their release from the plant matrix is affected by: solubility in GI fluids, permeability in intestinal epithelial cells, enzymatic and chemical reactions in the GI tract [136]. Four essential steps are required to absorb bioactive compounds effectively: (a) release from the food matrix; (b) incorporation into bile salt micelles; (c) absorption by epithelial cells; and finally; (d) incorporation into the cyclomicron secretion into the lymphatic system.

The biological effects of a BP depend on its capacity to survive until it reaches the target organ. Thus, the main requirement of a BP is its stability or resistance to gastrointestinal enzyme hydrolysis, brush border and serum peptidase. Experimental evidence shows that the length of the peptide chain determines the ability of BPs to pass through the intestinal epithelium in humans by different mechanisms. For small peptides, it is possible to transport through active basolateral, while for large peptides through a transport mechanism mediated by exocytotic-vesicles [137].

However, many peptides are biologically active but are unlikely to be absorbed in the gastrointestinal tract via local effects or receptors that release hormones and cell signalling in the gut. Such BPs affect gastric emptying, gastrointestinal transport, nutrient absorption (amino acids, glucose, lipids) and composition of the colon microflora. They may also regulate food intake [138].

In addition to the presence of specific residues, charge, and molecular weight, hydrogen bonding potential and amino acid hydrophobic tend to affect the bioavailability resistance of BPs to proteases and enzyme hydrolysing peptides [11, 139, 140]. Lunasin, a BP isolated from soybeans and cereal (wheat, barley and rice), has 43 amino acids (MW 5.4 kDa), displays a helical structure and contains nine aspartic acid residues in the C-terminal region. Lunasin is highly bioavailable, heat-stable (100°C, 10 min), and anti-cancer against carcinogenic chemicals. In vivo digestibility of lunasin-fortified soy protein was studied in mice fed for four weeks [141].

During transit in the central digestive tract, the structural properties of the peptide will influence the stability of BPs, including molecular weight, charge, amino acid sequence, and hydrophobicity [126]. Tests using Sprague Dawley rats showed that the highest absorption of ACE inhibitor BPs was in the jejunum [7]. The results showed that BPs with 2–6 amino acids were easy to absorb than proteins and free amino acids [142]. Small (di- and tripeptide) and large (10–51 amino acids) peptides can pass through the intestinal barrier and exhibit their biological function at the target tissue level. However, as the molecular weight of BPs increases, their chances of passing through the intestinal barrier decrease further [143]. The presence of proline and proline hydroxyl will result in resistance of BPs to digestive enzymes, especially a tripeptide with Pro-Pro at the C-terminal [144]. In another study, the number of peptides in human plasma increased depending on the dose of the BP administered. Thus, it concluded that the saturation of BP transporters could affect the number of peptides that can enter the peripheral blood [145].

Encryption of BPs in their natural protein structure may protect these BPs from gastric digestion. Another way to protect BPs is to modify structural proteins such as phosphorylation of serine, threonine, or tyrosine can prevent hydrolysis by digestive proteases. As a result, protein or peptides have a greater chance of being absorbed in target organs or tissues [146]. Stability also depends on the degree of hydrophobicity/hydrophilicity. The more hydrophobic the structure, the more difficult it is to attack by proteases [147].

Therefore, it explained that the difference in bioavailability of BPs between in vitro and in vivo tests (after oral consumption), which may be smaller or larger, occurs due to an increase or decrease in BPs after being catalysed by gastrointestinal proteases. A simulation test of the gastrointestinal digestion process of several tempe legumes (*Phaseolus lunatus* L, *Canavalia ensiformis* L, *Mucuna pruriens*) showed that the proteolysis process by the digestive enzyme pepsin-pancreatin increased ACE inhibitory activity [7, 31, 34]. Another example is that BPs' antioxidant activity in vivo is more significant than in vitro [148]. The shape of the molecular structure also influences the stability of BPs. For example, a small BP (YPI) isolated from ovalbumin has additional stability when tested in GI digestive system simulation. BPs (YPI) and peptides containing P at the C end (RADHP and ADHP) are stable. If their structure is slightly modified by adding one or two amino acids to the C end (e.g. RADHPF, RADHPFL, FRADHPFL), they become unstable in simulated GI hydrolysis [149].

Finally, the use of BPs in nutraceutical and pharmacology for human health is still limited. For that, it is necessary to evaluate: (1) degradation of BPs by proteases in the digestive tract, which can affect bioaccessibility, stability, and bioavailability; (2) the existence of technology that allows modification of the structure of BPs such as (a) phosphorylation of amino acids in BPs to make them more resistant to hydrolysis by digestive enzymes; or (b) increase the amino acid hydrophobic at the N-terminal or C-terminal [150].

5. Technology for bioactive peptides

In general, protein-rich foods that undergo processing involving protease enzymes will produce peptides. However, not all peptides resulting from protein hydrolysis of foodstuffs will become bioactive peptides (BPs) beneficial to body health. The structural properties of BPs (composition, amino acid sequence, hydrophobic amino acid content, and resistance to digestive enzymes) will determine their beneficial functional properties [126], such as, example anti-diabetic, antihypertensive, cholesterol-lowering, antioxidant, and other functional properties.

Food processing processes related to conventional BPs production include cooking, ripening, fermentation and germination. In principle, the processing involves protease enzymes, e.g., chymotrypsin, trypsin, papain, thermolysin, and others) [151], either in the form of free or immobilised enzymes. For food processing by fermentation, protease enzymes derived from microbes are used in the process, while for germination, the enzymes are from growing seeds. Production of BPs increased by regulating the types of enzymes, microbes used, and germination time. Combining these processes (enzymatic process followed by fermentation, or vice versa) will increase the production of BPs so that it is more optimal [152]. The conventional production of the BPs product was a low amount and purity, making it less effective for the industrial scale [153]. So this conventional method for producing BPs does not necessarily involve a separation and purification process, but the production of functional foods containing healthy BPs in the form of fermented food products [153].

The process technology used to produce functional or nutraceutical food will affect the functional, nutritional and biological properties of the protein in the food. Therefore, several things to pay attention to, namely: (1) the effect of using a thermal (or non-thermal) process on the components of the food produced, including its effect on its functional properties and preservation capabilities; (2) available extraction processes and formulations and their optimization; (3) innovative and sustainable applications that can be developed [127]. In addition, consideration of the choice of processing technology must also be based on the desired nutritional function and appearance and sensory properties (such as colour, texture, and taste in the mouth) to be attractive to consumers [154]. Thermal processes can encourage non-enzymatic Maillard reactions between amino groups and reducing sugars [155]. This process will produce colour, sensory properties that affect consumer acceptance and reduce the activity of BPs [155, 156]. The use of thermal processes (e.g. boiling, cooking, blanching, frying, and sterilising) for softening cell walls and inactivate microorganisms and enzymes to make the shelf life longer [127]. The development of non-thermal processes has several weaknesses; for example, the use of nanofiltration membranes requires energy [157]. Freeze-drying, encapsulation, and solvent extraction techniques are costly. To overcome this limitation, food technology experts must develop new alternative technologies (technology that can maintain bio-accessibility, stability, bio-availability and bio-activity of active components). Including BPs, processed food ingredients and the form of pure isolates (capsules or nanocapsules).

The production of BPs has become more accessible, faster, and more effective with the development of science and technology. Production of BPs on an industrial scale usually uses an enzyme hydrolysis process. So the BPs production process uses computer equipment and database search algorithms to predict target peptides and their properties. By selecting the correct protease enzyme through the database, it is possible to select the protein-enzyme combination, in-silico hydrolysis, and the nature of the peptide to be produced [152, 153, 158]. This in-silico hydrolysis method is a functional and widely practised approach for producing legume BPs (**Table 3**).

The legume or various food peptides resulting from enzymatic hydrolysis was then fractionated and purified using a combination of various chromatographic techniques [158–160]. Isoelectric focusing and ultrafiltration are separate macromolecular compounds (such as protein and pectin). Meanwhile, extraction techniques use solvents or supercritical solutions to isolate small molecule bioactive compounds such as antioxidants [157, 161, 162]. This extraction technique, combined with thermal technology (e.g. pasteurisation or spray drying), has been applied to functional foods. This conventional food processing technology is well documented and well established, but its application for the isolation of BPs still needs development and improvement.

The weakness of current technology is that there is still a need for studies on product safety for health. For example, advanced technologies such as cold plasma, nanotechnology, ultrasound, and others, are thought to affect advanced lipid oxidation processes and cause cell tissue damage. For this reason, the effect of this advanced technology on the safety and health of the food components produced needs to be studied to obtain a complete understanding. In this case, it is necessary to adapt the product and technology to the desired functional properties of the active ingredient. For example, modification or interaction with other macronutrients (e.g. dietary fibre) can increase the bio-availability of bioactive compounds [163].

On the other hand, encapsulation technology using legume protein ingredients as a material is also a technique for providing chemical compounds found naturally in plants and other nutraceutical compounds (such as vitamins, minerals, BPs, or others). Thus this encapsulation allowing these compounds (including BPs) to enter the body and undergo release and degradation by enzymes digestion [164]. Other technologies used to protect the active ingredients or nutraceuticals (such as BPs and others) are encapsulation, edible films and coatings, and vacuum impregnation. One may be promising is nutrigenomics, where the active ingredients are given to individuals on a Taylor-made basis according to the genetic characteristics of each individual [165].

Although several researchers have evaluated and characterised BPs that BPs isolated from food have potential bioactive activity and therapeutic functions, and have high bio-availability (bio-accessibility) (due to the support of excellent and modern processing technology), however, all of them can only have a positive impact on human health when combined with healthy living habits [4].

6. Conclusions

Legumes have various biological activities that are good for body health, such as antihypertensive, anti-diabetic, anti-cancer, antioxidant, and others, but legumes also contain anti-nutritional compounds. Food processing is an effective process to remove anti-nutritional compounds and, at the same time, can produce BP compounds that are healthy for the body. Although the number of active components in food is abundant, it is not necessarily able to prevent disease because it is very dependent on the amount available to function in target organs or tissues. One of the contributing factors is the BPs enzyme in holding the action while in the digestive tract. Some legumes showed that hydrolysis by these enzymes increased their bio-accessibility and bio-availability in the digestive tract of rats (in antihypertensive testing). Due to the diverse nature of BP, it is necessary to develop technology that is following the desired functional properties of BP, for example, to protect it so that it is stable while passing through the digestive tract using microencapsulation, edible film and coating technology. Further research still needs to be developed related to the study of safety separation technology for the products produced. From the excellent stability and bioavailability of BPs from legumes, it is likely to be more promising to develop alternative healthy functional food products containing BPs and sensory properties that attract consumers.

Author details

Retno Indrati
Department of Food and Agricultural Product Technology, Universitas Gadjah Mada, Yogyakarta, Indonesia

*Address all correspondence to: indrati@ugm.ac.id

References

[1] Sulistiani HR, Handayani S, Pangastuti A. Characterization of isoflavone bioactive compounds and antioxidant activity test of ethanol extract of tempe made from *Glycine soja*, *Lablab purpureus*, and *Phaseolus lunatus*. Biofarmasi. 2014; 12:62-72. DOI: 10.13057/biofar/f120203 [Indonesian]

[2] Li Y, Jiang B, Zhang T, Mu W, Liu J. Antioxidant and free radical-scavenging activities of chickpea protein hydrolysate (CPH). Food Chem. 2008; 106:444-450. DOI: 10.1016/j.foodchem.2007.04.067

[3] WHO. World Health Statistics 2020: A visual summary. World Health Organization. Available from: https://www.who.int/data/gho/whs-2020-visual-summary. [Accessed: 2021-05-15]

[4] Li Y, Yu J. Research progress in structure-activity relationship of bioactive peptides. J Med Food. 2015; 18:147-156. DOI: 10.1089/jmf.2014.0028

[5] Hernandez-Ledesma B, Amigo L, Ramos M, Recio I. Angiotensin converting enzyme inhibitory activity in commercial fermented products. Formation of peptides under simulated gastrointestinal digestion. J. Agric. Food Chem. 2004; 52:1504-1510. DOI: 10.1021/jf034997b

[6] Zhang JH, Tatsumi E, Ding CH, Li LT. Angiotensin I-converting enzyme inhibitory peptides in douchi, a Chinese traditional fermented soybean product. Food Chem. 2006; 98:551-557. DOI: 10.1016/j.foodchem.2005.06.024

[7] Pertiwi MGP, Marsono Y, Indrati R. In vitro gastrointestinal simulation of tempe prepared from koro kratok (*Phaseolus lunatus* L.) as an angiotensin-converting enzyme inhibitor. J Food Sci Technol. 2020; 57:1847-1855. DOI: 10.1007/s13197-019-04219-1

[8] Nout MJR, Kiers JL. Tempe fermentation, innovation, and functionality: update into the third millennium. J Appl Microb. 2005; 96:789-805. DOI: 10.1111/j.1365-2672.2004.02471.x

[9] Jakubczyk A, Karaś M, Złotek U, Szymanowska U. Identification of potential inhibitory peptides of enzymes involved in the metabolic syndrome obtained by simulated gastrointestinal digestion of fermented bean (*Phaseolus vulgaris* L.) seeds. Food Res Int'l, 2017; 100: 489-496. DOI:10.1016/j.foodres.2017.07.046

[10] Amigo L, Hernández-Ledesma B. Current Evidence on the Bioavailability of Food Bioactive Peptides. Molecules. 2020; 25: 4479. DOI: 10.3390/molecules25194479

[11] Carbonaro M, Maselli P, Nucara A. Structural aspects of legume proteins and nutraceutical properties. Food Res Int'l. 2015; 76: 19-30. DOI: 10.1016/j.foodres.2014.11.007

[12] Zhang B, Deng Z, Ramdath DD, Tang Y, Chen PX, et al. Phenolic profiles of 20 Canadian lentil cultivars and their contribution to antioxidant activity and inhibitory effects on α-glucosidase and pancreatic lipase. Food Chem. 2015; 172:862-872. DOI: 10.1016/j.foodchem.2014.09.144

[13] Ciabotti S, Silva ACBB, Juhasz ACP, Mendonça CD, Tavano OL, et al. Chemical composition, protein profile, and isoflavones content in soybean genotypes with different seed coat colors. Int'l Food Res J. 2016; 23:621-629

[14] Sridhar KR, Seena S. Nutritional and antinutritional significance of four unconventional legumes of the genus *Canavalia* - A comparative study. Food Chem. 2006; 99: 267-288. DOI: 10.1016/j.foodchem.2005.07.049

[15] Vadivel V. The Nutritional and Antioxidant Contents of Wild Jack Bean (*Canavalia ensiformis* L. Dc.): An Under-Exploited Legume From South India. Int'l J Recent Sci Res. 2019; 10:35502-35508. DOI: 10.24327/ijrsr.2019.1010.4116

[16] Baptista A, Pinho O, Pinto E, Casal S, Mota C, et al. Characterization of protein and fat composition of seeds from common beans (*Phaseolus vulgaris* L.), cowpea (*Vigna unguiculata* L. Walp) and bambara groundnuts (*Vigna subterranea* L. Verdc) from Mozambique. J Food Measurement and Characterization. 2017; 11: 442-450. DOI: 10.1007/s11694-016-9412-2

[17] Kalidass C, Mahapatra AK. Evaluation of the proximate and phytochemical compositions of an underexploited legume *Mucuna pruriens var. utilis* (*Wall ex* Wight) L.H.Bailey. Int'l Food Res J. 2014; 21: 303-308.

[18] Perazzini R, Leonardi D, Ruggeri S, Alesiani D, D'Arcangelo G, et al. Characterization of *Phaseolus vulgaris* L. Landraces cultivated in central Italy. Plant Foods for Human Nutrition (Dordrecht, Netherlands). 2008; 63:211-218. DOI: 10.1007/s11130-008-0095-7

[19] Blessing IA, Gregory IO. Effect of processing on the proximate composition of the dehulled and undehulled mungbean (*Vigna radiata* (L.) Wilczek) Flours. Pakistan J Nutr. 2010; 9:1006-1016. DOI: 10.3923/pjn.2010.1006.1016

[20] Li GH, Le GW, Liu H, Shi YH. Mung-bean protein hydrolysates obtained with alcalase exhibit angiotensin I-converting enzyme inhibitory activity. Food Sci. Technol. Int. 2005; 11:281-287. DOI: 10.1177/1082013205056781

[21] El-Adawy TA. Nutritional composition and antinutritional factors of chickpeas (*Cicer arietinum* L.) undergoing different cooking methods and germination. Plant Foods for Human Nutrition. 2002; 57:83-97. DOI:10.1023/A:1013189620528

[22] FAO. Food and Agriculture Organization of the united nations. 2021. Available from: http://www.fao.org/faostat/en/#data/QC. [Accessed: 2021-05-08]

[23] Sanjukta S, Rai AK. Production of bioactive peptides during soybean fermentation and their potential health benefits. Trends in Food Science and Technology. Elsevier Ltd. 2016; p. 1-10. DOI: 10.1016/j.tifs.2016.01.010

[24] Yan J, Zhao J, Yang R, Zhao W. Bioactive peptidas with antidiabetic properties: a review. Int'l J Food Sci Technol. 2019; 54:1909-1919. DOI: 10.1111/ijfs.14090

[25] Mojica L, Luna-Vital DA, de Mejia EG. Characterization of peptides from common bean protein isolates and their potential to inhibit markers of type-2 diabetes, hypertension and oxidative stress. J Sci Food Agric. 2017; 97:2401-2410. DOI: 10.1002/jsfa.8053

[26] Kalpanadevi V, Mohan VR. Effect of processing on antinutrients and in vitro protein digestibility of the underutilized Legume, *Vigna Unguiculata* (L.) Walp Subsp. Food Sci. Technol. 2013; 51:455-461. DOI: 10.1016/j.lwt.2012.09.030

[27] Sharma S, Goyal R, Barwal S. Domestic processing effects on physicochemical, nutritional and anti-nutritional attributes in soybean (*Glycine max* L. Merill). Int'l Food Res J. 2013; 20: 3203-3209.

[28] Doss A, Pugalenthi M, Vadivel VG, Subhashini G, Subash AR. Effects of processing technique on the nutritional composition and antinutrients content of under –utilized food legume *Canavalia ensiformis* L. DC. Int'l Food Res J. 2011; 18: 965-970.

[29] Vadivel V, Pugelanthi M. Effect of various processing methods on the levels of antinutritional constituents and protein digestibility of *Mucuna pruriens*

(l.) Dc. Var. *utilis* (wall. Ex wight) baker ex burck (velvet bean) seeds. J Food Biochem. 2008; 32:795-812. DOI: 10.1111/j.1745-4514.2008.00199.x

[30] Shimelis EA, Rakshit SK. Antinutritional factors and in vitro protein digestibility of improved haricot bean (*Phaseolus vulgaris* L.) varieties grown in Ethiopia. Int'l J Food Sci Nutr. 2005; 56:377-387. DOI: 10.1080/09637480500512930

[31] Puspitajati E, Cahyanto MN, Marsono Y, Indrati R. Jack bean as tempe ingredients: The safety study and fate of protein against gastrointestinal enzymes. IOP Conference Series: Earth Environ Sci. 2019; 346:012070. DOI: 10.1088/1755-1315/346/1/012070

[32] Thorn KA, Tinsley AM, Weber CW, Berry JW. Antinutritional factors in legumes of the Sonoran Desert. Ecology Food Nut. 1983; 13:251-256. DOI: 10.1080/03670244.1983.9990757

[33] Pebrianti SA, Cahyanto MN, Indrati R. Proteolysis and angiotensin-converting enzyme inhibitory activity of peptide fractions from pigeon pea tempe. Int'l J Sci Res. 2020; 9:306-311. DOI: 10.21275/SR20903093551

[34] Rizkaprilisa W, Marsono Y, Indrati R. Bioactive peptide tempe made from *Mucuna pruriens* (L) DC as an inhibitor of angiotensin-I-converting enzyme (ACE) in a digestion simulation. Prev. Nutr. Food Sci. 2020; 25:93-97. DOI: 10.3746/pnf.2020.25.1.93

[35] Sarmadi BH, Ismail A. Antioxidative peptides from food proteins: A review. Peptides. 2010; 31: 1949-1956. DOI: 10.1016/j.peptides.2010.06.020

[36] Chalamaiah M, Keskin-Ulug S, Hong H, Wu J. Regulatory requirements of bioactive peptides (protein hydrolysates) from food proteins. J Funct Foods. 2019; 58:123-129. DOI: 10.1016/j.jff.2019.04.050

[37] González-Montoya M, Hernández-Ledesma B, Mora-Escobedo R, Martínez-Villaluenga C. Bioactive peptides from germinated soybean with anti-diabetic potential by inhibition of dipeptidyl peptidase-IV, α-amylase, and α-glucosidase enzymes. Int. J. Mol. Sci. 2018; 19: 2883. DOI: 10.3390/ijms19102883

[38] Rocha TdS, Hernandez LMR, Chang YK, de Mejía EG. Impact of germination and enzymatic hydrolysis of cowpea bean (*Vigna unguiculata*) on the generation of peptides capable of inhibiting dipeptidyl peptidase IV. Food Res Int'l. 2014; 64:799-809. DOI: 10.1016/j.foodres.2014.08.016

[39] Mojica L, de Mejia EG, Granados-Silvestre MÁ, Menjivar M. Evaluation of the hypoglycemic potential of a black bean hydrolyzed protein isolate and its pure peptides using in silico, in vitro and in vivo approaches. J. Funct. Foods. 2017; 31:274-286 DOI: 10.1016/j.jff.2017.02.006

[40] Li GH, Wan JZ, Le GW, Shi YH. Novel angiotensin I-converting enzyme inhibitory peptides isolated from alcalase hydrolysate of mung bean protein. J Peptide Sci. 2006; 12:509-514. DOI: 10.1002/psc.758

[41] Gouda KGM, Gowda LR, Rao AGA, Prakash V. Angiotensin I-Converting Enzyme Inhibitory Peptide Derived from Glycinin, the 11S Globulin of Soybean (*Glycine max*). J Agric Food Chem. 2006; 54:4568-4573.DOI: 10.1021/jf060264q

[42] Capriotti AL, Caruso G, Cavaliere C, Samperi R, Ventura S, et al. Identification of potential bioactive peptides generated by simulated gastrointestinal digestion of soybean seeds and soy milk proteins. J Food Composition Anal. 2015; 44:205-213. DOI: 10.1016/j.jfca.2015.08.007

[43] Puchalska P, García MC, Marina ML. Identification of native angiotensin-I

converting enzyme inhibitory peptides in commercial soybean based infant formulas using HPLC-Q-ToF-MS. Food Chem. 2014; 157:62-69. DOI:10.1016/j.foodchem.2014.01.130

[44] Kuba M, Tana C, Tawata S, Yasuda M. Production of angiotensin I-converting enzyme inhibitory peptides from soybean protein with *Monascus purpureus* acid proteinase. Process Biochem. 2005; 40: 2191-2196. DOI:10.1016/j.procbio.2004.08.010

[45] Kodera T, Nio N. Identification of an angiotensin I-converting enzyme inhibitory peptides from protein hydrolysates by a soybean protease and the antihypertensive effects of hydrolysates in 4 spontaneously hypertensive model rats. J. Food Sci. 2006; 71:164-173. DOI: 10.1111/j.1365-2621.2006.tb15612.x

[46] Gu Y, Wu J. LC-MS/MS coupled with QSAR modeling in characterising of angiotensin I-converting enzyme inhibitory peptides from soybean proteins. Food Chem. 2013; 141:2682-2690. DOI: 10.1016/j.foodchem.2013.04.064

[47] Nakahara T, Sugimoto K, Sano A, Yamaguchi H, Katayama H, et al. Antihypertensive mechanism of a peptide-enriched soy sauce-like seasoning: the active constituents and its suppressive effect on renin-angiotensin-aldosterone system. J. Food Sci. 2011; 76:H201-H206. DOI: 10.1111/j.1750-3841.2011.02362.x.

[48] Wu J, Yuan L, Aluko RE. Restriction of the in vitro formation of angiotensin II by leucinyl-arginyl-tryptophan, a novel peptide with potent angiotensin I-converting enzyme inhibitory activity. Biosci Biotechnol Biochem. 2006; 70:1277-1280. DOI: 10.1271/bbb.70.1277.

[49] Nawaz KAA, David SM, Murugesh E, Thandeeswaran M, Gopikrishnan K, et al. Identifcation and in silico characterization of a novel peptide inhibitor of angiotensin converting enzyme from pigeon pea (*Cajanus cajan*). Phytomedicine. 2017; 36:1-7. DOI: 10.1016/j.phymed.2017.09.013

[50] Lammi C, Zanoni C, Arnoldi A. IAVPGEVA, IAVPTGVA, and LPYP, three peptides from soy glycinin, modulate cholesterol metabolism in HepG2 cells through the activation of the LDLR-SREBP2 pathway. J. Funct. Foods 2015; 14:469-478. DOI: 10.1016/j.jff.2015.02.021

[51] Lammi C, Zanoni C, Arnoldi A, Vistoli G. Two peptides from soy beta-Conglycinin Induce a hypocholesterolemic effect in HepG2 Cells by a statin-like mechanism: Comparative in vitro and in silico modeling studies. J. Agric. Food Chem. 2015; 63: 7945-7951. DOI: 10.1021/acs.jafc.5b03497

[52] Yoshikawa M, Fujita H, Matoba N, Takenaka Y, Yamamoto T, et al. Bioactive peptides derived from food proteins preventing lifestyle-related diseases. Bio Factors. 2000; 12:143-146. DOI: 10.1002/biof.5520120122

[53] Pak VV, Koo M, Kwon DY, Yun L. Design of a highly potent inhibitory peptide acting as a competitive inhibitor of HMG-CoA reductase. Amino Acids. 2012; 43:2015-2025. DOI:10.1007/s00726-012-1276-0

[54] Hernandez LMR, de Mejia EG. Bean peptides have higher in silico binding affinities than Ezetimibe for the N-terminal domain of cholesterol receptor Niemann-Pick C1 Like-1. Peptides. 2017; 90:83-89. DOI: 10.1016/j.peptides.2017.02.011

[55] Pina-Pérez MC, Ferrús-Pérez MA. Antimicrobial potential of legume extracts against foodborne pathogens: A review. Trends in Food Sci Technol. 2018; 72:114-124. DOI: 10.1016/j.tifs.2017.12.007

[56] López-Barrios L, Antunes-Ricardo M, Gutiérrez-Uribe JA. Changes in

antioxidant and anti-inflammatory activity of black bean (*Phaseolus vulgaris* L.) protein isolates due to germination and enzymatic digestion. Food Chem. 2016; 203:417-424. DOI: 10.1016/j.foodchem.2016.02.048

[57] Cam A., de Mejia E.G. RGD-peptide lunasin inhibits Akt-mediated NF-κB activation in human macrophages through interaction with the αVβ3 integrin. Mol. Nutr. Food Res. 2012; 56:1569-1581. DOI: 10.1002/mnfr.201200301

[58] Galvez AF, Chen N, Macasieb J, de Lumen BO. Chemopreventive property of a soybean peptide (Lunasin) that binds to deacetylated histones and inhibits acetylation. Cancer Res. 2001; 61: 7473-7478. PMID: 11606382

[59] Fang EF, Lin P, Wong JH, Tsao SW, Ng TB. A lectin with anti-HIV-1 reverse transcriptase, antitumor, and nitric oxide inducing activities from seeds of *Phaseolus vulgaris* cv. Extralong Autumn Purple Bean. J Agric Food Chem. 2010; 58:2221-2229. DOI: 10.1021/jf903964u

[60] Xue Z, Wen H, Zhai L, Yu Y, Li Y, et al. Antioxidant activity and anti-proliferative effect of a bioactive peptide from chickpea (*Cicer arietinum* L.). Food Res Int'l. 2015; 77: 75-81. DOI: 10.1016/j.foodres.2015.09.027

[61] Marshall T. Differences between in vitro, in vivo, and in silico studies. The Marshall Protocol Knowledge Base. 2018. Available from: https://mpkb.org/home/patients/assessing_literature/in_vitro_studies

[62] Chen HM, Muramoto K, Yamauchi F, Fujimoto K, Nokihara K. Antioxidative properties of histidine-containing peptides designed from peptide fragments found in the digests of a soybean protein. J Agric Food Chem 1998; 46:49-53. DOI: 10.1021/jf970649w

[63] Chen HM, Muramoto K, Yamauchi F, Nokihara K. Antioxidant activity of design peptides based on the antioxidative peptide isolated from digests of a soybean protein. J Agric Food Chem. 1996; 44:2619-2623. DOI: 10.1021/jf950833m

[64] Markland FS, Smith EL. Subtilisins: primary structure, chemical and physical properties. In: Boyer PD. (ed.), The Enzyme. New York: Academic Press, 1971; p. 561-608.

[65] Daliri EBM., Oh DH., Lee BH. Bioactive Peptides. Foods. 2017; 6:32. DOI: 10.3390/foods6050032

[66] Jia J, Maa H, Zhao W, Wang Z, Tian W, et al. The use of ultrasound for enzymatic preparation of ACE-inhibitory peptides from wheat germ protein. Food Chem. 2010; 119:336-342. DOI: 10.1016/j.foodchem.2009.06.036

[67] Mendis E, Rajapakse N, Kim SK. Antioxidant properties of a radicals scavenging peptide purified from enzymatically prepared fish skin gelatin hydrolysate. J Agric Food Chem. 2005; 53:581-587. DOI: 10.1021/jf048877v

[68] Zhong F, Liu J, Ma J, Shoemaker CF. Preparation of hypocholesterol peptides from soy protein and their hypocholesterolemic effect in mice. Food Res Int. 2007; 40: 661-667. DOI: 10.1016/j.foodres.2006.11.011

[69] McCann KB, Shiell BJ, Michalski WP, Lee A, Wan J, et al. Isolation and characterisation of a novel antibacterial peptide from bovine a S1-casein. Int Dairy J. 2006; 16:316-323. DOI: 10.1016/j.idairyj.2005.05.005

[70] Reyes-Díaz A, Del-Toro-Sánchez CL, Rodríguez-Figueroa JC, Valdéz-Hurtado S, Wong-Corral FJ, et al. Legume Proteins as a Promising Source of Anti-Inflammatory Peptides. Curr Protein and Peptide Sci. 2019; 20:1204-1217 DOI: 10.2174/1389203720666190430110647

[71] Oseguera-Toledo ME, de Mejıa EG, Dia VP, Amaya-Llano SL. Common bean (*Phaseolus vulgaris* L.) hydrolysates inhibit inflammation in LPS-induced macrophages through suppression of NF-jB pathways. Food Chem. 2011; 127:1175-85. DOI: 10.1016/j.foodchem.2011.01.121

[72] Galicia-Garcia U, Benito-Vicente A, Jebari S, Larrea-Sebal A, Siddiqi H, et al. Pathophysiology of Type 2 Diabetes Mellitus. Int J Mol Sci. 2020; 21:6275. DOI: 10.3390/ijms21176275

[73] Thong K, Gupta PS, Blann A, Ryder R. The influence of age and metformin treatment status on reported gastrointestinal side effects with liraglutide treatment in Type 2 diabetes. Diabetes Res. Clin. Pract. 2015; 109:124-129. DOI: 10.1016/j.diabres.2015.04.009

[74] Thulé PM, Umpierrez G. Sulfonylureas: A new look at old therapy. Curr. Diabetes Rep. 2014; 14:1-8. DOI: 10.1007/s11892-014-0473-5

[75] Dujic T, Causevic A, Bego T, Malenica M, Velija-Asimi Z, et al. Organic cation transporter 1 variants and gastrointestinal side effects of metformin in patients with Type 2 diabetes. Diabet. Med. 2015; 4:511-514. DOI: 10.1111/dme.13040

[76] Chakrabarti S, Jahandideh F, Wu J. Food-derived bioactive peptides on inflammation and oxidative stress. BioMed Res. Int. 2014; 608979. DOI: 10.1155/2014/608979

[77] Oseguera-Toledo ME, de Mejıa EG, Reynoso-Camacho R, Cardador-Mart!ınez A, Amaya-Llano SL. Proteins and bioative peptides: Mechanisms of action on diabetes management. Nutrafoods. 2015; 13:147-157. DOI: 10.1007/s13749-014-0052-z.

[78] Kwon DY, Hong SM, Ahn IS, Kim MJ, Yang HJ, et al. Isoflavonoids and peptides from meju, long-term fermented soybeans, increase insulin sensitivity and exert insulinotropic effects in vitro. Nutrition. 2011; 27:244-252. DOI: 10.1016/j.nut.2010.02.004

[79] Fuchs FD, Whelton PK. High Blood Pressure and Cardiovascular Disease. Hypertension. 2020; 75:285-292. DOI: 10.1161/HYPERTENSIONAHA.119.14240

[80] García-Tejedor A, Sánchez-Rivera L, Castelló-Ruiz M, Recio I, Salom JB, et al. Novel antihypertensive lactoferrin-derived peptides produced by *Kluyveromyces marxianus:* Gastrointestinal stability profile and in vivo angiotensin I-converting enzyme (ACE) inhibition. J. Agric. Food Chem. 2014; 62:1609-1616. DOI: 10.1021/jf4053868

[81] Jung WY, Jung JY, Lee HJ, Jeon CO. Functional Characterization of Bacterial Communities Responsible for Fermentation of Doenjang: A Traditional Korean Fermented Soybean Paste. Front Microbiol. 2016; 7:827. DOI: 10.3389/fmicb.2016.00827

[82] Shin ZI, Yu R, Park SA, Chung DK, Ahn CW, et al. His-His-Leu, an angiotensin I converting enzyme inhibitory peptide derived from Korean soybean paste, exerts antihypertensive activity in vivo. J. Agric. Food Chem. 2001; 49:3004-3009. DOI: 10.1021/jf001135r.

[83] Koyama M, Hattori S, Amano Y, Watanabe M, Nakamura K. Blood pressure-lowering peptides from neo-fermented buckwheat sprouts: A new approach to estimating ACE-inhibitory activity. PLoS ONE 2014; 9: e105802. DOI: 10.1371/journal.pone.0105802

[84] Nakamura Y, Yamamoto N, Sakai K, Okubo A, Yamazaki S, et al. Purification and characterization of Angiotensin I-converting enzyme inhibitors from sour milk. J. Dairy Sci. 1995; 78:777-783. DOI: 10.3168/jds.S0022-0302(95)76689-9

[85] Mancini GJ, Baker S, Bergeron J, Fitchett D, Frohlich J, et al. Diagnosis,

prevention, and management of statin adverse effects and intolerance: Canadian consensus working group update. Can. J. Cardiol. 2016; 32:S35–S65. DOI: 10.1016/j.cjca.2016.01.003

[86] Carter AA, Gomes T, Camacho X, Juurlink DN, Shah BR, et al. Risk of incident diabetes among patients treated with statins: Population Based Study. BMJ 2013; 346: f2610. DOI: 10.1136/bmj.f2610

[87] Chogtu B, Magazine R, Bairy KL. Statin use and risk of diabetes mellitus. World J Diabetes. 2015; 6:352-357. DOI: 10.4239/wjd.v6.i2.352

[88] Ahmad Z. Statin Intolerance. Am J Cardiology. 2014; 113:1765-1771. DOI: 10.1016/j.amjcard.2014.02.033

[89] Siow HL, Choi SB, Gan CY. Structure-activity studies of protease activating, lipase inhibiting, bile acid binding and cholesterol-lowering effects of pre-screened Cumin seed bioactive peptides. J. Funct. Foods. 2016; 27:600-611. DOI: 10.1016/J.JFF.2016.10.013

[90] Yusmarini, Indrati R, Utami T, Marsono Y. Ability of fermented soymilk by *Lactobacillus plantarum*-1 in bile acids binding. Majalah Farmasi Indonesia. 2010; 21:202-208. (in Indonesian)

[91] Zhang H, Yokoyama WH, Zhang H. Concentration-dependent displacement of cholesterol in micelles by hydrophobic rice bran protein hydrolysates. J. Sci. Food Agric. 2012; 92:1395-1401. DOI: 10.1002/jsfa.4713

[92] Marques MR, Freitas RAMS, Carlos ACC, Siguemoto ÉS, Fontanari GG, et al. Peptides from cowpea present antioxidant activity, inhibit cholesterol synthesis and its solubilisation into micelles. Food Chem. 2015; 168:288-293. DOI: 10.1016/j.foodchem.2014.07.049

[93] Lapphanichayakool P, Sutheerawattananonda M, Limpeanchob N. Hypocholesterolemic effect of sericin-derived oligopeptides in high-cholesterol fed rats. J. Nat. Med. 2017; 71: 208-215. DOI: 10.1007/s11418-016-1050-9

[94] Lammi C, Zanoni C, Calabresi L, Arnoldi A. Lupin protein exerts cholesterol-lowering effects targeting PCSK9: From clinical evidences to elucidation of the in vitro molecular mechanism using HepG2 cells. J. Funct. Foods. 2016; 23:230-240. DOI: 10.1016/j.jff.2016.02.042

[95] Duranti M, Lovati MR, Dani V, Barbiroli A, Scarafoni A, et al. The alpha'-subunit from soybean 7S globulin lowers plasma lipids and upregulates liver b-VLDL receptors in rats fed a hypercholesterolemic diet. J. Nutr. 2004; 134:1334-1339. DOI: 10.1093/jn/134.6.1334

[96] Wang WY, de Mejia EG. A new frontier in soy bioactive peptides that may prevent age-related chronic diseases. Compr Rev Food Sci Food Saf. 2005; 4:63-78. DOI: 10.1111/j.1541-4337.2005.tb00075.x

[97] Liu JR, Chen MJ, Lin CW. Antimutagenic and antioxidant properties of milk-kefir and soymilk-kefir. J Agric Food Chem. 2005; 53:2467-2474. DOI: 10.1021/jf048934k

[98] Wang D, Wang LJ, Zhu FX, Zhu JY, Chen XD, et al. In vitro and in vivo studies on the antioxidant activities of the aqueous extracts of Douchi (a traditional Chinese salt-fermented soybean food). Food Chem 2008; 107:1421-1428

[99] Barari L, Mosavi N, Asgharpour F, Asadi A, Moulana Z, et al. Antibacterial and antifungal effect of chickpea (*Cicer arietinum*) aqueous seed extract. World Journal of Pharmaceutical Sciences. 2015; 4:336-346.

[100] Yazdi MM, Asoodeh A, Chamani J. A novel antioxidant and antimicrobial

peptide from hen egg white lysozyme hydrolysates. J. Funct. Foods. 2012; 4:278-286. DOI: 10.1016/j.jff.2011.12.004

[101] Mansour SC, Pena OM, Hancock RE. Host defense peptides: Front-line immunomodulators. Trends Immunol. 2014; 35:443-450. DOI: 10.1016/j.it.2014.07.004

[102] Tomioka H, Nakagami H, Tenma A, Saito Y, Kaga T, et al. Novel anti-microbial peptide SR-0379 accelerates wound healing via the PI3 Kinase/Akt/mTOR pathway. PLoS ONE. 2014; 9:e92597. DOI: 10.1371/journal.pone.0092597

[103] Pane K, Durante L, Crescenzi O, Cafaro V, Pizzo E, et al. Antimicrobial potency of cationic antimicrobial peptides can be predicted from their amino acid composition: Application to the detection of "cryptic" antimicrobial peptides. J. Theor. Biol. 2017; 419:254-265. DOI: 10.1016/j.jtbi.2017.02.012

[104] Li Y, Xiang Q, Zhang Q, Huang Y, Su Z. Overview on the recent study of antimicrobial peptides: Origins, functions, relative mechanisms and application. Peptides. 2012; 37: 207-215. DOI: 10.1016/j.peptides.2012.07.001

[105] Shah P, Hsiao FSH, Ho YH, Chen CS. The proteome targets of intracellular targeting antimicrobial peptides. Proteomics. 2016; 16:1225-1237. DOI: 10.1002/pmic.201500380

[106] Guterstam P, Madani F, Hirose H, Takeuchi T, Futaki S, et al. Elucidating cell-penetrating peptide mechanisms of action for membrane interaction, cellular uptake, and translocation utilizing the hydrophobic counter-anion pyrenebutyrate. Biochim. Biophys. Acta (BBA)-Biomembr. 2009; 1788:2509-2517. DOI: 10.1016/j.bbamem.2009.09.014

[107] Aggarwal BB, Shishodia S, Sandur SK, Pandey MK, Sethi G. Inflammation and cancer: How hot is the link? Biochemical Pharmacology. 2006; 72:1605-1621. DOI: 10.1016/j.bcp.2006.06.029.

[108] Korniluk A, Koper O, Kemona H, Dymicka-Piekarska V. From inflammation to cancer. Irish J Med Sci. 2017; 186:57-62. DOI: 10.1007/s11845-016-1464-0

[109] Vendramini-Costa DB, Carvalho JE. Molecular link mechanisms between inflammation and cancer Current Pharmaceutical Design. 2012; 18:3831-3852. DOI: 10.2174/138161212802083707

[110] Bastiaannet E, Sampieri K, Dekkers OM, de Craen AJM, van Herk-Sukel MPP, et al. Use of aspirin postdiagnosis improves survival for colon cancer patients. Br J Cancer. 2012; 106:1564-1570. DOI: 10.1038/bjc.2012.101

[111] Crawford S. Anti-inflammatory/antioxidant use in long-term maintenance cancer therapy: A new therapeutic approach to disease progression and recurrence. Therapeutic Advances in Medical Oncology. 2014; 6:52-68. DOI: 10.1177/1758834014521111

[112] De Mejia EG, Dia VP. Lunasin and lunasin-like peptides inhibit inflammation through suppression of NFkB pathway in the macrophage. Peptides. 2009; 30:2388-2398. DOI: 10.1016/j.peptides.2009.08.005

[113] Lei Y, Zhang D, Yu J, Dong H, Zhang J, et al. Targeting autophagy in cancer stem cells as an anticancer therapy. Cancer Lett. 2017; 393:33-39. DOI: 10.1016/j.canlet.2017.02.012

[114] Hartman TJ, Albert PS, Zhang Z, Bagshaw D, Kris-Etherton PM, et al. Consumption of a legume-enriched, low-glycemic index diet is associated with biomarkers of insulin resistance and inflammation among men at risk for colorectal cancer. J Nutr. 2010; 140:60-67. DOI: 10.3945/jn.109.114249

[115] Wang Y, Wang Z, Fu L, Chen Y, Fang J. Legume consumption and

colorectal adenoma risk: a meta-analysis of observational studies. PLoS One. 2013; 8:e67335. DOI: 10.1371/journal.pone.0067335

[116] Chen Z, Li W, Santhanam RK, Wang C, Gao X, et al. Bioactive peptide with antioxidant and anticancer activities from black soybean [*Glycine max* (L.) Merr.] byproduct: isolation, identification and molecular docking study. Eur Food Res Technol. 2019; 245:677-689. DOI: 10.1007/s00217-018-3190-5

[117] Galvez AF, de Lumen BO. A soybean cDNA encoding a chromatin-binding peptide inhibits mitosis of mammalian cells. Nat. Biotechnol. 1999; 17:495-500. DOI: 10.1038/8676

[118] Kim SE, Kim HH, Kim JY, Kang YI, Woo HJ, et al. Anticancer activity of hydrophobic peptides from soy proteins. Biofactors. 2000; 12:151-155. DOI: 10.1002/biof.5520120124

[119] Lam SK, Ng TB. Apoptosis of human breast cancer cells induced by hemagglutinin from *Phaseolus vulgaris* cv. Legumi Secchi. Food Chem. 2011; 126:595-602. DOI: 10.1016/j.foodchem.2010.11.049

[120] Ben-Neriah Y, Karin M. Inflammation meets cancer, with NF-κB as the matchmaker. Nat Immunol. 2011; 12:715-723. DOI: 10.1038/ni.2060

[121] Walsh JG, Cullen SP, Sheridan C, Lüthi AU, Gerner C, et al. Executioner caspase-3 and caspase-7 are functionally distinct proteases. Proc Natl Acad Sci USA. 2008; 105:12815-12819. DOI: 10.1073/pnas.0707715105

[122] Fernández-Tomé S, Sanchón J, Recio I, Hernández-Ledesma B. Transepithelial transport of lunasin and derived peptides: inhibitory effects on the gastrointestinal cancer cells viability. J Food Composition Anal. 2017; 68:101-110. DOI: 10.1016/j.jfca.2017.01.011

[123] Hsieh CC, Martínezvillaluenga C, de Lumen BO, Hernándezledesma B. Updating the research on the chemo-preventive and therapeutic role of peptide lunasin. J Sci Food Agric. 2018; 98:2070-2079. DOI: 10.1002/jsfa.8719

[124] Kromhout D, Spaaij CJK, de Goede J, Weggemans RM, Brug J, et al. The 2015 Dutch food-based dietary guidelines. Eur J Clin Nutr. 2016; 70:869-878. DOI: 10.1038/ejcn.2016.52

[125] Papandreou C, Becerra-Tomás N, Bulló M, Martínez-González MA, Corella D, et al. Legume consumption and risk of all-cause, cardiovascular, and cancer mortality in the PREDIMED study. Clin Nutr. 2019; 38:348-356. DOI: 10.1016/j.clnu.2017.12.019

[126] Xu Q, Hong H, Wu J, Yan X. Bioavailability of bioactive peptides derived from food proteins across the intestinal epithelial membrane: A review. Trends in Food Sci Technol. 2019; 86:399-411. DOI:10.1016/j.tifs.2019

[127] Galanakis C, editor. Nutraceutical and Functional Food Components. Effects of Innovative Processing Technique. Elsevier Inc. 2017. 382.p

[128] Manach C, Williamson G, Morand C, Scalbert A, Remesy C. Bioavailability and bioefficacy of polyphenols in humans. I. Review of 97 bioavailability studies. Am J Clin Nutr. 2005; 81:230s-242s. DOI: 10.1093/ajcn/81.1.230S

[129] Blenford D. Bioavailability is key to nutrient effectiveness. Food Ingredients and Process Int'l. 1995; 17:28-30.

[130] Lesser S, Cermak R, Wolffram S. The fatty acid pattern of dietary fat influences the oral bioavailability of the flavonol quercetin in pigs. Br J Nutr. 2006; 96: 1047-1052. DOI: 10.1017/BJN20061953

[131] Carbonell-Capella JM, Buniowska M, Barba FJ, Esteve MJ,

Frigola A. Analytical methods for determining bioavailability and bioaccessibility of bioactive compounds from fruits and vegetables: A review. Comprehensive Rev Food Sci Food Safety, 2014; 13:155-171. DOI: 10.1111/1541-337.12049

[132] Manach C, Scalbert A, Morand C, Remesy C, Jimenez L. Polyphenols: Food sources and bioavailability. Am J Clin Nutr. 2004; 79:727-747. DOI: 10.1093/ajcn/79.5.727

[133] Wood RJ. Bioavailability: Definition, general aspects and fortificants. In: Caballero B, Prentice A, Allen L, editors. Encyclopedia of Human Nutrition. 2nd ed. Academic Press. 2005. 2000 p.

[134] Fairweather-Tait SJ. Bioavailability of nutrients. In: Macrae R, Robinson RK, Sadler MJ, editors. Encyclopaedia of food science, food technology and nutrition. London: Academic Press; 1993. p. 384-388.

[135] Schumann K, Classen HG, Hages M, Prinz-Langenhol R, Pietrzik K, et al. Bioavailability of oral vitamins, minerals and trace elements in perspective. Drug Res, 1997; 47:369-380. PMID: 9150856

[136] McClements DJ, Xiao H. Excipient foods: Designing food matrices that improve the oral bioavailability of pharmaceuticals and nutraceuticals. Food Funct. 2014; 5: 1320-1333. DOI: 10.1039/c4fo00100a.

[137] Segura-Campos M, Chel-Guerrero L, Betancur-Ancona D, Hernandez-Escalante VM. Bioavailability of bioactive peptides. Food Rev Int'l. 2011; 27:213-226. DOI: 10.1080/87559129.2011.563395

[138] Jahan-Mihan A, Luhovyy BL, Khoury DEL, Anderson GH. Dietary proteins as determinants of metabolic and physiologic functions of the gastrointestinal tract. Nutrients. 2011; 3: 574-603. DOI: 10.3390/nu3050574

[139] Carbonaro M, Grant G, Cappelloni M. Heat-induced denaturation impairs digestibility of legume (*Phaseolus vulgaris* L. and *Vicia faba* L.) 11S and 7S globulins in the small intestine of rat. J Sci Food Agric. 2005; 85:65-72. DOI: 10.1002/jsfa.1940

[140] Udenigwe CC, Aluko RE. Food protein-derived bioactive peptides: Production, processing, and potential health benefits. J Food Sci. 2012; 71:11-24. DOI: 10.1111/j.1750-3841.2011.02455.x

[141] Jeong HJ, Jeong JB, Kim DS, de Lumen BO. Inhibition of core histone acetylation by the cancer preventive peptide lunasin. J Agric Food Chem. 2007; 55: 632-637. DOI: 10.1021/jf062405u

[142] Grimble GK. The significance of peptides in clinical nutrition. Annu Rev Nutr. 1994; 14:419-447. DOI: 10.1146/annurev.nu.14.070194.002223

[143] Roberts PR, Burney JD, Black KW, Zaloga GP. Effect of chain length on absorption of biologically active peptides from the gastrointestinal tract. Digestion. 1999; 60:332-337. DOI: 10.1159/000007679

[144] FitzGerald RJ, Meisel H. Milk protein-derived peptide inhibitors of angiotensin-I-converting enzyme. Br J Nutr. 2000; 84:S33–S37. DOI: 10.1017/s0007114500002221

[145] Matsui T, Tamaya K, Seki E, Osajima K, Matsumoto K, et al. Val-Tyr as a natural antihypertensive dipeptide can be absorbed into the human circulatory blood system. Clin Exp Pharmacol Physiol. 2002; 29:204-208. DOI: 10.1046/j.1440-1681.2002.03628.x

[146] Boutrou R, Jardin J, Blais A, Tomé D, Léonil J. Glycosylations of κ–caseinderived caseinomacropeptide reduce its accessibility to endo- but not exointestinal brush border membrane peptidases. J Agric Food Chem. 2008; 56: 8166-8173. DOI: 10.1021/jf801140d

[147] [Hausch F, Shan L, Santiago NA, Gray GM, Khosla C. Intestinal digestive resistance of immunodominant gliadin peptides. Am J Physiology-Gastro-intestinal and Liver Physiology. 2002; 283:996-1003. DOI: 10.1152/ajpgi.00136.2002

[148] Li GH, Le GW, Shi YH, Shrestha S. Angiotensin I-converting enzyme inhibitorypeptides derived from food proteins and their physiological and pharmacological effects. Nutr Res. 2004; 24:469-486. DOI: 10.1016/j.nutres.2003.10.014

[149] Miguel M, Aleixandre A. Antihypertensive peptides derived from egg proteins. J Nutr. 2006; 136:1457-1460.

[150] Gianfranceschia GL, Gianfranceschia G, Quassintib L, Bramuccib M. Biochemical requirements of bioactive peptides for nutraceutical efficacy. J. Funct Foods. 2018; 47:252-263. DOI: 10.1016/j.jff.2018.05.034

[151] Kim SK, Wijesekara I. Development and biological activities of marine-derived bioactive peptides: a review. J Funct Foods. 2010; 2:1-9. DOI: 10.1016/j.jff.2010.01.003

[152] Agyei D, Danquah MK. Industrial-scale manufacturing of pharmaceutical-grade bioactive peptides. Biotechnology Advances. 2011; 29:272-277. DOI: 10.1016/j.biotechadv.2011.01.001

[153] Udenigwe CC. Bioinformatics approaches, prospects and challenges of food bioactive peptide research. Trends Food Sci Technol. 2014; 36:137-143. DOI: 10.1016/j.tifs.2014.02.00

[154] Day L, Seymour RB, Pitts KF, Konczak I, Lundin L. Incorporation of functional ingredients into foods. Trends Food Sci Technol. 2009; 20;388-395. DOI: 10.1016/j.tifs.2008.05.002

[155] Jiang Z, Wang L, Wu W, Wang Y. Biological activities and physicochemical properties of Maillard reaction products in sugar-bovine casein peptide model systems. Food Chem. 2013; 141:3837-3845. DOI: 10.1016/j.foodchem.2013.06.041

[156] Zhao CJ, Hu Y, Schieber A, G€anzle M. Fate of ACE inhibitory peptides during the bread-making process: quantification of peptides in sourdough, bread crumb, steamed bread and soda crackers. J Cereal Sci. 2013; 57:514-519. DOI: 10.1016/j.jcs.2013.02.009

[157] Galanakis CM. Recovery of high added-value components from food wastes: Conventional, emerging technologies and commercialized applications. Trends Food Sci Technol. 2012; 26:68-87. DOI: 10.1016/j.tifs.2012.03.003

[158] Tadesse S, Emire S. Production and processing of antioxidant bioactive peptides: A driving force for the functional food market. Heliyon. 2020; 6:e04765. DOI: 10.1016/j.heliyon.2020.e04765

[159] Lemes AC, Sala L, Ores JDC, Braga ARC, Egea MB, et al. A review of the latest advances in encrypted bioactive peptides from protein-rich waste. Int J Mol Sci. 2016; 17:950. DOI: 10.3390/ijms17060950.

[160] Piovesana S, Capriotti AL, Cavaliere C, Barbera GL, Montone CM, et al. Recent trends and analytical challenges in plant bioactive peptide separation, identification and validation. Anal Bioanal Chem. 2018; 410:3425-3444. DOI: 10.1007/s00216-018-0852-x

[161] Galanakis C.M. Emerging technologies for the production of nutraceuticals from agricultural b y-products: A viewpoint of opportunities and challenges. Food and Bioproducts Processing. 2013; 91:575-579. DOI: 10.1016/j.fbp.2013.01.004

[162] Heng WW, Xiong LW, Ramanan RN, Hong TL, Kong KW, et al.

Two level factorial design for the optimization of phenolics and flavonoids recovery from palm kernel by-product. Industrial Crops and Products. 2015; 63:238-248. DOI:10.1016/j.indcrop.2014.09.049

[163] Dupas C, Baglieri AM, Ordonaud C, Tom D, Maillard MN. Chlorogenic acid is poorly absorbed, independently of the food matrix: A Caco-2 cells and rat chronic absorption study. Mol Nut Food Res. 2006; 50:1053-1060. DOI: 10.1002/mnfr.200600034

[164] Maltais A, Remondetto GE, Subirade M. Soy protein cold-set hydrogels as controlled delivery devices for nutraceutical compounds. Food Hydrocolloids. 2009; 23:1647-1653. DOI: 10.1016/j.foodhyd.2008.12.006

[165] Betoret E, Betoret N, Vidal D, Fito P. Functional foods development: Trends and technologies. Trends Food Sci Technol. 2011; 22:498-508. DOI: 10.1016/j.tifs.2011.05.004

Chapter 4

Nutraceutical Properties of Legume Seeds: Phytochemical Compounds

Hai Ha Pham Thi and Thanh Luan Nguyen

Abstract

Legume seeds have an important role as nutraceuticals in human health (providing protein, carbohydrates, fiber, amino acids, and micronutrients) and act as sustainable food sources in livestock farming and aquaculture. Legume seeds contain a wide range of bioactive compounds that have significant health benefits, mainly classified under phenolic compounds, phytosterols, oligosaccharides, carbohydrates, and saponins. Some of these compounds play an important role in plant defense mechanisms against predators and environmental conditions. Heat-labile antinutritional factors (protease inhibitors and lectins) and heat-stable antinutritional factors (tannins and phytic acid) can be reduced by thermal treatment or postharvest to eliminate any potential negative effects from consumption. Substantial studies have demonstrated that these bioactive compounds possess multiple biological activities, including antioxidant properties, antibacterial, anticancer, anti-inflammatory, antidiabetic, cardiovascular protective. They also have various values for aquaculture, such as fishmeal alternative. In this review, the main bioactive compounds and important biological functions of legume seeds are summarized, and the mechanism of action is discussed.

Keywords: phytochemicals, bioactivities, antioxidant, antibacterial, anticancer, legume seeds

1. Introduction

Legumes, including pulses (dried seed legumes), belong to the Leguminosae family (also called Fabaceae), as shown in the Food and Agricultural Organization (FAO) of the United Nations. Leguminosae is an extensive family of plants with over 18,000 species of various types, and only a limited number are used as human resources or as animal feed. Seeds such as dry beans, broad beans, dry peas, chickpeas, cowpeas, lentils, and mung beans are listed by FAO as being consumed for their high nutrition source of proteins, minerals, vitamins, and bioactive compounds. In general, legumes are known for their high levels of bioactive compounds, such as phenolic compounds, phytosterols, bioactive carbohydrates, and saponins, which aid in the reduction in the risk of oxidant properties, bacteria, cancer, inflammation, and diabetes. Legumes have recently gained popularity as excellent sources of high nutrition and can be vital sources of ingredients for use in functional foods and other applications. Legumes are also a rich source of amino acids such as lysine and tryptophan but are low in sulfur-containing amino acids and may be a cost-effective ally in the fight against malnutrition. They are now regarded as a future superfood, capable of achieving zero hunger at a time when one in every five children under the age of five is chronically malnourished [1]. As a result, people consume legume seeds as a major source of protein worldwide. Legumes are more affordable, especially to low-income families, where consumption of animal protein may be restricted due to economic, social, cultural, or religious factors. In addition, the consumption of legumes has also been demonstrated to be connected with outstanding beneficial health, including hypocholesterolemic, antiatherogenic, anticarcinogenic, cardiovascular protection, and hypoglycemic properties [2].

The nutritional demand for legumes is increasing globally as consumers have become more aware of their nutritional and health benefits. Furthermore, in recent years, more people have substituted vegetable protein for animal protein, increasing demand for legumes, which are the primary source of plant proteins. Therefore, developing good extraction and isolation techniques to obtain a high content of bioactive compounds is critical for legumes to become a competitive source of phytonutrients.

The aim of this review is to concentrate on the phytochemicals and bioactivities of legume seeds on mechanisms of action. Furthermore, the quantities and compositions of these phytochemicals in a variety of legumes are shown. The information demonstrated in this study is helpful for the ingredient selection of legumes for the application of functional foods [1].

2. Bioactive compounds

In a conventional method for extracting bioactive compounds in legume seeds, the first step is to soak the dry seeds in water, followed by heat treatment such as boiling. It is effective in that it increases the nutritional value of legumes to some extent and diminishes the levels of phytates and tannins, leading to a higher starch digestibility [3]. However, phenolic acid and saponins can be destroyed in boiling water.

The isolation and purification of biomolecules, especially phenolic compounds, different types of solvents, such as methanol, hexane, and ethanol in a liquid–solid extraction method, are used based on the polarity of the solute of interest. In a study carried out by Xu BJ et al. [3, 4], 50% acetone can be used to extract bioactive constituents in chickpeas and soybeans, whereas black beans are treated with 70% acetone.

In recent years, new extraction techniques have been developed to provide a significant reduction in extraction time, solvent needed, and energy consumption, as well as to improve compounds recovery (**Table 1**). Microwave-assisted extraction (MAE) and ultrasound-assisted extraction (UAE) have attracted the attention of researchers for isolating bioactive compounds. MAE is an efficient method used in extracting bioactivities from legume seeds [22]. On the other side, UAE has also been used for the extraction by using ultrasound to disrupt legume seed cell walls. This method is regarded to be one of the simplest extraction techniques because it makes use of common laboratory equipment such as an ultrasonic bath [15].

2.1 Phenolic compounds

Phenolic compounds can be easily found in legume seeds. This is a vast group of bioactive compounds, chemically containing at least one benzene ring with one or more hydroxyl substituents, and ranges in complexity from simple phenolic

No.	Compound	Extracts and techniques	Content	Pulse	Unit	References
1	Total phenolic content	70% ethanol	21.9	Lentil	mg GAE/g	[5]
			18.8	Red beans		
			18.7	Soybeans		
			17.0	Mung beans		
2	Gallic acid	70% acetone and HPLC	479.26	Velvet beans	μg/g	[6–8]
			28.64	Black beans		
			24.55	Broad beans		
			12.26	Red kidney beans		
3	*p*-hydroxybenzoic acid	HPLC-DAD	19.2 - 60.5	Chickpea varieties	mg/kg	[9]
	Syringic acid		45.9			
	Gentisic acid		8.1–26.0			
	Protocatechuic acid		12.1–163.5	Pea varieties		
	p-hydroxybenzoic acid		45.5–101.7			
4	Kaempferol	—	6% of TPC	Raw and germination of dark common beans	—	[10]
	Quercetin		26% of TPC			
5	v	Methanol extraction and pressure cooking	Between 50 and 300	Kidney, pinto, black and borlotti beans	μg/g	[11]
			372.2–287.87	Black beans		
	Genistein		1166.15	Soybeans		
	Daidzein		1064.56			
6	Proanthocyanidins	T-25 ULTRA-TURRAX homogenizer	4.09–5.73	Black beans	mg CAE/g	[4, 9]
			13.8	Adzuki bean coats		
			3.73–10.20	Lentil cultivars		
7	Catechin and procyanidins	80% HCl-methanol and HPLC-MS	74.48	Lentil coats	μg/g	[12]
8	Soyasaponin I	Solid phase extraction and HPLC	630–900	Different raw legume samples	mg/kg	[13]
	Dehydrosoyasaponin I		650–1300			
9	Saponins	Ultrasound-assisted extraction (ethanol solvent)	4.55	Lupins	g/100 g	[14–17]
			10.63	Lentils		
			2.97	Chickpeas		
			4.08	Soybeans		
			12.90	Fenugreeks		
10	Total saponin content	70% acetone (0.5% acetic acid)	24.29 (bean hull) 2.20 (whole bean)	Mung beans	mg SbaE/g	[18]
			73.60 (bean hull) 10.82 (whole bean)	Adzuki beans		

No.	Compound	Extracts and techniques	Content	Pulse	Unit	References
11	β-sitosterol	Hexane/diethyl-ether (1:1), saponification and HPLC	15.4–24.2 (cooked)	Lentils	mg/100 g	[19, 20]
			123.4			
	Campesterol		2.18–2.58 (cooked)			
			15.0			
	Stigmasterol		2.60–2.63 (cooked)			
			20.0			
12	Resistant starch	Incubated with enzymes	0.6	Cowpeas	g/100 g	[7]
			3.4	Lentils		
			2.5	Peas		
			2	Kidney beans		
			4.2	White beans		
13	Raffinose	HPLC-HRMS analysis	3.3 (kernel)	Adzuki beans	g/kg	[21]
			2.9 (coat)			
			13.2 (kernel)	Peas		
			11.1 (coat)			
			4.8 (kernel)	Broad beans		
			4.3 (coat)			
			9.2 (kernel)	Green soybeans		
			10.1 (coat)			
			4.0 (kernel)	Mung beans		
			10.3 (coat)			
14	Lectins	—	2.4–5	Total protein in kidney bean seeds	%	[16]
			0.8	Total protein in soybean and lime bean protein		
			0.6	Total protein in pea seeds		

Table 1.
Content of bioactive compounds in legume seeds.

molecules to highly polymerized compounds [23]. The primary phenolic compounds found in legume seeds are phenolic acids, flavonoids, and condensed tannins (**Figure 1**). The distribution of these compounds differs in the cotyledon (mainly containing non-flavonoids, such phenolic acid and hydroxycinnamic) and primarily concentrates on the seed coats (mainly flavonoids) [3, 24, 25]. The phenolic compound may exist in a free, solubilized conjugated form or in an insoluble-bound form. Some free and conjugated phenolic compounds are thought to be absorbed in the small and large intestines; otherwise, the bound forms with associated non-digestive sugars are made bioavailable by the digestion of enzymes or microorganisms present in the intestine lumen [26].

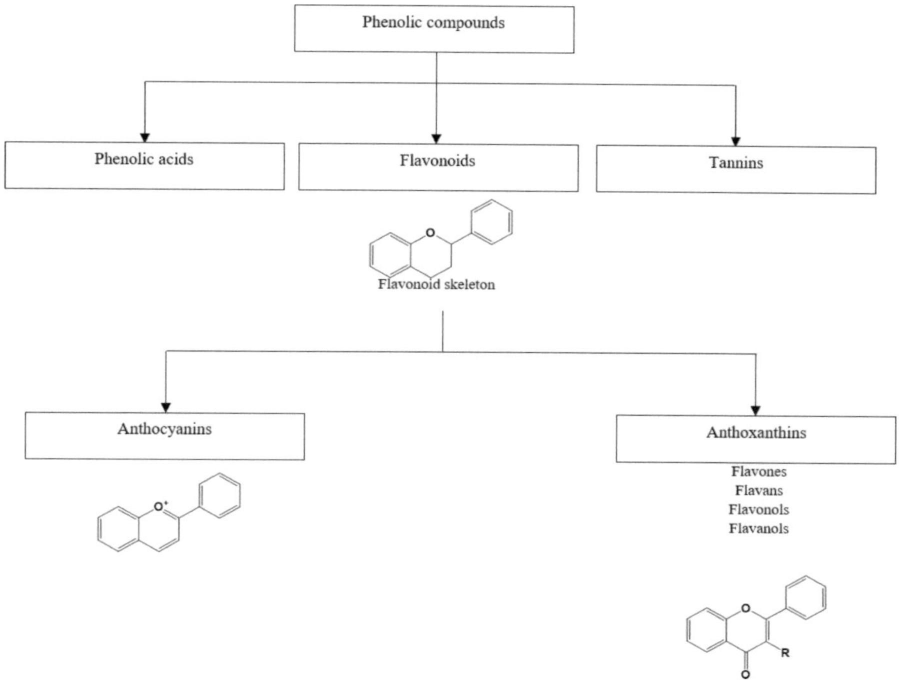

Figure 1.
Phenolic compounds in legume seeds.

Phenolic compounds found in important legumes, including lentil, pea, bean, and chickpea, are flavonoids, such as glycosides of flavonols, flavones, and isoflavonoids (primarily daidzein and genistein), and some hydroxybenzoic and hydroxycinnamic compounds [25]. Lentils were reported to have the highest amount of total phenolic content (TPC), which had 21.9 mg gallic acid equivalents (GAE)/g, compared with soybean, bean, and peas [5], and a slight decrease in red bean and soybean (18.8 and 18.7 mg GAE/g, respectively) and the lowest in mung bean (17.0 mg GAE/g).

In general, processing of legumes (including thermal processing, soaking, and roasting) usually affects the number of phenolic compounds. The study by Lafarga et al. [26] revealed that boiling methods retained more polyphenol than that of cooking broth. This explained that high temperature and the destruction of tissue structures of the cooking process caused the diffusion of phenolics and their leaching into water. By contrast, the germination process in legumes generally improved the nutritional quality, including phenolic compounds [27].

2.1.1 Phenolic acid

According to chemical structure, phenolic acids in legume seeds can be divided into hydroxybenzoic acids and hydroxycinnamic acids (**Figure 2**). Gallic, *p*-hydroxybenzoic, protocatechuic, vanillic, and syringic acids are the most common hydroxybenzoic acids in common beans and are mainly present in foods as glycosides. In addition, caffeic, ferulic, *p*-coumaric, and sinapic acids are the most frequently occurring hydroxycinnamic acids in legumes. The level of gallic acid in velvet beans was the highest (479.26 μg/g), followed by black, broad, and red kidney beans (28.64, 24.55, and 12.26 μg/g, respectively) [3, 28, 29]. Lopez et al. [20] documented that ferulic acid derivatives contained the highest percentage of TPC in

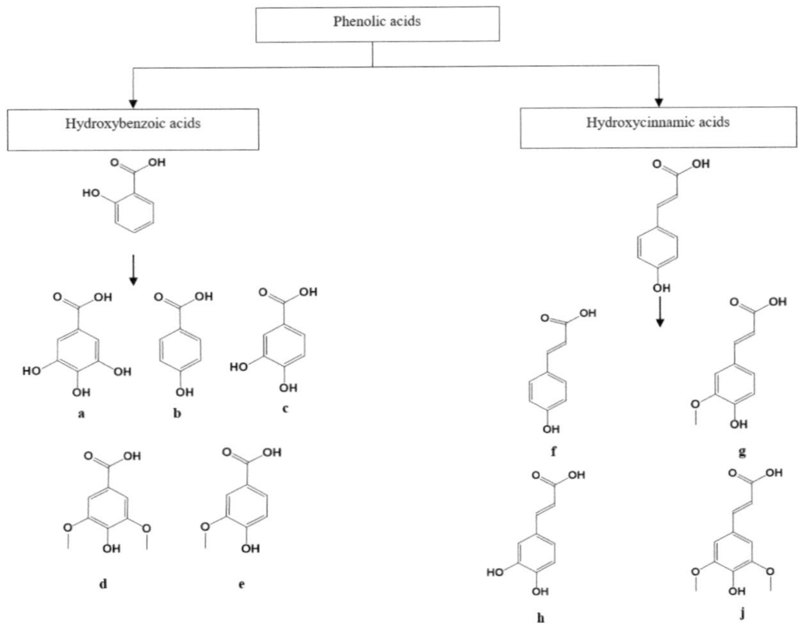

Figure 2.
Chemical structure of major phenolic acid compounds present in legume seeds: Gallic acid (a), p-hydroxybenzoic acid (b), protocatechuic acid (c), syringic acid (d), vanillic acid (e), ferulic acid (f), p-coumaric acid (g), caffeic acid (h), sinapic acid (i).

both raw and cooked dark beans (19 and 24%, respectively). The *p*-hydroxybenzoic acid (19.2 to 60.5 mg/kg), syringic acid (45.9 mg/kg), and gentisic acid (8.1 to 26.0 mg/kg) were presented in significant amounts in seeds of six chickpea varieties, while the six field pea seeds were found to contain protocatechuic acid of between 12.1 and 163.5 mg/kg, and *p*-hydroxybenzoic acid, which ranged from 45.5 to 101.7 mg/kg [7]. Yihan Liu et al. [30] reported that the differences among four types of cooking methods (traditional or boiling, pressure, microwave, and slow) and heating solution can affect the percentage of phenolic acids. Gallic acid content increased after processing in soybean (79.81 from 54.96 μg/g dry weight), while on the contrary, it decreased in black beans (36.02 from 67.88 μg/g dry weight) [31].

2.1.2 Flavonoids

Flavonoids are the main phenolic compound found in legumes, and their presence affects the flavor and color of common beans [7, 20]. They are low-molecular-weight compounds (approximately 300 g/mol), and their general structures are formed with two aromatic rings, joined by a three-carbon bridge, usually in the form of a heterocyclic ring C. Flavonoids are divided into two groups: anthocyanins (colored compounds) and anthoxanthins (colorless compounds) (**Figure 3**). According to the study by Amarowicz and Pegg [32], flavonols, flavan-3-ols (flavanols), flavones, and anthocyanins are the main flavonoids present in leguminous seeds. The presence of flavonols and flavones can impact the color of anthocyanins group [33]. Catechins are called flavan-3-ols, which are primarily identified in legumes as having colored seed coats, such as kidney, navy, and pinto beans. Catechins, along with procyanidins, are common in raw lentil coats and represent 69% of TPC (74.48 μg/g) [12], while other flavonoid glycosides, such as quercetin, myricetin, apigenin, and luteolin, are only found in trace amounts. Duenas et al. [20] also showed that

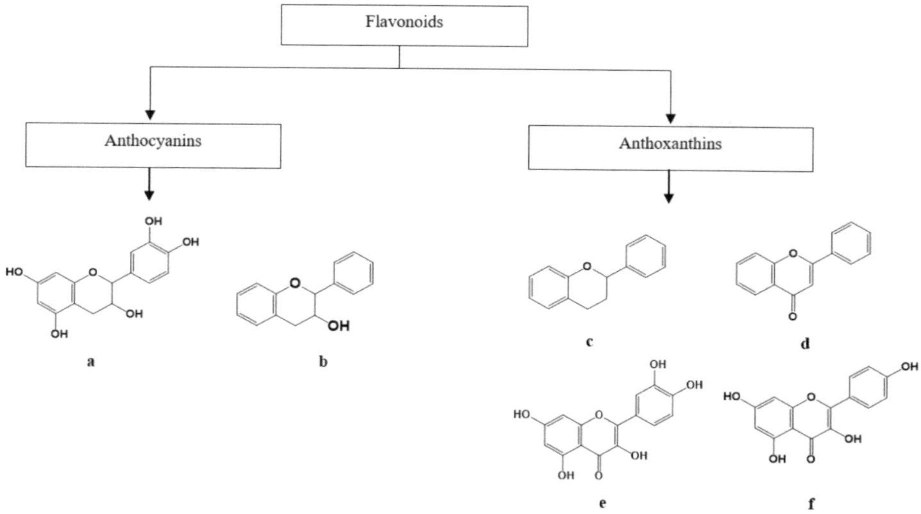

Figure 3.
Chemical structure of major flavonoid compounds: Flavanols (a), catechin (b), flavonols (c), flavones (d), quercetin (e), kaempferol (f).

kaempferol and quercetin in raw and germination of dark common beans contained approximately 6 and 26% of TPC, respectively.

A study by Teixeira-Guedes et al. [31] showed the effect of cooking methods on flavonoid profiles of different varieties of common beans, such as kidney, pinto, black, and borlotti bean. Pressure cooking increased the levels of catechins for all bean varieties, which ranged approximately from 50 to 300 μg/g, except in black beans, where the levels decreased from 372.2 to 287.87 μg/g. The catechin levels in genistein and daidzein content were also increased by this processing and were detected only in soybeans (go up to about 906 and 988 μg/g, respectively).

2.1.3 Condensed tannins

Tannins are polyphenols that are high in molecular weight with large numbers of hydroxyl groups in their structure, which have the ability to bind with carbohydrates and protein, but only to a limited extent. They are classified as hydrolyzable or condensed (non-hydrolyzable tannins), with flavonoids present at various levels of condensation [3]. Condensed tannins, also known as proanthocyanidins (PACs), are chemically oligomeric and polymeric flavonoids, which at high temperature release into catechins and anthocyanidins [7]. Mostly in the same class as flavonoids, PACs mainly distribute in common bean seed coats and play a crucial function in plant defenses that are susceptible to oxidative damage by many environmental factors. Lentils, black beans, and red beans were recorded to contain a high concentration of condensed tannins [34]. In the case of common beans varieties, PACs' content of black beans was in the range of 4.09 to 5.73 mg catechin equivalents (CAE)/g, for adzuki bean coats the content was 13.8 mg CAE/g, and for lentil cultivars, 3.73 to 10.20 mg CAE/g [9, 26].

2.2 Saponins

Saponins are bioactive compounds found in legumes, consisting of a triterpenoid aglycone (sapogenin) linked to one or more oligosaccharide moieties. They

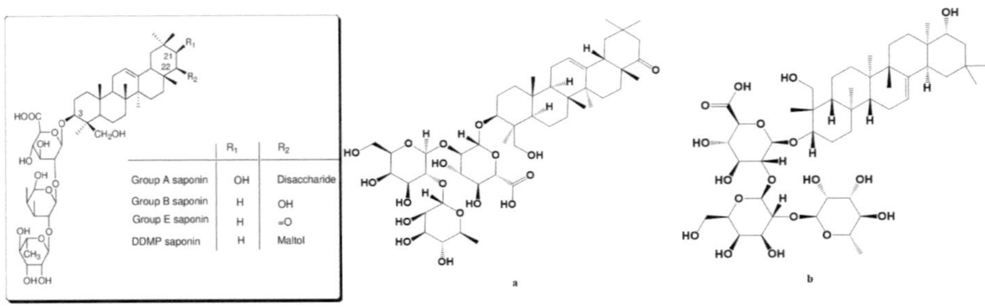

Figure 4.
Chemical structure of major saponins: Dehydrosoyasaponin I (a), soyasaponin I (b) [36].

have the ability to absorb free radicals and activate antioxidant enzymes. The most common saponins include the soyasaponins, which are divided into three groups, A, B, and E saponins, based on the chemical structure of aglycone. Saponins from the B group, which have been studied to be the primary compound in legume seeds [3, 35], contain soyasaponin I (approximately from 630 to 900 mg/kg) and dehydrosoyasaponin I (approximately from 650 to 1300 mg/kg) (**Figure 4**). In contrast to the seed coat or cotyledon part, the hilum portion of the seed has the highest concentration of saponins [37].

Saponins have been investigated in a variety of edible legumes, as well as the effect of solvent during UAE, which has been evaluated from lupins (4.55 g/100 g), lentils (10.63 g/100 g), chickpeas (2.97 g/100 g), soybeans (4.08 g/100 g), fenugreek (12.90 g/100 g), and various beans [14, 15, 17, 38]. The research by Wu et al. also showed that mung beans and adzuki beans contained the highest total saponins content in bean hull (24.29 mg saponins Ba equivalent (SbaE)/g and 73.60 mg SbaE/g, respectively) and whole bean (2.20 mg SbaE/g and 10.82 mg SbaE/g, respectively) [6].

2.3 Phytosterols

Phytosterols, primarily β-sitosterol, campesterol, and stigmasterol, are structurally similar to cholesterol and occur in a variety of plant types (**Figure 5**) [8, 10]. Phytosterols have a double bond at carbon-5 that can be saturated by enzymatic hydrogenation in plants or during food processing to form plant stanols. They are assumed to have a wide variety of biological potentials and are a rich source of grain legumes, vegetable oils, cereal grains, and nuts. β-sitosterol was identified in lentils to be a common component in plants, of which the level was 123.4 mg/100 g, followed by 20.0 mg/100 g of stigmasterol and 15.0 mg/100 g of campesterol [13]. Additionally, β-sitosterol was found as the predominant compound in cooked lentils, ranging from 15.4 to 24.2 mg/100 g [8].

2.4 Carbohydrates

Carbohydrates are an essential component of legume seeds and possess a bioactive property against chronic diseases. Chickpeas have been shown to be a good source of carbohydrates, such as dietary fiber, starch, and oligosaccharides [39, 40]. Pigeon peas have a high content of carbohydrates (57.6%), which is the same in cotyledons. Black gram beans have also been documented to possess soluble mucilaginous polysaccharides along with dietary fiber.

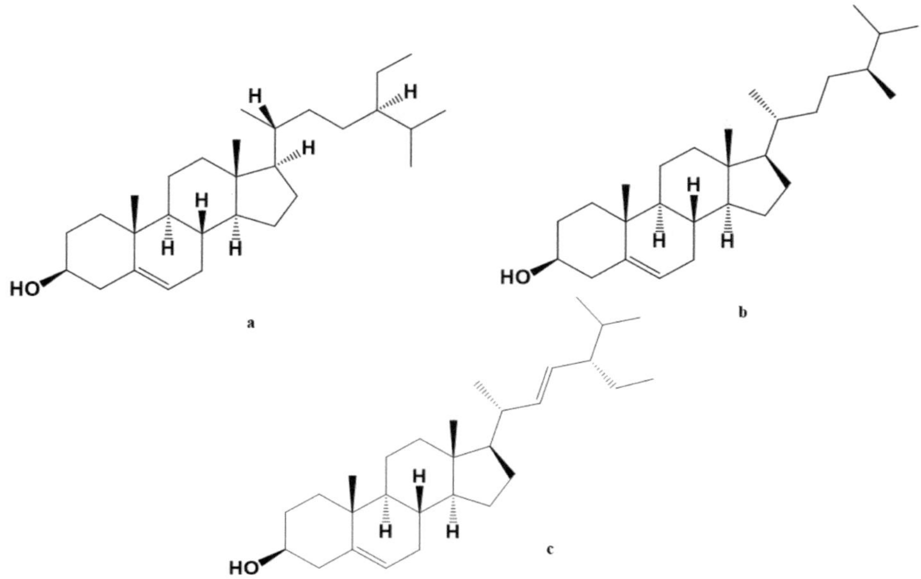

Figure 5.
Chemical structure of major phytosterols: β-sitosterol (a), campesterol (b), stigmasterol (c).

2.4.1 Dietary fiber

Dietary fiber has been demonstrated to be a beneficial food component and is made up of a combination of polymeric non-starch substances (such as cellulose, hemicellulose, and pectin) that are resistant to enzymatic digestion in the human gastrointestinal tract [41]. The dietary fiber contents of legume seeds vary according to the species, variety, and processing method. Dietary fiber, also called cell wall material, is of a lower level in cotyledon than testa [42]. Fiber concentration ranges from 8 to 27.5% and is between 3 and 14% of soluble fiber from almost gain legumes consumption. The gut bacteria metabolize and convert the solute fiber into fatty acids, which aid in the health of colonic cells. Guar beans have been identified to have the richest amounts of fiber, as well as soluble fiber (12.5%), among other legumes [41].

Some research projects have shown that dietary fiber can interact with other bioactive compounds such as phenolic compounds, which play an important role in health advantages. The interaction is accomplished through the formation of hydrogen, hydrophobic and covalent linkages between phenolic compounds, and components of legume cell wall fibers [43].

2.4.2 Resistant starch

Legumes are one of the best sources of resistant starch, and this component is not digested by humans. Resistant starch (RS) concentrations in legume seeds have been found to be higher than in cereals and many tubers (**Figure 6**) [45]. Cowpeas, lentils, peas, kidney beans, and white beans displayed the RS content of 0.6, 3.4, 2.5, 2, and 4.2 g/100 g of total seed material, respectively [3]. Legume seed processing has an effect on the RS content. Alonso et al. [46] showed that the formation of RS increased after refrigeration of legumes, which corresponded to the fact that cooling after gelatinization can develop the formation of RS.

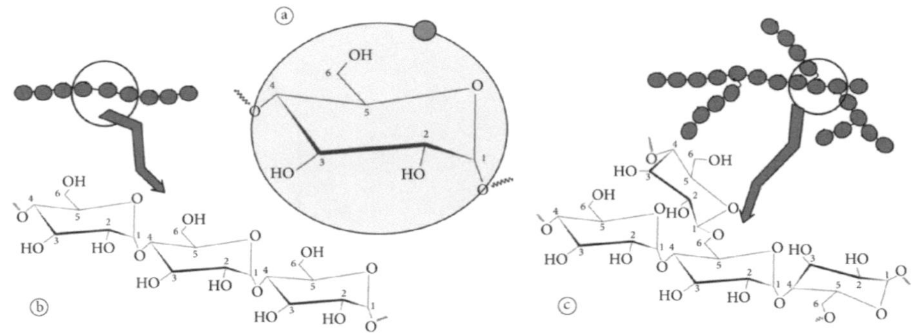

Figure 6.
Chemical structure of resistant starch: Glucose units (a), amylose (b), amylopectin (c) [44].

2.4.3 Oligosaccharides

Oligosaccharides, found in legumes, such as raffinose, stachyose, ciceritol, and verbascose (**Figure** 7), frequently cause flatulence in humans consuming legume seeds. They also induced discomfort and diarrhea. Nonetheless, oligosaccharides have recently been reported to have bioactive activities, particularly in small amounts. Raffinose was found in all parts of the legume plants, but it is built up in the seeds and roots during development. There were detectable amounts of these oligosaccharides in chickpeas, lentils, lupins, beans, peas, and faba beans (from 0.4 to 16.1% dry matter), with significant differences between the pulses studied [38]. In another study, Fan et al. [21] discovered that raffinose, stachyose, and verbascose were more concentrated in the seed kernels than in the seed coats (ratio of content in kernel/coat >1); the same was true for adzuki bean, pea, and broad bean, which showed the distribution ratio of oligosaccharides between 1 and 2

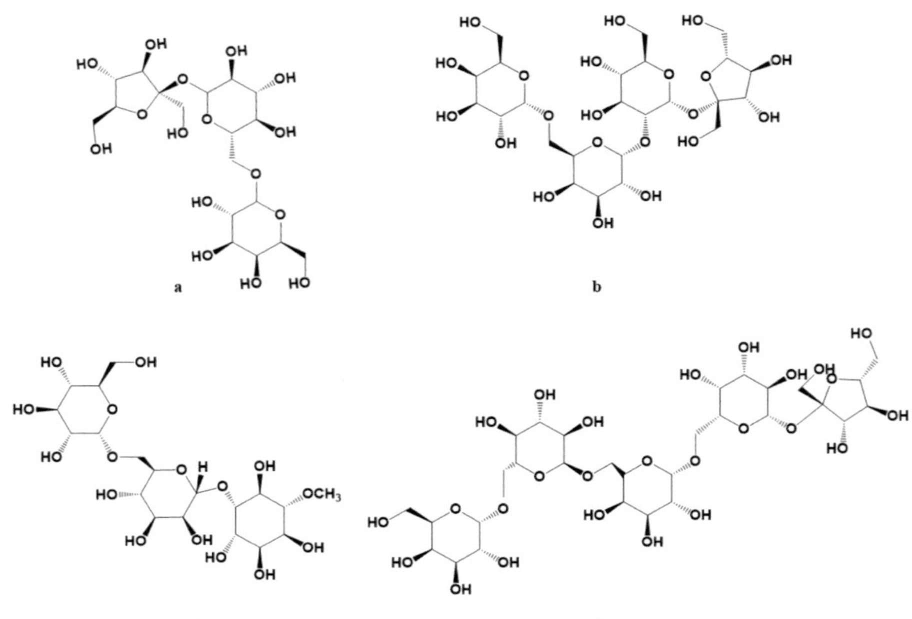

Figure 7.
Chemical structure of oligosaccharides: Raffinose (a), stachyose (b), ciceritol (c), verbascose (d).

in kernel and coat. Green soybean and mung bean, however, had higher levels of stachyose and raffinose in their seed coats than kernels (ratio of content in kernel/coat <1). Legumes generally have a decrease in total oligosaccharide content after soaking, which is most likely due to oligosaccharides leaching into the soaking water [43]. In chickpeas, lentils, yellow peas, green peas, and soybeans, it is observed that the oligosaccharide content is reduced after soaking in water with different factors (ultrasound and high hydrostatic pressure).

2.5 Antinutritional factors

Besides the nutritional compounds, legume seeds also contain some antinutritional factors (ANFs) that have been identified as bioactive constituents, which are lectins, phytic acid, alkaloids, amines, cyanogens, and other factors (**Figure 8**). Some of the ANFs that have unfavorable, undigested, or toxicological properties can be eliminated through plant genotype selection, postharvest, or thermal processing such as dehulling, soaking, germination, extraction, boiling, leaching, and/or fermentation [47].

Phytic acid is known widely as myo-inositol hexaphosphate (IP6), which is majorly stored in plants along with the salts (called phytates) [16, 38]. It is mostly considered an antinutrient due to its strong mineral, protein, and starch binding properties, which reduces bioavailability. Therefore, phytates affect enzyme activity, such as pepsin and trypsin; they also change the solubility, as well as digestibility. It has, however, been recognized for its antioxidant activity due to its ability to inhibit the formation of hydroxyl iron radicals. Another phytochemical of interest in legume seeds are lectins. Lectins are proteins or glycoproteins that are widely present in pulse and have the unique property of binding to carbohydrate-containing molecules. Lectins are hardly protein that does not degrade easily, and they can withstand stomach acid and digestive enzymes. Legumes have a wide range of lectin concentrations. They have been reported to contribute between 2.4 and 5% of total protein (17–23%) in kidney bean seeds, 0.8% of total protein in soybean and lime bean (34% and 21%, respectively), and approximately 0.6% of total protein (24–25%) in pea seeds [16].

Some of alkaloids are neurotoxins or neuromodulators. Quinolizidine alkaloids (QAs) are neurotoxin-secondary metabolites found in some Fabaceae, particularly in the genus Lupinus, including *Lupinus albus, L. mutabilis, L. angustifolius*, and others. QAs protect plants from insect pests; however, QA levels in food must be less than 0.02% when lupin is used as an ingredient. Pyrrolizidine alkaloids have the potential to cause mutations and even cancer in both animals and humans.

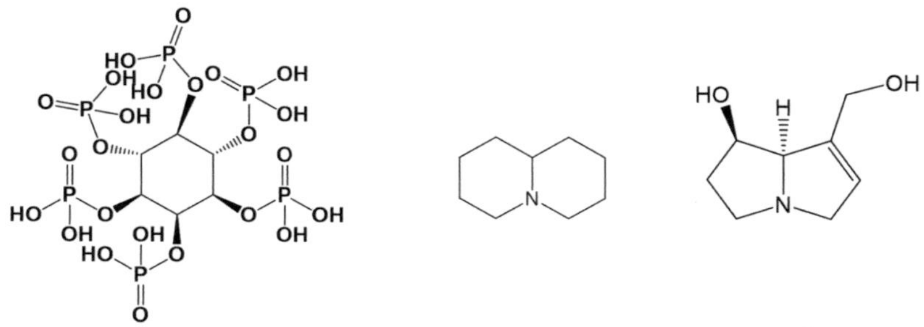

Figure 8.
Chemical structure of some ANFs: Phytic acid (a), quinolizidine alkaloids (b), pyrrolizidine alkaloids (c).

Furthermore, some toxins have long-term consequences by affecting species survival and reproductive fitness. The toxicity of alkaloids varies with concentration and is nontoxic at lower levels. Lupanine is the most toxic and higher in *L. albus* (700 mg/g total alkaloids), while sparteine (300 mg/g total alkaloids in *L. luteus*) and lupinine are the least toxic. After enzymatic hydrolysis, cyanogenic glucosides release HCN after wounding. HCN is a respiratory poison because it inhibits the mitochondrial respiratory chain and is lethal to most animals [48].

The endophytic fungus *Phomopsis leptostromiformis* is frequently found in *L. angustifolius*, an Australian forage plant used to feed sheep. Because of its antimitotic activity, this microorganism produces fomopsins and hepatocarcinogenic toxins that affect sheep [49].

Aside from lectins, protease inhibitors isolated from legume are divided into two main categories: the Kunitz inhibitor, which has a specificity aimed primarily against trypsin, and the Bowman-Birk inhibitor, which has the ability to inhibit chymotrypsin and trypsin at separate binding sites. They are found in common beans, lima beans, cowpeas, and lentils [16]. These hydrolyzed or modified proteins will become bioactive peptides (BPs), be commercialized as a nutraceutical product, and be involved in several body functions. BPs can be liberated from food proteins and exhibit bioactivity in both the small and large bowels [50].

3. Biological activities

3.1 Antioxidant activities

Natural products, such as mushrooms, vegetables, cereal, flowers, and wild fruits, have been extensively studied for their antioxidant properties. Antioxidant bioactive compounds have the ability to slow the oxidation process of important biomolecules found in human tissues and cells. It has been known that overproduction of free radicals plays a significant role in the onset of many chronic diseases such as Alzheimer's disease, various types of cancer, and diabetes. As can be seen from a review of some literature, the production of bioactive compounds is often less than 1% of the dry weight of the legume. Therefore, based on this consideration, even a technique such as chemical synthesis cannot yield large quantities of bioactive compounds.

Antioxidant activities of various legume species have been identified in numerous studies, with a positive association between antioxidant activities and total phenolic content. The chemical composition of phenolic compounds impacts their antioxidant function. The position and degree of hydroxylation on the B ring are the most significant factors in the activity of flavonoids, which are considered primary antioxidants [51]. Natural precursors of flavones and flavonols are chalcones, which have antioxidant potential (**Figure 9**) [25]. Several methods have been developed and utilized to evaluate the antioxidant activities in legumes, including *in vitro* assays for ferric-reducing antioxidant potential assay (FRAP),

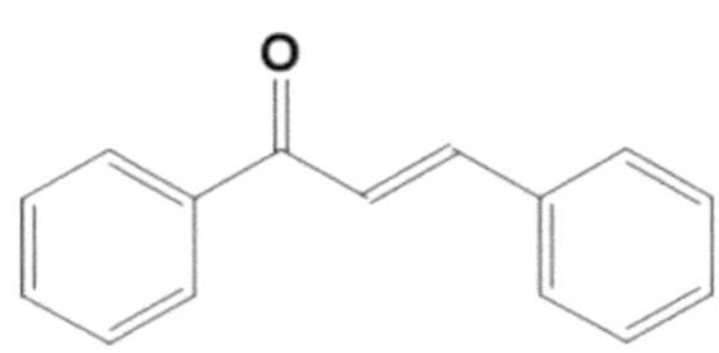

Figure 9.
Chemical structure of chalcone.

Trolox-equivalent antioxidant capacity (TEAC), 1-diphenyl-2-picrylhydrazyl free radical-scavenging assay (DPPH), the tests to measure values of oxygen radical absorbance capacity (ORAC), and total radical-trapping antioxidant parameter (TRAP). Xu and colleagues showed that dark-colored pulses had higher phenolic content and antioxidant activity than pale-colored pulses [19]; the same was true for anthocyanins, which attracted interest for their high antioxidant effects. Their study also reported that lentils had the highest DPPH and ORAC activity than green pea, yellow pea, and chickpea [52]. Similarly, lentils were observed to have the highest total antioxidant potential measured by FRAP and TRAP, among test pulses, but came in the second place by TEAC to broad beans [8], because seed coats contained mainly flavonoids, hydroxycinnamic, and hydroxybenzoic acids [25]. Red, brown, and black beans have been reported to have strong antioxidant activities in comparison with white beans [3]. More importantly, the antioxidant activity of the common bean seed coat was found to be higher than that of the cotyledon in several studies [20]. M. Dueñ et al. [25] demonstrated that the seed coats of lentils (with EC_{50} values were between 0.05 and 0.07 mg of sample) had a higher free radical-scavenging capacity than in cotyledon (with values from 21 to 29 mg of sample). Another research found that red kidney beans had the most antioxidant activity (15-µmol Trolox equivalents (TE)/g seed dry weight), while brown-eyed bean varieties had the least (6.22-µmol TE/g seed dry weight) [3].

In different circumstances, processes such as thermal processing, fermentation, and germination have a major impact on the antioxidant activities of common beans. Because of the increased level of total phenolic content, germination and fermentation may enhance the antioxidant properties of legume seeds even further. In some studies, it was noticeable that the antioxidant activities significantly increased in peas 4 days after germination and in the presence of light [27].

3.2 Antibacterial activities

Antimicrobial resistance has made the spread of bacterial, fungal, and viral infectious diseases a major public health concern. Natural compounds from plants are nowadays excellent candidates for use as alternative sources of antimicrobial substitutes. Legume seeds, which are high in phytochemical varieties, used these chemicals to defend themselves against microbes, pathogens, etc. The aforementioned antinutritional compounds, protease inhibitors, and polyphenols have been shown to be highly biologically antimicrobial agents [40]. Besides antioxidant agents, phenolics are also demonstrated in antibacterial potential against a wide spectrum of microorganisms. Polyphenols deplete critical essential mineral micronutrients (iron and zinc), disrupt the cytoplasmic membrane, inhibit microbial metabolism, and cause permeabilization of the cell membrane, resulting in microbe death [40]. Flavol-3-ols, flavonoids, and tannins (**Figure 10**) have received the most attention, because of their efficiency in resisting a variety of microbial virulence factors such as inhibition of biofilm formation, ligand adhesion reduction, and bacterial toxin neutralization [53]. Moreover, prenylated phenolics derived from legume seedlings indicated potent antibacterial activity against *Listeria monocytogenes* and methicillin-resistant *Staphylococcus aureus*; this compound has also served multiple goals, including providing health benefits and natural food preservation [54]. Protease inhibitors are also thought to have antimicrobial properties, and their mechanisms of action involve suppressing enzyme activities in response to attack by phytopathogenic microorganism-produced proteases. Methyl esterification of protein by methanol, which is isolated from broad bean, chickpea, and soybean, revealed efficient antibacterial activity against *Escherichia coli, S. aureus, Bacillus subtilis,* and *Pseudomonas aeruginosa* [40]. Methylate subunits interact with cell

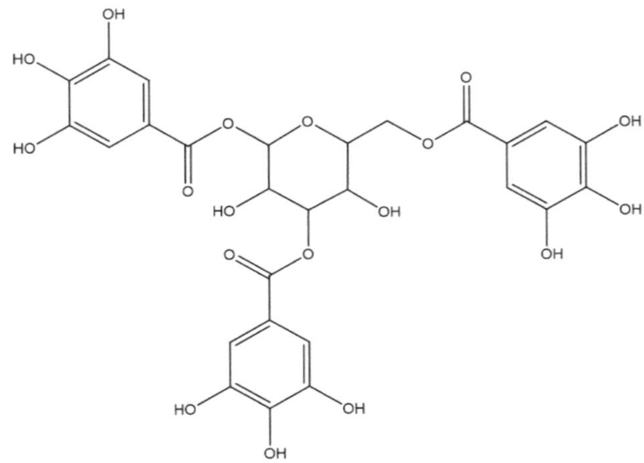

Figure 10.
Chemical structure of tannin.

walls and cell membrane, produce channels and pores and affect the integrity of bacteria cells, and finally achieve the lysis and death of the microorganism [55]. Antimicrobial peptides (AMPs) from natural sources of plants are generally effective against a wide range of microorganisms by interaction or disruption of the bacterial cell wall. Lectins are carbohydrate-binding proteins involved in plant's defense through the growth inhibition of bacteria, or disruption of the microbial cell wall by interacting with components on them such as teichoic and teichuronic acids, peptidoglycans, and lipopolysaccharides [53]. The seed extracts of lentils, fava beans, and peas show antibacterial activity (*P. aeruginosa* and *S. aureus*) [56]. AMPs derived from chickpeas, such as cicerin and arietin, have shown antifungal activity against *Botrytis cinerea, Mycosphaerella arachidicola, and Fusarium oxysporum*; and serine proteinase inhibitors that are found in chickpea seed extracts display antimetabolic activity against *Helicoverpa armigera* [40]. The effect of water extracts of colored azuki beans (such as green, black, and red) has been revealed to be more effective against *S. aureus, Aeromonas hydrophila,* and *Vibrio parahaemolyticus,* due to higher concentrations of polyphenols including proanthocyanidins, compared with the extracts of white azuki beans, which indicated no inhibition toward any of the bacteria examined [55].

In summary, the antibacterial properties of legumes are related to their variety and processing methods. These effects can be attributed primarily to the suppression of bacterial biofilm formation, cell wall disruption, and inhibition of microbial metabolism.

3.3 Anticancer activities

Cancer is recognized as the leading cause of death worldwide, and several research works have indicated that plant-derived secondary metabolites possess properties that fight against types of cancer varieties. According to the American Institute for Cancer Research (AICR), legumes contain a variety of compounds that may protect the human body against cancer, including lignans, saponins, resistant starch, and polyphenolic compounds [8]. Phenolic compounds, bioactive protein, and short-chain fatty acids extracted from legume seeds have several bioactive activities related to anticancer potentials, such as anti-inflammation, anti-proliferation, and pro-apoptotic effect [19]. Many studies reported that phenolic and flavonoids that are derived from plants exhibit potent anti-inflammatory activity by regulating the concentration of various inflammatory cytokines or mediators such as cyclooxygenase-2 (COX-2), tumor necrosis factor (TNF-α), and nuclear factor-kappa (NF-κB), interleukin 1, interleukin 6, interleukin 10, nitric oxide (NO), lipoxygenase (LOX), and iNOS [8]. Flavonoids isolated from black bean hulls can affect cell cycle by inducing cell cycle arrest at the S-phase and preventing progression to G2/M stages, as well as causing activation of apoptosis on OCI-Ly7 lymphoma cells in mouse [57]. In another experiment, phytosterol treatment reduces the development of production of carcinogens, inhibits cell growth, and also promotes apoptosis in cancer cells [10]. Saponins similar to those present in soybeans have been shown to have anticancer activity. Ginsenosides, a form of saponins isolated from ginseng, have been indicated to inhibit tumor cell proliferation and induce tumor cell differentiation and apoptosis in an *in vitro* assay, as well as *in vivo* to inhibit tumor invasion and metastasis [51]. In fact, several polypeptides have recently been researched and found to have powerful anticancer potential. Anticancer peptides derived from legumes can be found in the form of an intact long polypeptide chain, or they can be synthesized from their protein precursors through enzymatic hydrolysis [18]. Lunasin, a leader anticancer peptide derived from soybeans and other legumes, inhibits the chemical carcinogen-induced transformation of murine fibroblast cells to cancerous foci and induces selective apoptosis (**Figure 11**) [50].

Experimental studies have demonstrated that legume seeds and their active components can prevent and treat several types of cancers. These anticancer mechanisms mainly involve the regulation of carcinogen metabolism, inhibition of cell growth and proliferation, and induction of apoptosis.

3.4 Cardiovascular protection

Cardiovascular diseases have been considered to be a leading cause of premature death, of about 17.9 million people die per year. Dyslipidemia, hypertension, and type 2 diabetes are known to be risk factors for cardiovascular diseases, including stroke and coronary heart disease. A series of studies has shown that legume seeds can decrease the levels of blood lipids and blood pressure, contributing to protection from cardiovascular protection [11].

Legumes are rich in phytosterols, which have been demonstrated to inhibit the absorption of cholesterol in the intestine, followed by decreasing levels of low-density lipoprotein-cholesterol (LDL-C) and TAG, and enhanced the concentration of serum high-density lipoprotein-cholesterol (HDL-C) in the blood, a protective factor against coronary heart disease (CHD) [40, 58]. Furthermore, β-sterols, which are abundant in chickpeas, help to lower serum cholesterol, blood pressure, and

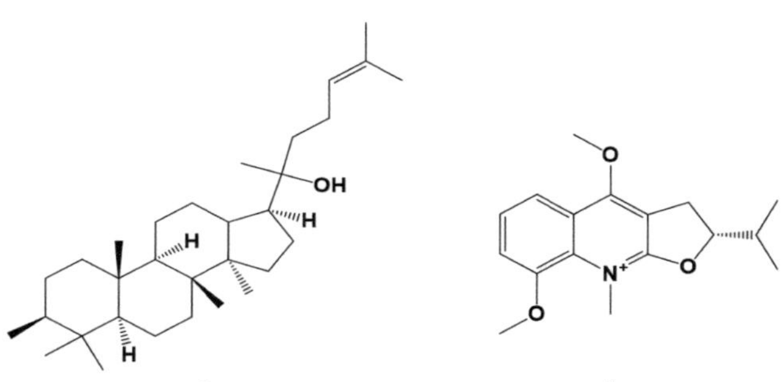

Figure 11.
Chemical structure of some anticancer constituents: Ginsenosides of Panax ginseng root (a), lunasin (b).

the risk of coronary heart disease [40]. Chickpeas also are high in dietary fiber, which help to lower total plasma cholesterol levels and can aid in weight loss and obesity reduction. These compounds are also thought to improve body metabolism and reduce chronic inflammation, serum lipid levels, blood pressure, and insulin resistance, as well as to affect fibrinolysis and coagulation, which may be essential in the plaque formation of existing atherosclerotic plaques [59].

Generally, legume has exhibited cardiovascular protective effects by attenuating hypertension and ameliorating dyslipidemia, such as in the improvement of HDL-C, reduction of LDL-C, TAG, and blood pressure.

3.5 Other bioactivities of legume

Apart from the bioactivities mentioned earlier, legume has other beneficial effects, such as anti-obesity and antidiabetic effects.

Diabetes mellitus is known as a severe metabolic disorder caused by insulin deficiency and/or insulin resistance, resulting in an abnormal increase in blood glucose. Legumes have been shown to regulate the levels of blood glucose and, in turn, provide protection against diabetes by resistant starch (NSPs). Moreover, short-chain fatty acids and the inhibition of α-amylase and α-glucosidase have been reported to induce hypoglycemic and hypocholesterolemic effects by suppressing glucose release and cholesterol production [17, 60]. Adzuki bean extracts reduce the final body weight of mice and adipose tissue accumulation and enhance lipolysis. This treatment also considerably decreases the serum triglyceride levels, total cholesterol, LDL-C, and liver lipids [61].

3.6 Benefits of legume seeds in aquaculture

Legumes contain large amounts of valuable protein. These proteins are not only abundant but also have a well-balanced amino acid profile, and may be used to substitute fish meal, which is an unsustainable resource. The substitution of plant protein sources for fish meals without compromising fish growth and physiology is a strategy for lowering feed costs and reducing aquafeed reliance on fish meals. Some studies reported that commercial hexane-extracted soybean meal with methionine supplement could replace 67% of the fish meal in the diet without negatively impacting milkfish growth and feed conversion ratio [62]. Another experiment showed that the substitution of up to 20% of fish meal protein with soybean meal protein in realistic diets for spotted rose snapper was an essential move for this high-value species [9]. Green mung bean in which the ANFs were inactivated by thermal processing was used as a replacement for fish meal in Asian sea bass and milkfish diets, with no negative effects on the fish's development [62, 63], and these studies were carried out on a 15-week feeding trial and were evaluated to measure growth, survival, FCR, PER, HSI, and liver and gut histology. Overall, legume is a promising alternative protein source for the aquaculture feed industry.

4. Conclusions

In conclusion, the utilization of legumes as food ingredients is of tremendous interest, not only for increasing the functionality of food items but also for developing functional foods with health advantages. The content, composition, and distribution of legumes, as well as their biological functions, are systemically outlined and analyzed in this review. However, the following aspects require additional investigation to fill knowledge gaps.

The extraction solvent has a significant impact on the extraction efficiency of bioactive substances. Mathematical modeling, such as response surface methodology, which is an ideal candidate for predicting the interactions between the target compound and solvent, has been successfully applied in the selection of a specific solvent for higher compound extraction yield in many plants. However, no study has used these modeling tools to optimize the extraction conditions of legume chemicals yet. As a result, future research can use these methods to reduce the time and effort required to identify solvents for common bean polyphenols in different kinds.

Furthermore, the research should seek more bioactive substances and investigate their benefits, as well as biologically active metabolites. Thus, future study can focus on extracting and purifying novel active chemicals from legumes, and clinical trials are also required to confirm the medicinal advantages of legumes. Moreover, legumes have the high potential to be a valuable substitute source of feed for the aquaculture industry.

Conflicts of interest

The authors declare that there is no conflict of interest.

Author details

Hai Ha Pham Thi[1] and Thanh Luan Nguyen[2*]

1 NTT Hi-Tech Institute, Nguyen Tat Thanh University, Ho Chi Minh City, Vietnam

2 Department of Science, Technology and International Affairs, HUTECH University, Ho Chi Minh City, Vietnam

*Address all correspondence to: nt.luan@hutech.edu.vn

References

[1] Martín-Cabrejas MA. Legumes: An overview. Legumes: Nutritional Quality, Processing and Potential Health Benefits. Croydon, UK: CPI Group (UK) Ltd. 2019:1-18

[2] Maphosa Y, Jideani VA. The role of legumes in human nutrition. Functional Food-Improve Health Through Adequate Food. 2017;**1**:13

[3] Singh JP, Singh B, Kaur A. Bioactive compounds of legume seeds. Phytochemistry Bioactive Compounds in Underutilized Vegetables and Legumes. 2020:1-21

[4] Khrisanapant P et al. A comprehensive characterisation of volatile and fatty acid profiles of legume seeds. Food. 2019;**8**(12):651

[5] Djordjevic TM, Šiler-Marinkovic SS, Dimitrijevic-Brankovic SI. Antioxidant activity and total phenolic content in some cereals and legumes. International Journal of Food Properties. 2011;**14**(1): 175-184

[6] Luo J et al. Phytochemical distribution in hull and cotyledon of adzuki bean (Vigna angularis L.) and mung bean (Vigna radiate L.), and their contribution to antioxidant, anti-inflammatory and anti-diabetic activities. Food Chemistry. 2016;**201**:350-360

[7] Singh B et al. Phenolic composition and antioxidant potential of grain legume seeds: A review. Food Research International. 2017;**101**:1-16

[8] Zhang B et al. Phytochemicals of lentil (Lens culinaris) and their antioxidant and anti-inflammatory effects. Journal of Food Bioactives. 2018;**1**:93-103

[9] Silva-Carrillo Y et al. The effect of substituting fish meal with soybean meal on growth, feed efficiency, body composition and blood chemistry in juvenile spotted rose snapper Lutjanus guttatus (Steindachner, 1869). Aquaculture. 2012;**364**:180-185

[10] Woyengo T, Ramprasath V, Jones P. Anticancer effects of phytosterols. European Journal of Clinical Nutrition. 2009;**63**(7):813-820

[11] Souza RG et al. Nuts and legume seeds for cardiovascular risk reduction: Scientific evidence and mechanisms of action. Nutrition Reviews. 2015;**73**(6):335-347

[12] Dueñas M, Hernández T, Estrella I. Changes in the content of bioactive polyphenolic compounds of lentils by the action of exogenous enzymes. Effect on their antioxidant activity. Food Chemistry. 2007;**101**(1):90-97

[13] Ryan E et al. Phytosterol, squalene, tocopherol content and fatty acid profile of selected seeds, grains, and legumes. Plant Foods for Human Nutrition. 2007;**62**(3):85-91

[14] Barakat H, Reim V, Rohn S. Stability of saponins from chickpea, soy and faba beans in vegetarian, broccoli-based bars subjected to different cooking techniques. Food Research International. 2015;**76**:142-149

[15] Del Hierro JN et al. Ultrasound-assisted extraction and bioaccessibility of saponins from edible seeds: Quinoa, lentil, fenugreek, soybean and lupin. Food Research International. 2018;**109**: 440-447

[16] Mohan V, Tresina P, Daffodil E. Antinutritional factors in legume seeds: characteristics and determination. Encyclopedia of Food and Health. 2016:211-220

[17] Moreno-Valdespino CA et al. Bioactive proteins and phytochemicals

from legumes: Mechanisms of action preventing obesity and type-2 diabetes. Food Research International. 2020;**130**: 108905

[18] Luna-Vital D, de Mejía EG. Peptides from legumes with antigastrointestinal cancer potential: Current evidence for their molecular mechanisms. Current Opinion in Food Science. 2018;**20**:13-18

[19] Rao S et al. Inhibitory effects of pulse bioactive compounds on cancer development pathways. Diseases. 2018;**6**(3):72

[20] Yang QQ et al. Polyphenols in common beans (Phaseolus vulgaris L.): Chemistry, analysis, and factors affecting composition. Comprehensive Reviews in Food Science and Food Safety. 2018;**17**(6):1518-1539

[21] Fan PH, Zang MT, Xing J. Oligosaccharides composition in eight food legumes species as detected by high-resolution mass spectrometry. Journal of the Science of Food and Agriculture. 2015;**95**(11):2228-2236

[22] Sutivisedsak N, Cheng HN, Willett JL, Lesch WC, Tangsrud RR, Biswas A. Microwave-assisted extraction of phenolics from bean (Phaseolus vulgaris L.). Food Research International. 2010;**43**(2):516-519

[23] Lin D et al. An overview of plant phenolic compounds and their importance in human nutrition and management of type 2 diabetes. Molecules. 2016;**21**(10):1374

[24] Amarowicz R. Legume seeds as an important component of human diet. Foods. 2020;9(12):1812

[25] Duenas M, Hernandez T, Estrella I. Assessment of in vitro antioxidant capacity of the seed coat and the cotyledon of legumes in relation to their phenolic contents. Food Chemistry. 2006;**98**(1):95-103

[26] Lafarga T et al. Bioaccessibility and antioxidant activity of phenolic compounds in cooked pulses. International Journal of Food Science & Technology. 2019;**54**(5):1816-1823

[27] López-Amorós M, Hernández T, Estrella I. Effect of germination on legume phenolic compounds and their antioxidant activity. Journal of Food Composition and Analysis. 2006;**19**(4): 277-283

[28] Malenčić D, Cvejić J, Miladinović J. Polyphenol content and antioxidant properties of colored soybean seeds from Central Europe. Journal of Medicinal Food. 2012;**15**(1):89-95

[29] Zhang Y et al. An investigation on phenolic and antioxidant capacity of under-utilized food legumes consumed in China. Food. 2020;**9**(4):438

[30] Liu Y et al. Composition of phenolic acids and antioxidant properties of selected pulses cooked with different heating conditions. Food. 2020;**9**(7):908

[31] Teixeira-Guedes CI et al. Impact of cooking method on phenolic composition and antioxidant potential of four varieties of Phaseolus vulgaris L. and Glycine max L. LWT. 2019;**103**: 238-246

[32] Amarowicz R, Pegg RB. Legumes as a source of natural antioxidants. European Journal of Lipid Science and Technology. 2008;**110**(10):865-878

[33] Wang M et al. Flavonoid content in different legume germplasm seeds quantified by HPLC. Plant Genetic Resources. 2008;**6**(1):62

[34] Sancho RAS, Pavan V, Pastore GM. Effect of in vitro digestion on bioactive compounds and antioxidant activity of common bean seed coats. Food Research International. 2015;**76**:74-78

[35] Sagratini G et al. Determination of soyasaponins I and βg in raw and

cooked legumes by solid phase extraction (SPE) coupled to liquid chromatography (LC)–mass spectrometry (MS) and assessment of their bioaccessibility by an in vitro digestion model. Journal of Agricultural and Food Chemistry. 2013;**61**(8): 1702-1709

[36] Dixit AK et al. 12. Soybean constituents and their functional benefits. Research Singpost. 2011;**37**(2):661

[37] Guajardo-Flores D et al. Characterization and quantification of saponins and flavonoids in sprouts, seed coats and cotyledons of germinated black beans. Food Chemistry. 2012;**134**(3):1312-1319

[38] Muzquiz M et al. Bioactive compounds in legumes: Pronutritive and antinutritive actions. Implications for nutrition and health. Phytochemistry Reviews. 2012;**11**(2):227-244

[39] Barman A et al. Nutraceutical properties of legume seeds and their impact on human health. Legume Seed Nutraceutical Research. IntechOpen. 2018:9-17

[40] Venkidasamy B et al. Indian pulses: A review on nutritional, functional and biochemical properties with future perspectives. Trends in Food Science & Technology. 2019;**88**:228-242

[41] Khan AR et al. Dietary fiber profile of food legumes. Sarhad Journal of Agriculture. 2007;**23**(3):763

[42] Guillon F, Champ M-J. Carbohydrate fractions of legumes: Uses in human nutrition and potential for health. British Journal of Nutrition. 2002;**88**(S3):293-306

[43] Han IH, Baik BK. Oligosaccharide content and composition of legumes and their reduction by soaking, cooking, ultrasound, and high hydrostatic pressure. Cereal Chemistry. 2006;**83**(4):428-433

[44] Alcázar-Alay SC, Meireles MAA. Physicochemical properties, modifications and applications of starches from different botanical sources. Food Science and Technology. 2015;**35**(2):215-236

[45] Nigudkar MR, Madan JG. Resistant starch content of traditional Indian legume preparations. Current Research in Nutrition and Food Science Journal. 2017;**5**(3):238-246

[46] García-Alonso A, Goni I, Saura-Calixto F. Resistant starch and potential glycaemic index of raw and cooked legumes (lentils, chickpeas and beans). Zeitschrift für Lebensmitteluntersuchung und-Forschung A. 1998;**206**(4):284-287

[47] Klupšaitė D, Juodeikienė G. Legume: Composition, protein extraction and functional properties. A review. Chemical Technology. 2015;**66**(1):5-12

[48] Frick KM, Kamphuis LG, Siddique KH, Singh KB, Foley RC. Quinolizidine alkaloid biosynthesis in lupins and prospects for grain quality improvement. Frontiers in Plant Science. 2017;**8**:87

[49] Gamboa-Gaitán MA. Hongos endófitos tropicales: Conocimiento actual y perpectivas. Acta Biológica Colombiana. 2006;**11**:3-20

[50] González-Montoya M, Cano-Sampedro E, Mora-Escobedo R. Bioactive peptides from le-gumes as anticancer therapeutic agents. Int J Cancer Clin Res. 2017;**4**:81

[51] Kerwin S. Soy saponins and the anticancer effects of soybeans and soy-based foods. Current Medicinal Chemistry-Anti-Cancer Agents. 2004;**4**(3):263-272

[52] Campos-Vega R et al. Common beans and their non-digestible fraction: Cancer inhibitory activity—an overview. Food. 2013;**2**(3):374-392

[53] Pina-Pérez MC, Pérez MF. Antimicrobial potential of legume extracts against foodborne pathogens: A review. Trends in Food Science & Technology. 2018;**72**:114-124

[54] Araya-Cloutier C et al. The position of prenylation of isoflavonoids and stilbenoids from legumes (Fabaceae) modulates the antimicrobial activity against Gram positive pathogens. Food Chemistry. 2017;**226**:193-201

[55] Sitohy M et al. Extent and mode of action of cationic legume proteins against Listeria monocytogenes and Salmonella Enteritidis. Probiotics and Antimicrobial Proteins. 2013;**5**(3):195-205

[56] El-Araby MM et al. Characterization and antimicrobial activity of lectins purified from three Egyptian leguminous seeds. AMB Express. 2020;**10**:1-14

[57] Aregueta-Robles U et al. Cytotoxic activity of a black bean (Phaseolus vulgaris L.) extract and its flavonoid fraction in both in vitro and in vivo models of lymphoma. Revista de Investigación Clínica. 2018;**70**(1):32-39

[58] Marventano S et al. Legume consumption and CVD risk: A systematic review and meta-analysis. Public Health Nutrition. 2017;**20**(2):245-254

[59] Satija A, Hu FB. Cardiovascular benefits of dietary fiber. Current Atherosclerosis Reports. 2012;**14**(6): 505-514

[60] Singhal P, Kaushik G, Mathur P. Antidiabetic potential of commonly consumed legumes: A review. Critical Reviews in Food Science and Nutrition. 2014;**54**(5):655-672

[61] Liu R et al. Saponins and flavonoids from adzuki bean (Vigna angularis L.) ameliorate high-fat diet-induced obesity in ICR mice. Frontiers in Pharmacology. 2017;**8**:687

[62] Apines-Amar, M. J. S., Coloso, R. M., Amar, M. N. G., Golez, M. S. M., Bunda, M. G. B., and Jaspe, C. J. Utilization of mung bean, Vigna radiata (Linnaeus) as a novel protein source in practical-type diets for juvenile milkfish, Chanos chanos (Forsskal): Effects on growth, feed efficiency, body composition, and histology of gut and liver. The Israeli Journal of Aquaculture-Bamidgeh. 2015;IJA_67.2015.1237

[63] Eusebio BP, Coloso RM. Nutritional evaluation of various plant protein sources in diets for Asian sea bass Lates calcarifer. Journal of Applied Ichthyology. 2000;**16**(2):56-60

Chapter 5

Phenolic Compounds in Legumes: Composition, Processing and Gut Health

Mayra Nicolás-García, Cristian Jiménez-Martínez, Madeleine Perucini-Avendaño, Brenda Hildeliza Camacho-Díaz, Antonio Ruperto Jiménez-Aparicio and Gloria Dávila-Ortiz

Abstract

Gut health is fundamental for human well-being and prevents chronic degenerative diseases and is influenced by the interaction between gut microbiota and food components. In recent years, interest in phenolic compounds has increased due to their health benefits such as antioxidant, antidiabetic, antimicrobial, anti-atherosclerotic, anti-inflammatory, anticarcinogenic, cardio- and neuro-protective properties. Legumes are an essential source of phytochemicals, particularly flavonoids and phenolic acids, distributed mainly in the seed coat, and have been reported to exhibit multiple biological effects. Flavonoids present in legumes have been shown to regulate metabolic stability and membrane transport in the intestine, thus improving bioavailability. Seed processing such as cooking allows the release of phenolic compounds, improving polyphenols digestion and absorption at the intestinal level, maintaining their protective capacity in the oxidative process at the cellular level, and modulating the gut microbiota. All these actions improve gut health, avoiding diseases like irritable bowel syndrome, inflammatory bowel disease, obesity, diabetes, colitis, and colorectal cancer. The effect of the consumption of legumes such as chickpea, pea, and bean, as well as the contribution of phenolic compounds to gut health, will be reviewed in this study.

Keywords: Legumes, biological effects, phenolic compounds, seed processing, chronic degenerative diseases, gut microbiota, gut health

1. Introduction

Eating habits are an important factor in the structure, formation, function, and modulation of the gut microbiota, which plays a crucial role in health; environmental factors, antibiotics, and lifestyle also contribute to the dysbiosis of the gut microbiota responsible for gastrointestinal diseases, like colon cancer. Several studies have shown that the gut has a greater impact than food processing and nutrient absorption. Gut health is a function of the gut barrier and gut microbiota essential elements for better health [1–6].

IntechOpen

Plants contribute diverse bioactive compounds to the diet [7, 8], including phenolic compounds. Legumes are part of the basic foods with great nutritional relevance due to their content of diverse phenolic compounds that promote health [5, 9]. These compounds are distributed in the whole seed and are mainly responsible for the seed coat color that depends on the composition and concentration [10–14]. The potential health benefits of phenolic compounds in the diet depend on their absorption and metabolism, which in turn are determined by their structure, including their conjugation with other phenols, degree of glycosylation/acylation, molecular size, and solubility [9, 15–17]. During seed processing, phenolic compounds can undergo various changes, altering their antioxidant activity [18, 19].

However, the presence of phenolic compounds and their mechanisms of action in preventing colon cancer or inflammation are probably mediated by the functional composition of the gut microbiota [20–22]. Epidemiological studies have confirmed that regular consumption of legumes has been associated with lower risk due to immunomodulatory effects and prevention of chronic and metabolic diseases, such as cardiovascular diseases, diabetes, cancer, and obesity, in addition to improving gut health [11, 16, 20, 23, 24].

During the absorption of phenolic compounds, like hydroxycinnamic acids (p-coumaric, caffeic, and ferulic) in free and conjugated forms, they are metabolized by the gut microbiota (e.g., genera *Bifidobacterium*, *Lactobacillus*, and *Escherichia*) are able to release them [8]. Moreover, flavonoid glycosides are deconjugated by microbial glucuronidases and sulfatases in the colon, releasing aglycones such as quercetin, myricetin, and kaempferol that can be metabolized by different genera of intestinal bacteria, including *Clostridium* sp., *Eubacterium* sp., *Enterococcus* sp., among others, to form hydroxyphenylacetic and hydroxyphenylpropionic acids as major metabolites leading to increased bioavailability of phenolic compounds [24].

Chickpea, pea, and bean seeds are among the most widely consumed food legumes worldwide. The genus *Cicer* comprises about 44 species. The chickpea (*Cicer arietinum* L.) has two commercial varieties 'Desi' and 'Kabuli' and their characteristics vary according to geographical distribution, shape, size, and color. The color of the Desi variety is dark in comparison with the Kabuli chickpea, which has a fine, light-colored covering and is the most widely consumed [20, 23].

Pisum sativum L., commonly known as pea, represents one of the oldest and most widespread cultivated legumes worldwide due to its wide availability, low cost, and high nutritional value [1, 16], they are small seeds with a green or yellow spherical shape. Quality characteristics depend on biological factors between the environment and genetics [25]. Another important legume in food is the genus *Phaseolus*, which includes species such as *P. vulgaris*, *P. lunatus*, *P. coccineus*, *P. acutifolius,* and *P. dumosus*; among these, the most cultivated in Mexico is the common bean (*P. vulgaris*) that has more than 70 varieties grouped according to their color in black, yellow, red, brown, white, purple, and pinto [14, 26, 27].

Phenolic compounds constitute an important group of secondary plant metabolites and influence the diversity and quantity of gut microbial species, allotting prebiotic effects to phenolic compounds, mainly flavonoids involved in modulating the taxonomic composition of the gut microbiota, increasing the relative abundance of beneficial species, and inhibiting the proliferation of bacterial species associated with negative implications [7, 24].

2. Phenolic compounds in chickpea, pea, and common bean: chemistry, distribution, and beneficial effects

Legumes are an excellent source of phytochemicals, including phenolic acids, flavonols, flavones, flavanols, flavanones, isoflavones, anthocyanins, tannins, and other phenolics [14, 16, 28, 29]. The structure of polyphenols and their composition and interaction in a food matrix are important determinants of their bioavailability and bioactivity [12, 15, 30]. **Figure 1** shows the structures of phenolic compounds present in chickpea, pea, and bean. Differences in the phenolic profile of various legumes influence the specific health benefits. The presence of phenolic acids and flavonoids in legumes such as chickpea, pea, and beans have been reported in different units of concentration and are presented in **Table 1**.

Phenolic compounds are present in soluble and insoluble forms. Therefore, it is very important to optimize the polyphenols extraction process [9, 10]. Most of the phenolic compounds associated with whole seed are in insoluble bound forms, mainly phenolic acids, linked covalently to cell wall structural components like cellulose, hemicellulose, lignin, and pectin [14, 20, 30, 45].

Chickpea contains several phenolic compounds, including lignans (secoisolariciresinol, pinoresinol, and lariciresinol), isoflavones, flavonoids, phenolic acids, and anthocyanins [20, 49]. Besides, it has significant amounts of flavonoids, especially isoflavones, the main ones being biochanin A and formononetin, to a lesser extent genistein and daidzein [5, 9, 12, 47].

The main compounds in peas are glycosylated flavonols, condensed tannins, as well as hydroxybenzoic and hydroxycinnamic acids, such as quercetin, kaempferol, luteolin, apigenin, flavan-3-ols, apigenin-7-glucoside, quercetin-3-rhamnoside, kaempferol-3-glucoside, flavonols, flavones, and stilbenes; the main compounds identified in the whole seed are hesperidin and catechin [26, 47]. In beans, phenolic acids and flavonoids represent 50% of the total content of phenolic compounds like vanillic, ferulic, 4-hydroxybenzoic, sinapic acids; quercetin, myricetin, and catechin are the major phenolic acids contained in bean seeds and determine the seed color [10, 14, 28].

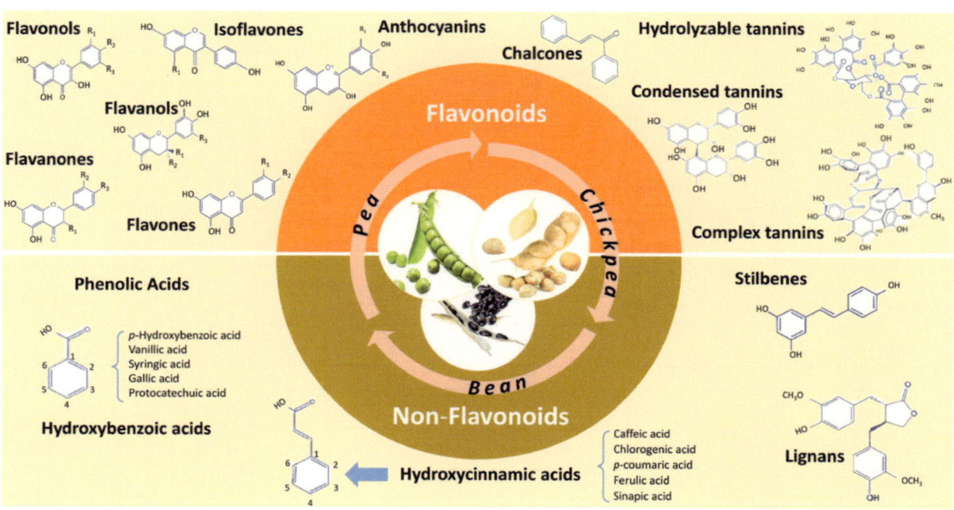

Figure 1.
Main phenolic compounds in legumes [20, 28, 31].

Compound	Concentration (DW)			References
	Chickpea	Pea	Bean	
Phenolic acids				
p-Hydroxybenzoic acid	10.5^{b}/0.08–1.63^{c}	588.94^{a}	0.30-16.30^{b}	[20, 31–34]
Protocatechuic acid	358.9^{b}/0.51^{c}	426.15^{a}/0.89-2.25^{d}	0.42-37.36^{b}/170-177^{c}	[20, 31–33, 35–37]
Syringic acid	222.1^{b}/0.63–1.95^{c}	nd	10.60-11.40^{c}	[20, 31, 34, 38]
Vanillic acid	80.8^{b}/0.34^{c}	536.67^{a}	618.44^{b}	[20, 31, 33, 36, 39]
Gallic acid	40.2^{b}/4.66^{c}	218.45^{a}/9.08-29.95^{d}	0.15–21.30^{c}	[20, 31, 33–35, 40]
Ellagic acid	0.43^{c}	433.87^{a}/899.19^{c}	4.3-18.08^{b}	[20, 33, 41–43]
Caffeic acid	0.11^{c}	146.11^{a}/0.20–0.92^{d}	3.08-11.70^{b}	[20, 33, 35, 36, 42]
Chlorogenic acid	nd	742.28^{a}/0.57–1.27^{d}	3.03-33.38^{b}/6.63-46.1^{c}	[33–37, 40, 42, 44]
p-coumaric acid	0.11^{c}	462.93^{a}/7.78^{c}/9.0-16.2^{d}	0.40-1.90^{b}/0.74^{c}	[20, 33, 40, 41, 44–46]
Ferulic acid	0.90^{b}/0.22^{c}	788.29^{a}/1.38-3.44^{d}	0.91-11.00^{b}	[20, 31, 33, 35, 36, 44, 46]
Sinapic acid	7.81^{b}/0.12^{d}	nd	2.9-86.27^{b}	[20, 31, 32, 39, 46]
Flavonoids				
Quercetin	nd	56.90^{c}/0.12-1.51^{d}	0.8-30.88^{b}/0.30-1.31^{c}	[33–35, 39, 40, 44, 46]
Quercetrin	nd	256.26^{c}	1.07^{c}	[33, 40]
Myricetin	nd	nd	1.99–5.98^{b}	[43, 44, 46]
Kaempferol	0.09^{b}	19.79^{c}	2.51^{b}/0.13^{c}	[33, 40, 44, 46, 47]
Rutin	0.101^{c}	83.01^{c}/0.26-1.31^{d}	0.20-119.70^{c}	[33–35, 37, 40]
Naringenin	nd	24.59^{c}	nd	[33]
Naringin	nd	201.41^{c}	2.35^{c}	[33, 40]
Hesperidin	nd	605.94^{c}	8.10^{c}	[33, 40]
Hespirtin	nd	158.29^{c}	0.56^{c}	[33, 40]
Catechin	nd	93-2303^{d}	10.05-78.34^{b}	[32, 39, 43–45]
Epicatechin	nd	1.03–13.02^{d}	10.90-34.48^{b}	[35, 42, 43]
Luteolin	1.56^{b}	3.24-8.57^{d}	2.41^{c}	[31, 35, 40]
Apigenin	nd	14.31^{c}	nd	[33]
Genistein	0.06^{e}	nd	3.64-4.74^{c}	[37, 48]
Daidzein	0.12^{e}	nd	nd	[48]
Formononetin	0.02^{b}/0.10^{e}	nd	35.94-163.34^{b}	[43, 47, 48]

Compound	Concentration (DW)			References
	Chickpea	Pea	Bean	
Biochanin A	0.78[b]	nd	nd	[47]
Biochanin glucoside	0.08[e]	nd	nd	[48]
Biochanin A derivative	3.31–5.25[b]	nd	nd	[47]

[a]ppm.
[b]µg/g.
[c]mg/100 g.
[d]mg/kg.
[e]mg/g.
DW: Dry weight, nd: not detected.

Table 1.
*Polyphenols reported in chickpea (*C. arietinum*), pea (*P. sativum*) and common bean (*P. vulgaris*) seeds.*

The phenolic composition of legumes has been particularly interesting for metabolic health because of their protection against oxidative damage [45]. Phenolic compounds constitute an important group of secondary plant metabolites, important for health by preventing multiple degenerative conditions in the body [16]. These compounds are biologically active and have been associated with antidiabetic, anticarcinogenic, antihypertensive, antimutagenic, antioxidant, antimicrobial, anti-inflammatory, anticholesterolemic, cardioprotective, immunostimulant, and anti-angiogenic properties [11, 14, 16, 20, 21, 29, 35, 41, 49, 50].

3. Impact of processing on phenolic compounds

Processing of legumes may result in an increase or decrease in the content of phenolic compounds. During processing, phenolic compounds may undergo various changes, altering the antioxidant activity of the products. Changes in phenolic content depend on the species, variety, and processing conditions [12, 18, 22]. Processes such as soaking, cooking, extrusion, germination, fermentation, and roasting improve the release of bound phenolic compounds, which influences the sensory properties of the seeds [51–54].

During processing, a reduction in the content of condensed tannins has been reported. In legumes, soaking has been found to decrease tannic acid content by approximately 20%, and germination reduces tannin content by 50%. The decrease of phenolic compounds during soaking and cooking may be due to several factors during the heat treatment, such as 1) polyphenol-protein interactions that decrease the extraction capacity, 2) the formation of tannin complexes with other water-soluble components, and 3) the lixiviation and thermal degradation of phenolic compounds [12, 14, 30]. However, unlike traditional processing or pressure cooking, the extrusion process is carried out in the absence of effluents, so the impact on phenolic content is less [52, 55, 56]. Arribas et al. [55] observed that extrusion does not affect the phenolic groups to the same extent; they reported that the anthocyanin content in extruded pea decreased from 4 to 50% as opposed to the flavonol content, which increased approximately three times.

On the other hand, the germination process increases bioactive compounds, like phenolic compounds, improving the seeds functionality. The increase is attributed to biosynthesis through the Shikimate pathway and the release of phenolic compounds. During germination, enzymatic reactions are activated, such as the enzyme phenylalanine ammonia lyase, which promote the phenolic compounds' biosynthesis. The endogenous esterases action allows the liberation of hydroxycinnamic acids linked to arabinoxylans and lignin in the cell wall [20, 57, 58]. Nevertheless, changes in isoflavones during this process may be related to genetic regulation. They may be induced by the metabolic pathways of naringenin chalcone and isoliquiritigenin, the precursors of isoflavonoids, present in legumes. Therefore, germination is an efficient alternative to increase antioxidant activity and has been used in legumes such as chickpeas, peas, and beans [9, 12, 48, 50]. Domínguez-Arispuro et al. [20] observed that the germination process in chickpeas induced an increase of 97 and 111% of the total phenolic and flavonoid content, respectively, as compared to the raw seed. Moreover, formononetin and biochanin contents of 0.10 and 0.18 mg/g, respectively, have been reported in raw chickpea; during a 10-day germination process, they increased to 1.42 and 2.10 mg/g respectively [48].

The fermentation process has been reported to cause an increase in free radical scavenging capacity. Changes in phenolic composition are associated with sensory, nutritional, and biochemical properties and depend on fermentation conditions such as optimal time and temperature to avoid a further reduction, mainly in tannin content [53, 59, 60]. Bulbula & Urga [53] reported the effect of different traditional processing methods on tannins in chickpea, noting that during boiling, toasting, and fermentation at 0 h, there are no differences from raw seed beans. However, during fermentation for 24, 48, 72 h and chickpea germination, tannin content decreased by 3.1, 14.4, 18.5, and 43.4%, respectively. The reduction of tannins during germination is generally attributed to enzymatic hydrolysis by polyphenolase.

4. Impact of phenolic compounds on the gut health and its relationship with human health

The gut microbiota plays an important role in food digestion, immunity, and other metabolic functions; its composition is influenced by endogenous and environmental factors such as age, diet, lifestyle, antibiotic intake, and xenobiotics. Optimal gut health depends on the microbial community structure, a balanced composition of gut microbiota, an epithelial barrier, and an intact host mucosa; therefore, a disorder of these components can lead to the development of intestinal diseases such as obesity, inflammatory bowel disease, and colon cancer [1–3, 6, 7, 21, 22, 24, 49, 61].

Legumes are composed of bioactive compounds, such as phenolic compounds, capable of modifying the physiological basal function within the intestinal microenvironment affecting the microbiota and epithelial barrier, improving metabolic and gastrointestinal health, enhancing resistance to colonization by pathogens, and exerting an impact on the gut microbiota. These actions lead to decrease the severity of diseases associated with the intestine due to their chemopreventive effects. However, not all polyphenols support gastrointestinal integrity equally, and their benefits depend on chemical structure and phenolic concentration [12, 21, 31, 49, 62]. Isoflavones, such as biochanin A, have been reported to improve gut health by exerting antioxidant and anti-inflammatory effects [12, 20, 21]. On the other hand, the effect of formononetin in an acute colitis model in mice induced by dextran sulfate sodium has been evaluated, observing an attenuation of colitis. This effect may be due to the inhibition of the NLRP3 immamasome pathway by the action of formononetin [9].

Bian et al. [2] suggest that kaempferol has a protective effect on the secretion of interleukin-8 (IL-8) and the barrier dysfunction of the Caco-2 monolayer in the lipopolysaccharide-induced epithelial-endothelial co-culture model. This effect is due to the inhibition of the nuclear factor-kappa B (NF-κB) signaling pathway, which allows the reduction of inflammatory bowel disease. Also, caffeic acid reduces the secretion of pro-inflammatory cytokines, including interleukin-6 (IL-6), tumor necrosis factor-alpha (TNF-α), and IFNγ, and colonic infiltration of CD3+ T cells, CD177+ neutrophils, and F4/80+ macrophages by inhibiting the activation of the NF-κB signaling pathway [63]. Naringin is a flavonoid that has a beneficial effect on intestinal disorders. Liu et al. [64] showed that naringin (50 μM) protects the integrity of rat intestinal microvascular endothelial cell monolayer barriers against TNF-α-induced disruption, preventing TNF-α-induced apoptosis and suppressing cell migration, and avoiding gut-vascular barrier disruption. Moreover, the proanthocyanidins or condensed tannins have been reported to have bioactive properties like anti-inflammatory and antimicrobial, causing a reduction in intestinal inflammation and promoting the growth of *Lactobacillus* spp. and *Bifidobacteria* spp. [1, 62].

Recent animal studies have shown that chickpeas consumption improves gut health by inhibiting the proliferation of cancer cells, attenuating inflammation, modulating the composition and activity of the microbiome, promoting epithelial barrier integrity, mucus production, and antimicrobial defenses [49, 65]. In addition, the quantitative structure–activity relationship on the cytotoxic effect of

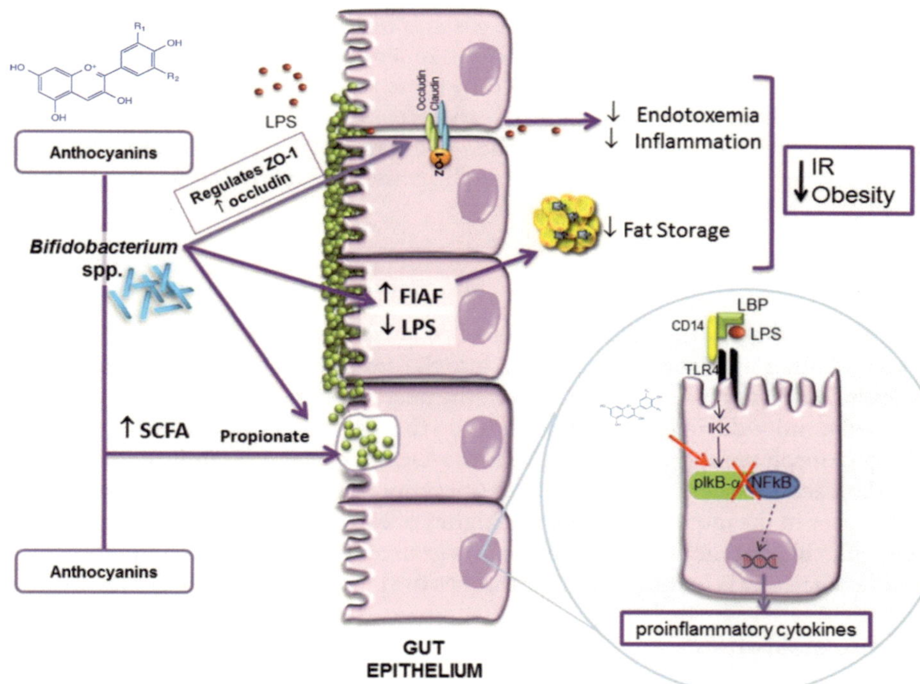

Figure 2.
The potential prebiotic effect of anthocyanins on gut microbiota and obesity. SCFA: Short-chain fatty acids; FIAF: Fasting-induced adipose factor; LPS: Lipopolysaccharide; ZO-1: Zonula occludens-1; IR: Insulin resistance. Anthocyanins and metabolites formed in the intestine change the composition of the gut microbiota. This is associated with restored tight-junction protein (ZO-1 and occludin) distribution, and localization. Hence, the gut permeability is decreased, and plasma lipopolysaccharide (LPS) levels (metabolic endotoxemia) are lowered, improving low-grade inflammation and obesity-related comorbidities. Anthocyanins decrease transcription factor NF-kB activity in the cell nucleus by decreasing gene expression of inflammatory cytokines, exerting their anti-inflammatory action. Anthocyanins have the ability to promote the growth of Bifidobacterium *spp., which increases the intestinal production of FIAF that inhibits fat storage in the host.* Bifidobacterium *spp. degrade SCFA; propionate stimulates mucus secretion and contributes to thickening of the mucus layer. At the same time, reduced mucus layer thickness favors microbiota encroachment. The figure is taken from [15].*

phenolic compounds have been studied. The *p*-hydroxybenzoic acid present in *P. sativum* exhibits strong cytotoxic activity in the colon carcinoma (HCT116) cell line [33]. Additional clinical evidence suggests that bean intake may reduce the incidence of developing advanced colorectal adenoma in humans [66]. Chen et al. [67] noted that cyanidin-3-*O*-glucoside protects against intestinal mucosal damage caused by 3-chloro-1, 2-propanediol (3-MCPD), a food contaminant. **Figure 2** shows the prebiotic effect of anthocyanins on the gut microbiota.

5. Phenolic compounds during gastrointestinal digestion: bioaccessibility and bioavailability

Bioaccessibility is defined from a nutritional point of view as the fraction of compounds liberated from the food matrix within the human gastrointestinal tract and available for intestinal absorption. The gastrointestinal tract is prone to oxidative stress due to its function as a primary digestive system and exposure to various stimuli [15, 29, 67]. The bioaccessibility and bioavailability of several phenolic compounds have been studied, noting that the aglycones in isoflavones are more bioavailable than their conjugated counterparts [9]. The absorption and bioavailability of phenolic compounds are commonly affected by low solubility, low permeability, and low stability in the gastrointestinal tract [29]. Some researchers have suggested that 5–20% of the total polyphenol content in legumes can be absorbed. The preventive action provided by these compounds depends on bioaccessibility. However, in the case of chronic diseases, such as stomach and colorectal cancer, they do not depend on the polyphenols bioaccessibility; still, gut microbiota can increase the bioavailability of the phenolic content of foods and quadruple their antioxidant activity [9, 12, 15, 29, 67].

5.1 Oral cavity absorption

During oral digestion, the food matrix is broken down, allowing phenolic compounds and other nutrients to be released into the environment due to enzymatic hydrolysis by salivary α-amylase. However, the decrease in polyphenols such as condensed tannins during mastication is due to the interaction with salivary proteins resulting in insoluble aggregates [50]. Luzardo-Ocampo et al. [68] indicated that the antioxidant activity of bean methanolic extracts during *in vitro* gastrointestinal digestion increases in the mouth stage due to a higher release of certain polyphenols such as catechin, chlorogenic acid, and vanillin. As gastrointestinal digestion progresses, antioxidant activity is increased until it reaches the large intestine, where it decreases.

5.2 Gastric absorption

Polyphenols are very poorly absorbed after ingestion and remain in the gastrointestinal tract, where they influence digestive enzymes activity, inhibiting crucial enzymes involved in the digestion of starch (α-amylase), lipids (pancreatic lipase), and protein digestibility (pepsin and trypsin). Digestibility is influenced by the polyphenol's interaction with food and endogenous proteins, like digestive enzymes, salivary proteins, gastric and intestinal mucosa, and other endogenous proteins on the luminal side of the intestinal tract [16, 69, 70]. Studies with *in vitro* simulation revealed that during gastric digestion (pH 1.2–2.0 in the presence of pepsin), there is a decrease in the recovery of phenolic compounds because they can interact with the pectin present in the food [29]. However, the presence of (+)-catechin has been evidenced in the stomach stage due to resistance to the acid environment [68].

5.3 Intestinal absorption

Polyphenols are not completely absorbed in the small intestine (5–10%). More than 90% enter the large intestine and are fermented by the human colon microbiota interacting with microorganisms (10–14 bacterial cells) and enzymes (α-L-rhamnosidase and β-D-glucosidase). Fermentation facilitates the liberation and absorption of insoluble bound phenolics involved in colorectal cancer prevention. The degradation of phenolic acids by enteric bacterial or chemical conversions may produce other metabolites, including protocatechuic acid, syringic acid, vanillic acid, phloroglucinol aldehyde, phloroglucinol acid, and gallic acid [3, 9, 15, 20, 21, 24, 30, 63, 69].

Phenolic compounds are catabolized by the gut microbiota, originate common phenolic (e.g., daidzein to equol, flavan-3-ols to valerolactones, and ellagitannins to urolithins) intermediates as in phenylpropionic, phenylacetic, and benzoic acids with different degrees of hydroxylation [17]. The flavonoids present in bound form (glycosides) and in free form (aglycones) are changed during digestion; a low fraction of these glycosides can be enzymatically hydrolyzed to aglycones in the small intestine, causing these aglycones to be more hydrophobic compared to the original glycosides. Aglycones are absorbed by epithelial cells through passive diffusion and then transported to the liver, where they will be metabolized. However, flavonoid glycosides are hardly absorbed in the small intestine due to their hydrophilic nature and reach the large intestine intact, where they will be metabolized by the gut microbiota [24].

During intestinal digestion, polyphenols such as anthocyanins, phenolic acids, catechin, quercetin, resveratrol, and rutin are unstable in the alkaline environment intestinal fluid (pH 6.8–8.0) due to their degradation [29]. The main phenolic acids absorbed in the small intestine are gallic, caffeic, and ferulic acids [63, 68]. Luzardo-Ocampo et al. [68] observed that phenolic acids in beans, like ferulic, chlorogenic, and vanillin, do not change their content during their passage through the small intestine, for 60–120 min, indicating their resistance to the intestinal enzymes and allowing them to arrive at the large intestine for fermentation. Milán-Noris et al. [12] observed that chickpea cooking increased intestinal absorption of the existent isoflavones. On the other hand, Cárdenas-Castro et al. [54] evaluated the bioaccessibility and *in vitro* release kinetics of phenolic compounds from two varieties of beans (Azufrado and Negro Jamapa). These authors reported that in cooked beans, the phenolic compounds showed 50% bioaccessibility, and 30% in cooked-fried beans, indicating that cooking did not modify the release kinetics of phenolic compounds during the first 60 min, being kaempferol-3-*O*-glucoside, quercetin-3-*O*-glucoside, and chlorogenic acid the main compounds released.

6. Interactions of phenolic compounds with the gut microbiota: metabolism and modulation

The interaction between the gut microbiota and the diet components is fundamental to the promotion of gut health. Lignans, flavonoids, and other phenolic compounds present in legumes participate in the modulation of the host's mucosal barrier integrity, attenuate the inflammatory process associated with colitis, and improve epithelial barrier integrity, aside from modulating fecal and cecal microbiota composition and providing beneficial effects against metabolic diseases like obesity. The interaction of gut microbiota and phenolic compounds, mainly anthocyanins, can implicate hydrolysis, demethylation, reduction, decarboxylation, dehydroxylation, or isomerization of compounds into simpler components to modulate absorption [15, 21, 49, 69]. Chlorogenic acid is poorly absorbed in the small intestine, but it has been shown that the bioavailability of this compound depends on the metabolism of the gut microflora. However, when this compound is metabolized in the colon, it modulates the colonic microbiota inducing a significant increase in the growth of *Bifidobacterium* spp. and *Clostridium coccoides-Eubacterium rectale*, revealing a potent antimicrobial activity by binding and permeabilizing the bacterial cell membrane [63]. On the other hand, it has been shown that phenolic compounds present in cooked chickpeas, like quercetin, daidzein, biochanin A, and formononetin, improved the integrity of the intestinal barrier by reducing its permeability and providing antioxidant and anti-inflammatory effects that promote gut health and decrease pathologies, like colitis [21, 49].

7. Conclusion

The inclusion of legumes such as chickpeas, peas and beans in the diet has increased consumer health benefits due to their content of bioactive compounds such as phenolic compounds and other nutrients. During digestion, these compounds are not completely absorbed in the intestinal tract and are metabolized in the colon, increasing their bioaccessibility and bioavailability. These compounds have been shown to participate in the modulation of the gut microbiota, the epithelial barrier and resistance to pathogen colonization, improving gut health by inhibiting the proliferation of cancer cells through their chemopreventive effects. The impact of phenolic compounds on the gut microbiota suggests that the incorporation of legumes into the diet and the design of novel functional foods may improve human health by preventing the development of metabolic and gastrointestinal disorders, including irritable bowel syndrome, inflammatory bowel disease, obesity, diabetes, colitis, and colorectal cancer. However, further research should be conducted to understand the impact of phenolic compounds during digestion and gut microbiota modulation.

Acknowledgements

The authors are grateful to the *Instituto Politécnico Nacional*, Mexico (SIP projects: 20181560, 20196640, and 20200291), and CONACyT project (241756).

Conflict of interest

The authors declare that they have no conflict of interest.

Author details

Mayra Nicolás-García[1], Cristian Jiménez-Martínez[1], Madeleine Perucini-Avendaño[1], Brenda Hildeliza Camacho-Díaz[2], Antonio Ruperto Jiménez-Aparicio[2] and Gloria Dávila-Ortiz[1]*

1 Departamento de Ingeniería Bioquímica, Escuela Nacional de Ciencias Biológicas, Instituto Politécnico Nacional (IPN), Mexico City, Mexico

2 Centro de Desarrollo de Productos Bióticos, Instituto Politécnico Nacional (IPN), Morelos, Mexico

*Address all correspondence to: gdavilao@yahoo.com

References

[1] Martens, L. G., Nilsen, M. M., & Provan, F. Pea hull fibre: Novel and sustainable fibre with important health and functional properties. ECNU. 2017; 10: 139-148.

[2] Bian, Y., Dong, Y., Sun, J., Sun, M., Hou, Q., Lai, Y., & Zhang, B. Protective effect of kaempferol on LPS-induced inflammation and barrier dysfunction in a coculture model of intestinal epithelial cells and intestinal microvascular endothelial cells. J. Agric. Food Chem. 2019; 68 (1): 160-167. https://doi.org/10.1021/acs.jafc.9b06294

[3] Moles, L., & Otaegui, D. The impact of diet on microbiota evolution and human health. Is diet an adequate tool for microbiota modulation?. Nutrients, 2020; 12 (6): 1654. https://doi.org/10.3390/nu12061654

[4] Wan, M. L., Ling, K. H., El-Nezami, H., & Wang, M. F. Influence of functional food components on gut health. Crit. Rev. Food Sci. Nutr. 2019; 59 (12): 1927-1936. https://doi.org/10.1080/10408398.2018.1433629

[5] Cid-Gallegos, M. S., Sánchez-Chino, X. M., Juárez Chairez, M. F., Álvarez González, I., Madrigal-Bujaidar, E., & Jiménez-Martínez, C. Anticarcinogenic activity of phenolic compounds from sprouted legumes. Food Res. Int. 2020; 1-16. https://doi.org/10.1080/87559129.2020.1840581

[6] Fassarella, M., Blaak, E. E., Penders, J., Nauta, A., Smidt, H., & Zoetendal, E. G. Gut microbiome stability and resilience: elucidating the response to perturbations in order to modulate gut health. Gut. 2021; 70 (3): 595-605. http://dx.doi.org/10.1136/gutjnl-2020-321747

[7] Danneskiold-Samsøe, N. B., Barros, H. D. D. F. Q., Santos, R., Bicas, J. L., Cazarin, C. B. B., Madsen, L., & Júnior, M. R. M. Interplay between food and gut microbiota in health and disease. Food Res. Int. 2019; 115: 23-31. https://doi.org/10.1016/j.foodres.2018.07.043

[8] Zmora, N., Suez, J., & Elinav, E. You are what you eat: diet, health and the gut microbiota. Nat. Rev. Gastroenterol Hepatol. 2019; 16 (1): 35-56. https://doi.org/10.1038/s41575-018-0061-2

[9] De Camargo, A. C., Favero, B. T., Morzelle, M. C., Franchin, M., Alvarez-Parrilla, E., de la Rosa, L. A., Vilar Geraldi, M., Maróstica Júnior, M. R., Shahidi, F., & Schwember, A. R. Is chickpea a potential substitute for soybean? Phenolic bioactives and potential health benefits. Int. J. Mol. Sci. 2019; 20 (11): 2644. https://doi.org/10.3390/ijms20112644

[10] Singh, B., Singh, J. P., Kaur, A., & Singh, N. Phenolic composition and antioxidant potential of grain legume seeds: A review. Food Res Int. 2017; 101: 1-16. https://doi.org/10.1016/j.foodres.2017.09.026

[11] Awika, J. M., Rose, D. J., & Simsek, S. Complementary effects of cereal and pulse polyphenols and dietary fiber on chronic inflammation and gut health. Food Funct. 2018; 9 (3): 1389-1409. https://doi.org/10.1039/C7FO02011B

[12] Milán-Noris, A. K., Gutiérrez-Uribe, J. A., Santacruz, A., Serna-Saldívar, S. O., & Martínez-Villaluenga, C. Peptides and isoflavones in gastrointestinal digests contribute to the anti-inflammatory potential of cooked or germinated desi and kabuli chickpea (*Cicer arietinum* L.). Food Chem. 2018; 268: 66-76. https://doi.org/10.1016/j.foodchem.2018.06.068

[13] Fahim, J. R., Attia, E. Z., & Kamel, M. S. The phenolic profile of pea (*Pisum sativum*): a phytochemical and pharmacological overview. Phytochem.

Rev. 2019; 18 (1): 173-198. https://doi.org/10.1007/s11101-018-9586-9

[14] Nicolás-García, M., Perucini-Avendaño, M., Jiménez-Martínez, C., Perea-Flores, M. J., Gómez-Patiño, M. B., Arrieta-Báez, D., & Dávila-Ortiz, G. Bean phenolic compound changes during processing: chemical interactions and identification. J. Food Sci. 2021; 83 (3): 643-655. https://doi.org/10.1111/1750-3841.15632

[15] Jamar, G., Estadella, D., & Pisani, L. P. Contribution of anthocyanin-rich foods in obesity control through gut microbiota interactions. BioFactors, 2017; 43 (4): 507-516. https://doi.org/10.1002/biof.1365

[16] Guo, F., Xiong, H., Wang, X., Jiang, L., Yu, N., Hu, Z., Sun, Y., & Tsao, R. Phenolics of green pea (*Pisum sativum* L.) hulls, their plasma and urinary metabolites, bioavailability, and *in vivo* antioxidant activities in a rat model. J. Agric. Food Chem. 2019; 67 (43): 11955-11968. https://doi.org/10.1021/acs.jafc.9b04501

[17] Pei, R., Liu, X., & Bolling, B. Flavonoids and gut health. Curr. Opin. Biotechnol. 2020; 61: 153-159. https://doi.org/10.1016/j.copbio.2019.12.018

[18] Minatel, I. O., Borges, C. V., Ferreira, M. I., Gomez, H. A. G., Chen, C. Y. O., & Lima, G. P. P. Phenolic compounds: Functional properties, impact of processing and bioavailability. In Soto-Hernández M, editor. Phenolic compounds—Biological activity. 1ra ed. IntechOpen; 2017. p. 1-24. http://dx.doi.org/10.5772/66368

[19] Mecha, E., Leitão, S. T., Carbas, B., Serra, A. T., Moreira, P. M., Veloso, M. M., & Bronze, M. R. Characterization of soaking process' impact in common beans phenolic composition: Contribute from the unexplored Portuguese germplasm. Foods. 2019; 8 (8): 296. https://doi.org/10.3390/foods8080296

[20] Domínguez-Arispuro, D. M., Cuevas-Rodríguez, E. O., Milán-Carrillo, J., León-López, L., Gutiérrez-Dorado, R., & Reyes-Moreno, C. Optimal germination condition impacts on the antioxidant activity and phenolic acids profile in pigmented desi chickpea (*Cicer arietinum* L.) seeds. J. Food Sci. Technol. 2018; 55 (2): 638-647. https://doi.org/10.1007/s13197-017-2973-1

[21] Monk, J. M., Wu, W., Hutchinson, A. L., Pauls, P., Robinson, L. E., & Power, K. A. Navy and black bean supplementation attenuates colitis-associated inflammation and colonic epithelial damage. J. Nutr. Biochem. 2018; 56: 215-223. https://doi.org/10.1016/j.jnutbio.2018.02.013

[22] Aranda-Olmedo, I., & Rubio, L. A. Dietary legumes, intestinal microbiota, inflammation and colorectal cancer. J. Funct. Foods. 2020; 64: 103707. https://doi.org/10.1016/j.jff.2019.103707

[23] Juárez-Chairez, M. F., Cid-Gallegos, M. S., Meza-Márquez, O. G., & Jiménez-Martínez C. Biological activities of chickpea in human health (*Cicer arietinum* L.). A review. Plant Foods Hum Nutr. 2020; 75 (2): 142-153. https://doi.org/10.1007/s11130-020-00814-2

[24] Neri-Numa, I. A., Cazarin, C. B. B., Ruiz, A. L. T. G., Paulino, B. N., Molina, G., & Pastore, G. M. Targeting flavonoids on modulation of metabolic syndrome. J. Funct. Foods. 2020; 73: 104132. https://doi.org/10.1016/j.jff.2020.104132

[25] Khan, M. I., Khan, S. M., Hussain, I., ur Rehman, S., Shah, S. H. R., Ayub, Q., & ul Haq, N. Qualitative analysis of pea (*Pisum sativum*) seeds procured from different sources and locations of district Haripur-Khyber Pukhtunkhwa Pakistan. PAB. 2019; 8 (2): 1782-1788. http://dx.doi.org/10.19045/bspab.2019.80121

[26] Bitocchi, E., Rau, D., Bellucci, E., Rodriguez, M., Murgia, M. L., Gioia, T., & Papa, R. Beans (*Phaseolus ssp.*) as a model for understanding crop evolution. Front. Plant Sci. 2017; 8: 722. https://doi.org/10.3389/fpls.2017.00722

[27] Capistrán-Carabarin, A., Aquino-Bolaños, E. N., García-Díaz, Y. D., Chávez-Servia, J. L., Vera-Guzmán, A. M., & Carrillo-Rodríguez, J. C. Complementarity in phenolic compounds and the antioxidant activities of *Phaseolus coccineus* L. and *P. vulgaris* L. Landraces. Foods. 2019; 8 (8): 295. DOI: 10.3390/foods8080295

[28] Dias, R., Oliveira, H., Fernandes, I., Simal-Gandara, J., & Perez-Gregorio, R. Recent advances in extracting phenolic compounds from food and their use in disease prevention and as cosmetics. Crit. Rev. Food Sci. Nutr. 2020; 1-22. https://doi.org/10.1080/10408398.2020.1754162

[29] Peanparkdee, M., & Iwamoto, S. Encapsulation for improving *in vitro* gastrointestinal digestion of plant polyphenols and their applications in food products. Food Rev. Int, 2020; 1-19. https://doi.org/10.1080/87559129.2020.1733595

[30] Lafarga, T., Villaró, S., Bobo, G., Simó, J., & Aguiló-Aguayo, I. Bioaccessibility and antioxidant activity of phenolic compounds in cooked pulses. Int. J. Food Sci. Technol. 2019; 54 (5): 1816-1823. DOI:10.1111/ijfs.14082

[31] Gupta, R. K., Gupta, K., Sharma, A., Das, M., Ansari, I. A., & Dwivedi, P. D. Health risks and benefits of chickpea (*Cicer arietinum*) consumption. J. Agric. Food Chem. 2019; 65 (1): 6-22. https://doi.org/10.1021/acs.jafc.6b02629

[32] Fan, G., & Beta, T. Discrimination of geographical origin of Napirira bean (*Phaseolus vulgaris* L.) based on phenolic profiles and antioxidant activity. J. Food Compos. Anal. 2017; 62: 217-222. https://doi.org/10.1016/j.jfca.2017.07.001

[33] El-Feky, A. M., Elbatanony, M. M., & Mounier, M. M. Anti-cancer potential of the lipoidal and flavonoidal compounds from *Pisum sativum* and *Vicia faba* peels. Egypt. J. Basic Appl. Sci. 2018; 5 (4): 258-264. https://doi.org/10.1016/j.ejbas.2018.11.001

[34] Giusti, F., Capuano, E., Sagratini, G., & Pellegrini, N. A comprehensive investigation of the behaviour of phenolic compounds in legumes during domestic cooking and *in vitro* digestion. Food Chem. 2019; 285: 458-467. https://doi.org/10.1016/j.foodchem.2019.01.148

[35] Stanisavljević, N. S., Ilić, M. D., Matić, I. Z., Jovanović, Ž. S., Čupić, T., Dabić, D. Č., Natić, M. M., & Tešić, Ž. L. Identification of phenolic compounds from seed coats of differently colored European varieties of pea (*Pisum sativu*m L.) and characterization of their antioxidant and *in vitro* anticancer activities. Nutr. Cancer. 2016; 68 (6):988-1000. https://doi.org/10.1080/01635581.2016.1190019

[36] Telles, A. C., Kupski, L., & Furlong, E. B. Phenolic compound in beans as protection against mycotoxins. Food Chem. 2017; 214: 293-299. https://doi.org/10.1016/j.foodchem.2016.07.079

[37] Yang, Q. Q., Gan, R. Y., Ge, Y. Y., Zhang, D., & Corke, H. Ultrasonic treatment increases extraction rate of common bean (*Phaseolus vulgaris* L.) antioxidants. Antioxidants, 2019; 8 (4): 83. https://doi.org/10.3390/antiox8040083.

[38] Mojica, L., Meyer, A., Berhow, M. A., & de Mejía, E. G. Bean cultivars (*Phaseolus vulgaris* L.) have similar high antioxidant capacity, *in vitro* inhibition of α-amylase and α-glucosidase while diverse phenolic composition and concentration. Food Res. Int. 2015; 69:

38-48. https://doi.org/10.1016/j.foodres.2014.12.007

[39] Brigide, P., Canniatti-Brazaca, S., Pereira, M. & Huber, K. Effect of cooking in common bean cultivars on antioxidant activity and phenolic compounds. Braz. J. Food Res. 2018; 9 (3): 1-7. DOI:10.3895/rebrapa.v9n3.4538

[40] Eshraq, B., Mona, A., Sayed, A., & Emam, A. Effect of soaking, cooking and germination on chemical constituents and bioactive compounds as well as their cytotoxic activities of black bean extracts. Nat. Prod. Chem. Res. 2016; 4. 1-7. DOI:10.4172/2329-6836.1000237

[41] Zilani, M. N. H., Sultana, T., Rahman, S. A., Anisuzzman, M., Islam, M. A., Shilpi, J. A., & Hossain, M. G. Chemical composition and pharmacological activities of *Pisum sativum*. BMC Complement Altern Med. 2017; 17 (1): 1-9. https://doi.org/10.1186/s12906-017-1699-y

[42] Lomas-Soria, C., Pérez-Ramírez, I. F., Caballero-Pérez, J., Guevara-Gonzalez, R. G., Guevara-Olvera, L., Loarca-Piña, G., & Reynoso-Camacho, R. Cooked common beans (*Phaseolus vulgaris* L.) modulate renal genes in streptozotocin-induced diabetic rats. J. Nutr. Biochem. 2015; 26 (7): 761-768. https://doi.org/10.1016/j.jnutbio.2015.02.006

[43] Ombra, M. N., d'Acierno, A., Nazzaro, F., Riccardi, R., Spigno, P., Zaccardelli, M., & Fratianni, F. Phenolic composition and antioxidant and antiproliferative activities of the extracts of twelve common bean (*Phaseolus vulgaris* L.) endemic ecotypes of southern Italy before and after cooking. Oxid. Med. Cell. Longev. 2016; 1-12. https://doi.org/10.1155/2016/1398298

[44] Teixeira-Guedes, C. I., Oppolzer, D., Barros, A. I., & Pereira-Wilson, C. Impact of cooking method on phenolic composition and antioxidant potential of four varieties of *Phaseolus vulgaris* L. and *Glycine max* L. LWT-Food Sci Technol. 2019; 103: 238-246. https://doi.org/10.1016/j.lwt.2019.01.010

[45] Maharjan, P., Penny, J., Partington, D. L., & Panozzo, J. F. Genotype and environment effects on the chemical composition and rheological properties of field peas. J. Sci. Food Agric. 2019; 99 (12): 5409-5416. https://doi.org/10.1002/jsfa.9801

[46] Madrera, R. R., & Valles, B. S. Development and validation of ultrasound assisted extraction (UAE) and HPLC-DAD method for determination of polyphenols in dry beans (*Phaseolus vulgaris*). J. Food Compos. Anal. 2020; 85. https://doi.org/10.1016/j.jfca.2019.103334

[47] Megías, C., Cortés-Giraldo, I., Alaiz, M., Vioque, J., & Girón-Calle, J. Isoflavones in chickpea (*Cicer arietinum*) protein concentrates. J. Funct. Foods. 2016; 21: 186-192. https://doi.org/10.1016/j.jff.2015.12.012

[48] Gao, Y., Yao, Y., Zhu, Y., & Ren, G. Isoflavone content and composition in chickpea (*Cicer arietinum* L.) sprouts germinated under different conditions. J. Agric. Food Chem. 2015; 63 (10): 2701-2707. https://doi.org/10.1021/jf5057524

[49] Monk, J. M., Wu, W., McGillis, L. H., Wellings, H. R., Hutchinson, A. L., Liddle, D. M., & Power, K. A. Chickpea supplementation prior to colitis onset reduces inflammation in dextran sodium sulfate-treated C57Bl/6 male mice. Appl. Physiol. Nutr. Metab. 2018; 43 (9): 893-901. https://doi.org/10.1139/apnm-2017-0689

[50] Gan, Y., Fu, Y., Yang, L., Chen, J., Lei, H., & Liu, Q. Cyanidin-3-*O*-glucoside and cyanidin protect against

intestinal barrier damage and 2, 4, 6-trinitrobenzenesulfonic acid-induced colitis. J. Med. Food. 2020; 23 (1): 90-99. https://doi.org/10.1089/jmf.2019.4524

[51] Dueñas, M., Sarmento, T., Aguilera, Y., Benitez, V., Mollá, E., Esteban, R. M., & Martín-Cabrejas, M. A. Impact of cooking and germination on phenolic composition and dietary fibre fractions in dark beans (*Phaseolus vulgaris* L.) and lentils (*Lens culinaris* L.). LWT-Food Sci Technol. 2016; 66: 72-78. https://doi.org/10.1016/j.lwt.2015.10.025

[52] López-Martínez, L. X., Leyva-López, N., Gutiérrez-Grijalva, E. P., & Heredia, J. B. Effect of cooking and germination on bioactive compounds in pulses and their health benefits. J. Funct. Foods. 2017; 38: 624-634. http://dx.doi.org/10.1016/j.jff.2017.03.002

[53] Bulbula, D. D., & Urga, K. Study on the effect of traditional processing methods on nutritional composition and anti nutritional factors in chickpea (*Cicer arietinum*). Cogent Food Agric. 2018; 4 (1): 1422370. https://doi.org/10.1080/23311932.2017.1422370

[54] Cárdenas-Castro, A. P., Pérez-Jiménez, J., Bello-Pérez, L. A., Tovar, J., & Sáyago-Ayerdi, S. G. Bioaccessibility of phenolic compounds in common beans (*Phaseolus vulgaris* L.) after *in vitro* gastrointestinal digestion: A comparison of two cooking procedures. Cereal Chem. 2020; 97 (3). https://doi.org/10.1002/cche.10283

[55] Arribas, C., Cabellos, B., Cuadrado, C., Guillamón, E., & Pedrosa, M. M. The effect of extrusion on the bioactive compounds and antioxidant capacity of novel gluten-free expanded products based on carob fruit, pea and rice blends. Innov. Food Sci. Emerg. Technol. 2019; 52: 100-107. https://doi.org/10.1016/j.ifset.2018.12.003

[56] Gu, B. J., Masli, M. D. P., & Ganjyal, G. M. Whole faba bean flour exhibits unique expansion characteristics relative to the whole flours of lima, pinto, and red kidney beans during extrusion. J. Food Sci. 2020; 85 (2): 404-413. https://doi.org/10.1111/1750-3841.14951

[57] Ferreira, C. D., Bubolz, V. K., da Silva, J., Dittgen, C. L., Ziegler, V., de Oliveira Raphaelli, C., & de Oliveira, M. Changes in the chemical composition and bioactive compounds of chickpea (*Cicer arietinum* L.) fortified by germination. LWT-Food Sci Technol. 2019; 111: 363-369. https://doi.org/10.1016/j.lwt.2019.05.049

[58] Xu, M., Jin, Z., Ohm, J. B., Schwarz, P., Rao, J., & Chen, B. Effect of germination time on antioxidative activity and composition of yellow pea soluble free and polar soluble bound phenolic compounds. Food Funct. 2019; 10 (10): 6840-6850. DOI: 10.1039/C9FO00799G

[59] Worku, A., & Sahu, O. Significance of fermentation process on biochemical properties of *Phaseolus vulgaris* (red beans). Biotechnol. Rep. 2017; 16: 5-11. https://doi.org/10.1016/j.btre.2017.09.001

[60] Uzeta, C. M., Cuevas-Rodriguez, E. O., Cervantes, J. L., Carrillo, J. M., Dorado, R. G., & Moreno, C. R. Improvement nutritional/antioxidant properties of underutilized legume tepary bean (*Phaseolus acutifolius*) by solid state fermentation. Agrociencia, 2019; 53 (7): 987-1003.

[61] Fan, Y., & Pedersen, O. Gut microbiota in human metabolic health and disease. Nat. Rev. Microbiol. 2020; 1-17. https://doi.org/10.1038/s41579-020-0433-9

[62] Forgie, A. J., Gao, Y., Ju, T., Pepin, D. M., Yang, K., Gänzle, M. G., Ozga, J. A., Chan, C. B., & Willing, B. P. Pea polyphenolics and hydrolysis processing alter microbial community structure and early pathogen colonization in mice.

J. Nutr. Biochem. 2019; 67: 101-110. https://doi.org/10.1016/j.jnutbio.2019.01.012

[63] Coman, V., & Vodnar, D. C. Hydroxycinnamic acids and human health: Recent advances. J. Sci. Food Agric. 2020; 100 (2): 483-499. https://doi.org/10.1002/jsfa.10010

[64] Liu, P., Bian, Y., Fan, Y., Zhong, J., & Liu, Z. Protective effect of naringin on *in vitro* gut-vascular barrier disruption of intestinal microvascular endothelial cells induced by TNF-α. J. Agric. Food Chem. 2020; 68 (1): 168-175. https://doi.org/10.1021/acs.jafc.9b06347

[65] Sánchez Chino, X. M., Jiménez Martínez, C., Vásquez Garzón, V.R., Álvarez González, I., Villa Treviño, S., Madrigal Bujaidar, E., Dávila Ortiz, G., & Baltiérrez Hoyos R. Cooked chickpea consumption inhibits colon carcinogenesis in mice induced with azoxymethane and dextran sulfate sodium. J. Am. Coll. Nutr. 2017; 36 (5): 391-398. https://doi.org/10.1080/07315724.2017.1297744

[66] Chen, P. X., Zhang, H., Marcone, M. F., Pauls, K. P., Liu, R., Tang, Y., Zhang, B., Renaud, J. B., & Tsao, R. Anti-inflammatory effects of phenolic-rich cranberry bean (*Phaseolus vulgaris* L.) extracts and enhanced cellular antioxidant enzyme activities in Caco-2 cells. J. Funct. Foods. 2017; 38: 675-685. https://doi.org/10.1016/j.jff.2016.12.027

[67] Chen, G., Wang, G., Zhu, C., Jiang, X., Sun, J., Tian, L., & Bai, W. Effects of cyanidin-3-*O*-glucoside on 3-chloro-1, 2-propanediol induced intestinal microbiota dysbiosis in rats. Food Chem. Toxicol. 2019; 133: 110767. https://doi.org/10.1016/j.fct.2019.110767

[68] Luzardo-Ocampo, I., Campos-Vega, R., Gaytán-Martínez, M., Preciado-Ortiz, R., Mendoza, S., & Loarca-Piña, G. Bioaccessibility and antioxidant activity of free phenolic compounds and oligosaccharides from corn (*Zea mays* L.) and common bean (*Phaseolus vulgaris* L.) chips during *in vitro* gastrointestinal digestion and simulated colonic fermentation. Food Res. Int. 2017; 100: 304-311. https://doi.org/10.1016/j.foodres.2017.07.018

[69] Edwards, C. A., Havlik, J., Cong, W., Mullen, W., Preston, T., Morrison, D. J., & Combet, E. Polyphenols and health: Interactions between fibre, plant polyphenols and the gut microbiota. Nutr. Bull. 2017; 42 (4): 356-360. https://doi.org/10.1111/nbu.12296

[70] Cirkovic Velickovic, T. D., & Stanic-Vucinic, D. J. The role of dietary phenolic compounds in protein digestion and processing technologies to improve their antinutritive properties. Compr. Rev. Food Sci. Food Saf. 2018; 17 (1): 82-103. https://doi.org/10.1111/1541-4337.12320

Chapter 6

Assessment of Secondary Metabolites with Different Uses of Fenugreek

Gulsum Yaldiz and Mahmut Camlica

Abstract

Fenugreek (*Trigonella foenum-graecum*) is an annual medicinal plant with trifoliate leaves, a branched stem, white flowers, rooted tubers, and golden yellow seed belonging to Fabaceae family. Fenugreek is used in different industries such as pharmaceutical, nutraceutical and food industries as an ancient crop plant. Fenugreek is grown as a medicinal herb in many countries and has antioxidant, hypoglycemic, hypercholesterolemia, stomach protective, chemopreventive, laxative and appetite stimulating properties. In recent years, many important studies have been conducted on the biological activities and therapeutic properties of fenugreek mainly secondary metabolites such as alkaloids, flavonoids, steroids and saponins. These compounds are used for multipurpose uses in different industries and also appreciated by scientists. Based on these several health usefulness as discussed in review, fenugreek might be a good candidate for a herbal drug and used for preparation of new drugs. In this review, secondary metabolites used in different industries of fenugreek will be discussed and general benefits of them will be expressed within the all significant aspect of fenugreek as clearly. This review also highlights the traditional uses and nutraceutical properties (antioxidant activity, antibacterial, antifungal, anticancer hypoglemic effects and anti-inflammatory and immunological activity) of fenugreek. These uses and effect properties of fenugreek have been discussed and researchable areas were implied to depending on the previous studies. In the future, studies on fenugreek are needed some important applications to increasing the popularity of fenugreek. In this context, researchers should be focused on secondary and primary metabolite studies in fenugreek seeds and leaves. In addition to these, fenugreek germplasm should be collected and subjected to intensive selection via modern breeding programs and new fenugreek genotypes with desired properties should be obtained.

Keywords: Fenugreek, *Trigonella foenum-graecum* L., ancestor plant, multipurpose uses, secondary metabolite

1. Introduction

Fenugreek (*Trigonella foenum-graecum* L.) is commonly grown in many parts of the world for both culinary purposes and health benefits. Fenugreek is rich in minerals, protein, vitamin A and C, and contains several bioactive compounds including proteins, protease inhibitors. Its seeds contain 23–43% protein, up to 58% carbohydrate, nearly 10–13% moisture, 5–6% lipid and less than 1% minerals.

IntechOpen

In particular, the plant is rich in soluble fiber, mucilage and galactomannan which decrease the uptake of bile salts and starch absorption [1].

The main secondary metabolites and seed contents of fenugreek were given in **Figure 1** [2–4].

Polyphenol compounds *viz.* rhaponticin and isovitexin were noted as major bioactive compounds in seeds of fenugreek [5]. In addition, fenugreek seed extracts have a number of phenolic constituents similar to beta-D-glucopyranoside, methyl, alpha-d-mannopyranoside, methyl, and diethyl phthalate. Fenugreek also represents a significant source of antioxidants [6].

A wide range of beneficial effects of fenugreek seeds has been reported by a number of researchers. In addition to antidiabetic effects, seeds have significant anti-atherosclerotic [7], anti-inflammatory [8], antinociceptive [9], and antiulcerogenic activities [10] which are essential for cure of diabetes and cancer disease. Antioxidant property helps in anti-aging. The phenolic antioxidants present in the extract of fenugreek show free radical scavenging activity which reduce oxidative stress in the body. This reduced oxidative stress reduces frequency of age-related disorders [11].

It was reported that fenugreek seeds had low amount essential oil and fatty oil [12]. According to the essential oil compounds findings; olfactometry diacetyl, 1-Octen-3-one, sotolone, acetic acid; 3-Isobutyl-2-methoxypyrazine, butanoic acid, isovaleric acid, 3-isopropyl2-methoxypyrazine, caproic acid, eugenol, 3-Amino-4,5-dimethyl-3, linalool, (Z)-1,5-Octadiene-3-one, 4-dihydro-2(5H)-furanone were determined as the main components [13]. It has been noted that sotolone is mostly found in fenugreek (5 s)-enantiomeric form (95%) among these essential oil components.

A study on human sweat was conducted by Meghwal and Goswami [3] and regarding essential oil components; pinene,3-octen-2-one, 2,5-dimethylpyrazine, b-camphor; terpinen-4-ol, 4-isopropylbenzaldehyde were found as the odor in sweat, while neryl acetate and b-caryophylene, 2,5-dimethylpyrazine has been observed to be the main component responsible for the compound contributing to sweat odor.

Many studies have been conducted on the therapeutic applications of various plant species on different diseases such as fungal, viral and bacterial contamination. Therefore, approximately one third of the world's population uses traditional/therapeutic plants and their extracts in their treatments [14]. A drug with both antidiabetic and antioxidant activity is much more beneficial in the treatment of diabetes. In addition, herbal medicines are more preferred due to the undesirable side effects of the existing antidiabetic medicine. As can be seen from the above explanations, fenugreek has the potential to be a versatile herbal medicine. Therefore, further studies are needed to provide detailed information

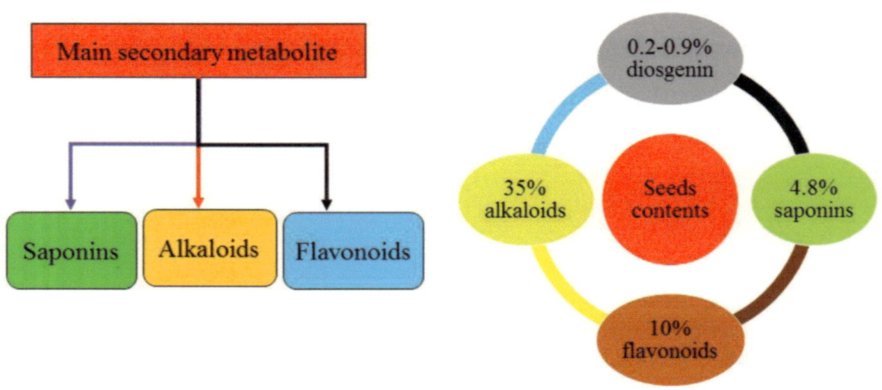

Figure 1.
Main secondary metabolites and seed context of fenugreek.

about the effects of fenugreek. So, in this review, it was aimed to inform about the studies conducted with the effects of fenugreek.

Generally, this study revealed the importance of focusing on the antioxidant, hypoglycemic, hypercholesterolemia, anticancer, antibacterial and antifungal properties, as well as on its medicinal properties, phytochemical and nutrient contents. In case, fenugreek is grown for bioactive secondary metabolites, concentrating on different activities such as polyphenol compounds, anti-inflammatory, antimicrobial properties.

2. Traditional uses of fenugreek

Fenugreek has been used in traditional cure treatments dates back to the 15th century. Different parts of fenugreek such as seeds and leaves were used to treatment of symptoms and ailments. For instance, a paste prepared with ground fenugreek seeds was used to treat eczema, local inflammations of the skin-as locally administered poultice or added to a hot bath [15].

Fenugreek is easily grown all over the world because of its wide adaptation and its usage varies significantly between countries.

Many studies were conducted to determine and confirm the traditional uses of fenugreek as herbal cure using different plant part (seeds and leaves) or pure phytochemicals (saponins, steroids, alkaloids). Fenugreek seeds have been reported to have an aphrodisiac effect in ancient times, but modern vaidyas have used it more for digestive and respiratory problems caused by phlegm and wind. In ancient Egypt, it was recorded that methi (fenugreek) was used to facilitate childbirth and increase milk flow, and modern Egyptian women still use fenugreek to make hilba tea to relieve menstrual cramps and relieve other abdominal pains. The Chinese call it hu lu ba and also use it to treat abdominal pain. While the fresh stems and leaves are mostly used as a winter vegetable, seeds are used as a flavor agent in different foods in India [16].

The seeds are also eaten raw as sprouts and used medicinally. While fenugreek is used in baking bread by Egyptian and Ethiopian, Switzerland uses it to flavor cheese. In the USA, it is mostly used in spice mixes for soups and stews [16]. Fenugreek has been used as a spice in cooking for centuries in European countries and remains a popular ingredient in curry powders, pickles, and spice mixes in India, Pakistan, Bangladesh, and other Asian countries. Fenugreek has been used in folk remedies to treat cellulite, boils and tuberculosis. Fenugreek remained a key ingredient in a 19th century patented drug for dysmenorrheal and post-menopausal symptoms [17].

Fenugreek was also used for ethnoveterinary applications such as the decrease of serum cholesterol in animals [18] and the increase of milk production in animals [19].

In the Ayurvedic and Unani systems of medicine, fenugreek is used to cure epilepsy, paralysis, gout, dropsy, chronic cough and piles. This crop has also known as potential of oleoresin and steroid production for oral contraceptives. In addition, ground seeds use as a control mechanism for the blood sugar and thereby checks the diabetes in human beings [20].

Fenugreek leaves are widely used for treatment of eye diseases in Iran [21] and gynecological disorders [22]. In traditional medicine, it is used to prepare infusions, water and alcohol extracts, tinctures, honeys, tonics with antidepressant and psychotonic properties, and muscle growth supplements. It is also used in the treatment of seborrhea, acne and dermatitis. The plant is widely used in cosmetology [23]. It was reported that fenugreek seeds have been used as an oral insulin substitute to decreasing blood sugar [17].

The aroma and taste of fenugreek has led to its use in imitation maple syrup [24]. Furthermore, fresh and dried leaves are used as vegetables in the diets. It was found that these leaves included calcium; zinc iron, phosphorous, riboflavin, carotene, thiamine, niacin and vitamin C. The leaves of fenugreek, which are stored in either in refrigeration conditions or dried in oven, are used to prepare in pressure cooker [25].

Aqueous solutions and softened fenugreek oils exert protective effects on mucous membranes in ulcer disease [26] and prevent colon cancer [27]. It is being utilized in the folk medicines for the treatment of cellulitis, tuberculosis and boils [28].

In Turkey, it has been found to be beneficial in healing internal wounds when taken with butter and sugar. It is used in healing hemorrhoids and it can be also used as a supplementary food supplement in the treatment of hyperthyroidism.

3. Nutraceutical properties of fenugreek

Pharmacological activities have been studied by Mehrafarin et al. [29] to explain the medicinal properties of fenugreek and its main metabolites. Many studies were conducted to increase the secondary metabolites of fenugreek by different applications (**Table 1**).

Part of fenugreek	Application	Secondary metabolites	References
Leaves	UV Radiation	Total phenols, Flavonoids, Alkaloids, Tannins, saponins, Anthocyanin content	El-Shora et al. [30]
Hypocotyls of the sprouts for callus initiation	Mannitol and Sodium Chloride	Total phenolics, Total flavonoids, Tannins,	Hussein and Aqlan, [31]
Seedling	Methyl Jasmonate	Diosgenin	Chaudhary et al. [32]
Fenugreek plant	Trichoderma strains	Trigonellin	Hosseini et al. [33]
Seeds and Callus	Cultured on MS medium	Trigonellin, Diosgenin	Altabtabaai et al. [34]
Leaf and stem	Collected from the local market Surat, Gujarat	Phenol content, Flavonoids content,	Varsha and Jain, [35]
Seeds	Collected from 50 regions of Iran	4-hydroxy isoleucine, Trigonellin	Rajabihashjin et al. [36]
Leaf	Water deficit, exogenous ethylene application and root symbioses	Trigonellin	Irankhah et al. [37]
Growth stages (Vegetative, full flowering and well-developed pods)	Rainfeed conditions	Proximate composition, total phenols, tannins, flavonoids and saponins	Abdouli et al. [38]
Fenugreek plants	Mycorrhizal fungal inoculum and exogenous methyl jasmonate; Water deficit	ABA, IAA, trigonelline, diosgenin	Irankhah et al. [39]
Seeds	Charcoal and drought stress	Trigonellin, diosgenin	Bitarafan et al. [40]
Plant tissues	Copper stress	Total phenols, Total flavanoids	Elleuch et al. [41]
Seeds	Gamma irradiation	Trigonelline, nicotinic acid, Diosgenin, Mucilage content	Parchin et al. [42]

Table 1.
Variation of fenugreek secondary metabolites in different cultural applications.

Fenugreek seeds contain the neuroprotective alkaloid trigonelline, which is one among the foremost alkaloids found in fenugreek seeds. Trigonelline consists of a methyl betaine derivative of nicotinic acid, aids in curing diabetes and treatment of neurodegenerative diseases. In addition, alkaloids like trimethylamine, neurin, choline, gentianine, carpine and betain are found in fenugreek. These alkaloids exhibit antibacterial, antiviral and memory improving activities [43].

Fenugreek also represents an important source of diosgenin, a saponin used as a precursor for the synthesis of steroid hormones. Diosgenin is a very valuable phytochemical due to its biological activities and pharmaceutical applications. In fact, this phytochemical has anticancer, anti-aging, cardioprotective and contraceptive properties [44–49] and antiviral, antimicrobial, antifungal and insecticidal activities [50, 51]. Anticancer effect of diosgenin has also been investigated in a number of preclinical studies, including growth inhibition and apoptosis induction in human colon cancer cells [27] and cell cycle in different cancer cell lines [52] has been documented.

So, fenugreek seeds have anti-inflammatory, hypoglycemic effects, anti-diabetes, cardioprotective, anticancer, antimicrobial properties, antipyretic and analgesic properties [53, 54]. Therefore, breeding strategies should be re-validated to increase the amounts of important active substances such as diosgenin and trigonellin in fenugreek seeds. Genetic variation in fenugreek should be introduced and a combination of traditional and molecular approaches should be used. In addition, the advantage of applying mutants to tissue cultures should be used in fenugreek.

3.1 Antioxidant activity of fenugreek

The fenugreek seeds contain polyphenolic compounds, which have been correlated to the beneficial health effects of fenugreek [55]. These polyphenolic compounds are known for several beneficial actions, such as antioxidant effect [56], cancer preventive activity [27], anti-diabetic effects [57] and hypocholesterolemic effect [12, 56].

In earlier studies Bors et al. [58] reported that the scavenging activities of phenolic substances are attributed to the active hydrogen donating ability of hydroxyl substituents. As an overall assessment, the presence of various phytochemicals, particularly naringenin and quercetin, may be responsible for the OH radical scavenging activity. Similarly, trigonellin isolated from ethanol extract of fenugreek seeds has been reported to reduce blood sugar and lipid profile in alloxane-diabetic rabbits [59]. This effect can be partially explained by the antioxidant properties of trigonellin due to its structural similarity to nicotinamide, which has an antioxidant effect [60].

Bukhari et al. [61] reported that fenugreek seed extract with methanol, ethanol, dichloromethane, acetone, hexane and ethyl acetate has a radical scavenging activity. In addition, Bhatia et al. [62] reported protective effect of fenugreek, on lipid peroxidation and on enzymatic antioxidants. Naidu et al. [63] reported that extracts of husk, fenugreek seed, and endosperm exhibited 72%, 64%, and 56% antioxidant activity respectively by free-radical scavenging method. Also, it was indicated that separation of fenugreek seeds into husk and endosperm could have advantage of process viability with respect to prior selective fractionation of bioactive components for their effective isolation.

In a similarly study, it was determined that fenugreek has a high phenolic content. Furthermore, antioxidant property was checked by reducing power, nitro blue tetrazolium (NBT) assay and H_2O_2 scavenging reported to show high superoxide and free radical scavenging [64].

Furthermore, Kaviarasan et al. [65] reported that 2,2′-Azinobis-(3-ethylbenzothiazoline-6-sulfonic acid) (ABTS) radicals are more reactive than 1,1-diphenyl-2-picrylhydrazyl (DPPH) radicals and unlike the reactions with DPPH radical which involve H atom transfer, the reactions with ABTS radicals involve electron transfer process. In addition, Shang et al. [66] identified five different flavonoids in fenugreek seeds, namely vitexin, tricin, naringenin, quercetin, tricin-7-O-beta-D-glucopyranoside, and fenugreek seed extract was found to have significant antiradical and antioxidant properties depending on the concentration.

In line with above researchers [65], they reported that an aqueous methanolic extract of fenugreek seeds was investigated for antiradical and antioxidant activity in different model systems, and antiradical activity was associated with the polyphenolic contained in the extract. As a result, it was determined that fenugreek seeds provide some important factors responsible for the antioxidant potential and provide evidence for numerous in vivo beneficial effects of seeds reported in the literature.

Similarly, Belguith-Hadriche et al. [67] investigated the hypocholesterolemic and antioxidant activities of various extracts of fenugreek seeds (water, methanol, ethyl acetate, hexane, dichloromethane) in rats fed cholesterol, and ethyl acetate only for rats fed a cholesterol-rich diet (HCD). It has been found that fenugreek extracts reduce triglycerides and low density lipoprotein cholesterol (LDL-C) and increase high density lipoprotein cholesterol (HDL-C). Based on these results, it was reported that ethyl acetate extract of fenugreek seeds had a significant hypocholesterolemic effect and antioxidant activity in cholesterol fed rats.

Furthermore, Liu et al. [68] determined the lipid peroxidation (LPO) and cyclooxygenase enzyme (COX) inhibitory activities of hexane, ethyl acetate, methanolic and water extracts to investigate the functional food use quality of fenugreek. They found that the extracts inhibited LPO by 55–95%, COX-1 by 6–87% and COX-2 by 36–70% at 250 lg/ml, respectively. Also, the isolates, excluding the saccharides, inhibited LPO and COX-1 and COX-2 enzymes between the ranges of 8–89%, 4–51% and 15–70%, respectively, at 25 lg/ml. The fenugreek seeds that were studied afforded 3.9 g of triglycerides and fatty acids, 6 g of polysaccharides and 233 mg of flavone C-glycosides per 100 g of seeds. The strong antioxidant activity in the LPO assays of the aqueous extract of fenugreek seed might be attributed to the flavones C-glycosides [69].

Likewise, the different solvent extracts of fenugreek seeds were used to examine the effects of extraction solvent on total phenolic content (TPC), DPPH and iron reducing antioxidant power (FRAP). It was observed that the extracts obtained using higher polar solvents were more effective than less ones, and the addition of 50% water to methanol, acetone or ethanol can enhance the extracting power and antioxidant activity estimation especially acetone and methanol. As a result, it was determined that the total phenolic content showed a good correlation with antioxidant activity FRAP and DPPH [70].

In similarly, Deshmukh et al. [71] reported that silver and iron oxide nanoparticles were successfully synthesized in a simple way at room temperature using an aqueous extract of fenugreek seeds. Then, all nanoparticles were characterized by various techniques to elucidate the stability and functionality of the nanoparticle. It has been determined that the nanoparticles synthesized with the assistance of ultrasound show higher stability and antibacterial and antioxidant activity due to the combined effect of ultrasound and biomolecules adhering to the surface of the nanoparticles.

Naidu et al. [6] also observed that the husk of fenugreek seeds contained higher total polyphenols (103.8 mg gallic acid equivalent/g and the total dietary fiber

(77.1 g/100 g), insoluble dietary fibers (31.9 g/100 g) and soluble dietary fibers (45.2 g/100 g). The bark, fenugreek seeds and endosperm extracts were reported to exhibit 72%, 64%, and 56% antioxidant activity, respectively, by the free radical scavenging method. As a result, separation of fenugreek seeds into husk and endosperms showed that the process viability advantage.

The antioxidant properties of germinated fenugreek seeds were examined in a study conducted by Dixit et al. [72]. Different fractions of germinated seeds were used at different levels to determine their antioxidant potential. Tests used are ferric reducing antioxidant power, DPPH, feriylmyoglobin / 2,2-azobis-3-ethylbenzthiazoline-6-sulfonic acid, pulse radiolysis, oxygen radical absorbance capacity and lipid peroxidation in rat liver mitochondrial preparations. An aqueous fenugreek fraction showed the highest antioxidant activity. Since the amount of phenolic and flavonoid compounds can be correlated with antioxidant activity, the contents of these extracts were measured and their polyphenols, flavonoids and other components were determined by HPLC analysis. This study reveals significant antioxidant activity in germinated fenugreek seeds, which may be due in part to the presence of flavonoids and polyphenols.

As seen from previous studies, the obtained different results may be attributed to different extraction methods and solvents used, different cultivars, growing conditions, maturity stage at harvest, or the storage conditions and time elapsed before the seeds were analyzed. Synthetic drugs used for the treatment of the diseases like cancer, diabetic and the antioxidants used for some treatment have side effects such as mutagenic and carcinogenic effects [73]. Some patients also have resistance to the synthetic drugs. To overcome this problem there is need to find effective natural drugs from traditional medicine. Therefore, fenugreek, which possesses phenolic compounds and antioxidant activity should have the ability to counteract these situations and might be a good candidate for a herbal drug.

3.2 Antibacterial and antifungal effect of fenugreek

The antibacterial activity of the plant extraction has been extensively investigated in many studies. Microbiological analyses revealed that fenugreek extracts exhibit antimicrobial activity against numerous bacteria [74]. Haouala et al. [75] determined the aqueous extracts obtained from various plant parts of fenugreek, different solvents such as methanol, petroleum ether and ethyl acetate fractions and their effects against fungal strains such as *Fusarium graminearum*, *Botrytis cinerea*, *Alternaria sp.*, *Rhizoctonia solani* and *Pythium aphanidermatum*. It was found that all parts of the fenugreek exhibited antifungal potential and the magnitude of the effect varied according to plant parts and fungal species. So, they suggested that fenugreek is an important source of biologically active compounds that are useful for developing better and new antifungal drugs.

Many studies have indicated the effectiveness of fenugreek extracts against *Helicobacter pylori* [76–80]. In a study conducted with honey produced with different plant pollens, the highest antibacterial activity against *Staphylococcus aureus*, *Pseudomonas aeruginosa*, and *Escherichia coli* was found in honey produced with fenugreek pollens [79].

Since cysteine-rich peptides, defensins have strong antifungal activity, the methanol-soluble fraction of fenugreek extract has been studied against nematodes and has been found to show nematicidal activity. It has also been reported to significantly cause the death of *Meloidogyne javanica* larvae [80]. Laroubi et al. [81] studied the prophylaxis effect of fenugreek seeds on renal stone formation in rats. They reported that, the fenugreek can be used in the treatment of patients with calcic urolithiasis.

In addition, Shaheed et al. [82] recorded that the inhibition results for each *Proteus mirabilis* and *E. coli* reached 10.5 and 10.0 mm, 9.0 and 9.5 mm, respectively at 50 mg/ml, and 13.0 and 11.5 mm, 10.5 and 7.5 mm, respectively at 100 mg/ml. Similarly, Hamadii [83] showed the highest percentage of inhibition against *Proteus mirabilis* and *E. coli* at 50 mg/ml from alcoholic fenugreek seed extract.

In another study, the antibacterial effects of fenugreek oil against *Escherichia coli, Samonella, typhimurium, Taphylococcus aureu, Aspergillus niger* microorganisms were investigated. These microorganisms have been chosen as common causes of some human and animal diseases and are contaminants that damage certain foods and are resistant to antibiotics. These results showed that fenugreek oil was stated to be suitable for human consumption and eating [84].

Similarly, it was determined that fenugreek essential oil showed the highest activity against *E. coli*, the inhibition zone reached 21 mm at 100% concentration, and against *Staphylococcus aureus*, it was found to reach 17 mm at 80% concentration [85]. Also, Sulieman et al. [86] found that the inhibition results for *Salmonella* reached 16 mm at 100% concentration. Thereby, these results are promising and may contribute to the future development of natural bio pesticides for the control of fenugreek for the microorganisms. Likewise, antibacterial effect was determined against *Escherichia coli* with an inhibition value of 20 mm at 100% concentration of fenugreek seed oil, while the antimicrobial activity was not detected at 50% and 90% concentrations [86]. As a result, fenugreek extracts show antibacterial activity against many gram positive and gram negative bacterial isolates. In addition, fenugreek seeds are potential sources of new antibacterial compounds as emphasized by the antibacterial activity of their different extracts. So, identification of different effective bacteria from crude plant extracts will assist in the development of drugs against pathogenic microorganism.

3.3 Anticancer effect of fenugreek

Cancer remains one of the leading causes of death worldwide. Flavonoids could also significantly contribute to fenugreek's anticarcinogenic properties. Fenugreek constitutes valuable raw material for the pharmaceutical industry that has long searched for effective cures for cancer.

Previous studies reported that fenugreek seeds have a preventive effect on cancer as in experimental models of cancer using cell lines or experimental animals.

Earlier studies revealed that seed extract of fenugreek importantly inhibits 7,12-dimethyl benz(a)anthracene-induced mammary hyperplasia and decreases its ratio in rats. It was also advised that anti-breast cancer preventive effect of fenugreek could be depending on the increasing apoptosis [87]. Furthermore, alcoholic whole plant extracts of fenugreek effected in vitro cytotoxicity against different human cancer cell lines such as IMR-32, a neuroblastoma cell line, and HT29, a cancer cell line [88].

A selective cytotoxic effect of fenugreek extract in vitro to a panel of cancer cell lines has been observed, including T-cell lymphoma by Alsemari et al. [89]. In addition, Sebastian and Thampan [90] and Prabhu and Krishnamoorthy [91] examined the growth of MCF-7 cells, which is an estrogen receptor positive breast cancer cell line, with ethanol extracts of fenugreek, and reported that the ethanol extract of fenugreek decreased cell viability and induced early apoptotic changes such as inversion of phosphatidyl serine and decreased mitochondrial membrane potential.

In a study conducted by Shabbeer et al. [92] treatment with fenugreek extract showed growth inhibitory effects on breast, pancreatic and prostate cancer cell lines but primary prostate or immortalized prostate cells remained unaffected.

In addition, in a dietary study involving fenugreek seed powder, it reduced colon tumor incidence and hepatic lipid peroxidation in rats treated with 1,2-dimethylhydrazine and also increased catalase, superoxide dismutase, glutathione S-transferase and glutathione peroxidase activities in the liver [93].

Li et al. [94] recorded that diosgenin modulates the STAT3 signaling pathway in hepatocellular carcinoma by suppressing the activation of c-Src, JAK1 and JAK2. They also noted that diosgenin reduced the expression of various STAT3-regulated genes, inhibited proliferation, and potentiated the apoptotic effects of paclitaxel and doxorubicin, which could be a new and potential treatment option for hepatocellular carcinoma and other cancers. Also many researchers reported that diosgenin exhibited anticancer and antiaging activities, as well as cardioprotective and contraceptive properties [44–49].

In addition, in different studies with diosgenin, it has antiproliferative activity such as prostate cancer (PC-3 and DU-145 cells) [95], colon cancer (HCT-116 and HT-29 cells) [96], erythroleukemia (HEL cells) [97], carcinoma (A431, Hep2 and RPMI 2650 cells) [98], stomach cancer (BGC-823 cells) [99], lung cancer (A549 [100], breast cancer (MCF-7) [101], hepatocellulacarcinoma (HepG2 and HCC cells) [102] and human chronic myeloid leukemia (CML) (K562 cells) [103].

As a result of the studies mentioned above, the role of fenugreek seeds and its main active ingredients as new supplements in diet-based preventive / therapeutic strategies to potentially alleviate human diseases remains an important area of study for future research [104].

3.4 Hypoglycemic effect, anti-inflammatory and immunological activity of fenugreek

Immunological changes include altered levels of cytokines and chemokines, changes in the numbers and activation states of various leukocyte populations, apoptosis, and fibrosis during diabetes. Therefore, treatment of diabetes and its complications may include pharmacological strategies to reduce inflammation [105]. Laroubi et al. [81] studied the prophylaxis effect of fenugreek seeds on renal stone formation in rats. And they said that the fenugreek can be used in the treatment of patients with calcic urolithiasis. Chauhan et al. [106] reported an antiinflammatory potential of fenugreek. Jung et al. [107] observed a reduction in the production of several inflammatory mediators, including NO and interleukins 1 and 6, in murine macrophages which had been pretreated with diosgenin and stimulated with lipopolysaccharide/interferon-γ.

In addition, Roberts [108] said that the gum, composed of galactose and mannose, is associated with reduced glycemic effect. Also, the hypoglycemic effect of fenugreek has been especially documented in humans and animals with type 1 and type 2 diabetes mellitus.

Xue et al. [109] reported that the fenugreek extract can lower kidney/body weight ratio and blood glucose and also improves hemorheological properties in experimental diabetic rats following repeated treatment for 6 weeks. A study on animals evaluated the hypoglycemic effects of the fenugreek seeds on dogs. The seeds lowered blood glucose levels, plasma glucagons and somatostatin levels; carbohydrate-induced hyperglycemia also was found to be reduced [110]. Most of the studies with polar fractions of fenugreek seeds point toward a strong anti-inflammatory and anti-arthritic activities mediated through anti-oxidant mechanisms [68, 111, 112].

In addition, Sharma et al. [113] recorded that guar gum of fenugreek prevents the rapid uptake of glucose in the small intestine, aids in blood sugar retention in diabetic patients and may also be effective in the treatment of hypercholesterolemia.

4. Conclusions

Especially recently, many of the beneficial properties of fenugreek have been experimentally proven and the potential of fenugreek's therapeutic applications has been demonstrated. It has high economic values because of including a lot of bioactive compounds. The important compounds can be listed steroidal sapogenins such as diosgenin, alkaloids as trigonelline, flavonoids, tannins, amino acids, steroidal glycosides, protein and others.

As can be seen from this review, the phenolic acids, dietary fiber, saponins and proteins contained in fenugreek are valuable additives to improve human nutrition. Additionally, in neurological studies conducted with fenugreek, the antidiabetic, antifertility, anticancer, antimicrobial, antiparasitic, lactation stimulating and hypocholesterolemic effects of fenugreek have been proven by many researchers. So, it has been also universally used as a spice in a conventional food or it has been interfered to prepare for some functional foods.

In the future, studies on fenugreek are needed some important applications to increasing the popularity of fenugreek. In this context, researchers should be focused quality criteria as primer or secondary metabolite, different cultural and molecular application, different techniques from sown to harvest times of fenugreek. Further research will be required to determine to know the molecules responsible the antioxidant properties in these extracts. Also, fenugreek germplasm can be collected and subjected to intensive selection via modern breeding programs. In addition to these, by selecting genotypes with superior characteristics, gene transfer can be made by determining suitable genes among these genotypes and new fenugreek genotypes with desired properties can be obtained. In addition, studies on fenugreek leaves are limited and generally focused on fenugreek seeds. In this context, researchers should be focused the quality criteria as primer or secondary metabolite in fenugreek leaves.

This study will help the researchers to obtained optimal fenugreek production, optimum biochemical components and adapt to difference environmental, other and specific farming conditions. In the future, when fenugreek is evaluated with all these aspects, its economic, industrial and medicinal value may increase and add value to the relevant sector.

Conflict of interest

The authors declare no conflict of interest.

Author details

Gulsum Yaldiz* and Mahmut Camlica
Agriculture Faculty, Department of Field Crops, Bolu Abant İzzet Baysal University, Bolu, Turkey

*Address all correspondence to: g_yaldiz@hotmail.com

References

[1] Madar Z, Shomer I. Polysaccharide composition of a gel fraction derived from fenugreek and its effect on starch digestion and bile acid absorption in rats. Journal of Agricultural and Food Chemistry. 1990;38(7):1535-1539. DOI: 10.1021/jf00097a023

[2] Çamlıca M, Yaldız G. Characterization of morphological and yield variation of fenugreek (*Trigonella foenum-graecum* L.) genotypes. Legume Research-An International Journal. 2019;42(4):500-504.

[3] Meghwal M, Goswami TK. A review on the functional properties, nutritional content, medicinal utilization and potential application of fenugreek. Journal of Food Processing and Technology. 2012;3(9):181. DOI: 10.4172/2157-7110.1000181

[4] Vaidya K, Ghosh A, Kumar V, Chaudhary S, Srivastava N, Katudia K, Tiwari T, Chikara SK. De Novo transcriptome sequencing in *Trigonella foenum-graecum* L. to identify genes involved in the biosynthesis of diosgenin. The Plant Genome. 2013;6(2):1-11. DOI: 10.3835/plantgenome2012.08.0021

[5] He Y, Wang X, Suo Y, Ding C, Wang H. Efficient protocol for isolation of rhaponticin and rhapontigenin with consecutive sample injection from fenugreek (*Trigonella foenum-graecum* L.) by HSCCC. Journal of Chromatographic Science. 2016;54(3): 479-485. DOI: 10.1093/chromsci/bmv169

[6] Naidu MM, Shyamala BN, Naik JP, Sulochanamma G, Srinivas P. Chemical composition and antioxidant activity of the husk and endosperm of fenugreek seeds. LWT-Food Science and Technology. 2011;44(2):451-456.

[7] Sharma RD, Sarkar A, Hazra DK, Misra B, Singh JB, Maheshwari BB. Toxicological evaluation of fenugreek seeds: a long term feeding experiment in diabetic patients. Phytotherapy Research. 1996; 10(6):519-520.

[8] Ahmadiani A, Javan M, Semnanian S, Barat E, Kamalinejad M. Anti-inflammatory and antipyretic effects of *Trigonella foenum-graecum* leaves extract in the rat. Journal of Ethnopharmaco logy. 2001;75(2-3): 283-286. DOI: 10.1016/S0378-8741(01)00187-8

[9] Puri D. Therapeutic potential of fenugreek. Indian Journal of Physiology and Pharmacology. 1998; 42:423-424.

[10] Suja Pandian R, Anuradha CV, Viswanathan P. Gastroprotective effect of fenugreek seeds (*Trigonella foenum graecum*) on experimental gastric ulcer in rats. Journal of Ethnopharmacology. 2002; 81:393-397.

[11] Rao AL, Bharani M, Pallavi V. Role of antioxidants and free radicals in health and disease. Advances in Pharmacology and Pharmacy. 2006;7:29-38.

[12] Sowmya P, Rajyalakshmi P. Hypocholesterolemic effect of germinated fenugreek seeds in human subjects. Plant Foods for Human Nutrition. 1999;53(4):359-365.

[13] Blank I, Lin J, Devaud S, Fumeaux R, Fay LB. The principal flavor components of fenugreek (*Trigonella foenum-graecum* L.). Nestle ́researchcenter. 211th ACS Symposium; 2011 p. 12-28. DOI: 10.1021/bk-1997-0660.ch003.

[14] Kumar AS, Reddy TS. Importance of traditional system of medicine- A review. International Journal of Phytotherapy. 2012;2:1-6.

[15] ESCOP Monographs. Matricariae flos. 2nd ed. New York: Thieme; 2003.

European Scientific Cooperative on Phytotherapy. 2003; pp. 312-9.

[16] Passano P. The many uses of methi. Manushi. 1995;2:31-34.

[17] Madar Z, Stark AH. New legume sources as therapeutic agents. British Journal of Nutrition. 2002;88:287-292.

[18] Sharma RD, Raghuram TC, Rao VD. Hypolipidaemic effect of fenugreek seeds. A clinical study. Phytotherapy Research. 1991;5(3):145-147.

[19] Bunney S. The illustrated book of herbs: their medicinal and culinary uses. Vol 1, No 2. London: Octopus Books. 1984; pp 320.

[20] Flammang A, Cifone M, Erexson G, Stankowski L. Genotoxicity testing of a fenugreek extract. Food and Chemical Toxicology. 2004;11:1769-1775.

[21] Miraldi E, Ferri S, Mostaghimi V. Botanical drugs and preparations in the traditional medicine of West Azerbaijan (Iran). Journal of Ethnopharmacology. 2001;2:77-87.

[22] Bashtian MH, Emami SA, Mousavifar N, Esmaily HA, Mahmoudi M, Poor AHM. Evaluation of Fenugreek (*Trigonella foenum-graceum* L.). Effects seeds extract on insulin resistance in women with polycystic ovarian syndrome. Iranian Journal of Pharmaceutical Research. 2013;12:475-448.

[23] Wijaya VH, Abdul Mun'im A, Djajadisastra J. Effectiveness test of fenugreek seed (*Trigonella foenum-graecum* L.) extract hair tonic in hair growth activity. International Journal of Current Research. 2013;5(11): 3453-3460.

[24] Basch E, Ulbricht C, Kuo G, Szapary P, Smith M. Therapeutic applications of fenugreek. Alternative Medicine Review. 2003;8:20-27.

[25] Betty R. The Many Healing Virtues of Fenugreek. Spice India, 2008; pp. 17-19.

[26] Pandian RS, Anuradha CV, Viswanathan P. Gastroprotective effect of fenugreek seeds (*Trigonella foenum graecum*) on experimental gastric ulcer in rats. Journal of Ethnopharmacology. 2002;81(3):393-397.

[27] Raju J, Patlolla JMR, Swamy MV, Rao CV. Diosgenin, a steroid saponin of *Trigonella foenum-graecum* (fenugreek), inhibits azoxymethane-induced aberrant crypt foci formation in F344 rats and induces apoptosis in HT-29 human colon cancer cells. Cancer Epidemiol Biomarkers Previous. 2004;13(8):1392-1398.

[28] Roberts KT. The potential of fenugreek (*Trigonella foenum-graecum*) as a functional food and nutraceutical and its effects on glycemia and lipidemia. Journal of Medicinal Food. 2011;14(12);1485-1489.

[29] Mehrafarin A, Rezazadeh SH, Naghdi BH, Noormohammadi GH, Zand E, Qaderi A. A review on biology, cultivation and biotechnology of fenugreek (*Trigonella foenum-graecum* L.) as a valuable medicinal plant and multipurpose. Journal of Medicinal Plants. 2011;10:6-24.

[30] El-Shora HM, El-Farrash AH, Kamal H, Aya A. Positive role of UV radiation in enhancing secondary metabolites production in fenugreek leaves. International Journal of Advanced Research. 2015;3(5):536-543.

[31] Hussein EA, Aqlan EM. Effect of mannitol and sodium chloride on some total secondary metabolites of fenugreek calli cultured in vitro. Plant Tissue Culture and Biotechnology. 2011;21(1):35-43.

[32] Chaudhary S, Chikara SK, Sharma MC, Chaudhary A, Syed BA,

Chaudhary PS, Mehta A, Patel M, Ghosh A, Iriti M. Elicitation of diosgenin production in *Trigonella foenum-graecum* (Fenugreek) seedlings by methyl jasmonate. International Journal of Molecular Sciences. 2015;16:29889-29899.

[33] Hosseini Z, Hassanloo T, Kowsari M, Majidian M. Trigonelline as an anti-diabetic metabolite increased in inoculated fenugreek by Trichoderma. Advanced Research in Microbial Metabolites and Technology. 2018;2:129-139.

[34] Altabtabaai TA, Alalwani BA, Abbas IS. HPLC analysis of the trigonelline and diosgenin in thecallus of fenugreek (*Trigonella foenum graecum* L.) seeds. Plant Archives. 2019;19(Supp. 2):2046-2050.

[35] Varsha T, Jain A. Comparative quantitative estimation of secondary metabolites and HPLC analysis in different plant parts of *Trigonella foenum graceum* (L.). International Journal of Research in Applied, Natural and Social Sciences. 2018;6(8):75-84.

[36] Rajabihashjin M, Asghari A, Zeinalabedini M, Ghaffari M. Evaluation of the effect of environmental condition on metabolites and morphology of seeds and prediction of 4-Hydroxy Isoleucine and trigonelline as two therapeutic metabolites of Persian fenugreeks. Preprints. 2019;1-26.

[37] Irankhah S, Sillo F, Nerva L, Ganjeali A, Balestrini R, Chitarra W. Combined effects of water deficit, exogenous ethylene application and root symbioses on trigonelline and aba accumulation in fenugreek. Applied Sciences. 2020;10(7): 2338-2348. DOI:10.3390/app10072338

[38] Abdouli H, Ayed MH, Elham M, Nabila B, Morencos MRA. Proximate composition, and total phenols, tannins, flavonoids and saponins, and in vitro ruminal fermentation activity of fenugreek cut at three maturity stages. Livestock Research for Rural Development. 2012;24:13.

[39] Irankhah S, Chitarra W, Nerva L, Antoniou C, Luminu E, Volpe V, Ganjeali G, Cheniany M, Mashreghi M, Fotopoulos V, Balestrini R. Impact of an arbuscular mycorrhizal fungal inoculum and exogenous MeJA on fenugreek secondary metabolite production under water deficit. Environmental and Experimental Botany. 2020;176: 104096.

[40] Bitarafan Z, Asghari HR, Hasanloo T, Gholami A, Moradi F, Khakimov B, Liu F, Andreasen C. The effect of charcoal on medicinal compounds of seeds of fenugreek (*Trigonella foenum-graecum* L.) exposed to drought stress. Industrial Crops and Products. 2019;131:323-329.

[41] Elleuch A, Chaâbene Z, Grubb DC, Drira N, Mejdoub H, Khemakhem B. Morphological and biochemical behavior of fenugreek (*Trigonella foenum-graecum*) under copper stress. Ecotoxicology and Environmental Safety. 2013;98: 46-53.

[42] Parchin RA, Ghomi AAN, Badi HN, Eskandari A, Navabpoura S, Mehrafarin A. Growth characteristics and phytochemical responses of Iranian fenugreek (*Trigonella foenum-graecum* L.) exposed to gamma irradiation. Industrial Crops and Products. 2019;139:111593.

[43] Murria S, Kaur N. Fenugreek alkaloids: A medicinal commodity. Rashtriya Krishi. 2018;13(2):13-14.

[44] Agarwal M, Agarwal M, Jain SC. In vitro regulation of bioactive compounds in Trigonella species by mutagenic treatments. Journal of Plant Sciences. 2015;3:40-44.

[45] Dias KLG, Correia N. de A, Pereira KKG, Barbosa Filho JM,

Cavalcante KVM, Araújo IGA, Silva DF, Guedes DN, Neto MDA, Bendhack LM, Medeiros, I. A. Mechanisms involved in the vasodilator effect induced by diosgenin in rat superior mesenteric artery. European Journal of Pharmacology. 2007;574(2-3): 172-178.

[46] Gong G, Qin Y, Huang W. Anti-thrombosis effect of diosgenin extract from Dioscorea zingiberensis C.H. Wright in vitro and in vivo. Phytomedicine. 2011;18(6): 458-463.

[47] Lee J, Jung K, Kim YS, Park D. Diosgenin inhibits melanogenesis through the activation of phosphatidylinositol-3-kinase pathway (PI3K) signaling. Life Science. 2007;81(3):249-254.

[48] Tada Y, Kanda N, Haratake A, Tobiishi M, Uchiwa H, Watanabe S. Novel effects of diosgenin on skin aging. Steroids. 2009;74(6):504-511. DOI: 10.1016/j.steroids.2009.01.006

[49] Yan LL, Zhang YJ, Gao WY, Man SL, Wang Y. In vitro and in vivo anticancer activity of steroid saponins of Paris polyphylla var. yunnanensis. Experimental Oncology. 2009; 31(1):27-32.

[50] Sautour M, Mitaine-Offer AC, Miyamoto T, Dongmo A, Lacaille-Dubois MA. Antifungal steroid saponins from Dioscorea cayenensis. Planta Medica. 2004;70(1):90-92. DOI: 10.1055/s-2004-815467

[51] Wang L, Ma T, Zheng Y, Lv S, Li Y, Liu S. Diosgenin inhibits IL-1β-induced expression of inflammatory mediators in human osteoarthritis chondrocytes. International Journal of Clinical and Experimental Pathology. 2015;8(5):4830-4836.

[52] Corbiere C, Liagre B, Terro F, Beneytout JL. Induction of antiproliferative effect by diosgenin through activation of p53, release of apoptosis-inducing factor (AIF) and modulation of caspase-3 activity in different human cancer cells. Cell Research. 2004;14:188-196.

[53] Malviya KG, Babhulkar MW, Mali P, Rangari VD. Evaluation of anti-inflammatory potential of *Trigonella foenum-graecum* (fenugreek) seed extracts by using carrageenan induced rat paw edema. Drug Invent. Today. 2010;2:109-111.

[54] Abdel-Nabey AA, Damir AA. Changes in some nutrients of fenugreek (*Trigonella foenum-graceum* L.) seeds during boiling. Plant Foods for Human Nutrition. 1990;40(4):267-274.

[55] Rayyan S, Fossen T, Andersen QM. Flavone C-Glycosides from seeds of fenugreek, *Trigonella foenum-graecum* L. Journal of Agricultural and Food Chemistry. 2010;58:7211-7217. DOI: 10.1021/jf100848c

[56] Ravikumar P, Amerada CV. Effect of fenugreek seeds on blood lipid peroxidation and antioxidants in diabetic rats. Phytotherapy Research. 1999;13:197-201.

[57] Broca C, Manteghetti M, Gross R, Baissac Y, Jacob M, Petit P, Sauveira Y, Ribes G. 4- Hydroxyisoleucine: Effects of synthetic and natural analogues on insulin secretion. European Journal of Pharmacolology. 2000;390:339-345.

[58] Bors W, Heller W, Michael C, Stettmaier K. Flavonoids and polyphenols: chemistry and biology. In: Cadenas E, Packer L (eds) Handbook of antioxidants. Marcel Dekker, New York, 1996:409-466.

[59] Adeeb AN, Hamadi SA, Al-Khateeb E. Isolation of trigonelline from Iraqi fenugreek seeds and studying its effects on blood glucose & lipid profile in normal and alloxan-diabetic rabbits. A master thesis, College of Pharmacy/University of Baghdad. 2002.

[60] Velikii NN, Obrosova IG, Efimoy AS. Nicotinamide coenzyme in the regulation of cellular metabolism in various types of diabetes. Vopr Med Khim. 1992;38(4):45-52.

[61] Bukhari SB, Muhammad IB, Shahabuddin M. Antioxidant activity from the extract of fenugreek seeds. Pakistan Journal of Analytical and Environmental Chemistry. 2008;9(2):78-83.

[62] Bhatia K, Kaur M, Atif F, Ali M, Rehman H, Rahman S. Aqueous extract of ameliorates additive urotoxicity of buthionine sulfoximine and cyclophosphamide in mice. Food and Chemical Toxicology. 2006;44:1744-1750.

[63] Naidu MM, Shyamala BN, Naik PJ, Sulochanamma G, Srinivas P. Chemical composition and antioxidant activity of the husk and endosperm of fenugreek seeds Food Science and Technology. 2010;44:451-456.

[64] Joglekar M, Mandal M, Somaiah MP, Murthy S.Comparative analysis of antioxidant and antibacterial properties of *Aegle marmelos*, *Coriandrum sativum* and *Trigonella foenum graecum*. Acta Biologica Indica. 2012;1(1):105-108.

[65] Kaviarasan S, Naik GH, Gangabhagirathi R, Anuradha CV, Priyadarsini KI. In vitro studies on antiradical and antioxidant activities of fenugreek (Trigonella foenum graecum) seeds. Food Chem. 2007;103:31-37.

[66] Shang M, Cai S, Han J, Li J, Zhao Y, Zheng J, Namba T, Kadota S, Tezuka Y, Fan W. Studies on flavonoids from Fenugreek (*Trigonella foenum graecum* L.). National Library of Medicine. 1998; 23(10):614-616, 639.

[67] Belguith-Hadriche O, Bouaziz M, Jamoussi K, Simmonds MS, El Feki A, Makni-Ayedi F. Comparative study on hypocholesterolemic and antioxidant activities of various extracts of fenugreek seeds. Food Chemistry. 138(2-3):1448-53.

[68] Liu Y, Kakani R, Nair MG. Compounds in functional food fenugreek spice exhibit anti-inflammatory and antioxidant activities. Food Chemistry. 2012;131:1187-1192.

[69] Musa KH, Abdullah A, Jusoh K, Subramaniam V. Antioxidant activity of pink-flesh guava (*Psidium guajava* L.): effect of extraction techniques and solvents. Food Analytical Methods. 2011;4(1):100-107.

[70] Mashkor IMAAL. Total phenol, total flavonoidsand antioxidant activity of pomegranate peel. International Journal of ChemTech Research. 2014;6(11):4656-4661.

[71] Deshmukh AR, Gupta A, Kim BS. Ultrasound assisted green synthesis of silver and iron oxide nanoparticles using fenugreek seed extract and their enhanced antibacterial and antioxidant activities. BioMed Research International. 2019; 1-14. DOI: 10.1155/2019/1714358

[72] Dixit P, Ghaskadbi S, Mohan H, Devasagayam TPA. Antioxidant properties of germinated fenugreek seeds. Phytotherapy Research. 2005;19(11):977-983. DOI: 10.1002/ptr.1769.

[73] Kaurinovic B, Popovic M, Vlaisavljevic S, Trivic S. Antioxidant capacity of *Ocimum basilicum* L. and *Origanum vulgare* L. extracts. Molecules. 2011; 16:7401-7414.

[74] Aqil F, Ahmad I. Broad-spectrum antibacterial and antifungal properties of certain traditionally used Indian medicinal plants. World Journal of Microbiology and Biotechnology. 2003;19:653-657.

[75] Haouala R, Hawala S, El-Ayeb A, Khanfir R, Boughanmi N. Aqueous and organic extracts of Trigonella foenum-graecum L. inhibit the mycelia growth of fungi. Journal of Environmental Sciences. 2008;20:1453-1457.

[76] O'Mahony R, Al-Khtheeri H, Weerasekera D, Fernando N, Vaira D, Holton J, Basset C. Bactericidal and anti-adhesive properties of culinary and medicinal plants against Helicobacter pylori. World Journal of Gastroenterology. 2005;11:7499-7507.

[77] Randhir R, Lin YT, Shetty K. Phenolics, their antioxidant and antimicrobial activity in dark germinated fenugreek sprouts in response to peptide and phytochemical elicitors. Asia Pacific Journal of Clinical Nutrition. 2004;13:295-307.

[78] Randhir R, Shetty K. Improved alpha-amylase and Helicobacter pylori inhibition by fenugreek extracts derived via solid-state bioconversion using Rhizopus oligosporus. Asia Pacific Journal of Clinical Nutrition. 2007;16:382-392.

[79] Mercan N, Guvensen A, Celik A, Katircioglu H. Antimicrobial activity and pollen composition of honey samples collected from different provinces in Turkey. Natural Product Research. 2007;21:187-195.

[80] Zia T, Hasnain SN, Hasan SK. Evaluation of the oral hypoglycaemic effect of *Trigonella foenum-graecum* L. (methi) in normal mice. Journal of Ethnopharmacology. 2001;75:191-195.

[81] Laroubi A, Touhami M, Farouk L, Zrara I, Aboufatima R, Benharref A. Prophylaxis effect of *Trigonella foenum graecum* L. seeds on renal stone formation in rats. Phytotherapy Research. 2007;21(10): 921-925.

[82] Shaheed KA, Alsirraj MA, Allaith SA, Noori NA, Obaid MH, Mouhsan ZM, Swedan SS. The biological activities of seeds extracts for fenugreek and black cumin and its inhibitory influences toward some pathogens. Iraq Medicine. 2018;2(2): 46-50.

[83] Hamadii RF. Effect of some plant extract on some bacterial isolated causing urinary tract infection (UTI). Anbar Journal. 2012;3.

[84] Han JS, Shin DH, Baek NI. Identification of growth inhibitory substance on foodborne microorganisms from *commiphora molmol* Engl. and its application to food products. Korean Journal of Food Science and Technology. 2001;33(4):401-408.

[85] Badreldin LE. Chemical composition of ginger (*Zingiber officinale* Rose) and detection of antimicrobial activity of its oil. M.Sc. Thesis, University of Gezira. 2006.

[86] Sulieman AME, Ahmed HE, Abdelrahim AM. The chemical composition of fenugreek (*Trigonella foenum graceum* L) and the antimicrobial properties of its seed oil. Gezira Journal of Engineering and Applied Sciences. 2008;3(2):52-71.

[87] Amin A, Alkaabi A, Al-Falasi S, Daoud SA. Chemopreventive activities of *Trigonella foenum-graecum* (Fenugreek) against breast cancer. Cell Biology International. 2001;29:687-694.

[88] Verma SK, Singh SK, Mathur A. In vitro cytotoxicity of *Calotropis procera* and *Trigonella foenum-graecum* against human cancer cell lines. Journal of Chemical and Pharmaceutical Research. 2010;2:861-865.

[89] Alsemari A, Alkhodairy F, Aldakan A, Al-Mohanna M, Bahoush E, Shinwari Z, Alaiya A. The selective cytotoxic anti-cancer properties and proteomic analysis of *Trigonella*

foenum-graecum. BMC Complement. Alternative Medicine. 2014;14:114.

[90] Sebastian KS, Thampan RV. Differential effects of soybean and fenugreek extracts on the growth of MCF-7 cells. Chemico-Biological Interactions. 2007;170:135-143.

[91] Prabhu A, Krishnamoorthy M. Anticancer activity of *Trigonella foenum-graecum* on *Ehrlich Ascites* carcinoma in Mus musculus system. Journal of Pharmacy Research. 2010;3:1181-1183.

[92] Shabbeer S, Sobolewski M, Anchoori RK, Kachhap S, Hidalgo M, Jimeno A, Davidson N, Carducci MA, Khan SR. Fenugreek: a naturally occurring edible spice as an anticancer agent. Cancer Biology and Therapy. 2009;8:272-278.

[93] Devasena T, Menon PV. Fenugreek seeds modulate 1,2-dimethylhydrazine-induced hepatic oxidative stress during colon carcinogenesis. The Italian journal of biochemistry. 2007;56:28-34.

[94] Li F, Fernandez PP, Rajendran P, Hui KM, Sethi G. Diosgenin, a steroidal saponin, inhibits STAT3 signaling pathway leading to suppression of proliferation and chemosensitization of human hepatocellular carcinoma cells. Cancer Letters. 2010;292:197-207.

[95] Chen PS, Shih YW, Huang HC, Cheng HW. Diosgenin, a steroidal saponin, inhibits migration and invasion of human prostate cancer pc-3 cells by reducing matrix metalloproteinases expression. PLoS ONE. 2011;6(5). DOI: 10.1371/journal.pone.0020164.e20164

[96] Lepage C, Léger DY, Bertrand J, Martin F, Beneytout JL, Liagre B. Diosgenin induces death receptor-5 through activation of p38 pathway and promotes TRAIL-induced apoptosis in colon cancer cells. Cancer Letters. 2011;301(2):193-202. DOI: 10.1016/j.canlet.2010.12.003

[97] Léger DY, Liagre B, Cardot PJP, Beneytout JL, Battu S. Diosgenin dose-dependent apoptosis and differentiation induction in human erythroleukemia cell line and sedimentation field-flow fractionation monitoring. Analytical Biochemistry. 2004;335(2):267-278. DOI: 10.1016/j.ab.2004.09.008

[98] Das S, Dey KK, Dey G, Pal I, Majumder A, MaitiChoudhury S, Kundu SC, Mandal M. Antineoplastic and apoptotic potential of traditional medicines thymoquinone and diosgenin in squamous cell carcinoma. PLoS ONE. 2012;7(10):e46641. DOI: 10.1371/journal.pone.0046641.e46641

[99] Mao ZJ, Tang QJ, Zhang CA, Qin ZF, Pang B, Wei PK, Liu B, Chou YN. Anti-proliferation and anti-invasion effects of diosgenin on gastric cancer BGC-823 cells with HIF-1α shRNAS. International Journal of Molecular Sciences. 2012;13(5):6521-6533. DOI: 10.3390/ijms13056521

[100] Mohammad RY, Somayyeh G, Gholamreza H, Majid M, Yousef R. Diosgenin inhibits hTERT gene expression in the A549 lung cancer cell line. Asian Pacific Journal of Cancer Prevention. 2013;14(11):6945-6948. DOI: 10.7314/APJCP.2013.14.11.6945

[101] Sung B, Prasad S, Yadav VR, Aggarwal BB. Cancer cell signaling pathways targeted by spice-derived nutraceuticals. Nutrition and Cancer. 2012;64(2):173-197. DOI: 10.1080/01635581.2012.630551

[102] Li Y, Wang X, Cheng S, Du J, Deng Z, Zhang Y, Liu Q, Gao J, Cheng B, Ling C. Diosgenin induces G2/M cell cycle arrest and apoptosis in human hepatocellular carcinoma cells. Oncology Reports. 2015;33(2):693-698. DOI: 10.3892/or.2014.3629

[103] Jiang S, Fan J, Wang Q, Ju D, Feng M, Li J, Guan JB, An D, Wang X, Ye L. Diosgenin induces ROS-dependent autophagy and cytotoxicity via mTOR signaling pathway in chronic myeloid leukemia cells. Phytomedicine. 2016;23(3):243-252. DOI: 10.1016/j.phymed.2016.01.010

[104] Abdelgawad MR, Mustafa, MMM, Kottb MKI. Phytochemical protection against diethylnitrosoamine induced hepatocarcinogenesis by *Trigonella foenum graecum* in female rats. Arab J. Nucl. Sci. Appl. 2012;45(2):523-536.

[105] Kothari V, Galdo JA, Mathews ST. Hypoglycemic agents and potential anti-inflammatory activity. Journal of Inflammation Research. 2016;9:27-38

[106] Chauhan G, Sharma M, Varma A, Khanrkwal H. Phytochemical analysis and anti-inflammatory potential of fenugreek, Medicinal plants. International Journal of Phytomedicines and Related Industries. 2010;2(1):39-44.

[107] Jung DH, Park HJ, Byun HE, Park YM, Kim TW, Kim BO, Um SH, Pyo S. Diosgenin inhibits macrophage-derived inflammatory mediators through downregulation of CK2, JNK, NF-κB and AP-1 activation. International Immunopharmacology. 2010;10:1047-1054.

[108] Roberts KT. The potential of fenugreek (*Trigonella foenum-graecum*) as a functional food and nutraceutical and its effects on glycemia and lipidemia. Journal of Medicinal Food. 2011;14(12):1485-1489.

[109] Xue WL, Li XS, Zhang J, Liu YH, Wang ZL, Zhang RJ. Effect of *Trigonella foenum-graecum* (fenugreek) extract on blood glucose, blood lipid and hemorheological properties in streptozotocin-induced diabetic rats. Asia Pacific Journal of Clinical Nutrition. 2007;1:422-426.

[110] Snehlata HS, Payal DR. Fenugreek (*Trigonella foenum-graecum* L.): an overview. International Journal of Current Pharmaceutical Research. 2012;2(4):169-187.

[111] Suresh P, Kavita CH, Babu SM, Reddy VP, Latha AK. Effect of ethanol extract of *Trigonella foenum-graecum* (Fenugreek) seeds on Freund's adjuvant induced arthritis in albino rats. Inflammation. 2012;35:1314-1321.

[112] Sindhu G, Ratheesh M, Shyni GL, Nambisan B, Helen A. Anti-inflammatory and antioxidative effects of mucilage of *Trigonella foenum graecum* (Fenugreek) on adjuvant induced arthritic rats. International Immunopharmacology. 2012;12:205-211.

[113] Sharma RD, Sarkar A, Hazra DK. Hypolipidaemic effect of fenugreek seeds: a chronic study in non-insulin dependent diabetic patients. Phytotherapy Research. 1996;10:332-334.

Chapter 7

Present and Future Perspective of Soybean Cultivation

Toshihiro Nakamori

Abstract

Soybeans have been cultivated as a traditional crop since ancient times in Japan, China, and other parts of Asia. Soybeans, as a source of protein, are rich in essential amino acids, but also contain a variety of functional and nutritional components. Their processed and fermented products support the maintenance of human health. Recently, new soybeans varieties containing superior nutritional components have been cultivated, and growing interest in plant-based foods has led to the establishment of new food products including dairy products such as butter and cream.

Keywords: soybeans, functional food, health benefits, protein, peptide, isoflavone, fat

1. Introduction: the origin of soybeans and their global production

Soybean (*Glycine max* (L.) Merr.) is a traditional crop that contributes significantly in Asia's food culture. Several theories regarding the origin and spread of soybean have been suggested. Nevertheless, it has been reported that soybean was being cultivated in north-eastern China even 3000 years ago [1]. Currently, soybean is one of the most essential plant resources cultivated in various regions worldwide. The global production of soybeans in 2019 was approximately 340 million tons (**Figure 1**), making it one of the most commonly produced plant resources globally [2]. Approximately 80% of cultivated soybeans are used in livestock feed, and its majority is used for non-dietary purposes (i.e., printing ink, binder and resin dispersion, etc).

Soybean is known as the "miracle crop" as its actual cultivation initiated after 1930, mainly in the United States, and its production increased rapidly and significantly [3]. Soybean is widely cultivated in regions extending from the equatorial tropics up to southern regions of Sweden and Canada in the north, and Argentina and southern Australia in the south, with a substantial number of soybean varieties appropriate for cultivation in different environments and uses appropriate for each region.

2. Functional components present in soybeans

The composition of soybeans varies marginally depending on the variety; however, it commonly consists of protein (35–45%), fat (18–20%), and carbohydrates (22–28%), an insoluble fiber called okara, a low percentage of starch and sucrose, and some oligosaccharides such as stachyose and raffinose [4]. The physiologically

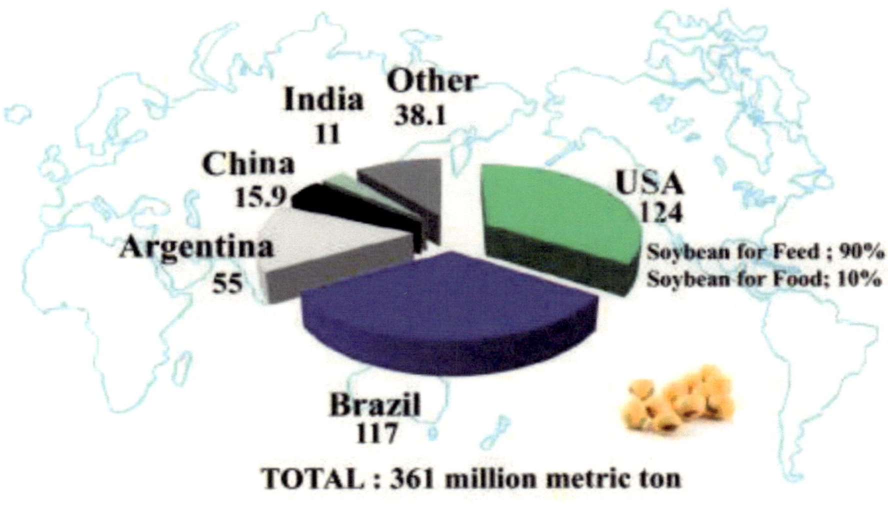

Figure 1.
World Soybean production.

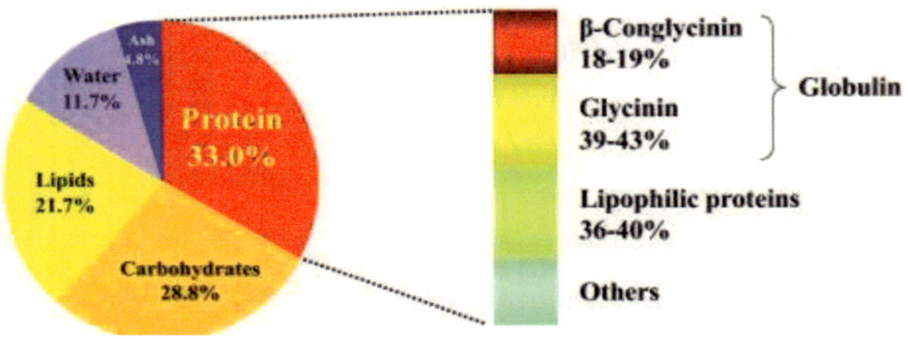

Figure 2.
Soybean composition.

functional components of soybeans include vitamins such as K, A, B_2, C, and D with a high content of vitamin E and B_1 while they also contain glycosides in the form of isoflavones and saponins as well as functional lipids including lecithin and sterols.

Besides lipid-associated proteins, two types of globulins, 11S globulin and 7S globulin represent the two major components of soybean proteins (**Figure 2**) [4, 5]. With respect to the physiological function of soybean proteins, Carroll and Hamilton [6] reported in 1975 that consumption of soybean proteins reduces plasma low-density lipoprotein concentration. Sugano et al. [7] and Kohno [8] further elucidated the mechanism through which the indigestible high-molecular-weight fraction of soybeans binds to excess bile acids in the intestinal tract and is excreted from the body. The U.S. Food and Drug Administration has approved the labeling of foods containing 25 g of soy protein per day as foods that reduce the risk of heart disease development due to the cholesterol-reducing effects of soy protein [9]. In Japan, soy protein represents a functional ingredient in food for special health-related use, and several correlated products are being commonly used [10]. It has been demonstrated that beta-conglycinin detected in 7S globulin reduces the visceral fat content and has been authorized for use in food for specific health-related use [11, 12].

Isoflavones, one of the essential components of soybeans, represent approximately 0.2–0.5% of seed weight, and their composition differ depending on the part of the soybean [13]. The isoflavone content per gram of protein in traditional soybean foods such as tofu is approximately 3.5 mg in aglycon units [14]. Isoflavones consist of genistein (~50%), daidzein (~40%), and glycitin (~10%). The physiological function (nonhormonal agent) of equal, a metabolite of daidzein ingested by the body and produced by intestinal microflora, is of particular scientific interest [15]. Isoflavones have been demonstrated to improve the blood lipid profile when consumed in combination with soy protein as a component of soy foods for a period of one to 3 months [16], and have shown to exert an anti-estrogenic effect on breast and prostate cancer that are hormone-related disorders [17, 18].

3. Improvement of soybean characteristics via breeding

Cultivation of soybean varieties suitable for different environments in various regions has been previously investigated, using high yield and pest resistance as the relevant development parameters. Development of soybeans with improved nutritional and functional characteristics has been pursued after utilizing wild species of soybean and employing genetic resources such as genetic mutations that are naturally generated during the cultivation process. Wild soybean (*Glycine soja* Sieb. et Zucc.) for instance, is characterized by its high protein and low-fat content. Yamada [19] reported that protein content (its main component) is adversely associated with its lipid content based on the examination of hybrids among this legume and soybean strains. Development of high protein soybeans by sacrificing a certain amount of the lipid content and agronomic traits would also be possible. Cultivation of soybeans including a higher protein content compared to conventional varieties is anticipated to contribute in the adaptation of soybean use products for protein intake, and for the improvement of productivity. Furthermore, it has been described that the 11S globulin content of soybeans increases when all 7S globulin genes are externally suppressed using microRNA [20]. Since 7S and 11S globulins are the main proteins in soybeans accounting for more than 50% of total protein content, changes on either protein levels result in significant effects on the gel water holding capacity, which is the main property of soybean protein. These proteins are thus expected to lead in the development of products that are completely different from conventional processed products from soybeans based on gelation properties.

Previous studies have attempted to enhance the content of alpha-tocopherol (α-Toc) and isoflavone through cultivation [21, 22]. Toc is a fat-soluble antioxidant commonly known as vitamin E. There are four naturally occurring homologs: α-, β-, γ-, and δ-Toc, with α-Toc being the most bioactive homolog. The total Toc content in soybean oil is relatively high among vegetable oils and fats. Nevertheless, the α-Toc content is low, ranging 5–7% [23] and as a result, vitamin E activity of soybean is not significant. Importantly, increasing the α-Toc content of soybeans has become a goal of cultivation. Dwiyanti et al. [21] demonstrated that gene expression related with α-Toc content was associated with the expression levels of one of the γ-Toc methyltransferase (E.C.2.1.1.95) isozyme (γ-TMT3) by genetic analysis of hybrids containing high α-Toc content using soybean genetic resources. In contrast, certain varieties have been identified in which α-Toc content increases from ×1.5 to ×5-fold during ripening in high temperature regions compared to that in standard regions [24], and an approach involving both genetic and cultivation environment is crucial when cultivating soybeans containing a high vitamin E content [24].

Various studies have also examined the potential of producing high-isoflavone soybeans [22, 25]. In isoflavone-related genetic analysis, it has been discovered that overexpression of the MYB transcription factor gene regulating the flavonoid biosynthetic pathway, increases the isoflavone and flavonol levels [26], suggesting the possibility of increasing isoflavone content via genetic engineering. In addition, a study has suggested that isoflavone content increases by ×3-fold when grown at a relatively lower temperature compared to that of standard growing conditions [26]. Isoflavone content may be thus increased by selecting varieties with genetic variation while considering the culturing conditions associated to the climate of the production area.

4. Traditional Japanese foods prepared from soybeans

Soybeans have been cultivated throughout Japan since the Yayoi period (1700–2300 years ago) [27, 28] and have contributed significantly in the preparation of Japanese food consumed nowadays. Currently, soybean is cultivated all over the country, from Hokkaido in the north to Okinawa in the south and there are varieties suitable for the cultivation conditions of each region. In the fiscal year of 2019, soybean demand in Japan was 3,670,000 tons, with food-related use accounting for 28% or 1,019,000 tons, while the remaining 72% accounted for non-food applications such as livestock feed. Currently, ~80% of soybeans intended for consumption are imported from overseas, primarily from the United States, Canada, and Brazil [29]. The main food-based applications of soybeans include the production of miso (soybean paste), soy sauce, tofu, fried bean curd, natto (fermented soybeans), dried bean curd, and soy milk (**Figure 3**). As an essential ingredient in Japanese cuisine, these traditional foods not only contribute to the health and nutrition of the Japanese population, but also in Japanese food culture. In December 2013, the

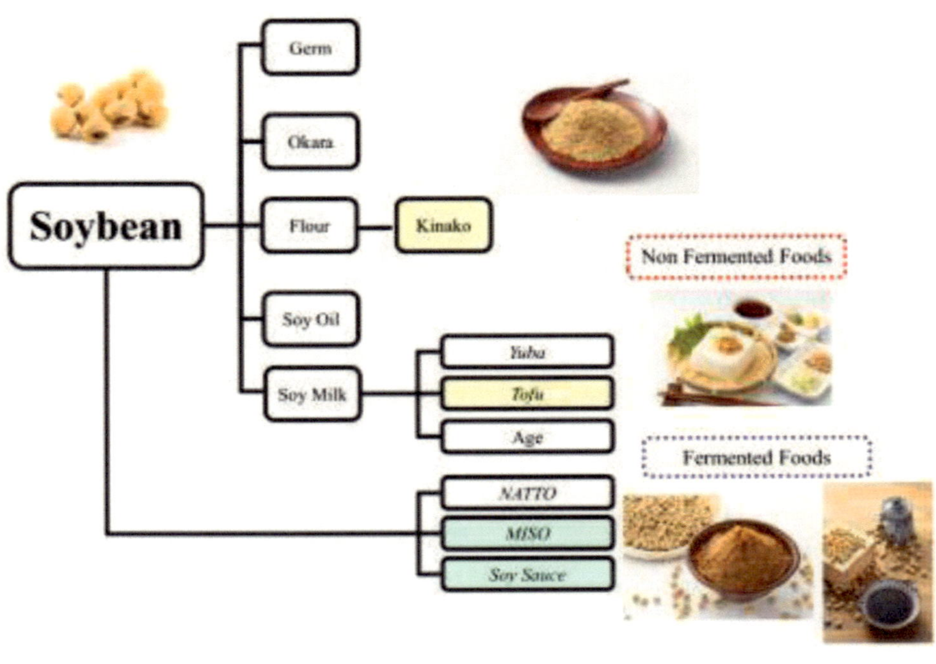

Figure 3.
The strain of Traditional Soybean Products.

"traditional food culture of the Japanese people" was registered as an UNESCO Intangible Cultural Heritage [30], and Japanese food culture is continuously gaining global attention due to the health benefits and taste of tofu, soy sauce, and miso [31].

The original form of miso is considered to have been established in Japan from the Chinese continent during the Nara period (710–794). Over the course of its 1400-year history, fermented soy sauce prepared from soybeans and grains mixed with mold-containing koji and salt has spread throughout the country and has been enhanced to fit each region's preferences. Soy sauce production is thought to have been established independently in each local environment [32]. Currently, various types of miso are available and they can be grouped into four main types: (1) rice miso prepared from soybeans, rice koji (malted rice) and salt; (2) barley miso prepared from soybeans, barley koji and salt; (3) soybean miso prepared from soybean miso balls, koji and salt; and (4) mixed miso prepared by mixing several types of miso. Rice miso is further classified based on different flavors and colors according to the koji ratio and salt concentration used, and is classified into sweet (white, red), sweet (light-colored, red), and dry (light-colored, red) rice miso while barley miso is classified as sweet and dry [33]. Among these, rice miso is the most widely produced miso and is manufactured in many regions of Japan. Rice miso is prepared by adding rice malt and salt to steamed soybeans and fermented. The salt inhibits bacterial growth as well as proteolytic enzymes, saccharolytic enzymes, and lipolytic enzymes produced by koji degradation to its ingredients; the aromatic flavor components that determine the taste of miso and miso-like color are mediated by the action of salt-resistant lactic acid bacteria and salt-resistant yeast during the fermentation and subsequent maturation processes [34]. Aroma extract dilution analysis (AEDA) is conducted to examine the aroma components in fermented soybean foods such as miso [35]. AEDA involves the detection of components eluted from a gas chromatography (GC) column based on the odor and can analyze the aroma of trace amounts of components that cannot be detected by GC [36]. The complex aroma of raw miso is attributed to major components such as 4-hydroxy-2(or 5)-ethyl-5(or 2)-methyl-3(2H)-furanone and 3-hydroxy-4,5-dimethyl-2(5H)-furanone. Furanone and other recently identified substantially low abundant components are considered to be key factors affecting the aroma of miso [37–39]. Moreover, it has been revealed that aroma alterations due to heat treatment during manufacturing and cooking procedures may be triggered by an increase in methional content, a major component, and a decrease in the levels of three components, 1-octen-3-one, (Z)-1,5-octadien-3-one, and trans-4,5-epoxy-(E)-2-decenal. Recently discovered trace components are important factors affecting soybean characteristics [40]. Miso contains numerous biological regulators, and it has been reported to inhibit melanin production, osteoporosis as well as to reduce cholesterol and blood pressure [41–43]. Miso consumption has considerably affected health maintenance of Japanese population during its long history, and is expected to improve accordingly the diets of the global population.

5. Development of new products prepared from soybeans

Soybeans contain an optimal balance of essential amino acids as suggested by their amino acid score of 100. Milk has the same amino acid score (100) and is an essential component of the human diet since ancient times. Throughout the history of food, consumption of milk can be summarized by the "dairy milk tree," that is characterized by the combination of fermentation and separation of protein-rich and fat-rich milk components [44]. A variety of dairy products ranging from butter and powdered milk to cream, condensed milk, ice cream, cheese, yogurt, and *Lactobacillus*-based beverages are prepared from milk, and they have become closely intertwined with the food culture of each region. Various studies have examined processing of soybeans to yield water-soluble protein components and oily components such as milk. Nevertheless, protein and fat components in soybeans are strongly associated, and effective classification of the two components represents a major challenge [45, 46].

Fuji Oil organization, which has been investigating soybeans for more than half a century, enhanced the process of separating lipophilic and hydrophilic proteins and developed an approach for separating soy milk into two parts: oily soy cream and low-fat soy milk. This new separation technology has been patented as the world's first Ultra Soy Separation (USS) method (**Figure 4**). The soymilk cream and low-fat soy milk separated using the USS process has been confirmed to contain various tastes and functions not detected in conventional soy milk. Soymilk cream has a deep richness of soybeans and improves the flavor of the ingredients combined with it and confers a delicious creamy texture to food. Low-fat soy milk is a healthy ingredient that although contains very limited oil amounts, it has a significantly strong soybean flavor and can be utilized as a soup stock. The first premium soy milk products developed after the USS manufacturing method were presented in the Japanese Pavilion of the Milan Expo in 2015, where food was the main theme and attracted particular attention as a food material that contains a new plant-based taste (**Figure 5**). A variety of desserts and beverages prepared using soy butter, soy cream, and low-fat soy milk that all benefit from plant-based taste of this novel ingredient, are now available in convenience stores in Japan and attract the attention of "flexitarians" who are interested in plant-based foods. In the near future, "dairy milk tree" based on soymilk cream and low-fat soy milk may further develop with the formation of a variety of new traditional foods. In summary, soybeans are anticipated to contribute in the amelioration of the health of the Japanese and global population.

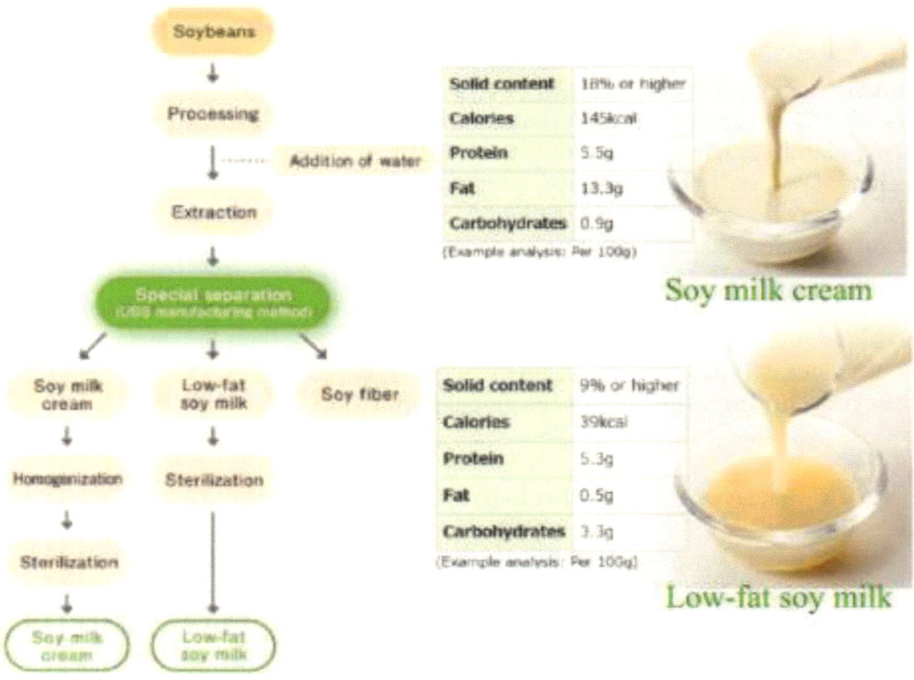

Figure 4.
USS (Ultra Soy Separation) method.

Figure 5.
Cheese-like soy milk products and their applications.

6. Conclusions

Soybean cultivated globally can be an important strategic crop that can help the resolution of the food shortage issue due to the increasing population based on its high protein content and nutritional value in the near future.

Author details

Toshihiro Nakamori
FUJI OIL Holdings Inc., Izumisano-shi, Osaka, Japan

*Address all correspondence to: nakamori.toshihiro@so.fujioil.co.jp

References

[1] Zhao TJ, Gai JY. The origin and evolution of cultivated soybean [*Glycine max* (L.) Merr.]. Scientia Agricultura Sinica. 2004;**37**(7):954-962

[2] USDA (United States Department of Agriculture). World Agricultural Production Report

[3] Manu SM, Halagalimath SP, Chandranath HT, Biradar BD. Effect of nutrient levels and plant growth regulators on harvest index and economics of soybean (glycine max). International Journal of Current Microbiology and Applied Sciences. 2020;**9**(3):890-897

[4] Perkins EG. Chapter 2—Composition of soybeans and soybean products. In: Erickson DR, editor. Practical Handbook of Soybean Processing and Utilization. USA: AOCS Press; 1995. pp. 9-28. Available from: https://www.aocs.org/stay-informed/books?SSO=True

[5] Samoto M, Maebuchi M, Miyazaki C, Kugitani H, Kohno M, Hirotsuka M, et al. Abundant proteins associated with lecithin in soy protein isolate. Food Chemistry. 2007;**102**: 317-322

[6] Carroll KK, Hamilton RMG. Effects of dietary protein and carbohydrate on plasma cholesterol levels in relation to atherosclerosis. Journal of Food Science. 1975;**40**:18-23

[7] Sugano M, Goto S, Yamada Y, Yoshida K, Hashimoto Y, Matsuo T, et al. Cholesterol-lowering activity of various undigested fractions of soybean protein in rats. The Journal of Nutrition. 1990;**120**:977-985

[8] Kohno M. Soybean protein and peptide as complementation medical food materials for treatment of dyslipidemia and inflammatory disorders. Food Science and Technology Research. 2017;**23**(6):773-782

[9] U.S. Food & drug. Code of Federal Regulations, Title 21, Volume 2, Revised as of April 1, 2020, CITE: 21CFR101.82

[10] Yamamoto T. Soybean Components and Food for Specified Health Uses (FOSHU). SOY in Health and Disease Prevention. 1st ed. Boca Raton, FL, USA: Taylor & Francis Group, CRC Press; 2005

[11] Kohno M, Hirotsuka M, Kito M. Decreases in serum triacylglycerol and visceral fat mediated by dietary soybean β-conglycinin. Journal of Atherosclerosis and Thrombosis. 2006;**13**(5):247-255

[12] Akao M, Abe R, Sato N, Hasegawa-Tanigome A, Kumagai H, Kumagai H. Prevention of osteoporosis by oral administration of phytate-removed and deamidated soybean β-conglycinin. International Journal of Molecular Sciences. 2015;**16**(1): 2117-2129

[13] Frank AA, Custer LJ, Cerns CM, Narala KK. Quantitation of phytoestrogens in legumes by HPLC. Journal of Agricultural and Food Chemistry. 1994;**42**(9):1905-1913

[14] Messina M, Nagata C, Wu AH. Estimated Asian adult soy protein and isoflavone intakes. Nutrition and Cancer. 2009;**55**:1-12

[15] Mruphy PA, Barua K, Hauck CC. Solvent extraction selection in the determination of isoflavones in soy foods. Journal of Chromatography B: Analytical Technologies in the Biomedical and Life Sciences. 2002;**777**(1-2):129-138

[16] Watanabe S, Noboru M, Yasunari M, Ideura T. A cross-sectional study on the effects of long term very low protein diets in patients with chronic kidney disease. Journal of Anti-Aging Medicine. 2010;**7**(2):7-13

[17] Taku K, Umegaki K, Sato Y, Taki Y, Endoh K, Watanabe S. Soy isoflavones lower serum total and LDL cholesterol in humans: A meta-analysis of 11 randomized controlled trials. The American Journal of Clinical Nutrition. 2007;**134**:1148-1156

[18] Morton MA, Arisaka O, Miyake N, Morgan LD, Evans BJ. Phytoestrogen concentrations in serum from Japanese men and women over forty years of age. The Journal of Nutrition. 2002;**132**: 3168-3171

[19] Yamada T. Molecular genetic study of soybean seed quality. The Japanese Society for Food Science and Technology. 2021;**68**(5):219-224

[20] Yamada T, Mori Y, Yasue K, Maruyama N, K3itamura, K., and Abe, J. Knock-down of the 7S globulin subunits shifts distribution of nitrogen sources to the residual protein fraction in transgenic soybean seeds. Plant Cell Reports. 2014;**33**:1963-1976

[21] Dwiyanti MS, Yamada T, Sato M, Abe J, Kitamura K. Genetic variation of γ-tocopherol methyltransferasegene contributes to elevated α-tocopherol content in soybean seeds. BMC Plant Biology. 2011;**11**:152

[22] Morrison MJ, Cober ER, Saleem MF, McLaughlic NB, Frégeau-Reid J, Ma BL, et al. Changes in isoflavone concentration with 58 years of genetic improvement of short-season soybean cultivars in Canada. Crop Science. 2008;**48**(6):2201-2208

[23] Ujiie A, Yamada T, Fujimoto K, Endo Y, Kitamura K. Identification of soybean varieties with high levels of α-tocopherol content. Breeding Science. 2005;**55**:123-125

[24] Yamada T, Maruyama K, Abe J. Molecular genetic study of temperature responsive γ-tocopherol methyltransferase genes in soybean seeds. Soy Protein Research. 2014;**17**:25-31

[25] Tsukamoto C, Shimada S, Igita K, Kudou S, Kokubun M, Okubo K, et al. Factors affecting isoflavone content in soybean seeds: Changes in isoflavones, saponins, and composition of fatty acids at different temperatures during seed development. Journal of Agricultural and Food Chemistry. 1995;**43**:1184-1192

[26] Yamada T, Kitamura K. Gene manipulation of the biosynthetic pathway for molecular breeding of soybean flavonoids. Soy Protein Research. 2010;**13**:37-42

[27] Van Huan N, Sugimoto H, Harada K. Genetic variation of local varieties of soybean in the western part of the Shikoku Mountains in Japan. Breeding Science. 2005;**55**:441-446

[28] Grawford GW. Advances in understanding early agriculture in Japan. Current Anthropology. 2011;**52**(4):S331-S345

[29] MAFF. Ministry of Agriculture, Forestry and Fisheries [Internet]. Available from: https://www.maff.go.jp/j/seisan/ryutu/daizu/

[30] MAFF. Ministry of Agriculture, Forestry and Fisheries [Internet]. Available from: https://www.maff.go.jp/j/keikaku/syokubunka/ich/pdf/leaflet_e2ok.pdf

[31] Suzuki N, Goto Y, Ota H, Kito K, Mano F, Joo E, et al. Characteristics of the Japanese diet described in epidemiologic publications: A qualitative systematic review. Journal of Nutritional Science and Vitaminology. 2018;**64**:129-137

[32] Ashkenazi M, Jacob J. Food Culture in Japan. USA, Santa Barbara, CA: Greenwood Publishing Group, ABC-CLIO, LLC; 2003

[33] Ebine H. Industrialization of Japanese Miso Fermentation. Industrialization of Indigenous Fermented Foods. 2nd ed Revised and

Expanded. ed. USA: CRC Press; 2004. pp. 99-147

[34] Homma N. On the aroma and aroma components of miso (part 2). Journal of Brewing Society of Japan. 1987;**82**:548-553

[35] Schieberle P. New developments in methods for analysis of volatile flavor compounds and their precursors. In: Goankar A, editor. Characterization of Food: Emerging Methods. Amsterdam: Elsevier; 1995. pp. 403-431

[36] Kumazawa K, Kaneko S, Nishimura O. Identification and characterization of volatile components causing the characteristic flavor in Miso (Japanese fermented soybean paste) and heat-processed Miso products. Journal of Agricultural and Food Chemistry. 2013;**61**(49):11968-11973

[37] Sugawara E. Identification of 4-hydoxy-2 (or 5)-ethyl-5 (or 2)-methyl-3(2H)-furanone as a flavor component in Miso. Nippon Shokuhin Kogyo Gakkaishi. 1991;**38**:491-493

[38] Ohata M, Kohama K, Morimitsu Y, Kubota K, Sugawara E. The formation mechanism by yeast of 4-hydroxy-2 (or 5)-ethyl-5(or 2)-methyl-3(2H)-furanone in miso. Bioscience, Biotechnology, and Biochemistry. 2007;**71**:407-413

[39] Inoue Y. Analysis of the cooked aroma and odorants that contribute to umami aftertaste of soy miso (Japanese soybean paste). Food Chemistry. 2016;**213**:521-528

[40] Kaneko S, Kumazawa K. Comparative aroma extract dilution analysis of changes in aroma components of seasoned soy sauce prepared from soy sauce and mirin during heating. Food Science and Technology. 2020;**26**(6):725-733

[41] Kamei H, Koide T, Hashimoto Y, Kojima T, Umeda T, Hasegawa M. Tumor cell growth-inhibiting effect of melanoidins extracted from Miso and soy sauce. Cancer Biotherapy Radiopharm. 1997;**12**(6):405-409

[42] Watanabe H. Biological efficacies of Miso. Journal of the Brewing Society of Japan. 2010;**105**:714-723

[43] Jayachandran M, Xu B. An insight into the health benefits of fermented soy products. Food Chemistry. 2019;**271**(15):362-371

[44] Wijesinha-Bettoni R, Burlingame B. Chapter 3 Milk and dairy product. In: Muehlhoff E, Bennett A, McMahon D, editors. Milk and Dairy Products in Human Nutrition. Rome, Italy: Food and Agriculture Organisation of the United Nations (FAO); 2013. pp. 41-102. Available from: https://www.fao.org/3/i3396e/i3396e.pdf

[45] Giri SK, Mangaraj S. Processing influences on composition and quality attributes of soymilk and its powder. Food Engineering Reviews. 2012;**4**: 149-164

[46] Han H, Choi JK, Park J, Im HC, Han JH, Huh MH, et al. Recent innovations in processing technologies for improvement of nutritional quality of soymilk. The Journal of Food. 2021;**19**(1):287-303

Chapter 8

Soybean Seed Compounds as Natural Health Protectors

Gabriel Giezi Boldrini, Glenda Daniela Martin Molinero, María Verónica Pérez Chaca, Nidia Noemí Gómez and Silvina Mónica Alvarez

Abstract

Glycine max (L) Merrill, better known as soy or soybean, is a legume of asian origin considered an excellent biotype, given the fact that it contains almost everything the human being needs for the diet. Its cultivation worldwide is one of the most important, and soy itself and its derivatives are highly on demand. The health effects of soy derived foods have been investigated for more than 25 years, and some of them remain controversial. On the other hand, we wondered if soy could be used to ameliorate the toxic effects of heavy metals. Therefore, in this chapter we review general characteristics of soy as well as its nutritional potential, and we compiled the newest information about the health effects of soy. In order to test our hypothesis, we developed a model of animals exposed to cadmium, and we gave them a soy based diet, comparing it with a casein-based diet as control. This allowed us to collect information about its effect on the respiratory and nervous system. Among the results of this review, we show that it reduces the cholesterol level and obesity while also having antidiabetic effects. We enumerate the benefits of soy-based diets on the respiratory system, such as protection against lung cancer and radiotherapy, better lung function in asthma patients and protection against cadmium intoxication. In the cardiovascular system it reduces the risk of coronary heart disease, improves blood pressure, glycemic control, and inflammation while it reduces not all but some of the alterations induced by cadmium exposure on the aorta and heart. It apparently promotes neurogenesis, improves cognitive functions, and reduces the oxidative stress and apoptosis induced by cadmium exposure in the cerebellum. Taken all together, this information let us conclude that soy consumption would exhibit numerous benefits for human health, although future studies should try to elucidate the best outcome considering variables such as gender, age, treatment duration and dosage of soy products consumption in the diet.

Keywords: Soybean, cadmium, oxidative stress, anti inflammatory

1. Introduction

Soybean has been consumed in many countries since before recorded history. The health effects of soy derived foods have been investigated for more than 25 years. Actually, more than 2000 soy related articles are published annually. Most of the research is conducted because there is evidence that soy has beneficial effects on health, specially preventing chronic diseases.

In our laboratory, we specialize in the effect of heavy metals intoxication on different organs and given the fact that soy is known for its antioxidant properties, we wanted to study if the addition of soy in the diets of exposed animals would ameliorate the deleterious effects of a heavy metal.

In this chapter, we begin providing some background information about soy characteristics and nutritional information; but the main intent of this review is to provide a compilation of the current understanding of the health effects of soybean derived foods mainly on the lipid profile and on different systems such as the respiratory system, the cardiovascular system and the central nervous system. We made a search compiling the most relevant works in the field and added our most outstanding results regarding the possible uses of soy to counteract some environmental intoxications.

2. *G. max* (L) Merrill "soy": general characteristics

The scientific name for soybeans is *G. max (L.) Merrill.* It is of Asian origin, native to north and central China [1]. It belongs to the *Fabaceae* family and the subfamily *Faboideae.* Depending on the country, it is popularly known by different names: soya (Portugal, France and England), soia (Italy) and sojabohne (Germany) [2]. It is an annual, herbaceous, shrubby species and its vegetative cycle ranges from three to seven months, depending on environmental conditions [3]. The optimal circumstances for its growth are the subtropical regions due to their permanently humid weather [4]. Morphologically, it is an erect and branched plant, reaching 80 to 100 cm in height. Almost all varieties show pubescence on the stems, leaves and pods. The basal leaves are unifoliate, oval, with a short half-life; from the second pair of leaves their folial development is alternate, trifoliate with oval or lanceolate leaflets, narrow or wide, depending on the variety, green in vegetative state and at maturity they turn yellow-brown.

After the vegetative growth period, the duration of which depends on the cultivar, latitude and environmental conditions (length of day and temperature) in addition to cultural practices, the soybean plant enters the reproductive phase. The flower is perfect (hermaphrodite), with axillary budding, developing a grouping of 2 to 35 flowers, which can be white or purple [5]. Palmer et al. [6] and Takahashi et al. [5] established that the color of soybean flowers is controlled mainly by six genes (W1, W2, W3, W4, Wm and Wp). Under the W1 genotype, the combination with W3W4 results in deep purple flowers, W3w4 has pale purple flowers or with purple coloration at the base of the petal, w3W4 produces purple flowers and w3w4 has almost white flowers [7].

The seed develops in pods 4 to 6 cm long; each pod has between 2 to 3 seeds. The seed has a shape that varies from round to subtly oval and can present different colors depending on the variety; they can be essentially yellow, black or green. The root system is pivotal and can reach a depth of 15–30 cm; it is capable of nodulation in symbiosis with bacteria of the genus *Rhizobium* [8].

Nutritionally, soy is an excellent biotype; since it contains almost everything the human being needs for his diet. It has between 38 and 40% protein, 18% fat due to its polyunsaturated nature, 15% carbohydrates and 15% fiber. It supplies most of the amino acids needed for protein synthesis, predominantly Lysine [9]. It is the only protein of vegetable origin with an amino acid score of 100%, when compared to proteins of animal origin, although it is limited in the amino acid methionine, so it is important that it is combined with a cereal (rice, quinoa, oats) or with animal proteins to be able to form complete proteins [10, 11]. It has a high concentration of potassium and is a good source of magnesium, phosphorus, iron, calcium, manganese and phosphates. It also provides vitamins such as vitamins E and B6 [9, 12].

Soy contains isoflavones, which are mainly found in roots and seeds [13]. The concentration ranges between 1.2 and 4.2 mg/g dry weight, depending on the characteristics of the soil in which it is grown, the climate, and the plant maturation at the time of harvesting the seeds [14]. Like other phenolic compounds, soy isoflavones are mainly found as glycosylated conjugates (> 80%), which are absorbed in the intestine and have low estrogenic activity [15]; only after hydrolysis do they acquire their maximum bioavailability and biological activity [16].

2.1 Soybean cultivation characteristics

The cultivation of soy (*G. max*) worldwide is one of the most important due to its profitability and high demand for the product and its derivatives [17]. To improve soybean crop yields, a set of variables such as genetic aspects, nutrient availability, as well as other factors (crop rotation) must be analyzed, with the aim of achieving sustainability in production.

Pérez [18] assures that in order for this variety to obtain adequate growth and yield, it needs a significant amount of available nutrients so that it is able to yield good grain production. The nutrient requirements per ton of harvested soybean grain exceeds that one needed by other field crops such as corn or wheat.

Among the macronutrients and micronutrients necessary for its development are: phosphorus (P), potassium (K), calcium (Ca), magnesium (Mg), zinc (Zn), sulfur (S), boron (B), copper (Cu) and organic matter (OM) [17]. According to Ortiz [19], soy contains high concentrations of nitrogen (N) in the grains and in the plant. Biological nitrogen fixation (BNF) increases as the crop develops, in this way the root nodules transfer between 30 and 50% of the fixed N to the vegetative stages, 80–90% between flowering and the formation of the fruit, reaching the maximum contribution during the filling of the grains [20].

2.2 Industrial use of soy

Soy is industrially used for the production of oil for both human and animal consumption. From the processed grain, 20% crude oil and 75% soy flour are obtained. Most of the soy flour is used as raw material for animals. Small percentages of protein are processed for human consumption in the form of soy milk, soy flour (SF), soy protein concentrate (SPC), tofu, and many retail food products [21].

Soy is also used for the production of industrial chemicals such as biodiesel, bio-composites, candles, ink, and wood. It is also used to produce synthetic wood floors and interior plywood. Soy products and other materials obtained from natural resources can be used as ingredients for the production of adhesives for wood due to their ability to bond with different materials [21]. Biodegradable soy-based adhesives are considerably less polluting for the environment than petrochemical adhesives [22].

Soybean residual biomass (a low-cost residue resulting from oil extraction) was functionalized with an industrial sulfur-based chelating agent, which is a precipitation agent used in industrial wastewater treatment. This biomass combination and chelating agent is used for removing heavy metals from aqueous solution, such as Pb (II), Ni (II) and Cu (II) [23].

Various derived by-products from the soy processing industry are used for the manufacture of food, cosmetics and cleaning products. Also, soy is one of the most widely used crops for the production of biodiesel [24].

Soy production represents a very important fraction of the PBI of the entire Southern Common Market (Mercosur) agribusiness, with great economic importance for these countries. Practically, the Mercosur countries make up 42% of the total soy planted in the world, compared to 33% planted in the US, satisfying together the growing world demand [25].

3. Soybean and health

Different studies have been conducted to determine which components of soy exert bioactive effects. Soy components include protein, lipids, fiber and phytochemicals including isoflavones. The three main isoflavones found in soybeans are genistein, daidzein and glycitein. Furthermore, genistein has been suggested to act as an inhibitor of oxidative stress, angiogenesis, and metastasis [26].

3.1 Anti-cancer soy property

It has been suggested that many of the health effects of soy may be related to estrogen receptors (ER), mediated by soy-associated phytoestrogens [27]. Epidemiologic studies have shown that isoflavones may exert their anticancer effects through an estrogen receptor (ER) signaling pathway in breast, endometrial, and ovarian cancers. In addition, some results from lung cancer studies link the decrease of lung cancer with soy intake. The mechanisms and active compounds in soybeans that may be responsible for this relationship will need to be elucidated in the future [28]. A research conducted in 2014 with separated women according to menopause status explored the association between breast cancer and soy isoflavone consumption and it concluded that soy isoflavone intake could reduce the risk of breast cancer for both pre- and post-menopausal women in Asian countries [29].

In men with benign prostatic hyperplasia (BPH), finasteride treatment induced epidermal growth factor receptor (EGFR) nuclear translocation, but, when finasteride was combined with soy isoflavones, EGFR remained on the cell membrane. Since nuclear EGFR is a predictor of poor outcome in prostate cancer, addition of ERβ agonists should be considered as a treatment that might offer some benefit to patients [30].

Other studies have demonstrated that high intake of total soybean and non-fermented soybean products could reduce the risk of gastric cancer, but fermented soybean products could increase that risk, indicating that the beneficial effect of soy-based food might be related to the non-fermented products [31]. Hua et al. [32] indicated that the isoflavone intake decreased the ovarian cancer risk by 33%, demonstrating that the intake of dietary isoflavones played a protective role against ovarian cancer. He et al. [33] showed that a daily diet rich in isoflavones might potentially decrease colorectal cancer incidence. In this sense, Yu et al. [34] results revealed that soy isoflavone consumption reduced the risk of developing colorectal cancer by 23%. Moreover, a higher intake of soy isoflavones was associated with a lower risk of mortality from lung, gastric, and colorectal cancers. Nachvak et al. [35] reported that an increase of 10 mg/day of soy isoflavones consumption was associated with a 7% lower risk of cancer mortality.

3.2 Soy enhances lipid profiles

The potential health effects of soybean in humans and animals remain controversial. One of the strongest health claims involves protection against coronary disease, based upon reductions in plasma cholesterol and triglycerides, and protection against atherosclerosis. However, the physiological mechanism by which soy may improve blood lipid profiles has been subject of numerous investigations. Additional health benefits involve antidiabetic effects, reduced weight gain, and improved body composition. Obesity clinically leads to nonalcoholic fatty liver disease (NAFLD), type 2-diabetes, and coronary heart disease. NAFLD is a potential risk factor for the development of both type 2-diabetes and metabolic syndrome. This illness is defined by excessive hepatic triglyceride (TG) content in the absence of excessive alcohol consumption. Moreover, NAFLD is the most common cause of chronic liver disease worldwide and it is a precursor of the more advanced liver disease nonalcoholic steatohepatitis (NASH). NAFLD is a condition that may progress to cirrhosis in up to 25% of the patients [36]. To date, the growing global disease burden of NAFLD has not been remedied by pharmaceutical treatment but soy derived foods in the diet have been proposed as a therapeutic tool for patients with NAFLD [37].

The amount of liver fat appears to be related to the amount of fat incorporated in the diet. Therefore, the excess of TG in the diet can favor fatty liver formation. Sources of hepatic TG were detected directly in NAFLD patients, as follows: a) TGs that come from non-esterified fatty acids, which flow into the liver through lipolysis of adipose tissue; b) TG in smaller amounts derived from *de novo* lipogenesis and dietary fat. Research shows that bioactive compounds in soy can prevent and treat NAFLD. Soy modulates lipid metabolism and regulates the expression of related transcription factors [36].

Hepatic lipid metabolism is partially controlled by transcription factors: sterol regulatory element binding proteins (SREBPs) and peroxisome proliferator-activated receptors (PPARs) [38]. Soybean intake decreases the expression of SREBP-1c and increases the expression of SREBP-2. SREBP-1c is a transcription factor that stimulates the expression of genes related to *de novo* lipogenesis. Insulin induces the expression of an inactive SREBP-1c precursor and likewise promotes its activation [37].

SREBP-1c stimulates the transcription of fatty acid biosynthesis genes such as acetyl-CoA carboxylase (ACC) and fatty acid synthase (FAS). In contrast, PPARα regulates genes involved in hepatic fatty acid oxidation. For example, carnitine palmitoyltransferase-1 (CPT-1) is an enzyme that regulates fatty acid β-oxidation [37, 38]. In conclusion, SREBP is associated with the regulation of hepatic lipogenesis and reduction of cholesterol synthesis [39].

Interactions between soy components are believed to improve fatty acid oxidation in the liver by increasing the expression of peroxisome proliferator activated receptor α (PPARα) regulated genes, thus decreasing lipid accumulation in the liver [38]. Another nuclear receptor involved in lipid metabolism is peroxisome proliferator-activated receptor (PPAR)γ2, whose expression is greatly increased in response to a high fat (HF) diet, especially a diet high in saturated and unsaturated fatty acids [36].

Japanese diet includes a high consumption of soy/isoflavones, fish/n-3 polyunsaturated fatty acids (PUFAs), salt/salted foods, and green tea, and a low consumption of red meat and saturated fat. Regarding this, inverse associations between soy/isoflavones or fish/n-3 PUFAs and diabetes have been found [36]. Several groups studied the effects of β-conglycinin, a soy protein, on the prevention and improvement of NAFLD. Soybean is a popular ingredient in Japan, used in foods such as tofu, miso, and natto. Ikaga et al. [36] found that in Japan the intake of soybeans and related foods was 58.6 g/day, and their protein intake was 5.1 g, while 30% of the soy protein was β-conglycinin.

In humans, supplementation with β-conglycinin is able to reduce and/or prevent intra-abdominal obesity; it is also necessary to emphasize that this protein is considered in numerous investigations as an allergen, although this information is controversial [40]. Small peptides released from β-conglycinin digestion may directly or indirectly affect lipid metabolism in adipose tissue and liver. In addition, digestion-released peptides were shown to activate low-density lipoprotein receptors and inhibit FAS biological activity. Therefore, β-conglycinin may be a promising dietary protein for NAFLD and intra-abdominal obesity amelioration. These results suggest that soy β-conglycinin could be a potentially useful dietary protein source for the prevention of hypertriglyceridemia, hyperinsulinemia, and hyperglycemia [36, 41].

Most studies on diet-induced obesity have always focused on the role of saturated fats, from animal fats, but there is enough evidence to suggest that polyunsaturated fatty acids (PUFAs), from vegetable fats also contribute to obesity [42–44]. On the other hand, there is strong evidence relating the cardiovascular health with soybean oil. Replacing high saturated FAs (SFAs) fats and oils or *trans*-FAs (TFAs) oils with either high–oleic acid soybean oil (H-OSBO) or oils high in n–6 PUFAs would have favorable effects on plasma lipid risk factors. Soybean oil (SBO) is the dominant edible oil in the marketplace and is a primary contributor of PUFAs in the US diet. (H-OSBO) is trait-enhanced oil high in oleic acid with superior functional properties, including high heat and oxidative stability for use in food manufacturing as an alternative to partially hydrogenated vegetable oils [45].

3.3 Soy and the respiratory system

It is known that nutrition is very important in lung health and defense [46–48]. Protection against lung cancer and the radioprotective effect of soy isoflavones are well described in the bibliography [49, 50]. Soy isoflavones seem to increase chemotherapy induced cell death in different types of lung cancer [51, 52]. They also improve survival of non-carcinogenic cells after chemo- or radiotherapy. Its mechanisms might be related to the inhibition of the infiltration and activation of macrophages and neutrophils [53–55].

Asthma is a chronic disease affecting a great number of people in the world and airway remodeling is one of the main pathological hallmarks of uncontrolled asthma. Remodeling is attributed to treatment failure with conventional medications and poor asthma control. Plasminogen activator inhibitor-1 (PAI-1) is associated with asthma severity due to its role in airway remodeling. Patients with asthma have demonstrated that consumption of soy isoflavones results in better lung function. Genistein induces reduction of PAI-1 generation from airway epithelial cells and may be part of the therapeutic mechanism [56]. Also, soy isoflavones recovered Th1/Th2 lymphocytes balance in bronchoalveolar lavage, inhibited eosinophil infiltration, collagen deposition and airway mucus production in murine models of allergic asthma [57, 58].

Hirayama *et al.* [59] provided evidence of a possible protective effect of the traditional japanese diet against tobacco carcinogen, showing significant reductions in the prevalence of COPD and breathlessness due to isoflavones. Additionally, Zhang *et al.* [60] reported that L-Lysine, an essential amino acid abundant in soy, was effective against sepsis-induced acute lung injury in a lipopolysaccharide induced murine model. Moreover, soy lecithins can be used for artificial pulmonary surfactant synthesis that are prepared at lower costs and they are useful for the treatment of respiratory distress, inflammatory pulmonary diseases, and dyspnea caused by asthma, among others [61].

It is known that the respiratory system is highly susceptible to environmental contamination. Airways are the access door to a number of chemical compounds, microorganisms or viruses that can injure lungs [62–66]. Cadmium is a heavy metal associated with lung damage or pulmonary cancer [67, 68] and the sources of cadmium contamination are related to industry applications, color pigments, fossil fuel combustion, Ni-Cd batteries, and the use of phosphate fertilizers [69]. The absorption of Cd takes place mainly through the respiratory tract and to a smaller extent via the gastro-intestinal tract. Cadmium induces oxidative stress and generates ROS, activates apoptosis, alters gene expression, inhibits respiratory chain complexes, and alters the inner mitochondrial permeability. Although inhalation is the most common source of intoxication, we recently demonstrated that ingested cadmium leads to pulmonary oxidative stress, cell death and fibrosis in rat lungs.

We developed a model of animals exposed to cadmium, and we gave them a soy based diet, comparing it with a casein-based diet as control. We showed that zinc and selenium levels in lungs were modified when cadmium intoxicated rats were fed with a soy-based diet. Zinc availability has an important role in lung redox-status maintenance [70]. The higher levels of Zn we found in the soy-fed Cd intoxicated animals suggests there is a protective effect exerted by the diet in cadmium intoxicated lungs. Selenium is an essential micronutrient associated with antioxidant defense, protection against cancers, and physiological functions in the nervous system [71]; it also is a cofactor of selenoproteins, such as glutathione peroxidases [72]. We observed a depletion of this element in animals fed with casein after 60 days of cadmium-induced pulmonary oxidative stress, however; Se concentration was maintained in soy-fed rats.

The enzymatic antioxidant system is composed by a group of proteins responsible for maintaining the redox state. Superoxide dismutase 2 (SOD-2) is in charge of transforming superoxide anions into hydrogen peroxide and oxygen, as McCord and Fridovich [73] well described. It is known that the expression of antioxidant enzymes is mediated by antioxidant response elements (ARE) [74]. SOD has been identified as Nrf2-regulated antioxidant enzyme [75, 76]. Our studies demonstrated that mRNA levels of Nrf-2 and SOD-2 increased only in soy fed Cd exposed animals, revealing an improvement in the antioxidant system when soy was included in the diet. Soy also reduces oxidative stress caused by other sources [77, 78]. It has been reported that flavonoids scavenge free radicals, chelate redox-active metal ions and increase metallothionein expression [79–81].

LDH activity measured in bronchoalveolar lavages is also an indicator of cell damage [82, 83]; we found a strong increase of its activity in casein-fed Cd exposed animals, which decreased in the soy-fed group. mRNA ratio of BAX (a pro-apoptotic protein) and Bcl-2 (anti-apoptotic) showed similar results to LDH activity. These data are in accordance with other studies that report that apoptosis is one of the main mechanisms involved in cell death induced by cadmium in the lungs [84–86]. Taking into account these data, we assume that a soy based diet exerts a protective effect against cadmium induced cell death in the lungs.

Pulmonary fibrosis is a common response to multiple pathologies [87–89]. Fibrosis is also related to aging [90], lung infection processes [91], and chemical intoxications [92–94]. Our study also showed that Cd caused advanced pulmonary fibrosis, capillary fragility, and numerous fused alveoli when casein was the protein source in the diet (**Figure 1A**). However, the animals that were fed with a soy diet showed less severe injuries in the lungs (**Figure 1B**).

3.4 Soy and the cardiovascular system

It is known that cardiovascular diseases (CVD) are the leading cause of death worldwide, which is affecting millions of people in both developing and developed countries. The association between CVD risk and dietary consumption has been investigated and Yan et al. [95] summarized several studies made in the last few years, showing that soy foods were associated with a reduced risk of CVD (including coronary heart disease [CHD] and stroke). Furthermore, the American Heart Association published a statement saying that daily consumption of ≥25 g of soy protein with its associated phytochemicals intact can improve lipid profiles in hypercholesterolemic humans and that soy protein without the isoflavones appears to be less effective [96].

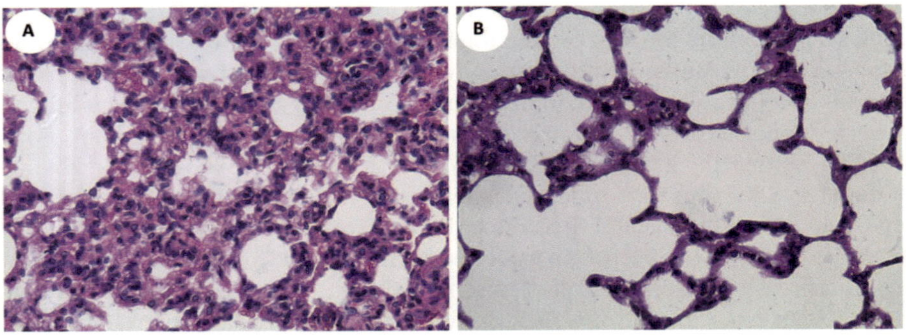

Figure 1.
Micrographs of lung parenchyma sections stained with H&E (40x). Casein-fed (A) and soy-fed (B) rat lungs exposed to 15 ppm Cd in drinking water for 60 days. (A) Presented numerous non-functional spaces, advanced pulmonary fibrosis, capillary fragility and fused alveoli. (B) Showed no evidence of capillary fragility and the number of fused alveoli was lesser; there also was evidence of pulmonary fibrosis but it was not homogeneous and the non-functional spaces were in the periphery of the lobes.

More recently, 3 prospective cohorts studies in the US showed that higher intake of isoflavones and tofu was associated with a moderately lower risk of developing CHD, and in women the favorable association of tofu were more pronounced in young women or postmenopausal women without hormone use [97]. All these results are consistent with what we mentioned above, about soy being able to decrease LDL-cholesterol levels. Besides, Ramdath et al. [98] reviewed the effects of soy on blood pressure, glucose levels, inflammatory markers, and obesity. They concluded that isoflavones and their metabolites may improve blood pressure, glycemic control, and inflammation.

On the other hand, epidemiological studies have shown that exposure to heavy metals is associated with high prevalence and incidence of cardiovascular diseases; therefore, Ferramola et al. [99] studied the effect of soy based-diets on animals exposed to cadmium in the myocardium. They found that those animals fed a soy-based diet while exposed to Cd had increased activity of CAT and SOD, suggesting an increased antioxidant response in the heart but it failed to protect the heart from Cd-induced histological alterations.

Working in the same model, Perez Diaz et al. [100] found that soy administered as a dietary protein source can modulate the pro-inflammatory and pro-apoptotic effects of a subchronic intoxication with Cd in rat's aortas. This was done by decreasing the expression of ICAM-1, which would result in a reduced leukocyte accumulation into the vasculature wall and by increasing the Bax/bcl-2 ratio respectively.

3.5 Soy and the central nervous system

It is known that soybean isoflavones influence neuronal proliferation *in vitro* and *in vivo* [101]. Besides, it has been reported that the *in vivo* administration of soybean protein hydrolysate increased the expression of brain derived neurotrophic factor (BDNF) in the cerebral cortex and the number of neurons in hippocampus and cerebral cortex, suggesting that it might promote neurogenesis [102]. Also, soy isoflavones, such as genistein, daidzein, and its metabolite, S-equol, have been informed to be an effective supplement to promote astrocyte migration in developing and/or injured adult brains, accelerating glial cell migration *via* GPER-mediated signal transduction pathway [103].

As we stated before, soy isoflavones are referred to as phytoestrogens because they bind to the estrogen receptor (ER) and as such, they affect estrogen mediated processes [104]. Soy isoflavones can exert both agonistic and antagonistic estrogenic effects [105], and have inhibitory effects on tyrosine kinase, topoisomerase and angiogenesis [106]. There are several studies that mentioned that soy isoflavones can improve cognitive function in both humans and rats [107, 108] and the SOPHIA study [109] observed the effects of soy isoflavone supplementation (110 mg/day) on cognitive function of postmenopausal women. A good performance was observed in this treatment. Duffy et al. [110] showed that postmenopausal women who received a daily treatment with 60 mg soy isoflavones along 12 weeks showed significantly better results in memorizing images and in sustained attention and planning tasks. Furthermore, isoflavone treatment showed no effect on menopausal symptoms. However, another study in healthy postmenopausal women aged 60 to 75 years showed no significant changes in cognitive function after 1 year of treatment [111]. Contradictory results have also been found in men. Some researchers have found that soy isoflavones appear to be detrimental to cognitive function in men [112], showing that middle-aged men with high tofu consumption had lower brain weight and greater cognitive decline compared to those who consumed less tofu. However, other studies demonstrated that supplementation with soy isoflavones improves cognitive function in men [108]. The treatment was carried out in young men, and consisted of the consumption of high doses of soy (100 mg of soy isoflavones/day), showing significant improvements in short- and long-term memory, as well as mental flexibility. Regardless, clinical trials indicate that soy isoflavones may improve cognitive function not only in postmenopausal women and young adults, but also in young adult men.

On the other hand, cognitive function was also evaluated in male rats and surprisingly the results were the same as in humans. Lund et al. demonstrated that the performance of a visual spatial memory test on male rats fed a soy isoflavone diet was less satisfactory than that executed on rats fed an isoflavone-free diet [113]. Other research showed that male rats that consumed soy isoflavones for life showed poorer performance in the radial arm maze test than male rats that switched to the isoflavone-free diet at 80 days of age and continued the same diet up to 120 days of age. However, another study reported that 10-month-old male rats fed 0.3 g/kg of isoflavone for 16 weeks had a significantly higher performance than male rats fed the isoflavone-free diet in the spatial water maze test [114]. The discrepancy between the two studies can be attributed to the difference in the age of the rats, dietary regimes, and maze tests.

Among the multiple mechanisms that soy has, it is also known to act at the tyrosine kinase level. Phosphorylation of this protein is a regulatory mechanism involved in numerous responses in the brain, including neuroregeneration [115], synaptic plasticity [116] and neuronal damage [117]. Protein tyrosine kinases are highly expressed in the brain and are reported to be involved in the induction of long-term potentiation (LTP) in the hippocampus [118], which is crucial for learning and memory. Genistein is known to inhibit tyrosine kinase [119], which is generally considered to be detrimental to a neuron. However, these effects appear to occur only when the level of genistein in a cell is high [120]. In a model of calcium ATPase inhibitor-induced apoptosis, low concentrations of genistein prevented apoptosis, however high concentrations reversed all phenotypes. High concentrations of genistein were also shown to induce apoptosis in primary cortical neurons [121], blocked tyrosine kinase activity, and contributed to H2O2-induced apoptosis in SH-SY5Y human neuroblastoma cells [122]. It can be postulated that LTP suppression by genistein is dose dependent, similar to its effects on neuroprotection.

On the other hand, Deol et al. [123] found that soy diets would produce significant dysregulation of gene expression in the hypothalamus of male mice, the most notable of which is the gene encoding Oxytocin (Oxt). These results demonstrated that different fatty diets can have differential effects on hypothalamic gene expression and increase the possibility that a rich soy diet could contribute both to increasing rates of metabolic disease as well as affecting neurological functions.

Also, just as we did with lungs, we analyzed the effect of a soy-based diet after chronic cadmium intoxication. Cadmium has been shown to have negative effects on the central nervous system, it has been observed that it is capable of crossing the blood–brain barrier (BBB), increasing its permeability in rats [124] allowing it to accumulate in the brain of developing and adult rats, thus affecting the integrity of the vascular endothelium, producing cellular dysfunction and cerebral edema [125–127]. These could induce cognitive dysfunctions [128, 129], leading to neurological alterations like olfactory dysfunctions, peripheral neuropathy, decreased ability to concentrate and neurodegenerative diseases [130, 131]. Moreover, it is known that oxidative stress is a factor involved in neurodegenerative disorders in adults, such as strokes, traumas and seizures [132], where the CNS is susceptible to free radicals damage [133].

In cerebellum, the motor coordination center, also involved in cognitive processing and sensory discrimination, the effects of Cd in experimental animals include cerebellar bleeding, cerebellar edema and hyperactivity [134, 135]. Our investigation revealed an increase in Cd concentrations in the cerebellum of the rats after 60 days of treatment while the animals fed a soy diet did not have high levels of the heavy metal. Soybean has an apparent defensive effect against Cd accumulation. Besides, some studies showed that a chronic ingestion of diet enriched in casein induced substantial BBB disruption in the cortex region resulting in non-specific blood-to-brain extravasation of plasma-derived proteins, while a diet enriched in soy protein showed no significant effects on BBB permeability [136]. Besides, it has been shown that certain amino acids found in protein-rich diets, specifically methionine, are associated with elevated plasma homocysteine levels [137, 138]. Tyagi et al. demonstrated that hyperhomocysteinemia induces structural and functional alterations of BBB by promoting oxidative stress and neuroinflammation in mouse models of cognitive dysfunction [139, 140].

Regarding the effect on the antioxidant defense system in cerebellum, we found a reduction of catalase activity in the intoxicated animals fed with soy, probably due to the isoflavones content, just like Halder [141] described in his study, where that an isoflavone like quercetin, helped to reduce Cd levels in brain tissue and modulated the antioxidant system of the cell by affecting expression of antioxidant enzymes at the transcription level. Furthermore, these animals showed increased expression of SOD-2, suggesting a protective, antioxidant effect of this protein source, as it was previously shown by Liu et al. [142].

It is known that the adult human brain contains about 20% of the body's total cholesterol [143], making cholesterol the main constituent of this tissue. Unlike cholesterol in other organs, in the brain it is primarily derived by *de novo* synthesis. The intact blood brain barrier (BBB) avoids the uptake of lipoproteins from the circulation [144]. Regarding the content of total cholesterol, it was significantly increased in both Cd exposed groups when compared to the respective controls, as well as the expression of HMGCoAR, the limiting enzyme in cholesterol synthesis. Additionally, we found an increase in total triglycerides (TG) and total phospholipids (PL) in the animals fed with casein, with no changes under a soy diet. The increased TG value was due to an increased expression of GPAT2, an enzyme that catalyzes the initial step in glycolipid biosynthesis. Similar results were found by Modi et al. [145] in the brain. Any of these modifications were found in the animals fed with soy, suggesting again a somewhat protective role of this diet.

Cd has been recognized as an inducer of apoptosis in a diversity of tissues including the brain [29]. In casein-fed cadmium intoxicated animals the Bax/Bcl-2 ratio increased. The replacement of casein for soy as dietary protein reduced this Bax increase but also increased Bcl-2 expression, changing the ratio. These data suggest that a soy diet inhibited Cd-induced apoptosis by increasing Bcl-2 expression, which may play an inducible cytoprotective role against Cd toxicity [146]. In this regard, it has been reported that a long-term intervention with genistein can lead to a decrease in apoptosis in hippocampal neurons of ovariectomized rats, upregulate the expression of Bcl-2, and downregulate the expression of Bax, playing genistein an anti-apoptotic and neuroprotective role [147]. This was also consistent with the structural changes observed, like a reduction in the number in the granular cells in the casein group and also with the morphometric results that indicated that Cd intoxication produced striking variations in the thickness of the granular layer in the casein group as reported by Mahmoud [148]. In the soy groups no significant differences were noticed, consistent with Farahmand who find that the antioxidant effect of *Salvia rhytidia* extract had an influence on the prevention of ischemia reperfusion injury of rat cerebellum and can reduce the ischemia injury confirmed by increase of the thickness of granular layer and size of Purkinje cells [149].

4. Conclusions

Soy (*G. max*) is a food of rich nutritional content, in which composition has proteins, oligosaccharides, dietary fiber, phytochemicals (especially isoflavones) and minerals.

Soy is used for the production of oil, flour, milk, protein concentrate and tofu, among others; products that can be used for both human and animal consumption. In addition, soybeans are used for the production of industrial chemicals such as biodiesel, in the treatment of industrial wastewater and in the manufacture of cosmetics and cleaning products.

In this chapter we summarized the benefits of adding soy to the daily diet. Its consumption exhibits numerous benefits for human health. It can reduce the risk of a variety of health problems, such as different cancers, cardiovascular diseases, stroke and it also acts as an anti-neuroinflammatory. What is more, it can ameliorate the toxic effect of cadmium intoxication in several organs.

Soy isoflavone may bind to some hormone receptors, such as estrogen receptors (ER). Therefore, the intake of phytoestrogens was found to be beneficial in breast, ovarian, endometrial and colorectal cancer and also in lung cancer. However, more research is needed to understand its effects, mechanisms and dosages. Several studies demonstrated that the intake of total soybean could also reduce the risk of gastric cancer. The consumption of soy protein and fiber reduces the risk of CVD and is associated with glycemic control. Soy is a high-quality protein that could be an alternative strategy to palliate pulmonary damage in several pathologies and to prevent the damage due to heavy metals intoxication.

Therefore, the use of soy in functional food is very interesting, but future studies should try to elucidate the best outcome considering variables such as genetic load, gender, age, treatment duration and the dosage of soy products consumption in the diet.

Conflict of interest

The authors report no conflicts of interest.

Author details

Gabriel Giezi Boldrini[1,2], Glenda Daniela Martin Molinero[1,2], María Verónica Pérez Chaca[3], Nidia Noemí Gómez[1,2,3] and Silvina Mónica Alvarez[1,2]*

1 Laboratory of Nutrition, Environment and Cell Metabolism, Faculty of Chemistry, Biochemistry and Pharmacy, National University of San Luis, Argentina

2 IMIBIO-SL CONICET, San Luis, Argentina

3 Laboratory of Morphophysiology, Faculty of Chemistry, Biochemistry and Pharmacy, National University of San Luis, Argentina

*Address all correspondence to: silvina.alvarez@gmail.com

References

[1] Gimenez AM. La soja, su evolución en Argentina [Internet]. 2018. Available from: https://www.monografias.com/trabajos82/soja-su-evolucion-argentina/soja-su-evolucion-argentina.shtml. [Accessed: 2021-05-01]

[2] Ridner E. Soja, propiedades nutricionales y su impacto en la salud. Sociedad Argentina de Nutrición, Buenos Aires. Grupo Q S.A. 2006; p 96.

[3] Singh G. The Soybean: Botany, Production and Uses; CABI. 2010; 1-494. DOI: 10.1079/9781845936440.0000

[4] Cortez JCO. Los rendimientos de cultivos y la agricultura en el estado plurinacional de Bolivia. Universidad Mayor de San Andrés. Revista de la Carrera de Ingeniería Agronómica, Apthapi. 2017;2(2):36-46.

[5] Takahashi R, Matsumura H, Oyoo ME, Khan NA. Genetic and Linkage Analysis of Purple–blue Flower in Soybean. Journal of Heredity. 2008;99(6):593-597. DOI:10.1093/jhered/esn041.

[6] Palmer RG, Pfeiffer TW, Buss GR, Kilen TC. Soybeans: improvement, production and uses Qualitative genetics. In: Boerma HR, Specht JE. 2004;editors. 3rd ed. 16: 137-233. https://DOI:.org/10.2134/agronmonogr16.3ed.c5.

[7] Mederos-Ramírez A, Ortiz-Pérez R. Analysis of the genotype environment interaction in the of soybean crop (Glycine max (L) Merrill). Cultivos Tropicales. 2021;42:1-10.

[8] Menjivar Menjivar A, Rivera Escobar SB, Vásquez Jovel KV. Evaluación fenológica y morfo-agronómica de ocho genotipos de soya (*Glycine max* L.) en la Estación Experimental y de Prácticas, Facultad de Ciencias Agronómicas, Universidad de El Salvador, San Luis Talpa, La Paz. [thesis], Universidad de El Salvador. 2017.

[9] Caicedo IM, Mejía-Gonzales A, Beltrán-Castro F. Industrialización del cultivo de soya. Revista Observatorio de la Economía Latinoamericana. 2019; ISSN: 1696-8352.

[10] Lope C, Ochoa X, Aguilera N. Soya. La oleaginosa de mayor importancia a nivel mundial [Internet]. 2018. Available from: http://www.oiapes.sagarhpa.sonora.gob.mx› paq-tec› paq-soya. [Accessed: 2021-04-27]

[11] Martín-Salinas C, López-Sobaler AM. Beneficios de la soja en la salud. Nutrición Hospitalaria. 2017;34:36-40. DOI:10.20960/nh.1569.

[12] Tosquy V, Alciva V, López R. Densidad de siembra en dos variedades de soya de temporal en Veracruz, México. Agronomía Mesoamericana. 2017;21: 63-72.

[13] Aguiar CL, Baptista AS, Alencar SM, Haddad R, Eberlin MN. Analysis of isoflavonoids from leguminous plant extracts by RPHPLC/DAD and electrospray ionization mass spectrometry. International Journal of Food Sciences and Nutrition 2007;58:116-124.

[14] Zhang J, Ge Y, Han F, Li B, Yan S, Sun J, Wang L. Isoflavone content of soybean cultivars from maturity group 0 to VI grown in northern and southern China. Journal of the American Oil Chemists' Society 2014;91:1019-1028.

[15] Nielsen ILF, Williamson G. Review of the factors affecting bioavailability of soy isoflavones in humans. Nutrition and Cancer 2007;57:1-10.

[16] de Cremoux P, This P, Leclercq G, Jacquot, Y. Controversies concerning the

use of phytoestrogens in menopause management: Bioavailability and metabolism. Maturitas 2010;65:334-339.

[17] Gerster G, Bacigaluppo S. Cómo alcanzar altos rendimientos en soja. Para mejorar la producción. INTA. 2009;vol. 42. ISSN: 2591-5088.

[18] Pérez B. Requerimientos nutricionales del cultivo de soya. IPNI. 2013;15(2),10-17.

[19] Ortiz, C. Calidad del grano de soya y su efecto en productos y subproductos industriales. Revista Soya noticias 2018;35(9):5-10.

[20] Hernández AJ, Guerra R, Tobía C, Villalobos E. Evaluación del potencial forrajero de diez cultivares de soya (*Glycine max* (L) Merr) en Venezuela. Agronomía Costarricense. 2013;37(2) (0377-9424):45-53. DOI: 10.15517/rac.v37i2.12761.

[21] Vnučec D, Kutnar A, Goršek A. Soy-based adhesives for wood-bonding – a review. Journal of Adhesion Science and Technology. 2016; 31:8, 910-931. DOI:10.1080/01694243.2016.1237278

[22] McDevitt JE, Grigsby W. Life Cycle Assessment of Bio- and Petro-Chemical Adhesives Used in Fiberboard Production. Journal of Polymers and the Environment. 2014; 22(4): 537-544. doi:10.1007/s10924-014-0677-4

[23] Bulgariu L,Bulgariu D. Functionalized soy waste biomass - A novel environmental-friendly biosorbent for the removal of heavy metals from aqueous solution. Journal of Cleaner Production. 2018;197(1): 875-885. DOI:10.1016/j.jclepro.2018.06.261

[24] Silva DR, Roa A, Quevedo RL, Quilodran BT. Production of biodiesel from soybean frying oil using native strains of Thraustochytrids. Chilean J. Agric. Anim. Sci. 2015;31(1): 29-41.

[25] Melgar R, Vitti G, de Melo Benites V. Fertilizando para altos rendimientos. Soja en Latinoamérica Instituto. 2011; 20. DOI: 10.3235/978-3-9523243-7-0.

[26] Mezei O, Banz WJ, Steger RW, Peluso MR, Winters TA, Shay N. Soy isoflavones exert antidiabetic and hypolipidemic effects through the PPAR pathways in obese Zucker rats and murine RAW 264.7 cells. J Nutr. 2003;133(5):1238-1243. DOI: 10.1093/jn/133.5.1238.

[27] Ronis MJ, Chen Y, Badeaux J, Badger TM. Dietary soy protein isolate attenuates metabolic syndrome in rats via effects on PPAR, LXR, and SREBP signaling. J Nutr. 2009;139(8):1431-1438. DOI: 10.3945/jn.109.107029.

[28] Yang WS, Va P, Wong MY, Zhang HL, Xiang YB. Soy intake is associated with lower lung cancer risk: results from a meta-analysis of epidemiologic studies. Am J Clin Nutr. 2011;94(6): 1575-1583. DOI: 10.3945/ajcn.111.020966.

[29] Chen M, Rao Y, Zheng Y, Wei S, Li Y, Guo T, Yin P. Association between soy isoflavone intake and breast cancer risk for pre- and post-menopausal women: a meta-analysis of epidemiological studies. PLoS One. 2014;9(2):e89288. DOI: 10.1371/journal.pone.0089288.

[30] Wu WF, Wang L, Spetsieris N, Boukovala M, Efstathiou E, Brössner C, Warner M, Gustafsson JA. Estrogen receptor β and treatment with a phytoestrogen are associated with inhibition of nuclear translocation of EGFR in the prostate. Proc Natl Acad Sci USA. 2021;118(13):e2011269118. DOI: 10.1073/pnas.2011269118

[31] Wang Y, Guo J, Yu F, Tian Y, Wu Y, Cui L, Liu LE. The association between soy-based food and soy isoflavone intake and the risk of gastric cancer: a systematic review and meta-analysis.

J Sci Food Agric. 2021 doi: 10.1002/jsfa.11334

[32] Hua X, Yu L, You R, Yang Y, Liao J, Chen D. Association among Dietary Flavonoids, Flavonoid Subclasses and Ovarian Cancer Risk: A Meta-Analysis. PLoS ONE. 2016;11:e0151134. DOI: 10.1371/journal.pone.0151134

[33] He X, Sun LM. Dietary intake of flavonoid subclasses and risk of colorectal cancer: Evidence from population studies. Oncotarget. 2016;7:26617-26627. doi: 10.18632/oncotarget.8562.

[34] Yu Y, Jing X, Li H, Zhao X, Wang D. Soy isoflavone consumption and colorectal cancer risk: A systematic review and meta-analysis. Sci. Rep. 2016;6:25939. doi: 10.1038/srep25939.

[35] Nachvak SM, Moradi S, Anjom-Shoae J, Rahmani J, Nasiri M, Maleki V, Sadeghi O. Soy, Soy Isoflavones, and Protein Intake in Relation to Mortality from All Causes, Cancers, and Cardiovascular Diseases: A Systematic Review and Dose-Response Meta-Analysis of Prospective Cohort Studies. J. Acad. Nutr. Diet. 2019;119:1483-1500. DOI: 10.1016/j.jand.2019.04.011

[36] Ikaga R, Li D, Yamazaki T. Dietary β-conglycinin prevents acute ethanol-induced fatty liver in mice. Biochem Biophys Res Commun. 2017;493(1):542-547. DOI: 10.1016/j.bbrc.2017.08.155.

[37] Seidemann L, Krüger A, Kegel-Hübner V, Seehofer D, Damm G. Influence of Genistein on Hepatic Lipid Metabolism in an In Vitro Model of Hepatic Steatosis. Molecules. 2021;26(4):1156. DOI: 10.3390/molecules26041156.

[38] Razzeto GS, Lucero López VR, Marra CA, Scardapane LA, Escudero NL, Gimenez MS. Soybean flour improves fatty acid profile and decreases hepatic triglyceride deposition in rats fed with normocaloric and hypercaloric diet. Food and Nutrition Sciences. 2015;6:1245-1257.

[39] Oliveira LP, de Jesús RP, Freire TO, Oliveira CP, Castro Lyra A, Lyra LG. Possible molecular mechanisms soy-mediated in preventing and treating nonalcoholic fatty liver disease. Nutr Hosp. 2012;27(4):991-998. DOI: 10.3305/nh.2012.27.4.5833. PMID: 23165534

[40] Wang T, Qin GX, Sun ZW, Zhao Y. Advances of research on glycinin and β-conglycinin: a review of two major soybean allergenic proteins. Crit Rev Food Sci Nutr. 2014;54(7): 850-862. DOI: 10.1080/10408398.2011.613534.

[41] Kohno M, Hirotsuka M, Kito M, Matsuzawa Y. Decreases in serum triacylglycerol and visceral fat mediated by dietary soybean beta-conglycinin. J Atheroscler Thromb. 2006;13(5): 247-255. DOI: 10.5551/jat.13.247.

[42] Costa, CAS;Carlos, AS;Santos, A;Monteiro, AM;Moura, EG;Nascimento-Saba, CC. Abdominal adiposity, insulin and bone quality in young male rats fed a high-fat diet containing soybean or canola oil. Clinics 2011;66(10),1811-1816. DOI: 10.1590/S1807-59322011001000022

[43] Deol P, Evans JR, Dhahbi J. Soybean oil is more obesogenic and diabetogenic than coconut oil and fructose in mouse: potential role for the liver. PloS One. 2015;10(7): e0132672. DOI: 10.1371/journal.pone.0132672. eCollection 2015.

[44] Deol P, Fahrmann J, Yang J. Omega-6 and omega-3 oxylipins are implicated in soybean oil-induced obesity in mice. Sci Rep. 2017;7(1):12488. DOI:10.1038/s41598-017-12624-9

[45] Huth PJ, Fulgoni VL 3rd, Larson BT. A systematic review of high-oleic

vegetable oil substitutions for other fats and oils on cardiovascular disease risk factors: implications for novel high-oleic soybean oils. Adv Nutr. 2015;6(6): 674-693. DOI: 10.3945/an.115.008979.

[46] Isaak M, Ulu A, Osunde A, Nordgren TM, Hanson C. Nutritional Factors in Occupational Lung Disease. Curr Allergy Asthma Rep. 2021;21(4):24. DOI: 10.1007/s11882-021-01003-0

[47] Damayanthi HDWT, Prabani KIP. Nutritional determinants and COVID-19 outcomes of older patients with COVID-19: A systematic review. Arch Gerontol Geriatr. 2021;95:104411. DOI: 10.1016/j.archger.2021.104411

[48] Nigro E, Perrotta F, Scialò F, D'Agnano V, Mallardo M, Bianco A, Daniele A. Food, Nutrition, Physical Activity and Microbiota: Which Impact on Lung Cancer? Int J Environ Res Public Health. 2021;18(5):2399. DOI: 10.3390/ijerph18052399

[49] Abernathy LM, Fountain MD, Joiner MC, Hillman GG. Innate Immune Pathways Associated with Lung Radioprotection by Soy Isoflavones. Front Oncol. 2017;7:7. DOI: 10.3389/fonc.2017.00007

[50] Yu Y, Xing Y, Zhang Q, Zhang Q, Huang S, Li X, Gao C. Soy isoflavone genistein inhibits has-circ-0031250/miR-873-5p/FOXM1 axis to suppress non-small-cell lung cancer progression. IUBMB Life. 2021;73(1):92-107. DOI: 10.1002/iub.2404

[51] Liu D, Yan L, Wang L, Tai W, Wang W, Yang C. Genistein enhances the effect of cisplatin on the inhibition of non-small cell lung cancer A549 cell growth in vitro and in vivo. Oncol Lett. 2014;8(6):2806-2810. DOI: 10.3892/ol.2014.2597

[52] Singh-Gupta V, Joiner MC, Runyan L, Yunker CK, Sarkar FH, Miller S, Gadgeel SM, Konski AA, Hillman GG. Soy isoflavones augment radiation effect by inhibiting APE1/Ref-1 DNA repair activity in non-small cell lung cancer. J Thorac Oncol. 2011;6(4):688-698. DOI: 10.1097/JTO.0b013e31821034ae

[53] Abernathy LM, Fountain MD, Rothstein SE, David JM, Yunker CK, Rakowski J, Lonardo F, Joiner MC, Hillman GG. Soy Isoflavones Promote Radioprotection of Normal Lung Tissue by Inhibition of Radiation-Induced Activation of Macrophages and Neutrophils. J Thorac Oncol. 2015;10(12):1703-12. DOI: 10.1097/JTO.0000000000000677

[54] Hillman GG, Singh-Gupta V, Lonardo F, Hoogstra DJ, Abernathy LM, Yunker CK, Rothstein SE, Rakowski J, Sarkar FH, Gadgeel S, Konski AA, Joiner MC. Radioprotection of lung tissue by soy isoflavones. J Thorac Oncol. 2013;8(11):1356-1364. DOI: 10.1097/JTO.0b013e3182a4713e

[55] Fountain MD, McLellan LA, Smith NL, Loughery BF, Rakowski JT, Tse HY, Hillman GG. Isoflavone-mediated radioprotection involves regulation of early endothelial cell death and inflammatory signaling in Radiation-Induced lung injury. Int J Radiat Biol. 2020;96(2):245-256. DOI: 10.1080/09553002.2020.1683642

[56] Bime C, Wei CY, Holbrook J, Smith LJ, Wise RA. Association of dietary soy genistein intake with lung function and asthma control: a post-hoc analysis of patients enrolled in a prospective multicentre clinical trial. Prim Care Respir J. 2012;21(4):398-404. DOI: 10.4104/pcrj.2012.00073

[57] Bao ZS, Hong L, Guan Y, Dong XW, Zheng HS, Tan GL, Xie QM. Inhibition of airway inflammation, hyperresponsiveness and remodeling by soy isoflavone in a murine model of allergic asthma. Int Immunopharmacol.

2011;11(8):899-906. DOI: 10.1016/j.intimp.2011.02.001

[58] Cho SH, Jo A, Casale T, Jeong SJ, Hong SJ, Cho JK, Holbrook JT, Kumar R, Smith LJ. Soy isoflavones reduce asthma exacerbation in asthmatic patients with high PAI-1-producing genotypes. J Allergy Clin Immunol. 2019;144(1):109-117.e4. DOI: 10.1016/j.jaci.2019.01.020.

[59] Hirayama F, Lee AH, Binns CW, Hiramatsu N, Mori M, Nishimura K. Dietary intake of isoflavones and polyunsaturated fatty acids associated with lung function, breathlessness and the prevalence of chronic obstructive pulmonary disease: possible protective effect of traditional Japanese diet. Mol Nutr Food Res. 2010;54(7):909-917. DOI: 10.1002/mnfr.200900316

[60] Zhang Y, Yu W, Han D, Meng J, Wang H, Cao G. L-lysine ameliorates sepsis-induced acute lung injury in a lipopolysaccharide-induced mouse model. Biomed Pharmacother. 2019;118:109307. DOI: 10.1016/j.biopha.2019.109307

[61] Yukitake K, Nakamura Y, Kawahara M, Nakahara H, Shibata O, Lee S. Development of low cost pulmonary surfactants composed of a mixture of lipids or lipids-peptides using higher aliphatic alcohol or soy lecithin. Colloids Surf B Biointerfaces. 2008;66(2):281-286. DOI: 10.1016/j.colsurfb.2008.07.007

[62] Flood-Garibay JA, Méndez-Rojas MÁ, Pérez-Cortés EJ. Sistema inmune respiratorio y consecuencias de contaminación aérea por materia particulada [Respiratory immune system and consequences due to particulate matter in air pollution]. Rev Med Inst Mex Seguro Soc. 2019 May 2;57(3):170-180.

[63] Harker JA, Lloyd CM. Overlapping and distinct features of viral and allergen immunity in the human lung. Immunity. 2021;54(4):617-631. DOI: 10.1016/j.immuni.2021.03.010

[64] Ather B, Mirza TM, Edemekong PF. Airborne Precautions. StatPearls [Internet]. 2021. Available from: https://www.ncbi.nlm.nih.gov/books/NBK531468. Accessed: [2021-05-03]

[65] Torres A, Cilloniz C, Niederman MS, Menéndez R, Chalmers JD, Wunderink RG, Van der Poll T. Pneumonia. Nat Rev Dis Primers. 2021; 7:25. DOI: 10.1038/s41572-021-00259-0

[66] Bracken-Clarke D, Kapoor D, Baird AM, Buchanan PJ, Gately K, Cuffe S, Finn SP. Vaping and lung cancer - A review of current data and recommendations. Lung Cancer. 2021;153:11-20. DOI: 10.1016/j.lungcan.2020.12.030

[67] Luevano J, Damodaran C. A review of molecular events of cadmium-induced carcinogenesis. J Environ Pathol Toxicol Oncol. 2014;33(3):183-194. DOI: 10.1615/jenvironpatholtoxicoloncol.2014011075

[68] Ganguly K, Levänen B, Palmberg L, Åkesson A, Lindén A. Cadmium in tobacco smokers: a neglected link to lung disease? Eur Respir Rev. 2018;27(147):170122. DOI: 10.1183/16000617.0122-2017

[69] Genchi G, Carocci A, Lauria G, Sinicropi MS, Catalano A. Nickel: Human Health and Environmental Toxicology. Int J Environ Res Public Health. 2020;17(3):679. DOI:10.3390/ijerph17030679

[70] Gomez NN, Fernandez MR, Zirulnik F, Gil E, Scardapane L, Ojeda MS, Gimenez MS. Chronic zinc deficiency induces an antioxidant adaptive response in rat lung. Exp Lung Res. 2003;29(7):485-502. DOI: 10.1080/01902140303776

[71] Burk RF. Selenium, an Antioxidant Nutrient. Nutrition in Clinical Care. 2002;5:75-79. DOI: 10.1046/j.1523-5408.2002.00006.x

[72] Labunskyy VM, Hatfield DL, Gladyshev VN. Selenoproteins: molecular pathways and physiological roles. Physiol Rev. 2014;94(3):739-777. DOI: 10.1152/physrev.00039.2013

[73] McCord JM, Fridovich I. Superoxide dismutase. An enzymic function for erythrocuprein (hemocuprein). J Biol Chem. 1969;244(22):6049-6055.

[74] Jaiswal AK. Nrf2 signaling in coordinated activation of antioxidant gene expression. Free Radic Biol Med. 2004;36(10):1199-1207. DOI: 10.1016/j.freeradbiomed.2004.02.074

[75] Hu R, Xu C, Shen G, Jain MR, Khor TO, Gopalkrishnan A, Lin W, Reddy B, Chan JY, Kong AN. Identification of Nrf2-regulated genes induced by chemopreventive isothiocyanate PEITC by oligonucleotide microarray. Life Sci. 2006;79(20):1944-1955. doi: 10.1016/j.lfs.2006.06.01

[76] Leonard MO, Kieran NE, Howell K, Burne MJ, Varadarajan R, Dhakshinamoorthy S, Porter AG, O'Farrelly C, Rabb H, Taylor CT. Reoxygenation-specific activation of the antioxidant transcription factor Nrf2 mediates cytoprotective gene expression in ischemia-reperfusion injury. FASEB J. 2006 Dec;20(14):2624-2626. doi: 10.1096/fj.06-5097fje.

[77] Aoki H, Otaka Y, Igarashi K, Takenaka A. Soy protein reduces paraquat-induced oxidative stress in rats. J Nutr. 2002;132(8):2258-2262. DOI: 10.1093/jn/132.8.2258

[78] Rancan L, Puig A, Balibrea JM, Paredes SD, García C, Jiménez L, Fernández-Tresguerres JA, Vara E. Protective effects of 17-β-oestradiol and phytoestrogen on age-induced oxidative stress and inhibition of surfactant synthesis in rat type II pneumocytes. Int J Food Sci Nutr. 2021;72(1):26-36. DOI: 10.1080/09637486.2020.1757044

[79] Sichel G, Corsaro C, Scalia M, Di Bilio AJ, Bonomo RP. In vitro scavenger activity of some flavonoids and melanins against O2-(.). Free Radic Biol Med. 1991;11(1):1-8. DOI: 10.1016/0891-5849(91)90181-2

[80] Morel I, Lescoat G, Cogrel P, Sergent O, Pasdeloup N, Brissot P, Cillard P, Cillard J. Antioxidant and iron-chelating activities of the flavonoids catechin, quercetin and diosmetin on iron-loaded rat hepatocyte cultures. Biochem Pharmacol. 1993;45(1):13-19. DOI: 10.1016/0006-2952(93)90371-3

[81] Kameoka S, Leavitt P, Chang C, Kuo SM. Expression of antioxidant proteins in human intestinal Caco-2 cells treated with dietary flavonoids. Cancer Lett. 1999;146(2):161-167. DOI: 10.1016/s0304-3835(99)00253-0

[82] Drent M, Cobben N, Henderson R, Jacobs J, Wouters E, van Dieijen-Visser M. BAL fluid LDH activity and LDH isoenzyme pattern in lipoid pneumonia caused by an intravenous injection of lamp oil. Eur Respir J. 1996;9:2416-2418. DOI: 10.1183/09031936.96.09112416

[83] Li XY, Gilmour PS, Donaldson K, MacNee W. Free radical activity and pro-inflammatory effects of particulate air pollution (PM10) in vivo and in vitro. Thorax. 1996;51(12):1216-1222. DOI:10.1136/thx.51.12.1216

[84] Kwon K, Jang J, Kwon S, Cho C, Oh H, Kim S. Cadmium induced acute lung injury and TUNEL expression of apoptosis in respiratory cells. Journal of Korean Medical Science. 2003;18:655-662. DOI: 10.3346/JKMS.2003.18.5.655

[85] Kiran Kumar KM, Naveen Kumar M, Patil RH, Nagesh R, Hegde SM, Kavya K, Babu RL, Ramesh GT, Sharma SC. Cadmium induces oxidative stress and apoptosis in lung epithelial cells. Toxicol Mech Methods. 2016;26(9):658-666. DOI: 10.1080/15376516.2016.1223240

[86] Xu YM, Wu DD, Zheng W, Yu FY, Yang F, Yao Y, Zhou Y, Ching YP, Lau AT. Proteome profiling of cadmium-induced apoptosis by antibody array analyses in human bronchial epithelial cells. Oncotarget. 2016;7(5):6146-58. DOI: 10.18632/oncotarget.6738

[87] Richeldi L, Collard HR, Jones MG. Idiopathic pulmonary fibrosis. Lancet. 2017;389(10082):1941-1952. DOI: 10.1016/S0140-6736(17)30866-8

[88] Meyer KC. Pulmonary fibrosis, part I: epidemiology, pathogenesis, and diagnosis. Expert Rev Respir Med. 2017;11(5):343-359. DOI: 10.1080/17476348.2017.1312346

[89] Klimova B, Kuca K, Novotny M, Maresova P. Cystic fibrosis revisited - A review study. Med Chem. 2017;13:102-109. DOI: 10.2174/157340641266616 0608113235

[90] Gulati S, Thannickal VJ. The Aging Lung and Idiopathic Pulmonary Fibrosis. Am J Med Sci. 2019;357(5):384-389. DOI: 10.1016/j.amjms.2019.02.008

[91] O'Hare M, Amarnani D, Whitmore HAB, An M, Marino C, Ramos L, Delgado-Tirado S, Hu X, Chmielewska N, Chandrahas A, Fitzek A, Heinrich F, Steurer S, Ondruschka B, Glatzel M, Krasemann S, Sepulveda-Falla D, Lagares D, Pedron J, Bushweller JH, Liu P, Arboleda-Velasquez JF, Kim LA. Targeting RUNX1 prevents pulmonary fibrosis and reduces expression of SARS-CoV-2 host mediators. Am J Pathol. 2021;S0002-9440(21)00161-9. DOI: 10.1016/j.ajpath.2021.04.006

[92] Della Latta V, Cecchettini A, Del Ry S, Morales MA. Bleomycin in the setting of lung fibrosis induction: From biological mechanisms to counteractions. Pharmacol Res. 2015;97:122-130. DOI: 10.1016/j.phrs.2015.04.012

[93] Hu X, Fernandes J, Jones DP, Go YM. Cadmium stimulates myofibroblast differentiation and mouse lung fibrosis. Toxicology. 2017;383:50-56. DOI: 10.1016/j.tox.2017.03.018

[94] Liu T, De Los Santos FG, Phan SH. The Bleomycin Model of Pulmonary Fibrosis. Methods Mol Biol. 2017;1627:27-42. DOI: 10.1007/978-1-4939-7113-8_2

[95] Yan Z, Zhang X, Li C, Jiao S, Dong W. Association between consumption of soy and risk of cardiovascular disease: a meta-analysis of observational studies. Eur J Prev Cardiol 2017, 24(7): 735-747. DOI: 10.1177/2047487316686441. Epub 2017 Jan 9.

[96] Erdman TW Jr. Soy protein and cardiovascular disease. A statement for healthcare professionals from the nutricion committee of the AHA. Circulation 2000;102: 2555-2559. DOI:.org/10.1161/01.CIR.102.20.2555

[97] Ma L, Liu G, Ding M, Zong G, Hu FB, Willett WC, Rimm EB, Manson JE, Sun Q. Isoflavone intake and the risk of coronary heart disease in US men and women. Results from 3 prospective cohort studies. Circulation 2020;141:1127-1137 DOI:.org/10.1161/CIRCULATIONAHA.119.041306

[98] Ramdath DD, Padhi EM, Sarfaraz S, Renwick S, Duncan AM. Beyond the cholesterol-lowering effect of soy protein: a review of the effects of dietary soy and its constituents on risk factors for cardiovascular disease. Nutrients. 2017;9(4):324. DOI: 10.3390/nu9040324.

[99] Ferramola ML, Perez Diaz MFF, Honoré SM, Sánchez SS, Antón RI, Anzulovich AC, Giménez MS. Cadmium-induced oxidative stress and histological damage in the myocardium. Effect of a soy-based diet. Toxicol Appl Pharmacol. 2012;265(3): 380-389. DOI: 10.1016/j.taap.2012.09.009.

[100] Perez Diaz MFF, Plateo Pignatari MG, Filippa VP, Mohamed FH, Marchevsky EJ, Gimenez MS, Ramirez DC. J Trace Elem Med Biol 2019;52: 239-246 DOI: 10.1016/j.jtemb.2019.01.007.

[101] Essawy AE, Abdou HM, Ibrahim HM, Bouthahab NM. Soybean isoflavone ameliorates cognitive impairment, neuroinflammation, and amyloid β accumulation in a rat model of Alzheimer's disease. Environ Sci Pollut Res Int. 2019;26(25):26060-26070. DOI: 10.1007/s11356-019-05862-z.

[102] Shimizu A, Mitani T, Tanaka S, Fujii H, Maebuchi M, Amiya Y, Tanaka M, Matsui T, Nakamura S, Katayama S. Soybean-Derived Glycine-Arginine Dipeptide Administration Promotes Neurotrophic Factor Expression in the Mouse Brain. J Agric Food Chem. 2018;66(30):7935-7941. DOI: 10.1021/acs.jafc.8b01581

[103] Ariyani W, Miyazaki W, Amano I, Hanamura K, Shirao T, Koibuchi N. Soy Isoflavones Accelerate Glial Cell Migration via GPER-Mediated Signal Transduction Pathway. Front Endocrinol. 2020;11:554941. DOI: 10.3389/fendo.2020.554941

[104] Molteni A, Brizio-Molteni L, Persky V. In vitro hormonal effects of soybean isoflavones. J Nutr. 1995;125(3 Suppl):751S-756S. DOI: 10.1093/jn/125.3_Suppl.751S.

[105] Miksicek RJ. Estrogenic flavonoids: structural requirements for biological activity. Proc Soc Exp Biol Med. 1995;208(1):44-50. DOI: 10.3181/00379727-208-43830.

[106] Adlercreutz H, Mazur W. Phyto-oestrogens and Western diseases. Ann Med. 1997 Apr;29(2):95-120. DOI: 10.3109/07853899709113696.

[107] Pan Y, Anthony M, Watson S, Clarkson TB. Soy phytoestrogens improve radial arm maze performance in ovariectomized retired breeder rats and do not attenuate benefits of 17h-estradiol treatment. Menopause. 2000;7:230 – 5. DOI: 10.1097/00042192-200007040-00004.

[108] File SE, Jarrett N, Fluck E, Duffy R, Casey K, Wiseman H. Eating soya improves human memory. Psychopharmacology. 2001;157:430 – 6. DOI: 10.1007/s002130100845.

[109] Kritz-Silverstein D, Von Muhlen D, Barrett-Connor E, Bressel MA. Isoflavones and cognitive function in older women: the SOy and Postmenopausal Health In Aging (SOPHIA) Study. Menopause. 2003;10:196 – 202.

[110] Duffy R, Wiseman H, File SE. Improved cognitive function in postmenopausal women after 12 weeks of consumption of a soya extract containing isoflavones. Pharmacol Biochem Behav. 2003;75:721 – 9. DOI: 10.1016/s0091-3057(03)00116-3.

[111] Kreijkamp-Kaspers S, Kok L, Grobbee DE, de Haan EH, Aleman A, Lampe JW. Effect of soy protein containing isoflavones on cognitive function, bone mineral density, and plasma lipids in postmenopausal women: a randomized controlled trial. JAMA. 2004;292:65 – 74. DOI: 10.1001/jama.292.1.65.

[112] White LR, Petrovitch H, Ross GW, Masaki K, Hardman J, Nelson J, et al. Brain aging and midlife tofu consumption. J Am Coll Nutr.

2000;19:242 – 55. DOI: 10.1080/07315724.2000.10718923.

[113] Lund TD, West TW, Tian LY, Bu LH, Simmons DL, Setchell KD. Visual spatial memory is enhanced in female rats (but inhibited in males) by dietary soy phytoestrogens. BMC Neurosci 2001;2:20. DOI: 10.1186/1471-2202-2-20.

[114] Lee YB, Lee HJ, Won MH, Hwang IK, Kang TC, Lee JY. Soy isoflavones improve spatial delayed matching-to-place performance and reduce cholinergic neuron loss in elderly male rats. J Nutr. 2004;134:1827 – 31. DOI:10.1093/jn/134.7.1827

[115] Bregman BS, Coumans JV, Dai HN, Kuhn PL, Lynskey J, McAtee M, Sandhu F. Transplants and neurotrophic factors increase regeneration and recovery of function after spinal cord injury. Prog Brain Res. 2002;137:257-273. doi: 10.1016/s0079-6123(02)37020-1

[116] Gurd JW. Protein tyrosine phosphorylation: implications for synaptic function. Neurochem Int. 1997;31:635 – 49. DOI:10.1016/S0197-0186(97)00022-3

[117] Phillis JW, Song D, O'Regan MH. Inhibition of tyrosine phosphorylation attenuates amino acid neurotransmitter release from the ischemic/reperfused rat cerebral cortex. Neurosci Lett. 1996;207: 151-154. DOI: 10.1016/0304-3940(96)12521-0.

[118] O'Dell TJ, Kandel ER, Grant SG. Long-term potentiation in the hippocampus is blocked by tyrosine kinase inhibitors. Nature. 1991;353:558 – 60. DOI: 10.1038/353558a0.

[119] Akiyama T, Ishida J, Nakagawa S, Ogawara H, Watanabe S, Itoh N. Genistein, a specific inhibitor of tyrosine-specific protein kinases. J Biol Chem. 1987;262:5592 – 5595.

[120] Linford NJ, Dorsa DM. 17h-Estradiol and the phytoestrogen genistein attenuate neuronal apoptosis induced by the endoplasmic reticulum calcium-ATPase inhibitor thapsigargin. Steroids 2002;67: 1029 – 1040. DOI: 10.1016/s0039-128x (02)00062-4.

[121] Linford NJ, Yang Y, Cook DG, Dorsa DM. Neuronal apoptosis resulting from high doses of the isoflavone genistein: role for calcium and p42/44 mitogen-activated protein kinase. J Pharmacol Exp Ther. 2001; 299:67 – 75.

[122] Bang OY, Hong HS, Kim DH, Kim H, Boo JH, Huh K, Mook-Jung I. Neuroprotective effect of genistein against beta amyloid-induced neurotoxicity. Neurobiol Dis. 2004;16: 21 – 8. DOI: 10.1016/j.nbd.2003.12.017

[123] Deol P, Kozlova E, Valdez M, Ho C, Yang EW, Richardson H, Gonzalez G, Truong E, Reid J, Valdez J, Deans JR, Martinez-Lomeli J, Evans JR, Jiang T, Sladek FM, Curras-Collazo MC. Dysregulation of Hypothalamic Gene Expression and the Oxytocinergic System by Soybean Oil Diets in Male Mice. Endocrinology. 2020;161(2):bqz044. DOI: 10.1210/endocr/bqz044.

[124] Antonio MT, Corredor L, Leret ML. Study of the activity of several brain enzymes like markers of the neurotoxicity induced by perinatal exposure to lead and/or cadmium. Toxicology Letters. 2003;143(3):331-340. DOI: 10.1016/s0378-4274(03)00194-2.

[125] Gabbiani G, Baic D, Déziel C. Toxicity of cadmium for the central nervous system. Exp Neurol. 1967;18(2):154-160. DOI: 10.1016/0014-4886(67)90037-4.

[126] Shukla A, Shukla GS, Srimal RC. Cadmium-induced alterations in blood-brain barrier permeability and its possible correlation with decreased

microvessel antioxidant potential in rat. Hum Exp Toxicol. 1996;15(5):400-405. DOI: 10.1177/096032719601500507.

[127] Mendez-Armenta M, Rios C. Cadmium neurotoxicity. Environmental Toxicology and Pharmacology. 2007;23(3):350-358. DOI: 10.1016/j.etap.2006.11.009

[128] Yokel, RA. Blood-brain barrier flux of aluminum, manganese, iron and other metals suspected to contribute to metal-induced neurodegeneration. Journal of Alzheimer's Disease. 2006;10(2-3), 223-253. DOI: 10.3233/jad-2006-102-309.

[129] Karri V, Schumacher M, Kumar V. Heavy metals (Pb, Cd, As and MeHg) as risk factors for cognitive dysfunction: a general review of metal mixture mechanism in brain. Environ Toxicol Pharm. 2016;48:203-213. DOI: 10.1016/j.etap.2016.09.016

[130] Monroe RK, Halvorsen SW. Cadmium blocks receptor-mediated Jak/STAT signaling in neurons by oxidative stress. Free Radic Biol Med. 2006;41(3):493-502. DOI: 10.1016/j.freeradbiomed.2006.04.023.

[131] Unsal C, Kanter M, Aktas C, Erboga M. Role of quercetin in cadmium-induced oxidative stress, neuronal damage, and apoptosis in rats. Toxicol Ind Health. 2015;31(12):1106-1115. DOI: 10.1177/0748233713486960

[132] Coyle JT, Puttfarcken P. Oxidative stress, glutamate, and neurodegenerative disorders. Science. 1993;262(5134):689-695. DOI: 10.1126/science.7901908.

[133] Cadet JL. Free radical mechanisms in the central nervous system: an overview. Int J Neurosci. 1988;40(1-2):13-18. DOI: 10.3109/00207458808985722.

[134] Abdalla FH, Schmatz R, Cardoso AM, Carvalho FB, Baldissarelli J, de Oliveira JS, Rosa MM, Gonçalves Nunes MA, Rubin MA, da Cruz IB, Barbisan F, Dressler VL, Pereira LB, Schetinger MR, Morsch VM, Gonçalves JF, Mazzanti CM. Quercetin protects the impairment of memory and anxiogenic-like behavior in rats exposed to cadmium: Possible involvement of the acetylcholinesterase and Na(+),K(+)-ATPase activities. Physiol Behav. 2014;135:152-167. DOI: 10.1016/j.physbeh.2014.06.008.

[135] Ferreira-Vieira TH, Guimaraes IM, Silva FR, Ribeiro FM. Alzheimer's disease: Targeting the Cholinergic System. Curr Neuropharmacol. 2016;14(1):101-115. DOI: 10.2174/1570159x13666150716165726

[136] Snelson M, Mamo JCL, Lam V, Giles C, Takechi R. Differential Effects of High-Protein Diets Derived from Soy and Casein on Blood-Brain Barrier Integrity in Wild-type Mice. Front Nutr. 2017;4:35. DOI:10.3389/fnut.2017.00035

[137] Ditscheid B, Funfstuck R, Busch M, Schubert R, Gerth J, Jahreis G. Effect of l-methionine supplementation on plasma homocysteine and other free amino acids: a placebo-controlled double-blind cross-over study. Eur J Clin Nutr. 2005;59:768-775. DOI: 10.1038/sj.ejcn.1602138

[138] Jakubowski H. Pathophysiological consequences of homocysteine excess. J Nutr. 2006;136:1741S–1749S. DOI: 10.1093/jn/136.6.1741S.

[139] Tyagi N, Kumar M, Pushpakumar SB, Lominadze D, Moshal KS, Sen U. Homocysteine attenuates blood brain barrier function by inducing oxidative stress and the junctional proteins. FASEB J. 2008 22:734.737. DOI: 10.3389/fnut.2018.00067

[140] Kalani A, Kamat PK, Givvimani S, Brown K, Metreveli N, Tyagi SC, et al. Nutri-epigenetics ameliorates

blood-brain barrier damage and neurodegeneration in hyperhomocysteinemia: role of folic acid. J Mol Neurosci. 2014;52:202-215. DOI: 10.1007/s12031-013-0122-5.

[141] Halder S, Kar R, Chakraborty S, Bhattacharya SK, Mediratta PK, Banerjee BD. Cadmium level in brain correlates with memory impairment in F1 and F2 generation mice: improvement with quercetin. Environ Sci Pollut Res Int. 2019;26(10):9632-9639. DOI: 10.1007/s11356-019-04283-2.

[142] Liu J, Chang SK, Wiesenborn D. Antioxidant properties of soybean isoflavone extract and tofu in vitro and in vivo. J Agric Food Chem. 2005;53(6):2333-2340. DOI: 10.1021/jf048552e

[143] Bjorkhem I, Meaney S, Fogelman AM. Brain cholesterol: Long secret life behind a barrier. Thromb Vasc Biol. 2004;24:806-815. DOI: 10.1161/01.ATV.0000120374.59826.1b.

[144] Pitas RE, Boyles JK, Lee SH, Hui D, Weisgraber KH. Lipoproteins and their receptors in the central nervous system. Characterization of the lipoproteins in cerebrospinal fluid and identification of apolipoprotein B,E (LDL) receptors in the brain. The Journal of Biological Chemistry. 1987;262(29):14352-14360.

[145] Modi HR, Katyare SS. Cadmium exposure-induced alterations in the lipid/phospholipids composition of rat brain microsomes and mitochondria. Neurosci Lett. 2009;464(2):108-112. DOI: 10.1016/j.neulet.2009.08.003.

[146] Kroemer G. The proto-oncogene Bcl-2 and its role in regulating. Nat Med. 1997;3(6):614-620. DOI: 10.1038/nm0697-614

[147] Peng Y, Jiang B, Wu H, Dai R, Tan L. Effects of genistein on neuronal apoptosis, and expression of Bcl-2 and Bax proteins in the hippocampus of ovariectomized rats. Neural Regen Res. 2012;7(36):2874-2881. DOI: 10.21203/rs.3.rs-258892/v1

[148] Mahmoud FY, El-Hakim AH, El Deen Amer AS, Fidal MW. Effect of exposure to cadmium on the hippocampus in adult albino rat and the possible role of L-carnitine. J Curr Med Res Pract. 2019;4:240-250. DOI: 10.4103/jcmrp.jcmrp_60_18

[149] Farahmand, M., Tadjali, M. Effect of Persian sage (Salvia Rhytidia) extract on histomorphometric changes of cerebellum following ischemia reperfusion injuries in rat. Comparative Clinical Pathology, 2014;24(5), 1129-1137. DOI: 10.1007/s00580-014-2046-z

Chapter 9

Soybean and Other Legume Proteins Exhibit Beneficial Physiological Effects on Metabolic Syndrome and Inflammatory-Related Disorders

Mitsutaka Kohno

Abstract

There is currently a trend in Western countries to increase the intake of plant proteins. In this chapter, the author explains that this is due to the beneficial physiological functions of plant proteins, based on the latest literature review and our own research results. Among plant proteins, soy protein has been reported to have many beneficial effects on the improvement and prevention of metabolic syndrome. This chapter outlines the excellent effects of soy protein on renal function [improvement of early symptoms of diabetic nephropathy], which is closely related to metabolic syndrome, and the effects of combining these effects as complementary medicine. In addition, recent findings about the anti-inflammatory and immune activation effects of soy protein as hydrolyzed peptides are outlined. A brief introduction of the recent results of other legume-derived proteins that have replaced soy proteins are also explained. By further deepening our understanding of the superior physiological functions of plant proteins, it is hoped that their use expands even further.

Keywords: soy protein, metabolic syndrome, chronic kidney disease, inflammatory disorder, pea protein, lupin protein, mung bean protein

1. Introduction

Protein is not only significant as an energy source, but also as a component of the body, such as muscle and connective tissue, and as a physiological function substance, such as enzymes, hormones, and immune antibodies.

On the other hand, the problem of food shortage (in particular, protein) due to global population growth is becoming increasingly serious. Because of the economic development of emerging countries, people who used to consume energy from "carbohydrates" such as bread and rice are now tending to consume "proteins" such as meat and seafood as a luxury item, and there are concerns about a shortage of protein supply on a global scale. Under these circumstances, the effective use of plant proteins as a protein source has been attracting attention. Plant proteins have been considered to be less adaptable to human tastes in terms of flavor and physical properties than animal proteins, but recent superior food processing technologies have led to the marketing of "delicious" plant protein foods that are at the same level as animal protein foods.

It has been reported that plant proteins, especially soy proteins, have beneficial functions to improve and prevent lifestyle-related diseases that cannot be overcome by animal proteins, which are currently prevalent all over the world. The US Food and Drug Administration [FDA] has approved the health claim for food labelling that the consumption of 25 g of soy protein per day reduces the risk of heart disease [1]. In Japan, the Consumer Affairs Agency [formerly the Ministry of Health, Labour and Welfare] has allowed the health labelling of soy protein as a food for specified health use "to people who are concerned about cholesterol levels" [FOSHU].

The beneficial physiological effects of soy protein are presumed to be due to anti-inflammatory properties. The anti-inflammatory effect of soy protein is enhanced by its processing into peptides. Indeed, it has been reported that soy peptides suppress muscle inflammation pain relief in rheumatoid arthritis and ameliorate inflammatory bowel disease.

Recently, it is being reported that not only soy proteins, but also some legume-derived proteins have excellent physiological effects that are similar to, or even absent from, soy proteins. In this chapter, the author 1] introduces the beneficial physiological effects of soy protein for MetS, CKD and inflammation; 2] reports that these effects acted complementarily when used in combination with drugs; and 3] suggests other legume-derived proteins as alternatives to soy protein as novel proteins from legumes with beneficial physiological functions.

By understanding these findings, it is hoped that plant proteins will be used more actively to contribute to the improvement of human health, as well as their value as protein nutrition, which is in short supply worldwide.

2. Soy protein and peptides

2.1 Soy protein for metabolic syndrome (MetS)

The concept of MetS has been proposed by several committees. The first formalized concept of MetS was proposed by a consultation group for the definition of diabetes for the World Health Organization (WHO); it was determined to have a high-risk status with multiple risk factors for cardiovascular disease. This group emphasized insulin resistance as the major underlying factor [2]. In 2001, a definition for MetS was devised by the National Cholesterol Education Program (NCEP) Adult Treatment Panel III (ATP III) [3]. The American Heart Association and the National Heart Lung and Blood Institute updated this definition in 2005 [4]. This updated definition is one of the most widely used criteria for MetS. The International Diabetes Foundation (IDF) published new criteria for MetS [5] in 2005. Although it includes the same general criteria as the other definitions, it requires that obesity, but not necessarily insulin resistance, be present. Although visceral obesity is now recognized as an important factor, the IDF definition has been criticized for its emphasis on obesity, rather than insulin resistance, in pathophysiology [6].

In Japan, in 2006, MetS was defined as a multiple risk factor clustering syndrome caused by visceral fat accumulation and insulin resistance that accompanies this accumulation [7]. In the MetS stage, it is advocated that lifestyle intervention to reduce visceral adiposity should be given priority over drug treatment. Subjects with multiple risk factor syndrome were diagnosed with MetS if their visceral fat areas determined by CT scan were over 100 cm^2.

The Japanese Committee for the Definition and Diagnosis of MetS aimed to select subjects with multiple risk factors in which lifestyle modification to reduce visceral adiposity has priority over drug treatment [8]. For this purpose, the Japanese government started a new health policy that provides a specific health check-up followed by specific counseling for subjects diagnosed with MetS according to the Japanese criteria from 2008.

Title	Number of articles	Number of total subjects	Outcome [significant difference]	Reference
Meta-analysis of the effects of soy protein intake on serum lipids.	38	743	Total-C↓, LDL-C↓, TG↓	[9]
Meta-analysis of the effects of soy protein containing isoflavones on the lipid profiles.	23	1,381	Total-C↓, LDL-C↓, TG↓, HDL-C↑	[10]
A meta-analysis of the effect of soy protein supplementation on serum lipids.	41	1,756	Total-C↓, LDL-C↓, TG↓, HDL-C↑	[11]
Hypocholesterolaemic effects of soya proteins: result of recent studies are predictable from Anderson meta-analysis data.	27	923	Total-C↓, LDL-C↓	[12]
Soy protein effects on serum lipoproteins: a quality assessment and meta-analysis of randomized, controlled studies.	43	1,946	LDL-C↓, TG↓, HDL-C↑	[13]
Effect of Plant Protein on Blood Lipids: A Systematic Review and Meta-Analysis of Randomized Controlled Trials.	112 (non-soy; 18)	5,774 (non-soy; 599)	LDL-C↓, Non-HDL-C↓, Apo-B↓	[14]
A Meta-Analysis of 46 Studies Identified by the FDA Demonstrates that Soy Protein Decreases Circulating LDL and Total Cholesterol Concentrations in Adults.	43	2,607	Total-C↓, LDL-C↓	[15]
The effects of isolated soy protein, isolated soy isoflavones and soy protein containing isoflavones on serum lipids in postmenopausal women: A systematic review and meta-analysis.	46		Total-C↓, LDL-C↓, TG↓, HDL-C↓, hypercholesterolemic subjects' Apo-A1↓	[16]

Note: ↓ and ↑ signs represent decrease and increase, respectively, after supplement of active compounds. Total-cholesterol (Total-C); low-density lipoprotein cholesterol (LDL-C); triglyceride (TG); high-density lipoprotein cholesterol (HDL-C); non-high-density lipoprotein cholesterol (non-HDL-C); apo-lipoprotein-B (Apo-B); apo-lipoprotein-AI (Apo-AI).

Table 1.
Meta-analysis on improving lipid metabolism in soy protein.

Soy protein exerts not only conventional nutritional value but also beneficial effects on human health. Many randomized controlled trials (RCTs) have assessed the effects of soy products on serum lipids. Systematic reviews and meta-analyses have reported improvements in lipid metabolism (**Table 1**) [9–16].

Soy protein isolate [SPI] is composed of three major components, glycinin [approx. 40%], β-conglycinin [approx. 20%], and lipophilic proteins (approx. 40%) [17]. Glycinin and β-conglycinin are storage proteins in soy, and lipophilic proteins consist primarily of membrane proteins. Among these components, β-conglycinin has the function of lowering serum triglycerides preferentially over serum cholesterol [18]. Digestive decomposition products of β-conglycinin were reported that lowering the activity of fatty acid synthase and increasing the activities of β-oxidation enzymes, and the fecal excretion of TG was high in β-conglycinin-fed mice and rats [19, 20]. Therefore, in the calculation based on the recommendation by the FDA, the same effect can be expected with 5 g of β-conglycinin. In clinical study, daily consumption of 5 g of β-conglycinin per subject significantly lowered serum TG concentrations, and the apo B and VLDL-TG concentrations were significantly decreased [21]. Hence, β-conglycinin consumption may specifically affect TG metabolism. In addition, the intake of 5 g of β-conglycinin per day decreased the body fat ratio and visceral fat [21, 22]. Additionally, serum adiponectin significantly increased with the consumption of β-conglycinin, and serum free fatty acids in the β-conglycinin group were significantly decreased. Tachibana et al. showed that β-conglycinin improves insulin sensitivity in rats [23]. β-conglycinin might be an important food component for the prevention and/or amelioration of visceral fat syndrome, which is also called MetS (**Table 2**) [21, 22, 24–27].

2.2 Soy protein for chronic kidney disease

Chronic kidney disease [CKD] is a major public health burden, with a global prevalence of ~11% in the general adult population [28]. If left untreated, CKD slowly progresses to end-stage renal disease, which requires dialysis or kidney transplant. Worldwide, a 31.7% increase in CKD mortality was observed over the last decade [29]. Effective interventions to prevent and delay the progression of CKD are well recognized. Prevention should start at the government level with the institution of multisectoral polices supporting sustainable development goals [SDGs] and ensuring safe and healthy environments.

CKD is bidirectionally associated with MetS and cardiovascular diseases [CVDs] [30, 31], and diabetic nephropathy [DN] is a complication of diabetes [32]. Moreover, it has been reported that 40% of patients undergoing dialysis are doing so because of DN [33], and approximately 50% of type II diabetes patients exhibit urinary albumin disease, which is an early stage of DN [34].

For CKD prevention, it is important to gain insight about commonly consumed foods and beverages in relation to kidney function. A report has been published in which PubMed was comprehensively searched for papers published until August 2019 describing prospective cohort studies and was supplemented by manual searches of reference lists from appropriate studies [35]. In this report, there was convincing evidence that a healthy dietary pattern may lower CKD risk. Red (processed) meat, poultry, fish, dairy, vegetables, legumes, nuts, and fruits were recommended foods for CKD patients. Dietary patterns were recommended adherence to the Dietary Approach to Stop Hypertension (DASH) diet, Mediterranean diet, and other healthy dietary patterns. As unhealthy diets, high-fat and high-sugar diets and high-acid-loaded diets were pointed out. In the Atherosclerosis Risk in Communities [ARIC] study of ~12,000 US participants with 23 years of follow-up,

Study title	Design of study	Number of subjects	Duration of study	Dose of β-conglycinin	Outcome [significant difference]	Reference
Decrease in serum triacylglycerol and visceral fat mediated by dietary soybean β-conglycinin*1.	Randomized, Double-Blind, Placebo-Controlled Study	Test1;138 Test2:102	Test1:12-wk Test2:20-wk*1	4.4 g/day	Test1; TG↓, Apo-B↓, VLDL-TG↓ Test2; Visceral fat↓	[21]
Effects of soybean beta-conglycinin on body fat ratio and serum lipid levels in healthy volunteers of female university students.	Randomized, single-blinded crossover design	41	8-wk	4.4 g/day	Body fat ratio↓	[22]
Serum triacylglycerol-lowering effect of soybean β-conglycinin in mildly hypertriacylglycerolemic individuals.	Randomized, Double-Blind, Placebo-Controlled Study	68	12-wk	2.3 g/day	TG↓, HDL-C↑, Apo C-II↓	[24]
Serum lipid-improving effect of soyabean β-conglycinin in hyperlipidaemic menopausal women.	Randomized, Double-Blind, Placebo-Controlled Study	100	12-wk	2.3 g/4.6 g	TG↓, LDL-C↓, Apo-B↓, NEFA↓	[25]
Improvement of Triglyceride Levels through the Intake of Enriched-β-Conglycinin Soybean (Nanahomare).	Randomized, Double-Blind, Placebo-Controlled Study	134	12-wk	38.8 g/week	TG↓	[26]
Effects of beta-conglycinin intake on circulating FGF21 levels and brown adipose tissue activity in Japanese young men.	Single-blinded randomized crossover trial	21	2-wk	9.2 g/day	FGF21↓, BAT act↑	[27]

*Note: *1; Test 1 is an examination of the serum triglyceride level and Test 2 is a measure of visceral fat by means of CT scanning. ↓ and ↑ signs represent decrease and increase, respectively, after supplement of β-conglycinin. Triglyceride (TG); apo-lipoprotein-B (Apo-B); very low-density lipoprotein triglyceride (VLDL-TG); high-density lipoprotein cholesterol (HDL-C); apo-lipoprotein-CII (Apo-CII); low-density lipoprotein cholesterol (LDL-C); non-esterified fatty acid (NEFA); fibroblast growth factor 21 (FGF21); brown adipo tissue activity (BAT act).*

Table 2.
Clinical studies for lipid metabolism improvements of β-conglycinin.

consumption of legumes was significantly associated with lower risks of CKD, with an HR of 0.83 [95% CI, 0.72; 0.95] for high versus low intakes [36]. Soy protein, which is representative of legumes, has been reported to suppress the progression of DN [37, 38]. The effects of soy protein on DN/CKD in clinical trials are summarized in **Table 3** [39–43].

Kidney disease patients are carefully monitored for protein intake, and restricted protein intake according to the progression of their condition by doctors and nutritionists. However, there are some reports showing that mild protein restriction does not suppress the progression of kidney disease [44–46]. Therefore, it is necessary to consider not only the quantity but also the quality of protein. Legumes, including soy protein, can be regarded as very significant proteins to help treat nephropathy.

Study title	Design of study	Number of subjects	Duration of study	Outcome [significant difference]	Reference
Soy protein intake, cardiorenal indices, and c-reactive protein in type 2 diabetes with nephropathy.	Longitudinal randomized clinical trial study	41	4-y	FPG↓, Total-C↓, LDL-C↓, TG↓, CRP↓, Proteinuria↓, Urinary creatinine↓	[39]
The effects of soy protein on chronic kidney disease: a meta-analysis of randomized controlled trials.	Meta-analysis [9 studies]	Total 197	6-wk~4-y	Serum creatinine↓, Phosphorus↓, TG↓	[40]
Soy-based renoprotection.	Single arm intervention (3 studies) Placebo-controlled chronic intervention [22 studies]	Total 634	4-wk~6-mo	Total-C↓, LDL-C↓, Urinary creatinine↓, [Urinary albumin↓]	[41]
Effects of soy protein containing isoflavones in patients with chronic kidney disease: A systematic review and meta-analysis.	Meta-analysis [12 studies]	Total 280		Serum creatine↓, Phosphorus↓, CRP↓, Proteinuria↓, BUN↓ [in predialysis subgroup]	[42]
Soy Protein and Chronic Kidney Disease: An Updated Review.	RCT (3 studies), DBRCT, CRCT, LRCT Total 6 studies	Total 335	1~24-wk	Urinary urea nitrogen↓, Proteinuria↓, Blood sodium↓, Serum Creatinine↓	[43]

Note: ↓ sign represents decrease, after supplement of soy protein. Fasting plasma glucose [FPG]; total-cholesterol (Total-C); low-density lipoprotein cholesterol (LDL-C); C-reactive protein [CRP], blood urea nitrogen (BUN).

Table 3.
Summary of clinical studies by soy protein for CKD.

2.3 Anti-inflammatory roles of soy protein and peptides

Inflammation can occur when infectious microorganisms such as bacteria, viruses, and fungi invade the body and circulate in the blood, and/or when they enter certain tissues [47, 48]. Inflammation can also occur during the course of pathologies such as tissue damage, cell death, cancer, ischemia, and degeneration [49–51].

There are reports of the anti-inflammatory effects of soy protein and its hydrolysate peptides [52]. Among them, lunasin is considered one of the most studied bioactive peptides. Since its discovery in soybean twenty years ago, many researchers around the world have focused their studies on demonstrating the chemo preventive and chemotherapeutic activity of lunasin [53–55]. Lunasin is a 44 amino acid peptide isolated from soy that has three domains implicated in anticancer activity: an RGD motif [Arg-Gly-Asp], a helical domain with a sequence conserved in chromatin binding proteins [Glu-Lys-His-Ile-Met-Glu-Lys-Ile], and a poly-aspartic acid tail [56]. Lunacin has been reported to have unique antioxidant, anti-inflammatory, and anti-cancer properties, and to play an important role in the regulation of cholesterol biosynthesis in the body [57]. Lunacin has potential as a dietary supplement by its high bioavailability and thermal stability.

Trypsin digests of soy proteins revealed that the sequence MITLAIPVNKPGR was able to stimulate phagocytosis in leukocytes. This peptide derived from β-conglycinin was named "Soymetide". The Met at its N-terminus was essential for its activity [58]. Four residues of the C-terminal residues of Soymetide-13 could be removed to form Soymetide-9 [MITLAIPVN], which had the highest activity. In these 9 residues [Soymetide-9], Soymetide-4 [MITL] is the minimal sequence required for its activity [58].

Soy protein with or without isoflavones was shown to reduce oxidative stress and have anti-inflammatory properties by inhibiting nuclear factor-kappa B [NF-κB] and blocking the secretion of pro-inflammatory cytokines in model rats and mouse. In clinical study by subjects with end-stage renal disease and healthy women over 70 years of age, their oxidative stress and inflammatory symptom were reduced [59]. The bioactive peptides RQRK and VIK were produced by digestion with pepsin and pancreatin from soy milk. These peptides inhibited lipopolysaccharide-induced inflammation in murine macrophages and the production of nitric oxide, interleukin [I])-1, nitric oxide synthase, and cyclooxygenase-2 [60].

Inflammatory bowel disease [IBD] is an intractable disease that causes inflammation of the gastrointestinal tract. Ulcerative colitis and Crohn's disease are the two major pathologies of IBD [61]. Ulcerative colitis is a non-specific inflammatory disorder that causes ulcers and erosion, primarily in the colonic mucosa. Young et al. revealed that soy peptides were effective in preventing dextran sulphate sodium[(DSS)-induced colitis in pigs [62]. The soy-derived tripeptide Val-Pro-Tyr [VPY] has been reported that anti-inflammatory effects in Caco-2 and THP-1 macrophages and inhibition of the secretion of IL-8 and TNF-α in a DSS-induced colitis model mouse [63]. They suggested that tripeptide VPY from soy peptides may be promising for the treatment of IBD.

Insulin resistance and diabetes has revealed to relate closely between nutrient excess and activation of the innate immune system in most organs pertinent to energy homeostasis by the research for a mechanism linking the pathogenesis of obesity over the past two decades [64–66]. Inflammation has been revealed to occur as a consequence of obesity, and to play a causative role in generating insulin resistance, defective insulin secretion [i.e., MetS), and disruption of other aspects of energy homeostasis by recent many studies. It has been reported that the suppressive effect of soy protein on the progression of CKD/DN, which is highly related to MetS, is also exerted by the anti-inflammatory effect in renal tubules [67]. From such a close relationship between MetS and inflammation, it is easy to predict that the beneficial effect of soy protein on MetS may be due to its anti-inflammatory effect.

3. Complementary effects of soy protein/peptide in combination with drugs

3.1 Effect of combined use with anti-hyperlipidaemic drugs

The mechanism by which soy protein lowers cholesterol differs from that of statins and fibrates. Soy protein lowers serum cholesterol levels by acting as a bile acid sequestrant, which binds bile in the gastrointestinal tract to prevent its reabsorption by performing the same anion exchange reaction as the resin cholestyramine [68, 69].

Statins and fibrates are drugs developed to improve blood lipid levels. Statins are known as the most efficient agents for reducing plasma cholesterol. Statins target hepatocytes and inhibit 3-hydroxy-3-methylglutaryl-coenzyme A [HMG-CoA] reductase in cholesterol metabolism. Accordingly, statin and soy protein are expected to act additively or synergistically to decrease cholesterol levels. There are known serious side effects from statins, including muscle symptoms, rhabdomyolysis [secondary renal failure due to destruction of specific muscle tissue], peripheral neuropathy, myopathy, liver dysfunction, and thrombocytopenia [70–73]. Rhabdomyolysis often induces sudden kidney failure [74]. Fibrates, which are antagonists of peroxisome proliferator-activated receptor α [PPARα], are used in adjunct therapy for hypertriglyceridemia and are usually used in combination with statins. As fibrate-related side effects, the slight gastric region discomfort and myopathy [myalgia with increased creatinine phosphokinase] have been reported. In addition, increasing of the gallstones risk has been known by fibrates because of increasing of cholesterol in the bile duct. Use in combination of statins and fibrates is reported to even more raise the risk of rhabdomyolysis. So, combination use of these two agents is contraindicated in principle.

Nabiki et al. examined the effects of SPI on weight loss, markers of diabetes, and parameters of dyslipidaemia in obese diabetic patients by treated with statins and/or fibrates because of high levels of LDL cholesterol and triglycerides [75]. As a result, body weights of these patients decreased significantly by approximately 1 kg and their waist circumferences got thinner significantly by approximately 2 cm. Total cholesterol, triglyceride, LDL cholesterol, apolipoprotein B, and glycated hemoglobin levels of these patients decreased significantly, and HDL cholesterol levels increased significantly. In addition, a lipid metabolism-improving effect was also observed in patients who did not decrease weight. Therefore, it was suggested that the improving effect of lipid metabolism-related factors in these patients was not only due to weight loss but also a direct effect of soy protein. Use of soy protein may help to reduce the drug dose for dyslipidaemia. SPI is recommended for patients with mild dyslipidaemia prior to drug therapy or for maladaptive disease patients, such as those who have side effects from medications.

Combination prescription of fibrates and statins for patients with renal dysfunction and dyslipidaemia is contraindicated. Thus, physicians are unable to adequately treat lipid abnormalities for chronic kidney disease patients. It has been reported that when chronic kidney disease patients with dyslipidaemia ingested β-conglycinin, a major component of soy protein, for 3 months, triglyceride and LDL cholesterol levels improved. Renal function during the consumption period of β-conglycinin showed a tendency to improve despite protein intake [76]. β-conglycinin may help improve lipid abnormalities in patients with renal dysfunction as a complementary medical food material without decreasing kidney function. Moreover, β-conglycinin may improve renal dysfunction as a direct and/or secondary effect of ameliorating lipid abnormalities.

3.2 Concomitant effect with rheumatoid arthritis drug

Rheumatoid arthritis is due to inflammation triggered by an immune response to autoantigens. Many of these patients have swelling and pain due to polyarticular arthritis. Their pain interferes with activities of daily living [ADLs], such as cleaning, washing, dressing, and undressing. These patients are anxious for more comfortable ADLs with reduced pain. The mechanisms of onset of rheumatoid arthritis have been reported in many studies. Based on these results, numerous new therapeutic agents have been developed.

As a specific case of improved inflammation, outpatients with rheumatoid arthritis consumed soy peptide with therapeutic drugs and the levels of IL-6 and IL-1β were significantly lower in the soy peptide group than the placebo group [77]. An increase in blood IL-6 levels is associated with extra-articular symptoms of rheumatoid arthritis, such as general malaise, loss of appetite, weight loss, and a slight fever. The Disease Activity Score 28 [DAS 28, objective assessment of rheumatoid arthritis disease activity by physicians] and the Clinical Disease Activity Index [CDAI, patient's own subjective indicator of rheumatoid arthritis disease activity] were calculated from the degree of ADLs' improvement, the severity of pain, and subjective symptoms recorded by visual analogue scale [VAS]. The DAS 28 score of the peptide group was markedly decreased, and the CDAI of the peptide group was significantly lower than that in the placebo group.

These effects on cytokines were also evident in a cell experiment using articular chondrocytes from patients with rheumatoid arthritis [78]. In this in vitro cell study, treatment with soy peptide significantly suppressed the mRNA levels of MMP-3 and ADAMTS-4 enhanced by IL-1β stimulation. This finding also suggests that soy peptides may prevent the degradation of articular cartilage.

4. Physiological effects of other legume proteins

Soy protein has excellent health benefits, but many soybeans grown in the world are genetically modified organisms [GMOs]. There is no problem with the safety of GMO soybeans. However, from the perspective of security, the use of soy protein in foods tends to be withheld. Recently, the use of pea and lupin proteins instead of soy protein has increased worldwide. Initially, pea protein was a substitute for soybean protein as an ingredient with physical characteristics functions, after that, its beneficial health function has been reported mainly in sports nutrition. Mung bean protein has a structure very similar to that of β-conglycinin. Mung bean protein has been reported to be responsible for the beneficial physiological functions reported for β-conglycinin.

4.1 Pea protein

Field pea [*Pisum sativum* L.] is grown in 84 different countries and constitutes the largest percentage [36%] of total pulse production worldwide [79]. Global pea production has continuously increased over the last 30 years. In 2008, field pea was cultivated on over 10 million hectares worldwide with a total world production of

12.13 million tons [80]. The top 5 countries for pea production are Canada, Russia, China, India and the USA. The global market for pea protein is expected to reach 34.8 million US dollars by 2020 [81]. The physical and chemical properties of pea protein can significantly influence its behaviors in food processing, storage and consumption [82, 83].

Life expectancy continues to increase worldwide. In the United States, adults 65 years of age and older are projected to more than double from 600 million to 1.6 billion worldwide between 2015 and 2050 [84]. Proper body composition, physical fitness, and a healthy appetite have been reported to lead to successful aging with higher performance [85, 86]. Skeletal muscle mass and strength begin to decline at age 30, and the rate of these losses accelerates at age 60 [87]. Protein ingestion strongly increases muscle protein synthesis rates [88]. Amino acids serve as precursors for de novo muscle protein synthesis and can act as strong signaling molecules activating translation initiation via the mechanistic/mammalian target of rapamycin complex-1 (mTORC1) pathway [89]. It was shown that BCAA ingestion increases myofibrillar protein synthesis rates during recovery from exercise only in young males [90]. Whey protein isolate [WPI] was used as the animal protein source because of its high concentration of BCAAs and its ability to increase satiety in response to a mixed meal [91]. While whey protein supplementation is known to enhance adaptations to resistance training, not all athletes are able or willing to consume whey or animal proteins. Vegetarian athletes who want to stick to a plant-based diet or those with restrictions on other animal foods often rely on other plant proteins as an equivalent alternative to whey protein [92]. Self-identify as vegetarian in just over 5% of U.S. adults aged 18–34 years and self-awareness as vegan in more than half of these respondents are reported in a 2016 Harris Poll conducted by the Vegetarian Resource Group [93]. Meat Free Mondays' movement and an upsurge of plant-based protein food products in the marketplace strongly reflect the recent acceptance of these lifestyles [94].

Field pea contains a well-balanced amino acid profile [95]. Because of its availability, low cost, nutritional value and health benefits, pea protein has been widely used as a substitute for soybean or animal proteins in various functional applications [96–99]. Pea protein can also be used as a nutritional supplement for sports and exercises. Pea protein is an excellent source of BCAAs and has high and balanced contents of leucine, isoleucine and valine. Indeed, there are reports that pea protein is as useful as whey protein in sports nutrition (**Table 4**) [100–103].

In the future, pea protein is expected to be widely used as a sports nutritional supplement as well as a physical and functional ingredient in place of soybean protein.

4.2 Lupin protein

Lupin (*Lupinus* L.) is an ancient pulse "bean" crop, and in the new genus of modern agriculture, the lupin seeds have great potential for high-protein food, animal feed, food potential, soil fertility improvement, plants as cover crops, crop residues as stable feed, and soil improvement [104, 105]. Lupin is well known for its ability to fix nitrogen and grow on infertile soils, and is further known to be valuable in terms of cropping rotations during the growing season in agriculture with cereals, hay, oilseeds, beans of other legumes, and disease break crops for pasture [104, 106]. Wild indigenous lupins have bitter alkaloids. All modern species of *L. angustifolius* have total alkaloid levels in seeds of up to 200 mg/kg [0.02%] or less, which is 100 times lower than the seed alkaloid levels of nearly wild types. Compared to almost all food crops, lupins have only recently become of interest in modern crop breeding.

Study title	Design of study	Number of subjects	Duration of study	Outcome [significant difference]	Reference
Pea proteins oral supplementation promotes muscle thickness gains during resistance training: a double-blind, randomized, Placebo-controlled clinical trial vs. Whey protein.	Randomized, Double-Blind, Placebo-Controlled Study	161	12-wk	Biceps brachii muscle thickness↑	[100]
The Effects of Whey vs. Pea Protein on Physical Adaptations Following 8-Weeks of High-Intensity Functional Training (HIFT): A Pilot Study.	Randomized, Double-Blind, Placebo-Controlled Study	15	8-wk	Result of the resistance training program↑	[101]
Effects of Whey and Pea Protein Supplementation on Post-Eccentric Exercise Muscle Damage: A Randomized Trial.	Randomized, Double-Blind, Placebo-Controlled Study	109	5-day	Creatinine kinase↓, Myoglobin↓	[102]
The Short-Term Effect of Whey Compared with Pea Protein on Appetite, Food Intake, and Energy Expenditure in Young and Older Men.	Randomized, single-blinded crossover design	30	One shot [postprandial data]	Appetite↑, Energy expenditure↑, 24-h energy intake↑	[103]

Note: ↓ and ↑ signs represent decrease and increase, respectively, after supplement of active compounds [pea protein only or, pea and whey proteins].

Table 4.
Clinical studies of pea protein for sports nutrition.

There has been considerable interest in lupin seeds recently, and as a human health food, the seeds are very high in dietary fiber, gluten-free, and virtually starch-free, and therefore have a very low glycemic impact [107]. What makes lupins even more valuable is that there are no genetically modified (GM) bean varieties under commercial cultivation. World production of lupin seed increased quickly in the 1970s and is dominated by Australian production.

Lupin seeds are high in protein, with levels similar to soybeans. Their grains are also known to be high in total dietary fiber, ~40 g/100 g dry matter, making lupins unique among ancient grains and beans. The main category of protein in lupin grains is globulin, with albumin making up the remainder. The major globulin categories are α-conglutin [35–37 g/100 g total protein], β-conglutin [44–45 g/100 g total protein], γ-conglutin [4–5 g/100 g total protein], and δ-conglutin [10–12 g/100 g total protein] [108–111]. Nutritionally, the limiting amino acids in lupin protein are the sulfur-containing amino acids methionine and cysteine [112]. Compared to soy protein, which have a more complete essential amino acid profile, the lupin protein was reported to be slightly below the required level of sulfur-containing amino acids needed by adults [113]. However, Singla et al. reported that the sulfur-containing amino acid levels of lupin protein were similar to those of soy [114]. This discrepancy is probably due to differences in lupin protein varieties and production environments. Carvajal-Larenas et al. reviewed in detail the amino acid composition of whole lupin seeds and concluded that it varies slightly among species. In vitro digestibility is ~98% high for uncooked lupin protein and is similar to soybean [115].

In vitro models of Lupinus albus γ-conglutin have shown the biological activity that enhances insulin and metformin activity on intracellular glucose consumption, indicating the potential for regulation on blood glucose by γ-conglutin [116]. As a possible improvement of lipid metabolism, an increase in LDL receptor activity has been demonstrated by HepG2 cells [117]. Furthermore, isolated lupin proteins of have been reported to have hyperlipidemic, anti-atherogenic, and hypocholesterolemic effects in rabbits, rats, and chickens [118, 119]. Several clinical human studies have shown that lupin protein decreases total and LDL cholesterol, as well as triglyceride and reduce the glycaemic response (**Table 5**) [120–127].

In general, the anti-nutrient factor of lupins is considered to be low compared to other legumes such as soybeans. Specifically, protease inhibitors are present at very low levels and are of minor importance in lupin crops. Trypsin inhibitor activity is described as "negligible" in Lupinus species, "very strong" at 43–84 trypsin inhibitor units [TIU/mg] in soybeans, and high [17–51 TIU/mg] in common beans [128]. Bitter lupin seed varieties contain quinolizine alkaloids, which may be toxic to humans. These toxic effects were recently reviewed by Carvajal-Larenes et al. [115]. Therefore, its maximum legal level of 0.02 g/100 g lupine powder and food has been legislated in several countries. There were no differences in alkaloids in grains among commercial *L. angustifolius* cultivars from western Australia in the same region and season, and all samples were below the levels permitted for maximum human food use.

Lupin protein, a legume, is a plant protein with similar attributes to soybean protein [129] and can be a substitute for soybean in the food industry [130, 131]. Further extensive research is expected due to the need for alternatives to animal proteins.

4.3 Mung bean protein

The mung bean (*Vigna radiata* L.) is one of the most important edible legume crops, grown on more than 6 million ha worldwide (approximately 8.5% of the global pulse area) and consumed by most households in Asia [132]. For individuals who cannot afford animal proteins or those who are vegetarian, mung bean is comparatively low cost and is a good source of protein. Furthermore, mung bean protein is more easily digestible than protein in other legumes [133]. In addition to the nutritional properties of mung bean, it has been known that mung beans have various physical regulation functions from ancient times. In the Compendium of Materia Medica (the "*Bencao Gangmu*"), a well-known Chinese pharmacopeia, mung beans have recorded to be utilized as a traditional Chinese medicine for its detoxification activities, recuperation of mentality, ability to alleviate heat stroke, and regulation of gastrointestinal upset.

Mung bean protein isolate (MuPI) dose-dependently reduced plasma lipid levels, such as total cholesterol, triglycerides, and non-high-density lipoprotein cholesterol [non-HDL-C] in hamsters [134, 135]. The mechanism underlying the cholesterol-lowering activity of mung bean protein was speculated to increase fecal bile acid and sterol excretion and decrease cholesterol absorption and synthesis. This mechanism is the same as that reported for SPI [68, 69]. In another study,

Title	Design of study	Number of total subjects	Outcome [significant difference]	Reference
Lupin protein compared to casein lowers the LDL cholesterol: HDL cholesterol-ratio of hypercholesterolemic adults	Randomized, double-blind, placebo-controlled, parallel trial	43	Total-C↓, LDL-C↓, LDL: HDL-C ratio↓	[120]
Hypocholesterolaemic effects of lupin protein and pea protein/ fiber combinations in moderately hypercholesterolaemic individuals	Randomized, double-blind, parallel group design	175	Total-C↓	[121]
Lupin protein positively affects plasma LDL cholesterol and LDL:HDL cholesterol ratio in hypercholesterolemic adults after four weeks of supplementation: a randomized, controlled crossover study	Randomized, controlled, double-blind crossover study	33 hypercholesterolemic subjects	LDL-C↓, HDL-C↑, LDL:HDL-C ratio↓	[122]
Consuming a mixed diet enriched with lupin protein beneficiallyaffects plasma lipids in hypercholesterolemic subjects: A randomized controlled trial	Randomized, controlled, double-blind three-phase crossover study	72 hypercholesterolemic subjects	LDL-C↓, Homocysteine↓, TG↓, Uric acid↓	[123]
Australian sweet lupin flour addition reduces the glycaemic index of a white bread breakfast without affecting palatability in healthy human volunteer	Randomized, single-blind, cross-over design	11 healthy subjects	Postprandial blood glucose↓	[124]
Lupin and soya reduce glycaemia acutely in type 2 diabetes	Randomized, cross-over trial	24 diabetic adults	Postprandial blood glucose↓, Insulin response↑	[125]
Hypoglycemic effect of lupin seed γ-conglutin in experimental animals and healthy human subjects	Randomized, double-blind, parallel group design	15 adult healthy volunteers	Postprandial blood glucose↓	[126]
Short-Term Effects of Lupin vs. Whey Supplementation on Glucose and Insulin Responses to a Standardized Meal in a Randomized Cross-Over Trial	Randomized, controlled, cross-over trial	12 healthy male and female volunteers	Postprandial blood glucose↓, Insulin response↑	[127]

Note: ↓ and ↑ signs represent decrease and increase, respectively, after supplement of active compounds. Total-cholesterol (Total-C); low-density lipoprotein cholesterol (LDL-C); triglyceride (TG); high-density lipoprotein cholesterol (HDL-C).

Table 5.
Clinical studies of lupin protein on improving lipid and glucose metabolisms.

MuPI was found to lower blood triglyceride levels in normal rats by inducing adiponectin and reducing triglyceride synthesis via insulin signaling [136]. This mechanism is the same as that reported for β-conglycinin [23]. From these findings, MuPI can be expected to be more effective in improving lipid metabolism. The main component of MuPI, accounting for over 80% of the protein, is 8S globulin. 8S globulin exhibited the highest degree of sequence identity [68%] and structural similarity with β-conglycinin [137, 138]. MuPI is expected to exhibit a four times stronger beneficial function on human health than SPI, in which β-conglycinin accounts for only 20% of the total protein.

The positive effects of MuPI on glucose metabolism in pre-diabetes patients was confirmed. In recent double-blind, placebo-controlled clinical trial, the test group subjects were instructed to consume a total of 2.5 g of MuPI twice daily for 12 weeks, with pre-diabetes [fasting plasma glucose level of 110–125 mg/dL or 2-h plasma glucose level of 140–200 mg/dL by the 75-g glucose tolerance test]. In this study, MuPI was shown to suppress to increase fasting plasma glucose and insulin levels compared to the placebo group. Triglyceride levels significantly decreased in subjects with hyperlipidaemia [139]. Another double-blind, placebo-controlled clinical trial of 44 healthy subjects showed that after consumption of 3.0 g/d MuPI for 8 weeks, insulin levels and homeostatic model assessment of insulin resistance values significantly decreased, and plasma glucose levels showed a downtrend, although it was not significant [140]. The lack of a beneficial effect of MuPI on blood glucose concentrations may be attributed to the exclusion of volunteers with abnormal blood glucose concentrations in this study. In this study, the body compositions of subjects were measured by dual-energy X-ray absorptiometry. As a result, a decrease in body fat mass and an increase in lean body mass in the test group were revealed. Conversely, in the control group, body fat mass increased and lean body mass decreased. The differences in body fat mass and lean body mass within each group and between the test and control groups were not statistically significant. However, the adiponectin level in the test group significantly increased, and it decreased in the control group. There was a significant difference between the net changes in the test and control groups [140]. These findings indicate that MuPI might improve insulin sensitivity by decreasing the accumulation of visceral fat.

Non-alcoholic fatty liver disease [NAFLD] represents a spectrum of liver diseases involving hepatocyte dysfunction caused by hepatic triglyceride accumulation in these cells. The prevalence of NAFLD has increased with the increased prevalence of obesity and metabolic syndrome. NAFLD is now a common disease, affecting 30% of the US population and 20% of Asian and European populations [141]. Rodent studies have shown that SPI intake reduces hepatic triglyceride accumulation [142, 143]. The detailed mechanism underlying the hepatic triglyceride-reducing effect of SPI remains to be elucidated, but β-conglycinin is likely to play an important role [135]. Indeed, the administration of purified β-conglycinin results in an even stronger reduction in hepatic triglycerides than SPI administration [18, 144]. From these results, it is expected that MuPI also has a preventive effect on NAFLD by preventing hepatic triglyceride accumulation. The effect of MuPI on hepatic triglyceride accumulation elucidated the potential ability of MuPI to prevent NAFLD onset and progression in experiments using an atherogenic diet-induced NASH mouse model in mice fed a normal-fat or high-fat diet [145]. In the abovementioned clinical trial [140], Alanine aminotransferase [ALT] levels increased slightly in the control group, whereas significantly decreased in the test group. Of the blood test items, ALT is one of important indicators of the degree of liver dysfunction.

The released peptides obtained from mung bean protein hydrolysate may exhibit bioactivity as angiotensin I-converting enzyme (ACE) inhibitors, antioxidants, and anti-cancer Asiatic acid carriers due to their sequence characteristics [146, 147]. A peptide [<3 kDa], with a small molecular weight isolated from MuPI hydrolysates, was reported to show high ACE inhibitory and antioxidant activities, including DPPH radical scavenging activity, hydroxyl radical scavenging ability, and metal-chelating activity [146]. Three kinds of novel peptides exerting high ACE inhibitory activity were isolated from the alcalase hydrolysate of MuPI, and the amino acid sequences of these peptides were identified to be Lys-Asp-Tyr-Arg-Leu, Val-Thr-Pro-Ala-Leu-Arg, and Lys-Leu-Pro-Ala-Gly-Thr-Leu-Phe [148].

The relationships between MuPI intake, strength, and lean body mass (LBM) in underactive vegetarians were examined, and the impact of MuPI supplementation on these indices was recorded utilizing an eight-week, randomized, controlled feeding trial. LBM significantly correlated with grams of protein consumed daily and was also significantly correlated with grip strength and lower body strength [149]. Mung beans are inadequate in threonine, tryptophan, and the sulfur-containing amino acids cysteine and methionine, but they contain high levels of essential amino acids, notably leucine, lysine, and phenylalanine [150]. Although it is necessary to consider the amino acid balance, it is expected that MuPI will be widely used in the field of sports nutrition in the future.

5. Conclusion

If the current pace of population growth continues, the global population is expected to surpass 9 billion by 2050. In addition to this increase in population, the change of dietary habits of emerging countries due to their increased GDP will require, in 2050, we will need twice as much protein as we had in 2005.

So far, we have been able to meet the increasing demand for protein by improving the productivity of agriculture. However, in the future, this growth alone will not be enough to absorb the increase, and the balance between supply and demand will begin to collapse as early as 2030. This prediction is called the "protein crisis," and has recently begun to attract attention, especially in Europe and the United States. To solve this protein crisis, it is essential to use highly productive plant proteins as food ingredients instead of animal proteins, which are less efficient in production.

WHO has called for the need to address the double burden of malnutrition. This means that we need to look not only at nutrient deficiencies, but also at nutrient excesses. Obesity caused by over-nutrition and the resulting lifestyle-related diseases are spreading around the world. In this regard, consumer demand for plant protein-based products is high and expected to grow considerably in the next decade. A variety of soy and other plant-based functional foods have been recommended by many health organizations worldwide.

Currently, contributions to the SDGs (Sustainable Developing Goals) are being appealed around the world. There is widespread recognition that the replacement of animal protein with vegetable protein not only contributes to human health, but also to the earth health. Wider and prudent use of plant proteins in the diet can help to supply adequate high-quality protein for the population and may reduce the potential for adverse environmental consequences.

This chapter focused on the recently reported physiological functions of legumes-derived plant proteins, including soybeans. Further research is expected to lead to more widely use of the legumes introduced in this chapter and to the discovery and use of legumes with new functionalities.

Acknowledgements

Of the research results presented in this chapter, our own research results were achieved in the laboratory at Fuji Oil Co. Ltd. and/or Fuji Oil Holdings Inc. to which I belonged until March of this year from 1986. I believe that I could not have done this without the cooperation of the researchers who belonged to that laboratory. I would like to take this opportunity to express my deepest gratitude.

Author details

Mitsutaka Kohno
Sports Nutrition Research Center, Kyushu Kyoritsu University, Fukuoka, Japan

*Address all correspondence to: m-kohno@kyukyo-u.ac.jp

References

[1] Food and Drug Administration: Food labelling health claims; soy protein and coronary heart disease. Fed Regist. 1999:64:57700-57733.

[2] Alberti KG, Zimmer PZ: Definition, diagnosis and classification of diabetes mellitus and its complication, part 1: Diagnosis and classification of diabetes mellitus provisional report of a WHO consultation. Diabet Med. 1998: 15:539-553.

[3] National Cholesterol Education Program (NCEP): Expert panel on detection and treatment on high blood cholesterol in adults: Third report of the National Cholesterol Education Program (NCEP) expert panel on detection, evaluation, and treatment of high blood cholesterol in adults (adult treatment panel III) final report. Circulation. 2002:106:3143-3421.

[4] Grundy SM, Cleeman JI, Daniels SR, Donato KA, Eckel RH, Franklin BA, Gordon DJ, Krauss RM, Savage PJ, Smith SC Jr, Spertus JA, Costa F, American Heart Association; National Heart, Lung, and Blood Institute: Diagnosis and management of the metabolic syndrome: An American Heart Association/National Heart, lung and blood institute scientific statement. Circulation. 2005:112:2735-2752.

[5] Zimmet P, Magliano D, Matsuzawa Y, Alberti G, Shaw J: The metabolic syndrome: A global public health problem and a new definition. J Atheroscle. Thromb. 2005:12: 295-300.

[6] Reaven GM: The metabolic syndrome: Is this diagnosis necessary? Am J Clin Nutr. 2006:83:1237-1247.

[7] The Examination Committee of Criteria for `Obesity Disease` in Japan: New criteria for `obesity disease` in Japan: Circulation J. 2006:66:987-992.

[8] Definition and the diagnostic standard for metabolic syndrome Committee to Evaluate Diagnostic Standards for Metabolic Syndrome: Nippon Naika Gakkai Zasshi. 2005:94:794-809 (in Japanese).

[9] Anderson JW, Johnstone BM and Cook-Newell MF: Meta-analysis of the effects of soy protein intake on serum lipids. N Eng J Med. 1995:333:276-282.

[10] Zhan S and Ho SC: Meta-analysis of the effects of soy protein containing isoflavones on the lipid profiles. Am J Clin Nutr. 2005:81:397-408.

[11] Reynolds K, Chin A, Lees KA, Nguyen A, Bujnowski D, He J: A meta-analysis of the effect of soy protein supplementation on serum lipids. Am J Cardiol. 2006:98(5):633-640.

[12] Sirtori CR, Eberini I, Arnoldi A: Hypocholesterolaemic effects of soya proteins: Result of recent studies are predictable from Anderson meta-analysis data. Brit J Nutr. 2007:97: 816-822.

[13] Anderson JW, Bush HM: Soy protein effects on serum lipoproteins: A quality assessment and meta-analysis of randomized, controlled studies. J Am Coll Nutr. 2011:30(2):79-91.

[14] Li SS, Mejia SB, Lytvyn L, Stewart SE, Viguiliouk E, Ha V, de Souza RJ, Leiter LA, Kendall CWC, Jenkins DA, Sievenpiper JL: Effect of plant protein on blood lipids: A systematic review and meta-analysis of randomized controlled trials. J Am Heart Assoc. 2017:6:e006659.

[15] Mejia SB, Messina M, Li SS, Viguiliouk E, Chiavaroli L, Khan TA, Srichaikul K, Mirrahimi A, Sievenpiper JL, Kris-Etherton P, Jenkins DJA: A meta-analysis of 46 studies identified by the FDA demonstrates that

soy protein decreases circulating LDL and Total cholesterol concentrations in adults. J Nutr. 2019:149:968-981.

[16] Moradi M, Daneshzad E, Azadbakht L: The effects of isolated soy protein, isolated soy isoflavones and soy protein containing isoflavones on serum lipids in postmenopausal women: A systematic review and meta-analysis. Crit Rev Food Sci Nutr. 2020:60(20): 3414-3428.

[17] Samoto M, Maebuchi M, Miyazaki C, Kugitani H, Kohno M, Hirotsuka M, Kito M: Abundant proteins associated with lecithin in soy protein isolate. Food Chem. 2007: 102(1):317-322.

[18] Aoyama T, Kohno M, Saito T, Fukui K, Takamatsu K, Yamamoto T, Hashimoto Y, Kito M: Reduction by phytate-reduced soybean β-conglycinin of plasma triglyceride level of young and adult rats. Biosci Biotechnol Biochem 2001:65:1071-1075.

[19] Moriyama T, Kishimoto K, Nagai K, Urade R, Ogawa T, Utsumi S, Maruyama N, Maebuchi M: Soybean β-conglycinin diet suppresses serum triglyceride levels in normal and genetically obese mice by induction of β-oxidation, downregulation of fatty acid synthase and inhibition of triglyceride absorption. Biosci Biotech Biochem 2004:68(2):352-359.

[20] Fukui K, Kojima M, Tachibana N, Kohno M, Takamatsu K, Hirotsuka M, Kito M: Effects of soybean β-conglycinin on hepatic lipid metabolism and fecal lipid excretion in normal adult rats. Biosci Biotech Biochem 2004:68(5): 1153-1155.

[21] Kohno M, Hirotsuka M, Kito M and Matsuzawa Y: Decrease in serum triacylglycerol and visceral fat mediated by dietary soybean β-conglycinin. J Atheroscler Thromb. 2006:13:247-255.

[22] Baba T, Ueda A, Kohno M, Fukui K, Miyazaki C, Hirotsuka M, Ishinaga M: Effects of soybean beta-conglycinin on body fat ratio and serum lipid levels in healthy volunteers of female university students. J Nutr Sci Vitaminol (Tokyo). 2004:50(1):26-31.

[23] Tachibana N, Iwaoka Y, Hirotsuka M, Horio F, Kohno M: Beta conglycinin lowers very-low-density lipoprotein-triglyceride levels by increasing adiponectin and insulin sensitivity in rats. Biosci Biotechnol Biochem 2010:74(6):1250-1255.

[24] Ohara S, Matsui Y, Tamesada M, Saitou M, Komatsu R, Yoshino T, Tsubaki T, Itakura H: Serum triacylglycerol-lowering effect of soybean b-conglycinin in mildly hypertriacylglycerolemic individuals. e-SPEN, Eur e-Journal Clin Nutr Met. 2007:2:12-16.

[25] Ma D, Taku K, Zhang Y, Jia M, Wang Y, Wang P: Serum lipid-improving effect of soyabean b-conglycinin in hyperlipidaemic menopausal women. Brit J Nutr 2013:110:1680-1684.

[26] Nishimura M, Ohkawara T, Sato Y, Satoh H, Takahashi Y, Hajika M, Nishihira J: Improvement of triglyceride levels through the intake of enriched-β-Conglycinin soybean (Nanahomare) revealed in a randomized, double-blind, Placebo-Controlled Study. Nutrients 2016:8:491.

[27] Taniguchi H, Shimizu K, Wada S, Nirengi S, Kataoka H, Higashi A: Effects of beta-conglycinin intake on circulating FGF21 levels and brown adipose tissue activity in Japanese young men: a single intake study and a randomized controlled trial. J Physiol Anthropol. 2020:39(1):17.

[28] Hill NR, Fatoba ST, Oke JL, Hirst JA, O'Callaghan CA, Lasserson DS, Hobbs FD: Global

prevalence of chronic kidney disease – A systematic review and meta-analysis. PLoS One. 2006:11(7):e0158765.

[29] GBD 2015 Mortality and causes of death collaborators: Global, regional, and national life expectancy, all-cause mortality, and cause-specific mortality for 249 causes of death, 1980-2015: A systematic analysis for the global burden of disease study 2015. Lancet. 2016:388: 1459-1544.

[30] Sun X, He J, Ji XL, Zhao YM, Lou HY, Song XX, Lou HY, Song XX, Shan LZ, Kang YX, Zeng WH, Pang XH, Zhang SZ, Ding Y, Ren YZ, Shan PF: Association of chronic kidney disease with coronary heart disease and stroke risks in patients with type 2 diabetes mellitus: An observational cross-sectional study in Hangzhou, China. Chin Med J. 2017:130:57-63.

[31] Hoogeveen EK, Geleijnse JM, Giltay EJ, Soedamah-Muthu SS, De Goede J, Oude Griep LM, Stijnen T, Kromhout: Kidney function and specific mortality in 60-80 years old post-myocardial infarction patients: A 10-year follow-up study. PLoS One 2017:12(2):e0171868.

[32] Agrawal V, Shah A, Rice C, Franklin BA, McCullough PA: Impact of treating the metabolic syndrome on chronic kidney disease. Nat Rev Nephrol. 2009:5:520-528.

[33] Yokoyama H, Kawai K, Kobayashi M, Japan diabetes clinical data management study group: Microalbuminuria is common in Japanese type 2 diabetic patients: A nationwide survey from the Japan diabetes clinical data management study group (JDDM 10). Diabetes Care 2007:30(4):989-992.

[34] Parving HH, Lewis JB, Ravid M, Remuzzi G, Hunsicker LG, DEMAND investigators: Prevalence and risk factors for microalbuminuria in a referred cohort of type II diabetic patients: A global perspective. Kidney Int. 2006:69:2057-2063.

[35] van Westing AC, Küpers LK, Geleijnse JM: Diet and Kidney Function: A Literature Review. Curr Hypertens Rep. 2020:22:14.

[36] Haring B, Selvin E, Liang M, Coresh J, Grams M, Petruski-Ivleva N, Steffen LM, Rebholz CM: Dietary protein sources and risk for incident chronic kidney disease: Results from the atherosclerosis risk in communities (ARIC) study. J Ren Nutr. 2017:27: 233-242.

[37] Asanoma M, Tachibana N, Hirotsuka M, Kohno M, Watanabe Y: Effects of soy protein isolate feeding on severe kidney damage in DOCA salt-treated obese Zucker rats. J Agric Food Chem. 2012:60:5367-5372.

[38] Jheng HG, Hirotsuka M, Goto Y, Shibata M, Matsumura Y, Kawada T: Dietary low-fat soy milk powder retards diabetic nephropathy progression via inhibition of renal fibrosis and renal inflammation. Mol Nutr Food Res. 2016:61(3):e2001600461.

[39] Azadbakht L, Atbak S, Esmaillzadeh A: Soy protein intake, cardiorenal indices, and c-reactive protein in type 2 diabetes with nephropathy. Diabetes Care 2008:31:648-654.

[40] Zhang J, Liu J, Su J, Tian F: The effects of soy protein on chronic kidney disease: A meta-analysis of randomized controlled trials. Eur J Clin Nutr 2014:68:987-993.

[41] McGraw NJ, Krul ES, Grunz-Borgmann E, Parrish AR: Soy-based renoprotection. World J Nephrol. 2016:5(3):233-257.

[42] Jing Z, Wei-Jie Y: Effects of soy protein containing isoflavones in patients with chronic kidney disease: A systematic review and meta-analysis. Clin Nutr. 2016:35(1):117-124.

[43] Rafieian-Kopaei M, Beigrezaei S, Nasri H, Kafeshani M: Soy protein and chronic kidney disease: An updated review. Int J Prev Med. 2017:8:105.

[44] Pan Y, Guo LL, Jin HM: Low-protein diet for diabetic nephropathy: A meta-analysis of randomized controlled trials. Am J Clin Nutr. 2008:88:660-666.

[45] Koya D, Haneda M, Inomata S, Suzuki Y, Suzuki D, Makino H, Shikata K, Murakami Y, Tomino Y, Yamada K, Araki SI, Kashiwagi A. Kikkawa R, Low-protein diet study group: Long-term effect of modification of dietary protein intake on the progression of diabetic nephropathy: A randomized control trial. Diabetologia. 2009:52(10):2037-2045.

[46] Rughooputh MS, Zeng R, Yao Y: Protein diet restriction slows chronic kidney disease progression in non-diabetic and in type 1 diabetic patients, but not in type 2 diabetic patients: A meta-analysis of randomized controlled trials using glomerular filtration rate as a surrogate. PLoS One. 2015:10(12): e0145505.

[47] Artis D, Spits H: The biology of innate lymphoid cells. Nature. 2015:517:293-301.

[48] Pedraza-Alva G, Pérez-Martínez L, Valdez-Hernández L, Meza-Sosa K.F, Ando-Kuri M: Negative regulation of the inflammasome: Keeping inflammation under control. Immunol Rev. 2015:265:231-257.

[49] Fernandes JV, Cobucci RN, Jatobá CA, Fernandes TA, de Azevedo JW, de Araújo JM: The role of the mediators of inflammation in cancer development. Pathol Oncol Res. 2015:21:527-534.

[50] Heppner FL, Ransohoff RM, Becher B: Immune attack: The role of inflammation in Alzheimer disease. Nat Rev Neurosci. 2015:16:358-372.

[51] Waisman A, Liblau RS, Becher B: Innate and adaptive immune responses in the CNS. Lancet Neurol. 2015:14: 945-955.

[52] Singh BP, Vij S, Hati S: Functional significance of bioactive peptides derived from soybean. Peptides 2014:54:171-179

[53] Jeong HJ, Park JH, Lam Y, de Lumen BO: Characterization of lunasin isolated from soybean. J Agric Food Chem. 2003:51:7901-7906.

[54] Hernández-Ledesma B, Hsieh CC, de Lumen BO: Lunasin, a novel seed peptide for cancer prevention. Peptides. 2009:30:426-430.

[55] Shidal C, Al-Rayyan N, Yaddanapudi K, Davis KR: Lunasin is a novel therapeutic agent for targeting melanoma cancer stem cells. Oncotarget. 2016:7:84128-84141.

[56] Lam Y, Galvez A, de Lumen BO: Lunasin suppresses E1A-mediated transformation of mammalian cells but does not inhibit growth of immortalized and established cancer cell lines. Nutr Cancer. 2003:47:88-94.

[57] Lule VK, Garg S, Pophaly SD, Hitesh, Tomar SK: Potential health benefits of lunasin: A multifaceted soy-derived bioactive peptide. J Food Sci. 2015:80:R485–R494.

[58] Tsuruki T, Kishi K, Takahashi M, Tanaka M, Matsukawa T, Yoshikawa M: Soymetide, an immunostimulating peptide derived from soybean beta-conglycinin, is an fMLP agonist. FEBS Lett. 2003:540:206-210.

[59] Draganidis D, Karagounis LG, Athanailidis I, Chatzinikolaou A, Jamurtas AZ, Fatouros IG: Inflammaging and skeletal muscle: Can protein intake make a difference? J Nutr. 2016:146:1940-1952.

[60] Dia VP, Bringe NA, de Mejia EG: Peptides in pepsin-pancreatin hydrolysates from commercially available soy products that inhibit lipopolysaccharide-induced inflammation in macrophages. Food Chem. 2014:152:423-431.

[61] Tontini GE, Vecchi M, Pastorelli L, Neurath MF, Neumann H: Differential diagnosis in inflammatory bowel disease colitis: State of the art and future perspectives. World J Gastr. 2015;21(1):21-46.

[62] Young D, Ibuki M, Nakamori T, Fan M, Mine Y: Soy-derived di- and tripeptides alleviate colon and ileum inflammation in pigs with dextran sodium sulfate-induced colitis. J Nutr. 2012:142:363-368.

[63] Kovacs-Nolan J, Zhang H, Ibuki M, Nakamori T, Yoshiura K, Turner PV, Matsui T, Mine Y: The PepT1-transportable soy tripeptide VPY reduces intestinal inflammation. Biocim Biophys Acta. 2012:1820:1753–1763.

[64] Hotamisligil GS: Inflammation and metabolic disorders. Nature. 2006:444(7121):860-867.

[65] Lumeng CN, Saltiel AR: Inflammatory links between obesity and metabolic disease. J Clin Invest. 2011:121(6):2111-2117.

[66] Lackey DE, Olefsky JM: Regulation of metabolism by the innate immune system. Nat Rev Endocrinol. 2016:12(1):15-28

[67] Ohta S, Asanoma S, Irie N, Tachibana N and Kohno M: Soy phospholipids exert a renoprotective effect by inhibiting the nuclear factor kappa B pathway in macrophages. 2021. in preparation.

[68] Sugano, M, Yamada Y, Yoshida K, Hashimoto Y, Matsuo T, Kimoto M: The hypocholesterolemic action of the undigested fraction of soybean protein in rats. Atherosclerosis. 1988:72:115-122.

[69] Sugano M, Goto S, Yamada Y, Yoshida K, Hashimoto Y, Matsuo T, Kimoto M: Cholesterol-lowering activity of various undigested fractions of soybean protein in rats. J Nutr., 1990:120(9):977-985.

[70] Kwak H-B: Statin-induced myopathy in skeletal muscle: The role of exercise. J Lifestyle Med. 2014:4(2):71-79.

[71] Lei Q, Peng WN, You H, Hu ZP, Lu W.: Statins in nervous system-associated disease: Angels or devils? Pharmazie. 2014:69(6):448-454.

[72] Laufs U, Scharnagi H, Halle M, Windler E, Endres M, Marz W: Treatment options for statin-associated muscle symptoms. Dtsch Arztebi Int. 2015:112:748-755.

[73] Haque T, Sasatomi E, Hayashi PH: Drug-induced liver injury: Pattern recognition and future directions. Gut Liver. 2016:10(1):27-36.

[74] Ambapkar SN, Shetty N, Dwivery A, Malve HO: Statin-induced rhabdomylysis in patients with renal failure and underlying undiagnosed hypothyroidism. Indian J Crit Care Med. 2016:20(5):305-307.

[75] Nabiki J, Yoshida M, Maeda A, Yamauchi F, Asayama S, Kohno M, Suzuki M, Harano A, Harano Y: Utility and significance of isolated soy protein snack for obesity and dyslipidemia. J Jpn Soc Clin Nutr. 2016:38(2):85-95 (in Japanese).

[76] Ueda F, Ohtoshi K, Kohno M, Harano Y: Effects and clinical significances of oral supplementary soybean β-conglycinin in subjects with dyslipidemia and renal dysfunction. 2021. in preparation.

[77] Yoshioka H, Yui N, Maegawa K, Kohno M, Misuya H and Yudoh K: Assessment of the effect of soy peptide in joint symptoms of rheumatoid arthritis patients. Pharmacometrics, 2015:88(3/4):39-46 (in Japanese).

[78] Arito M, Mitsui H, Kurokawa MS, Yudoh K, Kamada T, Niki H, Kato T: Effects of soy peptides on IL-1β-induced matrix-degrading enzymes in human articular chondrocytes. Integr Mol Med, 2016:3(3):661-665.

[79] Dahl WJ, Foster LM, Tyler RT: Review of the health benefits of peas (Pisum sativum L.). Brit J Nutr. 2012:108(S1):S3–S10.

[80] Schatz B, G Endres: Field pea production. NDSU Extension Service. 2009:A-1166 (Revised):1-8.

[81] Grand View Research 2015. Pea protein market by type (isolates, concentrates, textured), application, textured pea protein by form (dry, wet), by region – Global trends and forecast to 2020. http://www.researchandmarkets.com/research/b8x63n/pea_protein

[82] Shevkani K, Singh N, Kaur A, Rana JC: Structural and functional characterization of kidney bean and field pea protein isolates: A comparative study. Food Hydrocol. 2015:43:679-689.

[83] Lu ZX, He JF, Zhang YC, Bing DJ: Composition, physicochemical properties of pea protein and its application in functional foods. Crit Rev Food Sci Nutr. 2020:60(15):2593-2605.

[84] Roberts AW, Ogunwole SU, Blakeslee L, Rabe MA, editors. The population 65 years and older in the United States 2016: US Department of Commerce Washington (DC); 2018. United States Census Bureau.

[85] Cosco TD, Prina AM, Perales J, Stephan BC, Brayne C: Operational definitions of successful aging: A systematic review. Int Psychogeriatr 2014:26:373-381.

[86] Anton SD, Woods AJ, Ashizawa T, Barb D, Buford TW, Carter CS, Clark DJ, Cohen RA, Corbett DB, Cruz-Almeida Y, Dotson V, Ebner N, Efron PA, Fillingim RB, Foster TC, Gundermann DM, Joseph A-M, Karabetian C, Leeuwenburgh C, Manini TM, Marsiske M, Mankowski RT, Mutchie HL, Perri MG, Ranka S, Rashidi P, Sandesara B, Scarpace PJ, Sibille KT, Solber LM, Someya S, Uphold C, Wohlgemuth S, Wu SS, Pahor M: Successful aging: advancing the science of physical independence in older adults. Ageing Res Rev. 2015:24:304-327.

[87] Liguori I, Russo G, Aran L, Bulli G, Curcio F, Della-Morte D, Gargiulo G, Testa G, Cacciatore F, Bonaduce D, Abete P: Sarcopenia: Assessment of disease burden and strategies to improve outcomes. Clin Interv Aging. 2018:13:913-927.

[88] Rennie MJ, Bohe J, Wolfe RR: Latency, duration and dose response relationships of amino acid effects on human muscle protein synthesis. J Nutr, 2002:132:3225S–3227S.

[89] Kimball SR, Jefferson LS: Signalling pathways and molecular mechanisms through which branched-chain amino acids mediate translational control of protein synthesis. J Nutr. 2006:136:227S–231S.

[90] Jackman SR, Witard OC, Philp A, Wallis GA, Baar K, Tipton KD: Branched-chain amino acid ingestion stimulates muscle myofibrillar protein synthesis following resistance exercise in humans. Front Physiol. 2017:8:390. DOI: 10.3389/fphys.2017.00390.

[91] Hall WL, Millward DJ, Long SJ, Morgan LM: Casein and whey exert different effects on plasma amino acid profiles, gastrointestinal hormone

secretion and appetite. Br J Nutr. 2003:89:239-248.

[92] Fuhrman J, Ferreri DM: Fueling the vegetarian (vegan) athlete. Curr Sports Med Rep. 2010:9:233-241.

[93] How Many Adults in the U.S. Are Vegetarian and Vegan? 2019. Available from https://www.vrg.org/nutshell/Polls/2016_adults_veg.htm

[94] De Backer CJS, Hudders L: From meatless Mondays to meatless Sundays: Motivations for meat reduction among vegetarians and semi-vegetarians who mildly or significantly reduce their meat intake. Ecol Food Nutr. 2014:53:639-657.

[95] Nunes MC, Raymundo A, Sousa I: Rheological behaviour and microstructure of pea protein/k-carrageenan/starch gels with different setting conditions. Food Hydrocol, 2006:20(1):106-113.

[96] Iqbal A, Khalil IA, Ateeq N, Khan MS: Nutritional quality of important food legumes. Food Chem 2006:97(2):331-335.

[97] Maninder K, Sandhu KS, Singh N: Comparative study of the functional, thermal and pasting properties of flours from different field pea (*Pisum sativum* L.) and pigeon pea (*Cajanus cajan* L.) cultivars. Food Chem. 2007: 104(1):259-267.

[98] Aluko RE, Mofolasayo OA, Watts BM: Emulsifying and foaming properties of commercial yellow pea (*Pisum sativum* L.) seed flours. J Agri Food Chem, 2009:57(20):9793-9800.

[99] Barac, M, Cabrilo S, Pesic M, Stanojevic S, Zilic S, Macej O, Ristic N: Profile and functional properties of seed proteins from six pea (*Pisum sativum*) genotypes. Intr J Mol Sci 2010:11(12):4973-4990.

[100] Babault N, Païzis1 C, Deley G, Guérin-Deremaux L, Saniez M-H, Lefranc-Millot C, Allaert FA: Pea proteins oral supplementation promotes muscle thickness gains during resistance training: a double-blind, randomized, placebo-controlled clinical trial vs. whey protein. J Inter Soc Sports Nutr. 2015:12:3

[101] Banaszek A, Townsend JR, Bender D, Vantrease WC, Marshall AC, Johnson KD: The effects of whey vs. pea protein on physical adaptations following 8-weeks of high-intensity functional training (HIFT): A pilot study. Sports. 2019:7:12.

[102] Nieman DC, Zwetsloot KA, Simonson AJ, Hoyle AT, Wang X, Nelson HK, Lefranc-Millot C, Guérin-Deremaux L: Effects of whey and pea protein supplementation on post-eccentric exercise muscle damage: A randomized trial. Nutrients. 2020: 12:2382.

[103] Hawley AL, Gbur E, Tacinelli AM, Walker S, Murphy A, Burgess R, Baum BI: The short-term effect of whey compared with pea protein on appetite, food intake, and energy expenditure in young and old men. Curr Dev Nutr. 2020:4:nzaa009.

[104] Gladstones JS, Lupins as Crop Plants: Biology, Production and Utilization. edited by Gladstones JS, Atkins CA, Hamblin J: 1998. Wallingford, UK: CAB INTERNATIONAL, pp. 465.

[105] Lamber H, Clements JC, Nelson MN: How a phosphorus-acquisition strategy based on carboxylate exudation powers the success and agronomic potential of lupines (Lupinus, Fabaceae). Am J Bot. 2013:100(2):263-288.

[106] Seymour M, Kirkegaard J, Peoples M, White P, French R: Break-crop benefits to wheat in Western

Australia – Insights from over three decades of research. Crop Pasture Sci. 2012:63(1). DOI:10.1071/cp11320.

[107] Dove ER, Mori TA, Chew GT, Barden AE, Woodman RJ, Puddey IB, Sipsas S, Hodgson JM: Lupin and soya reduce glycaemia acutely in type 2 diabetes. Br J Nutr. 2011:106(7): 1045-1051.

[108] Blagrove RJ, Gillespie JM: Isolation, purification and characterization of the seed globulins of Lupinus angustifolius. Australian J Plant Physiol. 1975:2(1):13-27.

[109] Duranti M, Consonni A, Magni C, Sessa F, Scarafoni A: The major proteins of lupin seed: Characterization and molecular properties for use as functional and nutraceutical ingredients. Trends Food Sci Tech. 2008:19:624-633.

[110] Foley RC, Jimenez-Lopez JC, Kamphuis LG, Hane JK, Melser Su, Singh KB: Analysis of conglutin seed storage proteins across lupin species using transcriptomic, protein and comparative genomic approaches. BMC Plant Biology. 2015:15:106. DOI 10.1186/s12870-015-0485-6.

[111] Patterson DS. Composition and food uses of Lupins. In: Gladstones JS, Atkins CA, Hamblin J, editors. Lupins as a Crop Plants: Biology, Production and Utilization: CAB International London; 1998. p.353-384.

[112] Gulewicz P, Martı'nez-Villaluenga C, Frias J, Ciesiołka D, Gulewicz K, Vidal-Valverde C: Effect of germination on the protein fraction composition of different lupin seeds. Food Chem. 2008:107:830-844.

[113] Chew PG, Casey AJ, Johnson SK: Protein quality and physico-functionality of Australian sweet lupin (*Lupinus angustifolius* cv. *Gungurru*) protein concentrates prepared by isoelectric precipitation or ultrafiltration. Food Chem. 2003: 83(4):575-583.

[114] Singla P, Sharma S, Singh S: Amino acid composition, protein fractions and electrophoretic analysis of seed storage proteins in Lupins. Ind J Agric Biochem. 2017:30(1):33.

[115] Carvajal-Larenas FE, Linnemann AR, Robert Nout MJ: Lupinus mutabilis: composition, uses, toxicology and debittering. Crit Rev Food Sci Nutr. 2016: 56(9):1454-1487.

[116] Lovati MA, Manzoni C, Castiglioni S, Parolari A, Magni C, Duranti M: Lupin seed γ-conglutin lowers blood glucose in hyperglycemic rats and increases glucose consumption of HepG2 cells. Br J Nutr. 2012:107:67-73.

[117] Sirtori CR, Lovati MR, Manzoni C, Castiglioni S, Duranti M, Magni C, Morandi S, D'Agostina A, Arnoldi A: Proteins of white lupin seed, a naturally isoflavone-poor legume, reduce cholesterolemia in rats and increase LDL receptor activity in HepG2 cells. J Nutr. 2004:134:18-23.

[118] Marchesi M, Parolini C, Diani E, Rigamonti E, Cornelli L, Arnoldi A, Sirtori CR, Chiesa G: Hypolipidaemic and anti-atherosclerotic effects of lupin proteins in a rabbit model. Br J Nutr. 2008:100:707-710.

[119] Viveros A. Centeno C. Arija I. Brenes A: Cholesterol-lowering effects of dietary Lupin (*Lupinus albus* var *Multolupa*) in chicken diets. Poult Sci. 2007:86:2631-2638.

[120] Weiße K, Brandsch C, Zernsdorf B, Nembongwe GSN, Hofmann K, Eder K, Stangl GI: Lupin protein compared to casein lowers the LDL cholesterol: HDL cholesterol-ratio of hyper cholesterolemic adults. Eur J Nutr. 2010:49:65-71.

[121] Sirtori CR, Triolo M, Bosisio R, Bondioli A, Calabresi L, Vergori VD, Gomaraschi M, Mombell G, Pazzucconi F, Zacherl C, Arnoldi A: Hypocholesterolaemic effects of lupin protein and pea protein/fibre combinations in moderately hypercholesterolaemic individuals. Br J Nutr. 2012:107:1176-1183.

[122] Bähr M, Fechner A, Krämer J, Kiehntopf M, Jahreis G: Lupin protein positively affects plasma LDL cholesterol and LDL: HDL cholesterol ratio in hypercholesterolemic adults after four weeks of supplementation: A randomized, controlled crossover study. Nutr J. 2013:12:107.

[123] Bähr M, Fechner A, Krämer J, Kiehntopf M, Jahreis G: Consuming a mixed diet enriched with lupin protein beneficiallyaffects plasma lipids in hypercholesterolemic subjects: A randomized controlled trial. Clin Nutr. 2015:34:7-14.

[124] Hall RS, Thomas SJ, Johnson SK: Australian sweet lupin flour addition reduces the glycaemic index of a white bread breakfast without affecting palatability in healthy human volunteer. Asia Pac J Clin Nutr. 2005:14(1):91-97.

[125] Dove ER, Mori TA, Chew GT, Barden AE, Woodman RJ, Puddey IB, Sipsas S, Hodgson NM: Lupin and soya reduce glycaemia acutely in type 2 diabetes. Br J Nutr. 2011:106: 1045-1051.

[126] Bertoglio JC, Calvo MA, Hancke JL, Burgos RA, Riva A, Morazzoni P, Ponzone C, Magni C, Duranti M: Hypoglycemic effect of lupin seed γ-conglutin in experimental animals and healthy human subjects. Fitoterapia. 2011:82(5):933-938.

[127] Schopen K, Ewald AC, Johannes BW, Bloch W, Rittweger J, Frings-Meuthen P: Short-term effects of lupin vs. whey supplementation on glucose and insulin responses to a standardized meal in a randomized cross-over trial. Front Physiol. 2017:8:198.

[128] Guillamón E, Pedrosa MM, Burbano C, Cuadrado C, Sanchez MC, Muzquiz M: The trypsin inhibitors present in seed of different grain legume species and cultivar. Food Chem. 2008:107(1):68-74.

[129] Pihlanto A. Mattila P. Mäkinen S. Pajari A.M: Bioactivities of alternative protein sources and their potential health benefits. Food Funct. 2017:8:3443-3458.

[130] Jayasena V, Chih HJ, Nasar-Abbas S: Functional properties of sweet lupin protein isolated and tested at various pH levels. Res J Agric Biol Sci. 2010:6: 130-137.

[131] Kiosseoglou A, Doxastakis G, Alevisopoulos S, Kasapis S: Physical characterization of thermally induced networks of lupin protein isolates prepared by isoelectric precipitation and dialysis. Int J Food Sci Technol. 1999:34:253-263.

[132] Dahiya PK, Linnemann AR, Van Boekel MAJS, Khetarpaul N, Grewal RB, Nout MJR: Mung bean: Technological and nutritional potential. Crit Rev Food Sci Nutr. 2015:55:670-688.

[133] Mubarak AE: Nutritional composition and antinutritional factors of mung bean seeds (*Phaseolus aureus*) as affected by some home traditional processes. Food Chem. 2005:89: 489-495.

[134] Yao Y, Hao L, Shi Z. Wang L, Cheng X, Wang S, Ren G: Mung bean decreases plasma cholesterol by up-regulation of CYP7A1. Plant Foods Hum Nutr. 2014:69(2):134-136.

[135] Yao Y, Zhu Y, Ren G: Mung bean protein increases plasma cholesterol by

up-regulation of hepatic HMG-CoA reductase, and CYP7a1 in mRNA levels. J Food Nutr Res. 2014:2:770-775.

[136] Tachibana N, Wanezaki S, Nagata M, Motoyama T, Kohno M, Kitagawa S: Intake of mung bean protein isolate reduces plasma triglyceride level in rats. Funct Foods Health Dis. 2013:3(9):365-376.

[137] Maruyama N, Adachi M, Takahashi K, Yagasaki K, Kohno M, Takenaka Y, Okuda E, Nakagawa S, Mikami B, Utsumi S: Crystal structures of recombinant and native soybean beta-conglycinin beta homotrimers. Eur J Biochem. 2001:268:3595-3604.

[138] Itoh T, Garcia RN, Adachi M. Maruyama Y, Tecson-Mendoza EM, Mikami B, Utsumi S: Structure of 8Salpha globulin, the major seed storage protein of mung bean. Acta Crystallogr D Biol Crystallogr 2006:62:824-832.

[139] Kohno M, Motoyama T, Shigihara Y, Sakamoto M and Sugano H: Improvement of glucose metabolism via mung bean protein consumption: A clinical trial of glucodia tm isolated mung bean protein in Japan. Funct Foods Health Dis. 2017:7:115-134.

[140] Kohno M, Sugano H, Shigihara Y, Shiraishi Y, Motoyama T: Improvement of glucose and lipid metabolism via mung bean protein consumption: Clinical trials of GLUCODIA™ isolated mung bean protein in the USA and Canada. J Nutr Sci. 2018:7:e2.

[141] Anstee QM, Targher G, Day CP: Progression of NAFLD to diabetes mellitus, cardiovascular disease or cirrhosis. Nat Rev Gastro Hepatol 2013:10:330-344.

[142] Torre-Villalvazo I, Tovar AR, Ramos-Barragan VE, Cerbon-CervantesMA, Torres N: Soy protein ameliorates metabolic abnormalities in liver and adipose tissue of rats fed a high fat diet. J Nutr. 2008:138:462-468.

[143] Yang HY, Tzeng YH, Chai CY, Hsieh AT, Chen JR, Chang LS, Yang SS: Soy retards the progression of non-alcoholic steatohepatitis via improvement of insulin resistance and steatosis. Nutrition. 2011:27:943-948.

[144] Wanezaki S, Tachibana N, Nagata M, Saito S, Nagao K, Yanagita T, Kohno M: Soy beta-conglycinin improves obesity-induced metabolic abnormalities in a rat model of nonalcoholic fatty liver disease. Obes Res Clin Pract. 2015:9:168-174.

[145] Watanabe H, Inaba Y, Kimura K, Kaneko S, Asahara S-I, Kido Y, Matsumoto M, Kohno M, Tachibana N, Inoue H: Dietary mung bean protein reduces hepatic steatosis, fibrosis, and inflammation in male mice with diet-induced, non-alcoholic fatty liver disease. J Nutr. 2017:147:52-60.

[146] Xie J, Du M, Shen M, Wu T, Lin L: Physico-chemical properties, antioxidant activities and angiotensin-I converting enzyme inhibitory of protein hydrolysates from mung bean (*Vigna radiate*). Food Chem. 2019:270:243-250.

[147] Wongekalak L-o, Sakulsom P, Jirasripongpun K, Hongsprabhas P: Potential use of antioxidative mung bean protein hydrolysate as an anticancer asiatic acid carrier. Food Res Int. 2011:44:812-817.

[148] Li GH, Wan JZ, Le G-W, Shi Y-H: Novel angiotensin I-converting enzyme inhibitory peptides isolated from alcalase hydrolysate of mung bean protein. J Pept Sci. 2006:12:509-514.

[149] Bartholomae E, Incollingo A, Vizcaino M, Wharton C, Johnston CS: Mung bean protein supplement improves muscular strength in healthy,

underactive vegetarian adults. Nutrients. 2019:11:2423.

[150] Yi-Shen Z. Shuai S. FitzGerald R: Mung bean proteins and peptides: Nutritional, functional and bioactive properties. Food Nutr Res. 2018:62. DOI:10.29219.

Chapter 10

Unlocking Pharmacological and Therapeutic Potential of Hyacinth Bean (*Lablab purpureus* L.): Role of OMICS Based Biology, Biotic and Abiotic Elicitors

Krishna Kumar Rai, Nagendra Rai and Shashi Pandey-Rai

Abstract

Hyacinth bean also known as Indian bean is multipurpose legume crops consumed both as food by humans and as forage by animals. Being a rich source of protein, it also produces distinct secondary metabolites such as flavonoids, phenols and tyrosinase which not only help strengthened plant's own innate immunity against abiotic/biotrophic attackers but also play important therapeutic role in the treatment of various chronic diseases. However, despite its immense therapeutic and nutritional attributes in strengthening food, nutrition and therapeutic security in many developing countries, it is still considered as an "orphan crop" for unravelling its genetic potential and underlying molecular mechanisms for enhancing secondary metabolite production. Several lines of literatures have well documented the use of OMICS based techniques and biotic and abiotic elicitors for stimulating secondary metabolite production particularly in model as well as in few economically important crops. However, only limited reports have described their application for stimulating secondary metabolite production in underutilised crops. Therefore, the present chapter will decipher different dimensions of multi-omics tools and their integration with other conventional techniques (biotic and abiotic elicitors) for unlocking hidden genetic potential of hyacinth bean for elevating the production of secondary metabolites having pharmaceutical and therapeutic application. Additionally, the study will also provide valuable insights about how these advance OMICS tools can be successfully exploited for accelerating functional genomics and breeding research for unravelling their hidden pharmaceutical and therapeutic potential thereby ensuring food and therapeutic security for the betterment of mankind.

Keywords: Hyacinth bean, OMICS, biotic and abiotic elicitors, therapeutic, secondary metabolites, miRNAs

1. Introduction

Evolution, expansion and transformation of several wild crops via domestication and breeding have blessed the humans and animals with never ending wide varieties of plant-based foods around the globe [1]. Nevertheless, ~ 1 billion population around the world combat with hunger and malnutrition as they are unable to consume important vitamins/minerals thus affecting food and nutritional security in many developing countries [2]. These nutritional deficiencies could be due to the increased inclination towards consumption of specific crops as majority of the peoples relies on wheat, rice and maize for their food [1]. Several lines of literatures have documented that around 50% of total world population relies on above crops for catering while other crops like legumes are cultivated and consumed by marginal communities [3]. The crops cultivated by marginal communities are referred as neglected and underutilised crop species (NUCS) and rich source of vitamins, minerals and secondary metabolites having pharmaceutical properties [4]. These NUCS have the potential to counteract malnutrition by ensuring health/nutritional security, alleviate poverty by increasing resilience and sustainability to the farming systems [1]. However, compare to mainstream crops, less focused have been given towards the genetic improvement of NUCS.

Hyacinth bean (*Lablab purpureus* L.) which comes under the NUCS category is also an underutilised legume crops with great potential. The notable properties of hyacinth bean are (i) great adaptability under wide range of environmental conditions, (ii) rich source of protein, fibre and secondary metabolites like tyrosinase which is used in the treatment of various chronic diseases (iii) being stress tolerant, it require low cultivation/maintenance cost which could help smallholder farmers and their communities to generate more income [5, 6]. In Asia and Africa, hyacinth bean is also extensively consumed both as pulse and vegetable and is also exploited as green manure, forage/fodder for livestock, ornamental or medicinal herb [6]. The crop is ancient origin of which dates back to 1500 B.C. in Africa and later on introduce to India where it is mainly cultivated in tropical and subtropical regions and commonly known as "Sem" [7]. Besides, hyacinth bean possesses remarkable therapeutic potential for pharmaceutical application therefore, also considered as medicinal legume [5–7]. Consequently, hyacinth bean is tremendously used by therapeutic/pharmaceutical industries like for the large-scale production of drugs and skin ointments as it is rich source of aloe-emodin, rhein, chrysophanol, alkaloids, flavonoids and tyrosinase with broad spectrum pharmacological activities [8].

Furthermore, the seeds of hyacinth bean are also abundant in carbohydrate myoinositol that are exceptional in ovarian function in women by controlling oligomenorrhea and polycystic syndrome [8]. Additionally, the seeds also contain brassinolide which is a steroid which is clinically proven to cure prostate cancer in humans. The alkaloid spermidine found in hyacinth bean seeds is comprehensively used as a biomarker for the perception of skin cancer where as another alkaloid spermine is commonly used in the treatment of cancer/tumours [9]. Trigonelline another alkaloid found in its seed has demonstrated its role in the treatment of diabetes mellitus and also possess antimicrobial property against *Salmonella enterica* [8, 10]. Further, pantothenic acid obtained from hyacinth bean leaf have shown potential to combat aponeurosis thereby by stimulating fibroblast content in early postoperative period in rabbits [10, 11]. Flt3 receptor interacting lectin (FRIL) isolated from hyacinth bean seeds are exclusively employed as preservatives and has been used for preserving human cord blood for 1–2 months [12]. The hyacinth bean is also a rich source of flavonoids like isoflavone, kievitone and genistein which are phenomenal in reducing breast cancer progression in humans [12].

Despite of its pharmaceutical/therapeutic importance and catering food requirements of both humans and animals, hyacinth bean still lack focused research for its genomic improvement as compared to other mainstream crops (wheat, rice maize etc.). The genomic improvement through state-of -the art tools and techniques will not only reform its architectural growth but will also pave the way for rewiring the biosynthesis of imperative metabolites which will significantly impact its growth, yield and therapeutic potential. Therefore, this chapter provides valuable insights about the different state-of -the art tools and techniques that can be employed for the genetic improvement of hyacinth bean and how they can be exploited to inspire its therapeutic potential. Further, role of biotic and abiotic elicitors in stimulating the production of important metabolites in hyacinth bean has also been critically reviewed.

2. Medicinal/therapeutic properties of hyacinth bean

Being sessile in nature, plants have to withstand against various adverse climatic conditions to maintain their growth and developmental architecture. The plants are able to survive stressful conditions by synthesising diverse range of secondary metabolites and protease inhibitors that improve their adaptability [13]. Hyacinth bean for example, produces higher level of trypsin inhibitor (14–27 unit/mg) which is a unique property of this orphan legume crop compared to any other major legumes [14]. Like other serine inhibitors, trypsin inhibitor could also function as antifeedant or could also be responsible for strengthening growth, development and productivity by efficiently modulating proteolytic events with in hyacinth bean, mechanism of which has yet to be revealed [15]. Besides this, hyacinth bean also contains wide range of alkaloids, phenols and flavonoids which can be used in treatment of various chronic diseases essentially arthritis, nephritis sepsis as well as skin diseases thus significantly contributing towards human and animal health. All these nutritional and therapeutic properties make hyacinth bean a prime source of food, forage and cash crop in arid and semi-arid areas. However, till date the genes encoding these secondary metabolites are still ambiguous as the crop it self is considered as "orphan crop" for its genome revolution [1]. Further, both conventional and molecular breeding techniques also have been futile in the identification/linking of quantitative trait loci (QTLs) with production of these imperative secondary metabolites [4]. Therefore, all the above information's have reinstated the need to implement advance omics technology for unleashing the genetic constituent of hyacinth bean and to identify genes/proteins involved in the biosynthesis of important secondary metabolites.

2.1 Anti-inflammatory and analgesic properties

The phytochemicals or secondary metabolites synthesised by various under-utilised crops have the potential to boost innate immune response in humans as well as in animals thus providing immunity against infection, injury and irritation [7]. Several lines of literatures have strongly substantiated that various fruits, vegetables and food legumes synthesise various phytochemicals which are effectively exploited for the treatment of anti-inflammatory disorders, however their mechanism of action is still vague and needs to explored [13]. Various legumes such as soybean, mung bean, moth bean including hyacinth bean have diverted the attention of plant science community due to the presence of functional biological compounds which not only have health benefits and can also be simultaneously used for the treatment of various chronic diseases [8]. Researchers have analysed, tested and confirmed that the crude extracts of mung bean, hyacinth bean and soybean checks the synthesis of nitric oxide (NO) which is an inflammatory mediator thus significantly reducing the ear edema in mice caused by up-accumulation of arachidonic acid [16]. Likewise, another researcher evaluated crude extract of *Phaseolus vulgaris* and hyacinth bean was also effective in controlling the expression of 15-lipooxygenase (15-LOX) thus supressing the release of NO and prostaglandin E_2 (PGE_2). The limited release of NO and PGE_2 further downregulated the expression of inducible nitric oxide synthase (iNOS) and cyclooxygenase-2 (COX-2) with in the macrophages thus preventing the inflammatory disorder [17]. Overall, the result indicated that ethanolic extract of the above-mentioned beans can effectively modulate anti-inflammatory response by regulating the expressions of anti-inflammatory enzymes and transcription factors [17].

The phenols present in the dry seeds of legumes such as hyacinth bean have also been implicated to exaggerate anti-inflammatory response upon their adequate consumption [18]. A plethora of research have well indicated that seed and other ethanolic extract of food legumes is rich source of polyphenols and natural antioxidants capable of stimulating anti-inflammatory activity by suppressing the expression of 15-LOX as well as modulating the expression of cyclooxygenase −1 (COX-1) and COX-2 [19]. Similar findings have also been reported by Zhu et al. [16] in pinto bean, black bean and common bean where seed extract was effective in regulating the expression of interlukin-6 (IL-6), interferon-γ (IFN- γ) and IL-17A thus effectively ameliorating acute colitis in mice. In addition to phenolic compounds, these legumes also contain lectins which is protein capable of showing anti-inflammatory response after binding reversibly to carbohydrates [20]. For example, lectins isolated and purified from *Clitoria fairchildiana* showed enhanced anti-inflammatory activity against paw edema and reducing it up to 70% in affected mice [21]. Nonetheless, these phytochemicals have also been identified and documented in hyacinth bean which are capable of inducing multitude of immune response against chronic diseases, however, the functional genes/proteins responsible for their synthesis, transport and mechanisms by which they induce immune response needs an in-depth investigation.

2.2 Anti-diabetic properties

The flavonoids such as flavanones, flavanols, anthocyanidins, flavones and isoflavones present in fruits and vegetables have delineated themselves as key players in the treatment of cardiovascular diseases, diabetes and cancers [22]. Various legumes are also a rich source of dietary flavonoids that can regulate carbohydrate digestion, glucose uptake and insulin signalling via various signalling pathways [23]. Among all the flavonoids, dietary isoflavones *per se.,* daidzein and genistein are abundantly found in leguminous plants [24] found exclusively in soy foods. Increasing evidences have well suggested and evaluated the anti-diabetic properties of both the isoflavones i.e. daidzein and genistein in cell culture studies [24]. Dietary intake of both the isoflavones have been shown to modulate glucose metabolism and insulin levels in Type 1 & 2 diabetes thus exerting anti-diabetic effect by increasing lipid plasma composition and accordingly insulin sensitivity [25]. Concomitantly, another study used soy supplemented diet to control blood glucose level in mice [26]. The result of the study successfully demonstrated that soy food significantly improves lipid profile and glucose metabolism in the mice as direct result of increased phosphorylation of AMP activated protein kinase (AMPK) which in turn have caused favourable metabolic changes upon activation of genes/proteins involved in fatty acid oxidation pathway in peroxisome [27]. This report is consistent with the previous findings attributing exceptional role of soy food in the modulation of genes/proteins of AMPK pathway thus efficiently controlling blood glucose levels similar to other flavonoids [27].

Some researchers have also reported neutral to moderate effect of soy food and methanolic extract of hyacinth bean rich in isoflavones in controlling plasma lipid profile thereby confirming anti-diabetic effect of isoflavones could act differentially under *in-vitro* and *in-vivo* conditions [28]. For instance, dietary intake of isoflavones prevented the onset of diabetes in rats and improved blood glucose homeostasis by stimulating the function of pancreatic β-cells. Further, the researchers also noticed a sharp increase in insulin/glucagon ratio and C-peptide level in affected mice as compared to normal healthy rats [29]. The possible reason behind improved insulin/glucagon ratio in affected mice that consumed isoflavones could be due to the downregulation of gluconeogenic enzymes such as phosphoenolpyruvate carboxykinase and glucose-6-phosphatase beta oxidation. Furthermore, researchers have also indicated that genistein could modulate pancreatic β-cells via distinct metabolic pathways, Ca^{2+} signalling and calmodulin kinase II pathway [30] thus effectively regulating insulin synthesis in target cells/tissues. The isoflavones such as genistein and daidzein have also been shown to modulate Janus kinase/Signal transduce and activation of transcription (JAK/STAT), ERK-1/2 (serine/threonine protein kinase) and nuclear factor kappa-light chain enhancer of activated B cells (NF-κB) pathways thus stimulating the synthesis cytokinin's in response to pathological alterations [31].

Genistein have also been documented to stimulate the expression of protein kinase A and cAMP cascade which play important role in the proliferation of INS1 and pancreatic β-cells thus efficiently regulating glucose metabolism in mice [31]. However, in addition to isoflavones, anthocyanidin found in soybean seeds rich in cyanidin, delphinidin and petunidin have also demonstrated anti-diabetic effect in streptozotocin induced diabetic rats [32]. Researchers have used methanolic extract of anthocyanidin to diabetic mice and observed that the anthocyanidin effectively raised serum insulin concentration and glucose metabolism in rats. The possible reason behind the anti-diabetic effect of anthocyanidin could be due to the enhance translocation of GLUT4 (glucose transporter) which in turn have stimulated glucose uptake or anthocyanidin could have improved insulin signalling by causing phosphorylation of insulin receptor [33]. Similarly, in another study, researchers have also documented the beneficial effect of anthocyanidin by analysing it on diabetic animal model where they observe that diabetic animal treated with soybean anthocyanidin showed enhanced plasma insulin levels and low triglyceride content [34]. Furthermore, researchers continued their observation up to 12 weeks and observed that the diabetic mice exhibited reduced body weight, blood glucose level, triglyceride levels as revealed by lower expression of lipogenic gene expression in liver and fat [35]. Although, various studies have demonstrated the anti-diabetic effect of both isoflavones and anthocyanidin on animal system, there effect on controlling diabetes in humans are still limited. Therefore, efforts are needed to expand the dimension of research involving structural, biochemical and molecular characterisation of important therapeutic compounds obtained from underutilised legume crops for their efficient use in the human and animal welfare.

2.3 Anti-cancerous/tumour properties

The bioactive peptides found in certain legumes and cereals crops has been implicated to regulate growth and development of crops plants by imparting biotic and abiotic stress tolerance [36]. Further, researchers have also isolated and purified some of the plant bioactive peptides and demonstrated their pivotal impact on human health and immune response [37]. Lunasin, a 43 amino acid peptide initially identified and isolated from soybean has shown its tremendous competency in inhibiting cell division in tumour/cancer cells and protect DNA damage by delaying histone acetylation in mammalian cells under oxidative stress [38]. Later, lunasin was also identified in cereals and pseudo-cereals such as rice, wheat, barley and amaranth, however, its present in extremely low quantity as compared to soybean [39]. Being a rich source of lunasin, soybean has been extensively investigated in order to get valuable insight into its structure and function properties, mode action in preventing cancer and the ecological factors that can influence its biosynthesis and transport [37]. Initially, lunasin was identified as chemo-preventive agent but in-depth investigations by several researchers demonstrated that lunasin can effectively suppress skin tumorigenesis in mouse by delaying foci formation in DMBA NIH/3 T3 cells [40].

In addition, researchers have also well documented the chemo-preventive property of lunasin in breast cancer affected mice where they observed significant reduction (30–40%) in tumour cells after treating the mice with lunasin for two months [37]. However, not much research has been focused on lunasin therapeutic properties in soybean as well as in other underutilised legumes still researchers have hypothesised its broad-spectrum role in the treatment of lung cancer, colon cancer and leukaemia [36]. One of the possible mechanisms by which lunasin block cell division in cancer cells could be due to its ability inhibits cell cycle at G2 phase thereby inducing apoptosis in the affected cells [40]. Initial studies on lunasin's mode of action revealed that it can bind to hypoacetylated histone cores in cancer cells and inhibit acetylation in breast cancer cells and prostrate cancer cells [37]. Recently, researchers have made striking discovery claiming that lunasin binding can suppress the integrin signalling in cancer/tumour cells thereby inhibiting focal adhesion kinase/protein kinase B (FAK/AKT) and extracellular signal-regulated kinase 1 (ERK1) signalling in cancer cells [41]. Certain plant protease inhibitors such as Bowman-Birk inhibitors and flavonoids such as flavon-3-ols found in soybean and other legumes have also demonstrated their role in controlling breast and colon cancer [41]. However, detailed characterisation of their structural and functional properties in many legume crops is still ambiguous and need extensive research by employing advance omics technology for their potential application.

2.4 Anti-hypertension properties

Hypertension is one of the most important factors (apart from diabetes and high cholesterol level) causing cardiovascular disease in humans which is characterised by the increase in systolic/diastolic arterial pressure [42]. Studies have well documented that healthy diet/lifestyle i.e. reduce sodium intake and physical exercise are important factors controlling blood pressure, hypertension and ultimately risk of cardiovascular disease [42]. Various major and underutilised legumes are rich source of secondary metabolites, fibres and micronutrients thus forming an important framework of plant's bioactive compounds for healthy diet [43]. For example, some bioactive peptides from food as well as grain legumes have demonstrated their potential to combat both hyper and hypotensive effects. Peptides having Glu-Phe, Ile-Arg and Lys-Phe dipeptides identified form legume crops have shown anti-hypertensive effect by inhibiting the activity of Angiotensin-I- Converting Enzyme (ACE) [36, 44]. Similarly, proteins like tyrosinase and lupin present in legume crops have also shown their remarkable ability control both systolic/diastolic blood pressure in peoples suffering from hypertensive disorder [44].

The hypertensive property of both lupin and tyrosinase have also been extensively investigated under in vivo conditions in Goto-Kakizaki rats suffering from hypertension due to excessive consumption of Na rich diet. The researchers fed the hypertensive rats with both lupin and tyrosinase for two weeks and then observed significant reduction in the systolic/diastolic pressure in both the groups [45]. However, lupin treatment also significantly improved endothelium-dependent vasodilation in hypertensive rats more efficiently as compared to tyrosinase [45]. A large body of literatures have also indicated that these bioactive peptides/proteins do not only possess hypertensive and ACE inhibitory effect but are also actively involved in lowering cholesterol and lipid levels [44]. Researchers have also extensively studied hypocholesterolaemia by using bioactive peptides and proteins and identified that the peptide Ile-Ala-Val-Pro-Gly-Glu-Val-Ala was compellingly involved in lowering cholesterol and triglycerides levels by stimulating the activity of bile salts [46]. Furthermore, other studies have also well documented the role of soybean peptides/ proteins in efficiently controlling high cholesterol and lipid levels by efficiently modulating ratio of low density/high density lipo-proteins and expression of beta-very-low-density lipoprotein (β-VLDL) receptors thus minimising risk factors for cardiovascular disease [46]. Researchers have identified and evaluated several of these bioactive peptides from other legumes crops as well, however efforts are needed for in-depth characterisation of their function and mode of action in other underutilised legumes such as in hyacinth bean.

2.5 Antioxidative properties

Reactive oxygen species (ROS) generated as a consequence of oxidative stress are concomitantly involved in the onset and progression of various chronic diseases. Increased level of ROS has been shown to instigate severe damage to nucleic acids, cause membrane damage via lipid peroxidation and inhibit protein synthesis thus causing cell death or apoptosis [47]. Several crop plants including legumes contains various bioactive compounds such as flavonoids, phenols and some peptides that can efficiently scavenge ROS thus ameliorating stress induced oxidative damages [48]. Flavonoids such as flavanones and flavon-3-ols present in the seeds of certain leguminous plants such soybean and hyacinth bean have been reported to have antioxidative effect as demonstrated by both animal and cell culture studies [49]. In a study conducted by Babu et al. [50] oral infusion of flavanones and flavon-3-ols to an alloxan induced insulin dependent diabetic mouse, significantly enhanced the activity of hepatic catalase, superoxide dismutase and glutathione reductase enzymes thereby confirming their function as antioxidants. Similarly, in another study, researchers orally administered a flavonoid rich compound apigenin to streptozotocin-induced diabetic rats that significant reduction in the triglyceride levels which could be due the antioxidative effect of apigenin that effectively maintained ion/osmotic homeostasis [51]. Moreover, like apigenin, researchers also used anthocyanidin and luteolin treatment to diabetic rats which ultimately protected rat cells from oxidative damage via controlling the synthesis of iterlukin-1β and interferon-γ [52]. Like other flavonoids, anthocyanidin is also extensively present in legume plants which have received significant recognition owing to their health benefits and potential antioxidative properties [52].

Antioxidant peptides like His-Trp-Tyr-Lys have demonstrated to play exceptional role in ameliorating stress induce oxidative damage by efficiently regulating the scavenging of ROS [53]. Moreover, several studies have shown that thiol group of Cys residue can efficiently chelate metallic ions thus lowering the effect of oxidative stress and minimising the onset of chronic disease [53]. A study conducted by Morales-Medina et al. [54] reported that Val and Leu residues present at N-terminus of a peptide and Tyr and Trp residues present at C-terminus of same peptide have antioxidative properties that were effective in minimising lipid peroxidation and strengthening ion homeostasis. Furthermore, it is also well documented that seeds and leaves of legume plants are rich source of other bioactive compounds such as anthocyanins, polyphenols with antioxidative properties and are also capable of performing metal sequestration and stimulate the expression of other stress responsive genes [55]. Additionally, Zhu et al. [56] evaluated various other peptides from soybean and wheat having Leu-Leu-Pro-His-His repeat for its antioxidative activity by using distinct experimental procedure and conditions. The results indicated that the peptide was effective in stimulating 1,1-diphenyl-2-picrylhydrazyl (DPPH) activity, expression of enzymatic and non-enzymatic antioxidants such as catalase, superoxide dismutase, peroxidase and ascorbate thus controlling the level of ROS generation and minimising the chance of severe disease. Lunasin peptide found in soybean and other legume has also been extensively investigated for its antioxidative properties where the researchers documented that lunasin was effective in scavenging both hydrogen peroxide and superoxide anion thereby protecting cell from oxidative damage [42].

2.6 Cytotoxic properties

Since ancient times legumes have been ascribed to have pharmaceutical/therapeutic values far beyond than providing essential nutrition in the form of amino acids [57]. In recent years, various proteins/peptides form several legume species have been included in the category of nutraceuticals i.e., food or products derived from them having medicinal or therapeutic role in the prevention of disease along with nutritional benefits [57]. Various legume-based bioactive proteins/peptides have been isolated and characterised for their functional role such as Bowman-Birk inhibitors (BBIs), Kunitz inhibitors (KIs) and alpha amylase inhibitors (AAIs) which are also commonly known as anti-nutritional compounds [58]. Several researchers during their early epidemiological studies observed that the protease such as Bowman-Birk inhibitors isolated from soybean seeds were highly effective in the counteracting tumour growth under both *in vitro* and *in vivo* conditions [58]. Later, these inhibitors also demonstrated their involvement in the treatment of hypocholesteraemia, cell toxicity, lowering of blood glucose level and pressure. The BBIs are distributed across many plant species including fruits and vegetables which are characterised by the presence of conserved pattern of 14 cysteine residues forming disulphide linkages having multigene origin. An exceptional property of these protease inhibitors is that they are structurally and functionally stable under changing environment conditions and can effectively bind to IgE thus embarking their anti-proliferative effect in gut mucosa and colon cancer thereby keeping cellular toxicity at a bay [59]. Additionally, researchers have isolated and purified these inhibitors from liver, kidney and lungs to understand their mechanism of action however, their course of action is still under debate [60].

Furthermore, AAIs have demonstrated themselves has a suitable candidate for controlling triglyceride levels thus keeping obesity under check whereas lectins obtained from the seeds of legumes have also shown immense therapeutic potential displaying cytotoxic and anti-cancer activity [60]. For example, concanavalin lectin obtained from the seeds of *Canavalia ensiformis* L. are structurally stable and are highly resistant to denaturation and in vivo proteolysis displaying strong anti-hepatoma activity under acidic conditions [61]. The cytotoxic and anti-cancer activity of lectins have also been demonstrated using animal model where internalisation of lectins in small intestine showed stimulation in immune and hormonal activity thus confirming their role as therapeutic agents. Researchers in last few years have identified and characterised AAIs from soybean and *Phaseolus vulgaris* for clinical studies and obtained interesting outcome as AAIs were effective in controlling obesity and blood glucose level in hypercholesterolemic rat model [60]. Similarly, α-subunit of soybean 7S globulin protein has been well ascribed to stimulate the transcription of HepT9A4 hepatic cells thus increasing low density lipoprotein (LDL) uptake in HepG2 cells in hypercholesterolemic rat model [62]. However, the involvement of these protease inhibitors and lectins in the treatment of various chronic diseases are mainly confined to in vitro studies or animal model, therefore, efforts are needed to increase the dimension of their application by performing more human clinical trials [63]. Furthermore, efforts are also being diverted towards the identification of these medicinally important bioactive compounds in underutilised legumes such as hyacinth bean for increasing the bioavailability of these bioactive compounds for the benefit of mankind.

2.7 Anti-microbial properties

Several major and underutilised legumes are rich source of bioactive phenolic compounds or polyhydroxylated compounds with immense anti-nutritional and therapeutic potential [46]. These phenolic compounds also play significant role in the stimulation of immune response, protect cells from oxidative damage and other pathogenic diseases [64]. Several studies have documented that some phenolic compounds isolated from seeds of legume crops are indispensably involved in the treatment of cancer disease, microbial/pathogenic attack, inflammatory disease thus providing potential health benefits [46]. Phenolic compounds are large group of compounds comprising phenolic acids, flavonoids, tannins and stilbenes [64]. Several recent studies have well documented the anti-microbial activity of phenolic compounds obtained from Faba bean, broad bean, adzuki bean and Dolichos bean in their crude methanolic extract of leaf and seeds [65]. The total antioxidant activity (TAA) of methanolic extract of various phenolic compounds and tannins obtained from adzuki bean and lectins have been shown to exhibit anti-microbial activity against several bacterial strains showing average zone of inhibition of 8–20 mm [66]. In addition, these methanolic extract have also shown potential anti-fungal activity against *Saccharomyces cerevisiae, Candida albicans* and *Aspergillus niger* [65]. Plethora of research have well documented that phenolic compounds are actively involved in the termination of ROS signalling as well as metal sequestration thus strengthening anti-microbial activity against various pathogenic micro-organisms [66].

In addition to phenolic compounds, several bioactive peptides have also been instigated to play important role in regulating various biological activities along with antimicrobial and anti-inflammatory effects [57]. Studies have well reported that several of the ACE-inhibitory peptides containing Arg-Lys residues at C-terminus have shown enhanced anti-microbial activities against pathogenic microorganism [44]. Similarly, peptide containing Leu-Leu-His-His residues also have been shown to posses anti-microbial and anti-oxidative properties. Moreover, a group of researchers working on legume bioactive proteins attempted to used bioactive peptides in conjunction with phenolic compounds and ascertain that the amalgamation of both stimulate the defence mechanisms of plants against pathogenic attack [44]. Similarly, a protein dolichin extracted from hyacinth bean exhibited strong anti-microbial activity against *Fusarium oxysporum, Rhizoctonia solani and Coprinus comatus* [10]. Likewise, a 36 KDa AAIs from hyacinth bean showed significantly inhibited conidial germination and hyphal growth of *A. flavus* [10]. In addition, several studies have well documented to inhibit the progression of human immunodeficiency virus (HIV) by regulating the expression of reverse transcriptase and alpha/beta glucosidase enzyme as well as it also demonstrated to have low ribonuclease and translational inhibitory activities [44].

3. Integrated "OMICS" techniques for enhancing its therapeutic potential

Plants act as factories that synthesises wide array of nutritional and secondary metabolites with complex structure and functions. Essentially, therapeutic/pharmaceutical secondary metabolites are often extensively isolated and purified from wild plant species or under-utilised crops as compared to cultivated species. However, the chemical synthesis of these medicinally important metabolites is a daunting challenge and is not economically feasible. Recent advancement in the system biology tools have pave the way to exaggerate their synthesis in tissue culture plants, but still their applications are limited to certain plant species because of the complex nature of technological standardisation in respective crops and lack of proper understanding of biosynthetic pathway. In this section, we will be discussing recent advancements made in the system and synthetic biology tools to provide detailed glimpse of the biosynthetic pathways and to explore the unprecedented possibilities of their application for the human welfare. These cutting-edged technologies can be successfully exploited for the improvement/enhancement of secondary metabolites production or could also help in the identification of novel metabolites in cultivated plants as well.

3.1 Phenomics based imaging and analytical toolkits

The phenotype exhibited by plants at certain stages of growth/developments are the function of gene × environment interaction that govern a peculiar trait of interest expressed from the plant's genome [67]. The term "phenotype" corresponds to precise and rigorous recording of the distinct phenotypic parameters from single cell to whole plant level, which if conducted explicitly can help facilitate identification/classification of novel traits in several plant species. Phenomics is a sub-discipline of plant biology that deals with phenotyping under controlled green-house conditions as well as field experimentation using advanced imaging technologies and imaging tools [67]. Phenomics study is a three-step process involving (i) setting up experimental plot, light intensities, nutrition acquisition and temperature (ii) rigorous monitoring/phenotyping such as growth, stress response, photosynthesis, chlorophyll and secondary metabolite contents etc. using advanced imaging tools and (iii) computer-assisted data visualisation, interpretation and storage [68]. Recent technological advancements have paved the way for the development of high-resolution imaging platforms aided with advanced bio-informatic tools for the phenotyping several important traits in plants for cellular and functional analysis [69]. Therefore, phenomics has now been recognised as an indispensable tool that can provide valuable insights into plant's morphology and physiology which can be further integrated with functional genomics data for analysing key traits such as secondary metabolites production and other economically important traits [68].

Several informatorily databases and analytical toolkits have been developed to facilitate phenomics and taxonomic studies in various cultivated and under-utilised crops at a greater pace. For example., PlantCLEF (2019) is an online repository that contain wide variety of images of plant's organs with the sole purpose to facilitate identification and classification of underutilised crop plants having distinct features [70]. PlantCLEF act like a real-life computerised program that can identify and classify plant species using raw images by extracting similar traits/characteristics and matching them defined plant species and family [70]. Similarly, MPID (Medicinal plant images database) which is a premium database maintained by Hong Kong Baptist University that is known to accommodate vast range of phenotypic data related to medicinal and therapeutically important plants [71]. Furthermore, in addition to phenotypic data, it also acts as a repertoire of scientific/botanical names, therapeutic values, physiological and ecological parameters of more than 1000 medicinal plants. Likewise, MPDB (Medicinal plant database of Bangladesh) database is also specifically dedicated to store phenotypic and physiological data associated with medicinal and aromatic plants found in Bangladesh [72].

Apart from databases, several computer-based analytical tools and techniques have also been developed and implemented for recording high-resolution images and morpho-physiological parameters in selected plants [70]. Plant computer vision (PlantCV) is a freeware software package written explicitly in python language that provide valuable algorithms for analysing phenotypic data [71]. It can analyse phenotypic data for multiple plant species and compare them with in the database for identification of novel traits/characteristics in genetically un-explored crops [67]. Similarly, ImageJ is a Java based program equipped with various algorithms such as image enhancer, graphic correction, segmentation and measurement to facilitate accurate phenotyping of plant species [73]. HTPheno is an algorithm of ImageJ that allows monitoring of plant's growth and development in terms colour spectrum. It captures image related to growth and fitness by various angles, time period and temperature/light conditions in the form high-resolution images [74]. However, despite of these technological breakthroughs, the implantation of these state-of-the art techniques are limited certain plant species. Therefore, efforts are needed to establish, standardise and implement these advanced phenomics techniques in various under-utilised medicinally important crops in order to facilitate comprehensive analysis of their physiological, morphological and cellular functions.

3.2 Functional genomics approach

Identification of hereditary determinants governing morphological, physiological and biochemical properties are of astute importance to uncover genetic potential of plant species. With the advent of next-generation sequencing techniques it has now become possible to perform in-depth studies on economically/therapeutically important under-utilised crops [75]. Till date whole genome sequencing projects has led to the development of draft genomes and chloroplast genomes of various medicinally important plants which can be efficiently exploited in-conjunction with advanced bio-informatic tools to obtain information about gene families, gene regulatory networks, miRNA and non-coding RNAs involved in gene regulation in those plants whose genome sequence is not available [76]. Furthermore, it can also result in the development of DNA markers for DNA fingerprinting and DNA barcoding to facilitate efficient taxonomic identification of plant under study using specific region of DNA [77]. Several DNA fingerprinting/barcoding primers such as 18-S-rRNA, 5S-rRNA, rupture of the cranial cruciate ligament (rccl), maturase K (matK), internal transcribed spacer (ITS), intergenic spacer (trnH-psbA) have been successfully implemented for identification and classification of medicinal plants. In addition, several dominant and co-dominant markers such as single nucleotide polymorphism (SNP), sequence characterised amplified region (SCAR), amplified fragment length polymorphism (AFLP), inter simple-sequence repeat (ISSR) and random amplified polymorphic DNA (RAPD) have also facilitated identification and authentication of medicinal plants [76].

Transcriptome-wide profiling of genes of regulatory pathways can help researchers gain valuable insight into the functional mechanisms of plant's biosynthetic pathways. In the recent years, researchers have exploited expressed sequence tags (ESTs) for transcriptome wide analysis of important medicinal plants [77]. Later, the scientists began to use microarray which is probe hybridization-based technique for studying regulation of gene expression and candidate gene discovery [78]. Recently, various transcriptome-wide analysis studies have been conducted in several medicinally important plants and their sequencing and expression profiling data are available in various online databases such as GarlicESTdb (garlic EST database), GEO (gene expression omnibus), ArrayExpress, RASP (RNA atlas of structure probing), AgriSeqDB (RNA sequence database), EGENES

(EST database) that can help expedite transcriptomic research in those plants in which transcriptome wide analysis has yet not been completed [79]. Likewise, several toolkits have also been designed that explicitly analyse microarray data and can also be used in conjunction with other phenomics, transcriptomics, proteomics and epigenomics for the identification of functional biological pathways liked with secondary metabolite synthesis [79]. Notably used toolkits are iArray, BRB-Arraytools, KEGG (Kyoto encyclopaedia for genes and genomes), GENEVESTIGATOR, PLEXdb, ExPath are the ones which offers various features for microarray data analysis, visualisation, interpretation and annotation in the form of heat map, graph and tables [80].

In addition, few databases have also been developed such as CroFGD (*Catharanthus roseus* functional genomic database), TeaCon (database of gene co-expression network), PlaNet (plant co-expression network), AraNet (*Arabidopsis* co-expression network) for functional analysis and study of co-expression networks to identify functional biosynthetic pathways [81–83]. Furthermore, several non-coding RNAs (ncRNAs) such as small interfering RNAs (siRNA) and microRNAs (miRNAs) have also been discovered and are thought to play pivotal role in the regulation of secondary metabolite synthesis in medicinal and crop plants [84]. Intriguingly, several transcriptome-wide analyses in medicinal plants have well indicated that these ncRNAs whether siRNA or miRNA indeed have therapeutic properties which if harnessed systematically can help in the prevention of various chronic diseases such as cancer and influenza A virus infection [84]. In this context, a group of researchers have developed a miRNA database (MepmiRDB; medicinal plant microRNA database) devoted specifically for medicinal plants that provide plethora of information regarding gene sequence, expression levels and target miRNA for 30 different medicinal plants [85]. Besides, several software packages have also been developed such as sRNA-Seq-data, NATpipe, PLncPRO and CNIT that can greatly facilitate the identification ncRNAs, siRNAs and miRNAs in various medicinal plants as well as in crop plants that specifically involved in the regulation of secondary metabolites of therapeutic importance [85].

Several protein-coding genes have also been qualitatively and quantitatively analysed for their corresponding products to generate a profile of their proteome to help researchers gain valuable insights into the mechanisms underlying cellular and metabolic pathways in medicinal plants [86]. Fewer studies have been conducted to develop a complete proteome map in the medicinal plants describing the proteins involved in the regulation of secondary metabolite synthesis. For example, a study conducted by Jacobs et al. [87] identified various proteins involved in alkaloid biosynthesis in *C. roseus* using 2D gel electrophoresis and mass spectrometry. Likewise, Chin [88] also performed in-depth proteomic study using Matrix-Assisted-Laser Desorption and Ionisation (MALDI) Time of Flight (TOF) analysis to unravel proteins involved in the secondary metabolite production in the germinating seeds of orchid plants. In addition, several online toolkits such as STRING (search tools for retrieval of interacting genes), PAIR (predicted Arabidopsis interactome resource), UniProt, Pfam (protein families), IntAct (molecular interaction database) can also be exploited in non-model crop plants such as hyacinth bean to gain functional insight into proteins involved in the secondary metabolite productions [79]. A list of putative genes/TFs involved in the regulation of bioactive metabolites in legumes are presented in **Table 1**.

3.3 Metabolomics approach

Metabolomics is also a functional genomics tool with the sole purpose to provide in-depth understanding of different cellular and metabolic pathways in various

S. No	Legumes	Genes/Transcription factors (TFs)	Secondary metabolites	Pathway involved	Technique used	References
1.	*Glycine max*	Fatty acid desaturase 2 (FAD2)	Linoleic acid	Octadecanoid pathway	Generation of mutant followed by LC–MS analysis	Liu et al. [89]
2.	*Medicago truncatula*	Cytochrome 72A67 (CYP72A67), lateral organ boundaries domain TFs	Saponins	Isoprenoid pathway	Generation of mutant followed by GC–MS	Biazzi et al. [90]
3.	*Lupinus angustifolius* L.	Apetella 2/ ethylene responsive factor (AP2/ERF TF)	Quinolizidine alkaloids	Decarboxylation of lysine	Transcriptome de-novo assembly and QTL mapping	Kroc et al. [91]
4.	*Lupinus angustifolius* L.	13-hydroxylupanine O-tigloyltransferase (HMT/HLT), Lysine/ornithine decarboxylase (LDC) and 4-hydroxy-tetrahydrodipicolinate synthase (DHDPS)	Quinolizidine alkaloids	Decarboxylation of lysine	Transcriptome de-novo assembly and QTL mapping	Kroc et al. [91]
5.	*Trifolium repens* and *Medicago sativa*	R2R3-MYB TF (TaMYB14)	Proanthocyanidin	Flavonoid biosynthetic pathway	Gene silencing followed by LC–MS analysis	Hancock et al. [92]
6.	*Medicago truncatula*	MYB TF and MtLAR and MtANR	Proanthocyanidin	Flavonoid biosynthetic pathway	Targeted mutagenesis by HPLC analysis	Cañas and Beltrán [93]
7.	*Medicago truncatula*	Uridine diphosphate glucosyltransferases (UGT73K1 and UGT71G1)	Saponins and isoflavonoids	Triterpenoid saponins biosynthetic pathway	Localization through prokaryotic expression system followed by microarray analysis	Achnine et al. [94]
8.	*Leucaena leucocephala*	Hyp 1 – Hyp 7	β-amyrin and mimosine	Triterpenoid saponins biosynthetic pathway	Microarray analysis	Honda and Borthakur [95]
9.	*Lotus japonicus*	TM1624.23	Phenylpropanoid derivatives and pro-anthocyanidin metabolism	Phenylpropanoid pathway	Gas chromatography coupled to electron impact ionisation/time-of-flight mass spectrometry	Sanchez et al. [96]
10.	*Cicer arietinum*	CaUGT	Isoflavonoids	methylerythritol phosphate pathway	Next generation sequencing followed by marker assisted breeding	Jha et al. [97]

S. No	Legumes	Genes/Transcription factors (TFs)	Secondary metabolites	Pathway involved	Technique used	References
11.	*Medicago truncatula*	Trc genes	Trigonelline,	Tryptophane-kynurenine pathway.	Gene cloning and mutagenesis	Boivin et al. [98]
12.	*Medicago truncatula* and *Glycine max*	mtPAR, isoflavone synthase (IFS), mtTT8 and mtWD40 1	proanthocyanidin	Flavonoid biosynthetic pathway	Cloning, gene expression and microarray analysis	Li et al. [99]
13.	*Glycine max*	GmF3H1, GmF3H2 and GmFNSII-1	Isoflavones	Flavonoid biosynthetic pathway	CRISPR/Cas9-mediated metabolic engineering	Zhang et al. [82, 83]
14.	*Glycine max, Cicer arietinum*	No apical meristem-Arabidopsis transcription activator factor-Cup shaped cotyledon (NAC TF) NAC 4, NAC 29, NAC 25 and NAC 72	Abscisic acid and secondary metabolite synthesis	Biosynthetic pathway	Multi-OMICS platform	Jha et al. [100]
15.	*Glycine max*	GmCHS1–GmCHS9	Flavonoids and isoflavonoids	Flavonoid biosynthetic pathway	Cloning, gene expression and microarray analysis	Dastmalchi and Dhaubhadel [101]
16.	*Glycyrrhiza uralensi*	2-hydroxyisoflavanone synthase (CYP93C), 2,7,4 0-trihydroxyisoflavanone 4 0-O-methyltransferase/isoflavone 4 0-O-methyltransferase (HI4OMT) and isoflavone-7-O-methyltransferase (7-IOMT)	Flavonoids and isoflavonoids	Flavonoid biosynthetic pathway	Whole genome sequencing, assembly and gene expression	Mochida et al. [102]

Table 1.
List of putative genes/transcription factors and functional genomics tools involved in regulating biosynthesis of secondary metabolites in legumes.

organisms. Metabolomics is an advanced system biology tool with improved analytical methodologies, sensitivity and resolution that has been successfully exploited to understand biosynthesis of important metabolites in various plant species [103]. Several researchers have used this technique to discover candidate genes/proteins involved biosynthesis of specialised metabolites [104]. Furthermore, it has also provided great depth of understanding about the structural properties and diversity that exists among different metabolites as well as has facilitated to gain valuable insight into the type active ingredients that gives each metabolites its specific nutritional and medicinal properties [103]. Recent decades have witnessed the detailed characterisation of various medicinally important metabolites such as paclitaxel, artemisinin, vincristine, vinblastine, camptothecin and accuminata etc. from Pacific yew tree, *Artemisia annua*, *C. roseus, Camptotheca acuminata* and *Papaver somniferum* having anti-cancer and anti-malarial properties using this approach [104]. The identification of these medicinally important metabolites in above mentioned plants has served as model for studying the biosynthesis of specialised metabolites in other crop plants as well, which however, could not be possible by phenomics, genomics, transcriptomics and proteomics studies [105]. Metabolite profiling studies have been conducted in various transgenic plants by generating over-expression and gene-insertion based mutants to track the regulation of flavonoid biosynthesis (Nguyen et al. 20). In addition, metabolomics-based reverse genetic approach has also led to the identification of putative genes involved in the regulation of flavonoid synthesis driven by the conjugative action of post-translational modifications such as acetylation, phosphorylation, methylation, ubiquitination and biotinylation [105].

Several metabolomic studies have been conducted in model as well as crop legumes such as *Medicago truncatula*, *Lotus japonicus*, *Glycine max* and *Pisum sativum* to identify functional metabolites that are involved in the imparting biotic and abiotic stress tolerance for thus improving their growth and productivity [106]. However, fewer studies have conducted for the identification of medicinally important metabolites in legume plants compared to other model and medicinal plants [93]. Nonetheless, efforts are being made to revamp, standardise and implicate these advanced system biology tools for the identification, characterisation and quantification of important metabolites in various under-utilised crops as well [106]. The techniques like gas-chromatography mass spectrometry (GC–MS), liquid chromatography-mass spectrometry (LC–MS), nuclear magnetic resonance (NMR), capillary electrophoresis-mass spectrometry (CE-MS) and high-performance liquid chromatography-time of flight-mass spectrometry (HPLC-TOF-MS) have been successfully used for the assessment of medicinal constituents of functional metabolites [106]. With the advent of technological breakthroughs, metabolomics has also facilitated the generation of various protein reference maps of various model plant species and legume crops which can expedite the functional genomic analysis of genes/proteins in those plant species whose genome sequence is not available [93]. Nevertheless, efforts are needed to generate more protein reference maps to unravel cellular and biochemical signalling pathways and to identify novel genes and their product through comparative proteomic approach. The integrative analysis of “OMICS” datasets is crucial for the implementation of system biology tools for identification and mapping of secondary metabolite pathways in medicinal plants as well as legume crops and mechanism by which it can be achieved is depicted in **Figure 1**. Therefore, it has now become imperative to generate resourceful OMICS database that can help in the advancement integrative omics technology for precise understanding of molecular mechanisms and their possible application in legume improvement through breeding programs.

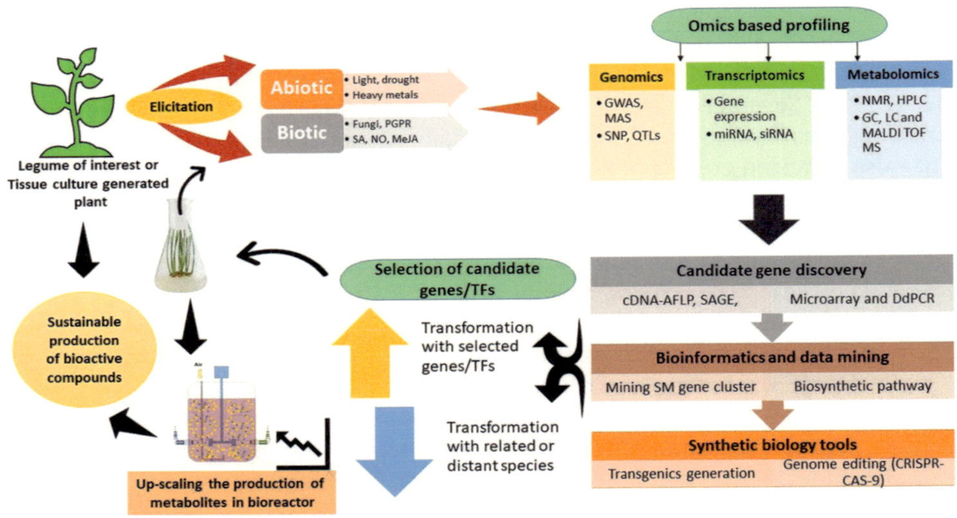

Figure 1.
Systemic workflow depicting the application of integrated OMICS tools as well as role of different biotic/abiotic elicitors for improving the biosynthesis known bioactive compounds and identification of novel therapeutic metabolites in legume crops. In this process, tissue culture plants or plants grown in field are treated with different biotic or abiotic elicitors either independently of in combination. The plants are then analysed for the differential expression of genes involved in the regulation of secondary metabolites using integrated OMICS techniques. Candidate genes are discovered using various techniques such as cDNA-AFLP, SAGE, analysed by bioinformatics tools and are rewired using synthetic biology tool. The transformed plants are then exploited for sustainable production of important bioactive metabolites. GWAS: Genome wide association studies; MAS: Marker assisted selection; SNP: Single nucleotide polymorphism; QTLs: Quantitative trait loci's; miRNA: microRNA; siRNA: Small interfering RNA; NMR: Nuclear magnetic resonance; HPLC: High performance liquid chromatography; GC: Gas chromatography; LC: Liquid chromatography; MALDI-TOF-MS: Matrix assisted laser desorption ionisation-time of flight-mass spectrometry; cDNA AFLP: Complementary DNA amplified fragment length polymorphism (RNA finger printing technique); SAGE: Serial analysis of gene expression; DdPCR: Differential display PCR; SM: Secondary metabolites; CRISPR-CAS 9: Clustered regulatory interspaced short palindromic repeat, CRISPR associated protein 9; TFs: Transcription factors.

4. Role of biotic and abiotic elicitors for enhancing its therapeutic potential

In plants, increase synthesis and accumulation of secondary metabolites occur upon their exposure to adverse climatic conditions which not only strengthen their growth but also revamp their innate immune response [107]. Several studies have indicated that distinct physical, chemical and microbial factors could act as abiotic/biotic elicitors for stimulating genes of metabolic pathways which will in turn result in the increase production important/specialised metabolites [108]. Now a day's elicitation is extensively used as a biotechnological tool to induce the biosynthesis of medicinally important metabolites in various tissues and organs of tissue cultured plants. The functional mechanism behind elicitor's elicitation involves signal perception by the receptors designed specifically to initiate signal transduction of the genes/transcription factor involved in the biosynthetic pathway (**Figure 2**) which in turn enhance the production and accumulation of different metabolites [107]. This section briefly describes different biotic/abiotic elicitors that can be employed for enhancing the production of secondary metabolites in medicinal as well as crop plants. A list of different biotic/abiotic elicitors involved in the regulation of bioactive metabolites in legumes are presented in **Table 2**.

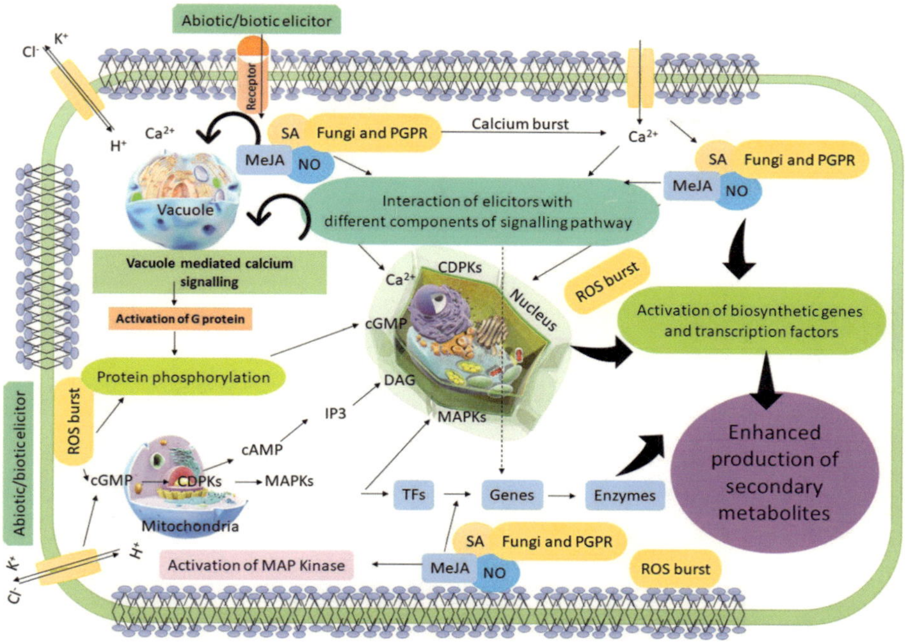

Figure 2.
Schematic representation of mechanisms by which biotic/abiotic elicitors signalling pathways involved in the biosynthesis of important bioactive metabolites in plants. The elicitors are perceived by the receptors which then interacts with various components of signal atransduction pathway. This interaction activates certain transcription factors which in turn regulate the expression of various biosynthetic genes/proteins thus stimulating enhanced production of important therapeutic metabolites. SA: Salicylic acid; NO: Nitric oxide; MeJA: Methyl jasmonate; ROS: Reactive oxygen species; CDPKs: Calcium dependent protein kinases; MAPKs: Mitogen activated protein kinases; IP3: Inositol triphosphate; DAG: Diacyglycerol; TFs: Transcription factors.

4.1 Biotic elicitors

Biological materials such as proteins, carbohydrates, inactivated enzymes, and polysaccharides etc. whether of plant, fungi or bacterial origin either in crude or purified form is used to induce the synthesis of secondary metabolites are termed as biotic elicitors [120]. Researchers have well indicated that proteins/enzymes are being explicitly used to stimulate the defence system of plants by increasing the synthesis of secondary metabolites involved in the regulation of stress responsive genes [121]. In tissue culture generated plants, several glycoprotein elicitors have been shown to elicit the production of phytoalexin, lectins and agglutinins that tremendously ameliorate the stress-induced oxidative damage [122]. Similarly, various fungal elicitor proteins such as PebC and PevD1 from *Botrytis cinerea* from and *Ss*Cut from *Sclerotinia sclerotiorum* elicited multiple defence response tomato and cotton plants in response to various biotic and abiotic stresses [123]. Furthermore, these elicitors have also been shown to activate G-proteins which in turn can also act as elicitor in stimulating secondary metabolite synthesis in plants such as stimulation of flavonoids and isoflavanones in soybean, benzophenanthridine alkaloids in bloodroot and β-thujaplicin in *Cupressus lusitanica* [122]. Ca^{2+} signalling also play pivotal role in the activation of various protein kinases such as calcium dependent protein kinase (CDPKs) and mitogen activated protein kinases (MAPKs) that have stimulated the sesquiterpenes biosynthesis in tobacco and French bean plants [120].

Polysaccharides such as xyloglucans, oligogalacturonides, hemicellulose and pectin derived from plant, bacterial or fungal cell wall could also be exploited as an

S. No.	Legumes	Abiotic/biotic elicitor used	Secondary metabolite elicited	Pathway involved	Probable role	References
1.	*Lupinus luteus*	Chitosan (0.12%), Salicylic acid (800 μM) and potassium cyanide (400 μM)	Isoflavonoid genistein	Phenylpropanoid pathway	Treatment of cancer, osteoporosis, and ischemic heart disease	Kneer et al. [109]
2.	*Vicia faba*	UV light (30-50 W for 5, 10 and 15 hr)	Phenolics and L-Dopamine	Pentose phosphate pathway	Act as neuromodulator and used for treatment of Parkinson's disease	Shetty et al. [110]
3.	*Medicago truncatula*	UV light (5.5 min at 8000 J m^{-2}), Methyl jasmonate (50 mM), Yeast	Triterpene saponins and other primary metabolites	Phenylpropanoid pathway	Act as anti-tumour, anti-mutagenic, anti-inflammatory, anti-viral and cardiac activities	Broeckling et al. [111]
4.	*Glycine max*	Methyl jasmonate (at 0.1 kg/m^3)	Genistein and Daidzein, and β-glycosidic type isoflavonoids	Phenylpropanoid pathway	Treatment of cancer, osteoporosis, and ischemic heart disease	Gueven and Knorr [112]
5.	*Lupinus luteus*	cadmium (at 10 mg/l) and lead (at 150 mg/l)	2′-hydroxygenistein glucoside and 2′-hydroxygenistein 7-O-glucoside	Phenylpropanoid pathway	Treatment of cancer, osteoporosis, and ischemic heart disease	Pawlak-Sprada et al. [113]
6.	*Lupinus angustifolius*	Fungal spore suspension (2×10^6 spores/ml, approximately 5 ml/plant)	Isoflavone phytoalexins or their precursors	Phenylpropanoid pathway	Treatment of cardiovascular disease, osteoporosis, hormone-dependent cancer and loss of cognitive function	Wojakowska et al. [114]
7.	*Phaseolus vulgaris*	Ascorbic acid (500 μM) Folic acid (50 μM) and glutamic acid (5 mM)	Phenolic composition and angiotensin I converting enzyme (ACE)	—	Treatment of hypertension and cardiovascular disease, inhibition of cholesterol synthesis	Dueñas et al. [115]
8.	*Phaseolus vulgaris*	Sucrose, gibberellins and proline	Quercetin-3O-glucoside, malvidin-3O-glucoside, and soyasaponins	Phenylpropanoid pathway	Act as anti-tumour, anti-mutagenic, anti-inflammatory, anti-viral and cardiac activities	Díaz-Sánchez et al. [116]

DOI: http://dx.doi.org/10.5772/intechopen.99345

S. No.	Legumes	Abiotic/biotic elicitor used	Secondary metabolite elicited	Pathway involved	Probable role	References
9.	*Glycine max*	$AgNO_3$ and H_2O_2	Glyceollin and Isoflavones	Phenylpropanoid pathway	Act as anti-tumour, treatment of cardiovascular disease and other chronic diseases	Kalli et al. [117]
10.	*Glycine max*	*B. subtilis and Rhizopus*	6-Prenyl daidzein and phaseol	Phenylpropanoid pathway	Menopausal relief, treatment of osteoporosis, blood cholesterol, and lowering the risk of some hormone-related cancers, and heart disease	Kalli et al. [117]
11.	*Trigonella foenum*	Arbuscular mycorrhizal fungal inoculum and exogenous methyl jasmonate	Trigonelline and diosgenin	Acetyl coenzyme A through the mevalonate pathway	A novel multitarget based chemo-preventive or therapeutic agent neuroprotective, anti-diabetic	Irankhah et al. [118]
12.	*Lens culinaris*	Sodium silicate	Flavonoids and phenolic acids	Phenylpropanoid and shikimic acid pathway	Act as anti-tumour, anti-mutagenic, anti-inflammatory, anti-viral and cardiac activities	Dębski et al. [119]
13.	*Trigonella foenum*	Sodium silicate + Fe EDTA	Flavonoids and phenolic acids	Phenylpropanoid and shikimic acid pathway	Act as anti-tumour, treatment of cardiovascular disease and other chronic diseases	Dębski et al. [119]
14.	*Medicago sativa*	Sodium silicate	Flavonoids and phenolic acids	Phenylpropanoid and shikimic acid pathway	Act as anti-tumour, anti-mutagenic, anti-inflammatory, anti-viral and cardiac activities	Dębski et al. [119]

Table 2.
List of different abiotic/biotic elicitors used for eliciting secondary metabolites production in legume crops.

elicitor to stimulate secondary metabolite synthesis in plants [124]. For instance, a polysaccharide derived from *Trichoderma atroviride* D16 was successfully regulated the genes involved in the production of tanshinone diterpene in *Salvia miltiorrhiza* and also increased the production of hairy roots up to 60% compared to control plants [125]. Likewise, oligosaccharide derived from *Fusarium oxysporum* efficiently stimulated the production of artemisinin in *Artemisia annua* and flavonoids in *Fagopyrum tataricum* plants suggesting definitive role of biotic elicitors in the stimulation of therapeutically important metabolites [126]. In another study, researchers also successfully exploited oligogalacturonides as biotic elicitor to stimulate the synthesis of phytoalexins in soybean plants [125]. The oligogalacturonides was also further utilised efficiently to stimulate stress defence response in *Nicotiana tabacum* plants by stimulating the biosynthesis of nutraceuticals [126]. Furthermore, the oligogalacturonides based elicitation of phytoalexins was also confirmed by Ferrari et al. [127] in soybean cell cultures.

Various phytohormones/signalling molecules such as salicylic acid (SA), nitric oxide (NO) jasmonic acid (JA), ethylene (ET) and abscisic acid (ABA) which can serve as an elicitor to elicit secondary metabolites production and stress-induced defence response in various plant species [127]. Among all, the role of SA, NO and JA have been extensively investigated for the elicitation of secondary metabolites synthesis and imparting resistance against biotic/abiotic stress induced oxidative damage in plants [128]. Methyl-jasmonate a derivative of jasmomic acid precisely activated the production of indole glucosinolate, β-thujaplicin and terpenes indole alkaloids in *Arabidopsis*, *C. roseus* and *C. lusitanica* plants [129, 130]. Similarly, studies have also indicated that both methyl jasmonate and salicylic acid either alone or in combination significantly enhanced the therapeutic attributes of *Hemidesmus indicus* by stimulating the synthesis of 2-hydroxy 4-methoxy benzaldehyde [131]. Moreover, Gai et al. [132] also observed SA and methyl jasmonate based elicitation of pharmacologically active alkaloids in the hairy root cultures of *Isatis tinctoria* L. Apart from plants, researchers have also widely used SA and methyl jasmonate to elicit the pharmaceutical alkaloids biosynthesis in microalgae *Arthrospira platensis* suggesting their robust application in prokaryotic system [133].

Various plant growth promoting rhizobacteria (PGPR) and arbuscular mycorrhizal fungal inoculum in conjunction with methyl jasmonate have been shown to enhance production of various secondary metabolites in *Rauvolfia serpentine,* fenugreek and *Solanum khasianum* [118, 134]. Bacterial and fungal based elicitation of therapeutically important compounds is being commonly used by various researchers because apart from increasing secondary metabolites production they also significantly improve growth and developments of plant exposed to various biotic and abiotic stress conditions [118]. In a study, Gorelick and Bernstein [135] observed differential effect of fungal elicitation in *Cannabis sativa* plants where fungal elicitation increases the production of cannabinoid and 3-deoxyanthocyanidin but causes significant reduction in anthocyanin content. Furthermore, researchers have also used arbuscular mycorrhizal fungal inoculum along with foliar application of chitosan that significantly boosted the biosynthesis of menthol and essential oils in *Mentha* × *Piperita* L. [136].

4.2 Abiotic elicitors

Elicitation of secondary metabolite synthesis by using substance of non-biological such as inorganic salts of heavy metals ($VOSO_4$, $NiSO_4$, $CdCl_2$, $AgNO_3$, $CuCl_2$), UV-radiation, heat, light etc. is known as abiotic elicitation and the substance used are known as abiotic elicitors. Abiotic elicitors such as high temperature, salt, drought, light and heavy metals etc. have also been successfully used as physical and chemical stimuli to elicit the biosynthesis of medicinally important metabolites in various plants [122]. These abiotic elicitors have been successfully used either independently or in combination either by foliar spray, irrigation or as hydroponics under both open field or controlled conditions for secondary metabolite production in medicinally important plants [122]. Present section deciphers the functional mechanism by which these different abiotic based elicitors elicit the production of therapeutically important compounds.

Drought is one of the most prevalent abiotic stress that alter plant growth and productivity around the globe [108]. Researchers have also indicated that in order to cope up with drought induced oxidative stress, plants synthesise certain metabolites such as glycine betaine and proline as mean to strengthen their defence system [122]. Based on this notion, researchers are using mannitol, calcium chloride and polyvinyl pyrrolidone (chemical which are used to induce drought stress) as a physical elicitor to induce the production of terpene indole alkaloids up to 2-fold in treated *C. roseus* plants compared to non-treated control [107]. Likewise, researchers have also observed increased synthesis of anti-inflammatory metabolite saikosaponins in *Bupleurum chinense* plants exposed to mild drought stress [137]. Furthermore, exposure of drought stress also significantly enhanced the production of rosmarinic ursolic and oleanolic acid in *Prunella vulgaris* [122]. Similarly, researchers have also observed sharp increase in the glycyrrhizic acid and betulinic acid content in *Glycyrrhiza uralensis* and *Hypericum brasiliense* plants upon exposure to drought stress conditions [137].

Salinity is also known to affect wide array of physiological and biochemical properties in plants thus affecting their growth and development [137]. Prolong exposure to salinity stress causes cellular dehydration and generation of oxidative stress in plants thus limiting their ion/osmotic homeostasis [122]. However, in order to withstand to salinity stress, plants synthesised various secondary metabolites like phenols, alkaloids and terpenes as an ameliorative mechanism to overcome oxidative damage. For instance, researchers observed significant increase in the biosynthesis of terpene indole alkaloids (TIAs) in *C. roseus* plants exposed to mild salt stress as compared to control plants. Similarly, another group of researchers also enhanced production of vincristine alkaloids and anthocyanin in *C. roseus* and *Grevillea ilicifolia* plants exposed to salt stress [137]. Salinity induced elicitation of secondary metabolites are also reported in *Datura innoxia, Oryza sativa, Triticum aestivum* and *Trifolium repens* where researchers have observed enhance synthesis of various alkaloids, polyamine and glycine betaine using NaCl as an elicitor [138].

High light intensity and temperature are also able to alter the course of secondary metabolites production in plants [138]. Prolong exposure of both high light and temperature can induce oxidative stress in plants that can have adverse effect growth, ontology and development. High temperature can also lead to the induction of premature leaf senescence, stomatal closure and can stimulate transpiration rate to a greater extent [137]. Nonetheless, despite affecting plant's growth these physical factors have also been reported to stimulate the biosynthesis of important secondary metabolites in the root of *Panax quinquefolius* [138]. Likewise, researchers have stimulated the production of gingerol and zingiberene metabolites by culturing *Zingiber officinale* explants under high light conditions. Moreover, exposing plants to short-term UV-B radiations have also been reported to stimulate the secondary metabolite synthesis. For example., Klein et al. [139] observed increase biosynthesis of betacyanin and betaxanthin metabolites in *Alternanthera sessilis* and *Alternanthera brasiliana* by exposing them to 10–40 J cm^{-2} of UV-B radiation.

Similarly, UV-B radiation (up to 30–90 min) and low temperature treatment significantly improved hypericin biosynthesis in *Hypericum perforatum* adventitious roots and enhances the synthesis of total hydroxycinnamic acids (HCAs) and some sesquiterpenes in *Crepidiastrum denticulatum* [140]. Furthermore, exposure of both high/low temperature have also been shown to improve biosynthesis of ginsenoside, hypericin and hyperforin metabolites in *Panax ginseng* and *Hypericum perforatum* plants [137].

Increasing bioaccumulation of heavy metals such as As, Cd, Cu, Ni, Co and Ag have significantly impacted the agricultural lands and productivity. These heavy metals when presence in excess amount adversely affects plant growth and development [137]. However, at low levels these heavy metals act as co-enzymes/co-factors in various cellular and metabolic pathways thus stimulating secondary metabolite production in plants [122]. Several researchers have well documented the role of heavy metals in stimulating oil content, shikonin/digitalin levels in *Brassica juncea* and production of betalains in *Beta vulgaris* [107]. Likewise, stimulatory effect of Cu^{2+}, Co^{2+}, $CdCl_2$ and $AgNO_3$ in stimulating lepidine in cultures of *Lepidium sativum*, betacyanins in callus of *Amaranthus caudatus*, tanshinone in root culture of *Perovskia abrotanoides* and various sesquiterpenoids in *Datura stramonium* [122]. In the recent years, researchers have synthesised various nanoparticles using various metals that have induce compelling impact on the plant, secondary metabolite production. For example., CdO nanoparticles not only induced the biosynthesis of phenolic compounds but also significantly improve growth and productivity of barley plants [141]. Likewise, Tripathi et al. [142, 143] also reported development of silver nanoparticles in *Withania coagulans* possess strong antibacterial, cytotoxic and antioxidative properties and was also able to enhance the production of withanolides. A large body of literature have well characterised the functional mechanism by which the nanoparticles can induce secondary metabolite synthesis is by stimulating ROS bursts [122]. These ROS act as signalling molecules at lower levels but can impose severe repercussion on the growth and development of plants when their level reaches beyond their antioxidative defence system. Thus, these nanoparticles presented themselves as most efficient mean to strengthen secondary metabolite production that will not only improve plant performance under stress conditions but can also improve the therapeutic/pharmacological potential of plants.

5. Conclusion

In the present era, hyacinth bean has been recognised as an omnipotent legume crop which has the ability to conquer malnutrition, food/hunger index and several chronic diseases all around the globe. Being rich source of genetic and genome resources, the information's reviewed here can significantly contributes towards unravelling its structural, biochemical and molecular genomics which can lead to the identification of signalling pathways involve in the biosynthesis of important therapeutic metabolites/compounds. Furthermore, the implementation of multi "OMICS" techniques are the need of the hour which can transform hyacinth bean and other underutilised legume crops from being "orphan" to "model crop" by exploiting them in the breeding programs. These underutilised legumes hold the potential for developing sustainable agriculture which can lead to hunger and disease-free world in the era of global warming/pandemic. Therefore, synergistic use of multi OMICS tools are of ultimate requirement for expanding the current horizons of underutilised legume crops to address important problems relevant to Nations be it on health, nutrition and environment.

Acknowledgements

The authors are also thankful to DST (Department of Science and Technology), and DBT-ISLS facility, BHU, Govt. of India (Grant No. DST/BHU/PURSE 2017-2018 and DST/BHU/FIST 2016-2017) for financial support. KKR is also thankful for financial support in the form of Research Associate by National Academy of Sciences India, Prayagraj is gratefully acknowledged.

Author details

Krishna Kumar Rai[1,2]*, Nagendra Rai[2] and Shashi Pandey-Rai[1]*

1 Laboratory of Morphogenesis, Department of Botany, Institute of Science, Banaras Hindu University, Varanasi, Uttar Pradesh, India

2 Indian Institute of Vegetable Research, Varanasi, Uttar Pradesh, India

*Address all correspondence to: raikrishna16@gmail.com and shashi.bhubotany@gmail.com

References

[1] Prasad, M., 2020. Omics of neglected and underutilized crop species: one small step for NUCS, one giant leap for addressing global hunger. Nucleus 63: 213-215.

[2] Fanzo, J., 2019. Healthy and sustainable diets and food systems: the key to achieving sustainable development goal 2? Food Ethics 4(2): 159-174.

[3] Kole, C., Muthamilarasan, M., Henry, R., Edwards, D., Sharma, R., Abberton, M., Batley, J., Bentley, A., Blakeney, M., Bryant, J. and Cai, H., 2015. Application of genomics-assisted breeding for generation of climate resilient crops: progress and prospects. Front. Plant Sci. 6: 563.

[4] Singh, R.K., Prasad, A., Muthamilarasan, M., Parida, S.K. and Prasad, M., 2020. Breeding and biotechnological interventions for trait improvement: status and prospects. Planta 252(4): 1-18.

[5] Maass, B.L., Knox, M.R., Venkatesha, S.C., Angessa, T.T., Ramme, S. and Pengelly, B.C., 2010. *Lablab purpureus*—A crop lost for Africa? Trop. Plant Biol. 3(3): 123-135.

[6] Rai, K.K., Rai, N. and Rai, S.P., 2018. Recent advancement in modern genomic tools for adaptation of *Lablab purpureus* L to biotic and abiotic stresses: present mechanisms and future adaptations. Acta Physiol. Plant 40(9):1-29.

[7] Maass, B.L., 2016. Origin, domestication and global dispersal of *Lablab purpureus* (L.) Sweet (Fabaceae): current understanding. Seed 10: 14.

[8] Naeem, M., Shabbir, A., Ansari, A.A., Aftab, T., Khan, M.M.A. and Uddin, M., 2020. Hyacinth bean (*Lablab purpureus* L.)–An underutilised crop with future potential. Sci. Hortic. 272: 109551.

[9] Chauhan, D., Shrivastava, A.K. and Patra, S., 2018. Secondary metabolites and antioxidants screening of *Lablab purpureus* (l.) sweet in different solvents. Advances in Pharmacology & Toxicology 19(1).

[10] Al-Snafi, A.E., 2017. The pharmacology and medical importance of *Dolichos lablab* (*Lablab purpureus*)-A review. IOSR J. Pharm. 7(2): 22-30.

[11] Rahman, S.A. and Akhter, M.S., 2018. Antibacterial and cytotoxic activity of seeds of white hyacinth bean (*Lablab purpureus* L. sweet 'white'). J. Adv. Biotech. Experimental Therap. 1(2): 49-54.

[12] Liu, Y.M., Shahed-Al-Mahmud, M., Chen, X., Chen, T.H., Liao, K.S., Lo, J.M., Wu, Y.M., Ho, M.C., Wu, C.Y., Wong, C.H. and Jan, J.T., 2020. A carbohydrate-binding protein from the edible Lablab beans effectively blocks the infections of influenza viruses and SARS-CoV-2. Cell Rep. 32(6): 108016.

[13] Ku, Y.S., Contador, C.A., Ng, M.S., Yu, J., Chung, G. and Lam, H.M., 2020. The effects of domestication on secondary metabolite composition in legumes. Front. Genet. 11.

[14] Guretzki, S. and Papenbrock, J., 2014. Characterization of *Lablab purpureus* regarding drought tolerance, trypsin inhibitor activity and cyanogenic potential for selection in breeding programmes. J. Agron. Crop Sci. 200(1): 24-35.

[15] Subagio, A., 2006. Characterization of hyacinth bean (*Lablab purpureus* (L.) sweet) seeds from Indonesia and their protein isolate. Food Chem. 95(1): 65-70.

[16] Zhu, F., Du, B. and Xu, B., 2018. Anti-inflammatory effects of phytochemicals from fruits, vegetables, and food legumes: A review. Crit. Rev. Food Sci. Nutr. 58(8): 1260-1270.

[17] Shi, Y., Mandal, R., Singh, A. and Pratap Singh, A., 2020. Legume lipoxygenase: Strategies for application in food industry. Legume Sci. 2(3): e44.

[18] Zhang, Y., Pechan, T. and Chang, S.K., 2018. Antioxidant and angiotensin-I converting enzyme inhibitory activities of phenolic extracts and fractions derived from three phenolic-rich legume varieties. J. Funct. Foods 42: 289-297.

[19] Turco, I., Ferretti, G. and Bacchetti, T., 2016. Review of the health benefits of Faba bean (*Vicia faba* L.) polyphenols. J. Food Nutr. Res. 55(4): 283-293.

[20] He, S., Simpson, B.K., Sun, H., Ngadi, M.O., Ma, Y. and Huang, T., 2018. *Phaseolus vulgaris* lectins: A systematic review of characteristics and health implications. Crit. Rev. Food Sci. Nutr. 58(1): 70-83.

[21] Leite, J.F.M., Assreuy, A.M.S., Mota, M.R.L., Bringel, P.H.D.S.F., e Lacerda, R.R., Gomes, V.D.M., Cajazeiras, J.B., Do Nascimento, K.S., Pessôa, H.D.L.F., Gadelha, C.A.D.A. and Delatorre, P., 2012. Antinociceptive and anti-inflammatory effects of a lectin-like substance from *Clitoria fairchildiana* R. Howard seeds. Molecules 17(3): 3277-3290.

[22] Buer, C.S., Imin, N. and Djordjevic, M.A., 2010. Flavonoids: new roles for old molecules. J. Integr. Plant Biol. 52(1): 98-111.

[23] Samanta, A., Das, G. and Das, S.K., 2011. Roles of flavonoids in plants. Carbon 100(6): 12-35.

[24] Tucak, M., Horvat, D., Cupic, T., Krizmanic, G., Tomas, V., Ravlic, M. and Popovic, S., 2018. Forage legumes as sources of bioactive phytoestrogens for use in pharmaceutics: a review. Curr. Pharm. Biotechnol. 19(7): 537-544.

[25] Gętek, M., Czech, N., Muc-Wierzgoń, M., Grochowska-Niedworok, E., Kokot, T. and Nowakowska-Zajdel, E., 2014. The active role of leguminous plant components in type 2 diabetes. Evidence Based Alternative Med. 1-12.

[26] Roopchand, D.E., Kuhn, P., Rojo, L.E., Lila, M.A. and Raskin, I., 2013. Blueberry polyphenol-enriched soybean flour reduces hyperglycemia, body weight gain and serum cholesterol in mice. Pharmacol. Res. 68(1): 59-67.

[27] Dihingia, A., Bordoloi, J., Dutta, P., Kalita, J. and Manna, P., 2018. Hexane-Isopropanolic Extract of Tungrymbai, a North-East Indian fermented soybean food prevents hepatic steatosis via regulating AMPK-mediated SREBP/FAS/ACC/HMGCR and PPARα/CPT1A/UCP2 pathways. Sci. Rep. 8(1): 1-12.

[28] Chen, H., Liu, L.J., Zhu, J.J., Xu, B. and Li, R., 2010. Effect of soybean oligosaccharides on blood lipid, glucose levels and antioxidant enzymes activity in high fat rats. Food Chem. 119(4): 1633-1636.

[29] Panneerselvam, S., Packirisamy, R.M., Bobby, Z., Jacob, S.E. and Sridhar, M.G., 2016. Soy isoflavones (*Glycine max*) ameliorate hypertriglyceridemia and hepatic steatosis in high fat-fed ovariectomized Wistar rats. J. Nutr. Biochem. 38: 57-69.

[30] Fu, Z., Zhang, W., Zhen, W., Lum, H., Nadler, J., Bassaganya-Riera, J., Jia, Z., Wang, Y., Misra, H. and Liu, D., 2010. Genistein induces pancreatic β-cell proliferation through activation of multiple signaling pathways and prevents insulin-deficient diabetes in mice. Endocrinology 151(7): 3026-3037.

[31] Nagaraju, G.P., Zafar, S.F. and El-Rayes, B.F., 2013. Pleiotropic effects

of genistein in metabolic, inflammatory, and malignant diseases. Nutri. Rev. 71(8): 562-572.

[32] Zhang, Y., Wang, P., Xu, Y., Meng, X. and Zhang, Y., 2016. Metabolomic analysis of biochemical changes in the plasma of high-fat diet and streptozotocin-induced diabetic rats after treatment with isoflavones extract of *Radix puerariae*. Evidence Based Alternative Med. 1-12.

[33] Sabahi, Z., Khoshnood-Mansoorkhani, M.J., Rahmani Namadi, S. and Moein, M., 2016. Antidiabetic and synergistic effects study of anthocyanin fraction from *Berberis integerrima* fruit on streptozotocin-induced diabetic rats' model. Trends Pharmacol. Sci. 2(1): 43-50.

[34] Huang, P.C., Wang, G.J., Fan, M.J., Asokan Shibu, M., Liu, Y.T., Padma Viswanadha, V., Lin, Y.L., Lai, C.H., Chen, Y.F., Liao, H.E. and Huang, C.Y., 2017. Cellular apoptosis and cardiac dysfunction in STZ-induced diabetic rats attenuated by anthocyanins via activation of IGFI-R/PI3K/Akt survival signaling. Environ. toxicol. 32(12): 2471-2480.

[35] Koh, E.S., Lim, J.H., Kim, M.Y., Chung, S., Shin, S.J., Choi, B.S., Kim, H.W., Hwang, S.Y., Kim, S.W., Park, C.W. and Chang, Y.S., 2015. Anthocyanin-rich Seoritae extract ameliorates renal lipotoxicity via activation of AMP-activated protein kinase in diabetic mice. J. Transl. Med. 13(1): 1-12.

[36] Malaguti, M., Dinelli, G., Leoncini, E., Bregola, V., Bosi, S., Cicero, A.F. and Hrelia, S., 2014. Bioactive peptides in cereals and legumes: agronomical, biochemical and clinical aspects. Int. J. Mol. Sci. 15(11): 21120-21135.

[37] González-Montoya, M., Cano-Sampedro, E. and Mora-Escobedo, R., 2017. Bioactive Peptides from Legumes as anticancer aherapeutic agents. Int. J. Cancer Clin. Res. 4: 081.

[38] De Mejia, E.G., Bradford, T. and Hasler, C., 2003. The anticarcinogenic potential of soybean lectin and lunasin. Nutr. Rev. 61(7): 239-246.

[39] Alaswad, A.A. and Krishnan, H.B., 2016. Immunological investigation for the presence of lunasin, a chemopreventive soybean peptide, in the seeds of diverse plants. J. Agric. Food Chem. 64 (14): 2901-2909.

[40] Jeong, H.J., Lam, Y. and de Lumen, B.O., 2002. Barley lunasin suppresses ras-induced colony formation and inhibits core histone acetylation in mammalian cells. J. Agric. Food Chem. 50(21): 5903-5908.

[41] Price, S.J., Pangloli, P., Krishnan, H.B. and Dia, V.P., 2016. Kunitz trypsin inhibitor in addition to Bowman-Birk inhibitor influence stability of lunasin against pepsin-pancreatin hydrolysis. Food Res. Int. 90: 205-215.

[42] Arnoldi, A., Zanoni, C., Lammi, C. and Boschin, G., 2015. The role of grain legumes in the prevention of hypercholesterolemia and hypertension. Crit. Rev. Plant Sci. 34(1-3): 144-168.

[43] Polak, R., Phillips, E.M. and Campbell, A., 2015. Legumes: Health benefits and culinary approaches to increase intake. Clin. Diabetes 33(4): 198-205.

[44] Maleki, S. and Razavi, S.H., 2020. Pulses' germination and fermentation: Two bioprocessing against hypertension by releasing ACE inhibitory peptides. Crit. Rev. Food Sci. Nutr. 1-18.

[45] El-Sayed, S.T., Al-Azzouny, R.A. and Ali, O.S., 2020. Evaluation of bile acids-binding capacity and tyrosinase inhibitor potency of novel peptides prepared from agricultural wastes via enzymatic hydrolysis. Scientific Study

& Research: Chemistry & Chemical Engineering, Biotechnology, Food Industry, 21(2): 141-154.

[46] Conti, M.V., Guzzetti, L., Panzeri, D., De Giuseppe, R., Coccetti, P., Labra, M. and Cena, H., 2021. Bioactive compounds in legumes: Implications for sustainable nutrition and health in the elderly population. Trends Food Sci. Technol. https://doi.org/10.1016/j.tifs.2021.02.072.

[47] Ray, P.D., Huang, B.W. and Tsuji, Y., 2012. Reactive oxygen species (ROS) homeostasis and redox regulation in cellular signaling. Cell. Signal. 24(5): 981-990.

[48] Pitzschke, A., Forzani, C. and Hirt, H., 2006. Reactive oxygen species signaling in plants. Antioxid. Redox Signal. 8(9-10): 1757-1764.

[49] Miller, N.J. and Ruiz-Larrea, M.B., 2002. Flavonoids and other plant phenols in the diet: Their significance as antioxidants. J. Nutr. Environ. Med. 12(1): 39-51.

[50] Babu, P.V.A., Liu, D. and Gilbert, E.R., 2013. Recent advances in understanding the anti-diabetic actions of dietary flavonoids. J. Nutr. Biochem. 24(11): 1777-1789.

[51] Chang, Y.Q., Tan, S.N., Yong, J.W. and Ge, L., 2012. Determination of flavonoids in *Costus speciosus* and *Etlingera elatior* by liquid chromatography-mass spectrometry. Anal. Lett. 45(4): 345-355.

[52] Mestry, S.N., Dhodi, J.B., Kumbhar, S.B. and Juvekar, A.R., 2017. Attenuation of diabetic nephropathy in streptozotocin-induced diabetic rats by *Punica granatum* Linn. leaves extract. J. Tradit. Complement Med. 7(3): 273-280.

[53] Matemu, A., Nakamura, S. and Katayama, S., 2021. Health benefits of antioxidative peptides derived from legume proteins with a high amino acid score. Antioxidants 10(2): 316.

[54] Morales-Medina, R., Tamm, F., Guadix, A.M., Guadix, E.M. and Drusch, S., 2016. Functional and antioxidant properties of hydrolysates of sardine (*S. pilchardus*) and horse mackerel (*T. mediterraneus*) for the microencapsulation of fish oil by spray-drying. Food chem. 194: 1208-1216.

[55] Peters, K., Delong, H. and Hossain, K., 2011. Anthocyanin, total polyphenols and antioxidant activity of common bean. Am. J. Food Technol. 6(5): 885-394.

[56] Zhu, X.L., Watanabe, K., Shiraishi, K., Ueki, T., Noda, Y., Matsui, T. and Matsumoto, K., 2008. Identification of ACE-inhibitory peptides in salt-free soy sauce that are transportable across caco-2 cell monolayers. Peptides 29(3): 338-344.

[57] Cornara, L., Xiao, J. and Burlando, B., 2016. Therapeutic potential of temperate forage legumes: a review. Crit. Rev. Food Sci. Nutr. 56: S149-S161.

[58] Etxeberria, U., de la Garza, A.L., Campión, J., Martinez, J.A. and Milagro, F.I., 2012. Antidiabetic effects of natural plant extracts via inhibition of carbohydrate hydrolysis enzymes with emphasis on pancreatic alpha amylase. Expert Opin. Ther. Targ. 16(3): 269-297.

[59] Clemente, A. and del Carmen Arques, M., 2014. Bowman-Birk inhibitors from legumes as colorectal chemopreventive agents. World J Gastroenterol. 20(30): 10305.

[60] Srikanth, S. and Chen, Z., 2016. Plant protease inhibitors in therapeutics-focus on cancer therapy. Front. pharmacol. 7: 470.

[61] Li, W.W., Yu, J.Y., Xu, H.L. and Bao, J.K., 2011. Concanavalin A: a potential

anti-neoplastic agent targeting apoptosis, autophagy and anti-angiogenesis for cancer therapeutics. Biochem. Biophys. Res. Commun. 414(2): 282-286.

[62] Consonni, A., Lovati, M.R., Manzoni, C., Pizzagalli, A., Morazzoni, P. and Duranti, M., 2010. Cloning, yeast expression, purification and biological activity of a truncated form of the soybean 7S globulin α′ subunit involved in Hep G2 cell cholesterol homeostasis. J. Nutr. Biochem. 21(9): 887-891.

[63] Chatterjee, C., Gleddie, S. and Xiao, C.W., 2018. Soybean bioactive peptides and their functional properties. Nutrients 10(9): 1211.

[64] Singh, B., Singh, J.P., Kaur, A. and Singh, N., 2017. Phenolic composition and antioxidant potential of grain legume seeds: A review. Food Res. Int. 101: 1-16.

[65] Araya-Cloutier, C., den Besten, H.M., Aisyah, S., Gruppen, H. and Vincken, J.P., 2017. The position of prenylation of isoflavonoids and stilbenoids from legumes (Fabaceae) modulates the antimicrobial activity against Gram positive pathogens. Food chem. 226: 193-201.

[66] Pina-Pérez, M.C. and Pérez, M.F., 2018. Antimicrobial potential of legume extracts against foodborne pathogens: A review. Trends Food Sci. Tech. 72: 114-124.

[67] Furbank, R.T. and Tester, M., 2011. Phenomics–technologies to relieve the phenotyping bottleneck. Trends plant Sci. 16(12): 635-644.

[68] Houle, D., Govindaraju, D.R. and Omholt, S., 2010. Phenomics: the next challenge. Nat. Rev. Genet. 11(12): 855-866.

[69] Egea-Cortines, M. and Doonan, J.H., 2018. Phenomics. Front. plant Sci. 9: 678.

[70] Goëau, H., Bonnet, P. and Joly, A., 2019, September. Overview of lifeclef plant identification task 2019: diving into data deficient tropical countries. In CLEF 2019-Conference and Labs of the Evaluation Forum (Vol. 2380, pp. 1-13). CEUR.

[71] Gao, L. and Lin, X., 2018. A method for accurately segmenting images of medicinal plant leaves with complex backgrounds. Comput. Electron. Agric. 155: 426-445.

[72] Uddin, S.B., 2010. Medicinal plants database of Bangladesh. Retrieved from mpbd. info.

[73] Rueden, C.T., Schindelin, J., Hiner, M.C., DeZonia, B.E., Walter, A.E., Arena, E.T. and Eliceiri, K.W., 2017. ImageJ2: ImageJ for the next generation of scientific image data. BMC Bioinformatics 18(1): 1-26.

[74] Hartmann, A., Czauderna, T., Hoffmann, R., Stein, N. and Schreiber, F., 2011. HTPheno: an image analysis pipeline for high-throughput plant phenotyping. BMC Bioinformatics 12(1): 1-9.

[75] Somerville, C. and Somerville, S., 1999. Plant functional genomics. Science 285(5426): 380-383.

[76] Goossens, A., Häkkinen, S.T., Laakso, I., Seppänen-Laakso, T., Biondi, S., De Sutter, V., Lammertyn, F., Nuutila, A.M., Söderlund, H., Zabeau, M. and Inzé, D., 2003. A functional genomics approach toward the understanding of secondary metabolism in plant cells. Proc. Natl. Acad. Sci. USA 100(14): 8595-8600.

[77] Modolo, L.V., Blount, J.W., Achnine, L., Naoumkina, M.A., Wang, X. and Dixon, R.A., 2007. A functional genomics approach to (iso) flavonoid glycosylation in the model legume *Medicago truncatula*. Plant Mol. Biol. 64(5): 499-518.

[78] Yang, D., Du, X., Yang, Z., Liang, Z., Guo, Z. and Liu, Y., 2014. Transcriptomics, proteomics, and metabolomics to reveal mechanisms underlying plant secondary metabolism. Eng. Life Sci. 14(5): 456-466.

[79] Robinson, A.J., Tamiru, M., Salby, R., Bolitho, C., Williams, A., Huggard, S., Fisch, E., Unsworth, K., Whelan, J. and Lewsey, M.G., 2018. AgriSeqDB: an online RNA-Seq database for functional studies of agriculturally relevant plant species. BMC Plant Biol. 18(1): 1-8.

[80] Chien, C.H., Chow, C.N., Wu, N.Y., Chiang-Hsieh, Y.F., Hou, P.F. and Chang, W.C., 2015, December. EXPath: a database of comparative expression analysis inferring metabolic pathways for plants. In BMC genomics BioMed Central 16: 1-10.

[81] She, J., Yan, H., Yang, J., Xu, W. and Su, Z., 2019. croFGD: Catharanthus roseus functional genomics database. Front. Genet. 10: 238.

[82] Zhang, P., Du, H., Wang, J., Pu, Y., Yang, C., Yan, R., Yang, H., Cheng, H. and Yu, D., 2020a. Multiplex CRISPR/Cas9-mediated metabolic engineering increases soya bean isoflavone content and resistance to soya bean mosaic virus. Plant Biotech. J. 18(6): 1384-1395.

[83] Zhang, R., Ma, Y., Hu, X., Chen, Y., He, X., Wang, P., Chen, Q., Ho, C.T., Wan, X., Zhang, Y. and Zhang, S., 2020b. TeaCoN: a database of gene co-expression network for tea plant (*Camellia sinensis*). BMC Genomics 21(1): 1-9.

[84] Bulgakov, V.P. and Avramenko, T.V., 2015. New opportunities for the regulation of secondary metabolism in plants: focus on microRNAs. Biotechnol. Lett. 37(9): 1719-1727.

[85] Yu, D., Lu, J., Shao, W., Ma, X., Xie, T., Ito, H., Wang, T., Xu, M., Wang, H. and Meng, Y., 2019. MepmiRDB: a medicinal plant microRNA database. Database 2019.

[86] Allen, E., Xie, Z., Gustafson, A.M. and Carrington, J.C., 2005. microRNA-directed phasing during trans-acting siRNA biogenesis in plants. Cell 121(2): 207-221.

[87] Jacobs, D.I., Gaspari, M., van der Greef, J., van der Heijden, R. and Verpoorte, R., 2005. Proteome analysis of the medicinal plant *Catharanthus roseus*. Planta 221(5): 690-704.

[88] Chin, C.F., 2016. The proteome of orchids. In Agricultural Proteomics Vol. 1, pp. 127-135, Springer, Cham.

[89] Liu, Z., Li, H., Wen, Z., Fan, X., Li, Y., Guan, R., Guo, Y., Wang, S., Wang, D. and Qiu, L., 2017. Comparison of genetic diversity between Chinese and American soybean (Glycine max (L.)) accessions revealed by high-density SNPs. Front. Plant Sci. 8: 2014.

[90] Biazzi, E., Carelli, M., Tava, A., Abbruscato, P., Losini, I., Avato, P., Scotti, C. and Calderini, O., 2015. CYP72A67 catalyzes a key oxidative step in *Medicago truncatula* hemolytic saponin biosynthesis. Mol. Plant 8(10): 1493-1506.

[91] Kroc, M., Czepiel, K., Wilczura, P., Mokrzycka, M. and Święcicki, W., 2019. Development and validation of a gene-targeted dCAPS marker for marker-assisted selection of low-alkaloid content in seeds of narrow-leafed lupin (*Lupinus angustifolius* L.). Genes 10(6): p.428.

[92] Hancock, K.R., Collette, V., Fraser, K., Greig, M., Xue, H., Richardson, K., Jones, C. and Rasmussen, S., 2012. Expression of the R2R3-MYB transcription factor TaMYB14 from *Trifolium arvense* activates proanthocyanidin biosynthesis in the legumes *Trifolium repens* and *Medicago sativa*. Plant physiol. 159(3): 1204-1220.

[93] Cañas, L.A. and Beltrán, J.P., 2018. Model legumes: functional genomics tools in *Medicago truncatula*. Functional Genomics in *Medicago truncatula*. 11-37.

[94] Achnine, L., Huhman, D.V., Farag, M.A., Sumner, L.W., Blount, J.W. and Dixon, R.A., 2005. Genomics-based selection and functional characterization of triterpene glycosyltransferases from the model legume *Medicago truncatula*. Plant J. 41(6): 875-887.

[95] Honda, M.D. and Borthakur, D., 2019. Mimosine concentration in *Leucaena leucocephala* under various environmental conditions. Tropical Grasslands-Forrajes Tropicales, 7(2):164-172.

[96] Sanchez, D.H., Szymanski, J., Erban, A., Udvardi, M.K. and Kopka, J., 2010. Mining for robust transcriptional and metabolic responses to long-term salt stress: a case study on the model legume *Lotus japonicus*. Plant Cell Environ. 33(4): 468-480.

[97] Jha, U.C., Jha, R., Bohra, A., Parida, S.K., Kole, P.C., Thakro, V., Singh, D. and Singh, N.P., 2018. Population structure and association analysis of heat stress relevant traits in chickpea (*Cicer arietinum* L.). 3 Biotech, 8(1): 43.

[98] Boivin, C., Barran, L.R., Malpica, C.A. and Rosenberg, C., 1991. Genetic analysis of a region of the *Rhizobium meliloti* pSym plasmid specifying catabolism of trigonelline, a secondary metabolite present in legumes. J. bacterial. 173(9): 2809-2817.

[99] Li, Y., Chen, X., Chen, Z., Cai, R., Zhang, H. and Xiang, Y., 2016. Identification and expression analysis of BURP domain-containing genes in *Medicago truncatula*. Front. Plant Sci. 7: 485.

[100] Jha, U.C., Bohra, A., Jha, R. and Parida, S.K., 2019. Salinity stress response and 'omics' approaches for improving salinity stress tolerance in major grain legumes. Plant Cell Rep. 38(3): 255-277.

[101] Dastmalchi, M. and Dhaubhadel, S., 2015. Soybean chalcone isomerase: evolution of the fold, and the differential expression and localization of the gene family. Planta, 241(2): 507-523.

[102] Mochida, K., Sakurai, T., Seki, H., Yoshida, T., Takahagi, K., Sawai, S., Uchiyama, H., Muranaka, T. and Saito, K., 2017. Draft genome assembly and annotation of *Glycyrrhiza uralensis*, a medicinal legume. Plant J. 89(2): 181-194.

[103] Breitling, R., Ceniceros, A., Jankevics, A. and Takano, E., 2013. Metabolomics for secondary metabolite research. Metabolites 3(4): 1076-1083.

[104] Nguyen, Q.T., Merlo, M.E., Medema, M.H., Jankevics, A., Breitling, R. and Takano, E., 2012. Metabolomics methods for the synthetic biology of secondary metabolism. FEBS Lett. 586(15): 2177-2183.

[105] Nakabayashi, R. and Saito, K., 2013. Metabolomics for unknown plant metabolites. Anal. Bioanal. Chem. 405(15): 5005-5011.

[106] Ramalingam, A., Kudapa, H., Pazhamala, L.T., Weckwerth, W. and Varshney, R.K., 2015. Proteomics and metabolomics: two emerging areas for legume improvement. Front. Plant Sci. 6: 1116.

[107] Ramakrishna, A., Ravishankar, G.A., 2011. Influence of abiotic stress signals on secondary metabolites in plants. Plant Signal. Behav. 6 (11): 1720 –1731.

[108] Naik, P.M., Al-Khayri, J.M., 2016. Abiotic and biotic elicitors –role in secondary metabolites production

through in vitro culture of medicinal plants. Abiotic and Biotic Stress in Plants - Recent Advances and Future Perspectives. https://doi.org/10.5772/61442

[109] Kneer, R., Poulev, A.A., Olesinski, A. and Raskin, I., 1999. Characterization of the elicitor-induced biosynthesis and secretion of genistein from roots of *Lupinus luteus* L. J. Exp. Bot. 50(339): 1553-1559.

[110] Shetty, P., Atallah, M.T. and Shetty, K., 2003. Stimulation of total phenolics, L-DOPA and antioxidant activity through proline-linked pentose phosphate pathway in response to proline and its analogue in germinating fava beans (*Vicia faba*). Process Biochem. 38(12): 1707-1717.

[111] Broeckling, C.D., Huhman, D.V., Farag, M.A., Smith, J.T., May, G.D., Mendes, P., Dixon, R.A. and Sumner, L.W., 2005. Metabolic profiling of *Medicago truncatula* cell cultures reveals the effects of biotic and abiotic elicitors on metabolism. J. Exp. Bot. 56(410): 323-336.

[112] Gueven, A. and Knorr, D., 2011. Isoflavonoid production by soy plant callus suspension culture. J. Food Eng. 103(3): 237-243.

[113] Pawlak-Sprada, S., Stobiecki, M. and Deckert, J., 2011. Activation of phenylpropanoid pathway in legume plants exposed to heavy metals. Part II. Profiling of isoflavonoids and their glycoconjugates induced in roots of lupine (*Lupinus luteus*) seedlings treated with cadmium and lead. Acta Biochim. Pol. 58(2).

[114] Wojakowska, A., Muth, D., Narożna, D., Mądrzak, C., Stobiecki, M. and Kachlicki, P., 2013. Changes of phenolic secondary metabolite profiles in the reaction of narrow leaf lupin (*Lupinus angustifolius*) plants to infections with *Colletotrichum lupini* fungus or treatment with its toxin. Metabolomics 9(3): 575-589.

[115] Dueñas, M., Sarmento, T., Aguilera, Y., Benitez, V., Mollá, E., Esteban, R.M. and Martín-Cabrejas, M.A., 2016. Impact of cooking and germination on phenolic composition and dietary fibre fractions in dark beans (*Phaseolus vulgaris* L.) and lentils (*Lens culinaris* L.). LWT - Food Sci. Technol. 66: 72-78.

[116] Díaz-Sánchez, E.K., Guajardo-Flores, D., Serna-Guerrero, D., Gutierrez-Uribe, J.A. and Jacobo-Velázquez, D.A., 2018. The application of chemical elicitors improves the flavonoid and saponin profiles of black beans after soaking. Rev. Mex. Ing. Quim. 17(1): 123-130.

[117] Kalli, S., Araya-Cloutier, C., de Bruijn, W.J., Chapman, J. and Vincken, J.P., 2020. Induction of promising antibacterial prenylated isoflavonoids from different subclasses by sequential elicitation of soybean. Phytochemistry 179: 112496.

[118] Irankhah, S., Chitarra, W., Nerva, L., Antoniou, C., Lumini, E., Volpe, V., Ganjeali, A., Cheniany, M., Mashreghi, M., Fotopoulos, V. and Balestrini, R., 2020. Impact of an arbuscular mycorrhizal fungal inoculum and exogenous MeJA on fenugreek secondary metabolite production under water deficit. Environ. Exp. Bot. 176: 104096.

[119] Dębski, H., Wiczkowski, W. and Horbowicz, M., 2021. Effect of elicitation with iron chelate and sodium metasilicate on phenolic compounds in legume sprouts. Molecules 26(5): 1345.

[120] Patel, H., Krishnamurthy, R., 2013. Elicitors in plant tissue culture. J. Pharmacogn. Phytochem. 2: 2278 –4136.

[121] Picard, K., Ponchet, M., Blein, J.P., Rey, P., Tirilly, Y., Benhamou, N., 2000. Plant Physiol. 124: 379 –395.

[122] Halder, M., Sarkar, S. and Jha, S., 2019. Elicitation: A biotechnological tool for enhanced production of secondary metabolites in hairy root cultures. Eng. Life Sci. 19(12): 880-895.

[123] Zhai, X., Jia, M., Chen, L., Zheng, C.J., Rahman, K., Han, T., Qin, L.P., 2016. The regulatory mechanism of fungal elicitor-induced secondary metabolite biosynthesis in medical plants. Crit. Rev. Microbiol. 43 (2): 238-261.

[124] Bi, F., Iqbal, S., Arman, M., Ali, A. and Hassan, M.U., 2011. Carrageenan as an elicitor of induced secondary metabolites and its effects on various growth characters of chickpea and maize plants. J. Saudi Chem. Soc. 15(3): 269-273.

[125] Peng, W., Ming, Q.L., Zhai, X., Zhang, Q., Rahman, K., Wu, S.J., Qin, L.P. and Han, T., 2019. Polysaccharide fraction extracted from endophytic fungus *Trichoderma atroviride* D16 has an influence on the proteomics profile of the *Salvia miltiorrhiza* hairy roots. Biomolecules 9(9): 415.

[126] Zhong, L., Niu, B., Tang, L., Chen, F., Zhao, G. and Zhao, J., 2016. Effects of polysaccharide elicitors from endophytic *Fusarium oxysporum* Fat9 on the growth, flavonoid accumulation and antioxidant property of *Fagopyrum tataricum* sprout cultures. Molecules 21(12): 1590.

[127] Ferrari, S., Savatin, D.V., Sicilia, F., Gramegna, G., Cervone, F. and De Lorenzo, G., 2013. Oligogalacturonides: plant damage-associated molecular patterns and regulators of growth and development. Front. Plant Sci 4: 49.

[128] Hayat, Q., Hayat, S., Irfan, M., Ahmad, A., 2010. Effect of exogenous salicylic acid under changing environment: a review. Environ. Exp. Bot. 68: 14 –25.

[129] Malekpoor, F., Salimi, A., Pirbalouti, A.G., 2016. Effect of jasmonic acid on total phenolic content and antioxidant activity of extract from the green and purple landraces of sweet basil. Acta Pol. Pharm. 73 (5): 1229 –1234.

[130] Złotek, U., Michalak-Majewska, M., Szymanowska, U., 2016. Effect of jasmonic acid elicitation on the yield, chemical composition, and antioxidant and anti-inflammatory properties of essential oil of lettuce leaf basil (*Ocimum basilicum* L.). Food Chem. 213: 1 –7.

[131] Nandy, S., Hazra, A.K., Pandey, D.K., Ray, P. and Dey, A., 2021. Elicitation of industrially promising vanillin type aromatic compound 2-hydroxy 4-methoxy benzaldehyde (MBAlD) yield in the in-vitro raised medicinal crop *Hemidesmus indicus* (L) R. Br. by methyl jasmonate and salicylic acid. Ind. Crops Prod. 164: 113375.

[132] Gai, Q.Y., Jiao, J., Wang, X., Zang, Y.P., Niu, L.L. and Fu, Y.J., 2019. Elicitation of Isatis tinctoria L. hairy root cultures by salicylic acid and methyl jasmonate for the enhanced production of pharmacologically active alkaloids and flavonoids. Plant Cell Tissue Organ Cult. 137(1): 77-86.

[133] Hadizadeh, M., Ofoghi, H., Kianirad, M. and Amidi, Z., 2019. Elicitation of pharmaceutical alkaloids biosynthesis by salicylic acid in marine microalgae *Arthrospira platensis*. Algal Res. 42: 101597.

[134] Srivastava, M., Sharma, S. and Misra, P., 2016. Elicitation based enhancement of secondary metabolites in *Rauwolfia serpentina* and *Solanum khasianum* hairy root cultures. Pharmacogn. Mag. 12: S315.

[135] Gorelick, J. and Bernstein, N., 2017. Chemical and physical elicitation for enhanced cannabinoid production in

cannabis. In *Cannabis sativa* L.-botany and biotechnology, pp. 439-456, Springer, Cham.

[136] Goudarzian, A., Pirbalouti, A.G. and Hossaynzadeh, M., 2020. Menthol, balance of menthol/menthone, and essential oil contents of mentha× piperita l. under foliar-applied chitosan and inoculation of arbuscular mycorrhizal fungi. J. Essent. Oil-Bear. Plants 23(5): 1012-1021.

[137] Thakur, M., Bhattacharya, S., Khosla, P.K. and Puri, S., 2019. Improving production of plant secondary metabolites through biotic and abiotic elicitation. J. Appl. Res. Med. Aromat. Plants 12: 1-12.

[138] Gorelick, J. and Bernstein, N., 2014. Elicitation: an underutilized tool in the development of medicinal plants as a source of therapeutic secondary metabolites. Adv. Agron. 124: 201-230.

[139] Klein, F.R.S., Reis, A., Kleinowski, A.M., Telles, R.T., Amarante, L.D., Peters, J.A. and Braga, E.J.B., 2018. UV-B radiation as an elicitor of secondary metabolite production in plants of the genus Alternanthera. Acta bot. bras. 32(4): 615-623.

[140] Park, S.Y., Lee, M.Y., Lee, C.H. and Oh, M.M., 2020. Physiologic and metabolic changes in *Crepidiastrum denticulatum* according to different energy levels of UV-B radiation. Int. J. Mol. Sci. 21(19).

[141] Večeřová, K., Večeřa, Z., Dočekal, B., Oravec, M., Pompeiano, A., Tříska, J. and Urban, O., 2016. Changes of primary and secondary metabolites in barley plants exposed to CdO nanoparticles. Environ. Pollut. 218: 207-218.

[142] Tripathi, D., Modi, A., Narayan, G. and Rai, S.P., 2019. Green and cost-effective synthesis of silver nanoparticles from endangered medicinal plant *Withania coagulans* and their potential biomedical properties. Mater. Sci. Eng. C 100: 152-164.

[143] Tripathi, D., Rai, K.K. and Pandey-Rai, S., 2021. Impact of green synthesized WcAgNPs on in-vitro plant regeneration and withanolides production by inducing key biosynthetic genes in *Withania coagulans*. Plant Cell Rep. 40(2): 283-299.

Chapter 11

Vigna unguiculata (L.) Walp: A Strategic Crop for Nutritional Security, Well Being and Environmental Protection

M. Duraipandian, K.E. Poorani, H. Abirami and M.B. Anusha

Abstract

Cowpea is the common legume crop plant widely cultivated in all over the world for human consumption and animal feed. The global biological name of cowpea is *Vigna unguicultа* (L.) Walp. The crop is cultivated globally in all warm-seasons, semiarid or specifically tropical regions by even poor farmers but originated from anciently to Africa. Cowpea able to grow in even variety of polluted soils and able to grow vigorously withstand in both biotic and abiotic stress conditions. Morphologically herbaceous dicotyledonous plant, grow annually. The entire plant parts of *Vigna unguiculata* (L) Walp such as dry seeds, leaves, roots and pods consumed by all humans and animals Cowpea has rich in nutrition, vitamins and minerals so preferred by many farmers cultivated as intercrop with other cereals. In this review able to discuss the nutritional, medicinal as well as ecological significance. The seeds of cowpea have high content of proteins, fiber foods like carbohydrates, low cholesterol, minerals and vitamins. The nutrient value is higher and delay hungry sense in cowpea when compare to other cereals or pulses. The cowpea used for various medical aspects to lower cholesterol, promote body growth, iron source to increase blood cells, improve gall bladder function, maintain good circulatory system, increase insulin production, decrease body weight and do excellent antioxidant mechanism. The black eye pea also maintains good health to reduce both communicable and non-communicable diseases. Cowpea to form symbioses with variety of beneficial soil microorganisms to increase soil fertility, soil-root aeration, improve humidity and do fix atmospheric nitrogen to reduce global warming, climate change and also increase activity of biogeochemical cycle in the environment. In future people focus on to cultivate cowpea with to increase global production in all countries.

Keywords: legume, tropical, biotic & abiotic, intercrop, nutrition, Fiber food, Symbiosis and biogeochemical cycle

1. Introduction

Legumes are third largest family of flowering plants which consist of over 20,000 species, under the family of Fabaceae, amoung them *Vigna unguiculata* (L.)

Walp is an annual, herbaceous, commonly cultivated throughout the landscape of the earth, with high values in its nutrients, medicinal value and also helps in ecological balances. This crop is cultivated worldwide. The cultivar group such as *Sesquipedal* is wildly known as long or sneak or asparagus bean with have sixteen ovules present and the seeds arranged within the pod compactly and it is suggested as a sub species in Ref. with molecular level [1].

It is an important crop in the semiarid regions across India, Africa and Russia, naturally it is the farmers friendly crop due to crop survival, yield, resistance towards pest, tolerance in sandy soil, low rainfall, whole plant is used as cattle feed and also well-suited to intercropping with other crops as the plant fixes nitrogen with *Agrobacterium*.

There are many species can exist but the common sub-species are includes as *V. unguiculata, Viola biflora, V. textilis and V. Sesquipedalis* are recognized. Among the subspecies, the *V. unguiculata, V. biflora* and *V. textilis* are cultivated worldwide. The bean plants have more morphological difference and found in same species with high differences in the shape, structure and size, shape, and structure. Cowpeas are growing by in semi-erect, erect and climbing manner.

The bean crop is cultivated for the purpose of cattle feed and its seeds have high nutrient values, in terms of rich protein, carbohydrates, vitamins and minerals. The cow pea plant green stems and leaves and seed coats are high nutrients, so consider as cattle feed which increase cattle milk production for many application for in the dairy industry. Many varieties cowpeas were traditionally cultivated in some states India, but in Africa consider as primary cultivar. The countries such as United States of America, United Kingdom, Europe and some parts of Asia.

The cowpea was emerging crop for food and Nigeria considered as largest production country approximately 3.67 million tons and the second largest cultivar Niger global production of cowpea are approximately 2.6 million tons and other countries cowpea production is approximately 2 million tons in 2020. The production of cowpea is marginally reduced in globally approximately 8.2 million tons in 2020. Cow pea seeds are usually cooked and made into stews and curries, or ground into flour or paste.

Most cowpeas are grown on the African continent, particularly in Nigeria and Niger and also in India which account for 66% of world production. Recently the estimate suggests that cowpeas are cultivated on 12.5 million hectares of land. The cowpea has been cultivated in tons of three million and two hundred million humans consume in everyday. The large number of production of cowpea in many places should drastically reduction by the involvement of many pests and insects, particularly causing over ninety percentage loss in crop yield.

2. Taxonomy of cowpea

Kingdom	Plantae
Sub kingdom	Trcheobionta
Super division	Spermatophyta
Division	Magnoliophyta
Class	Magnoliopsida
Sub class	Rosidae
Order	Rosales

DOI: http://dx.doi.org/10.5772/intechopen.103025

Family	*Fabaceae*
Genus	*Vigna*
Species	*V. unguiculata* (L.) Walp
Sub species	*Sesquipedalis* (L.) Verdc

Source: Wolfram Alpha Knowledge Base, 2011.

Figure 1.
Morphology of Cowpea.

Figure 2.
V. unguiculata *(L.) Walp Floral Diagram with Floral formula.*

Figure 3.
Variety of cowpea grown by the many countries of people.

The most common important cultivar of cowpea is *V. unguiculata* is globally accepted. The *V. unguiculata* is belongs to tracheobionta sub-kingdom means have well developed xylem for water and ions transport and prominent phloem for storage function. The super division of cow pea is having starchy seeds.

The Magnoliopsida class of this plant describes the ovule enclosed with ovary. The order of the plant os Rosales means the petals separated. This group crops family Fabaceae is particularly bean or pea plants. There are many sub species such as *Sesquipedalis*, *Tenuis, Dekindtiana* and *Stenophylla* characterized in 1993.

Cowpea are annual, small, sub erect herbaceous plant, have tap root system with root nodules, Glabrous stems and leaves are alternate and compound in nature. The fruits have pods and have cross pollination and epigeal seed germination. Inflorecence is receme, flowers bisexual, have five fused sepals and five free petals. It also has polyandrocium and monocarpellary superior ovary (**Figures 1** and **2**).

The word 'cowpea' primarily mentioned in America on 1798. The bean plant fodder was eagerly eating the cow animals, so the name as cowpea. The common name *V. unguiculata* plant is easily identified by the presence of black-eye on the hilum of bean like seeds. These plants cultivated in southern states in the earth. The pea seeds characteristically by pods are tightly closed so called as crowder pea. Their pods based leading to the other common names of southern pea or crowder pea. The cow pea plants varied in different place to place and different varieties of seeds present across the india and other countries (**Figure 3**). The black eyed seeds of the cowpea are common. The sub species *Sesquipedalis* have long pods in Asia so it called Chinese long-bean.

3. Multidimensional applications of cow pea

The cow pea does efficiently fix nitrogen to enhance to fertility of the different kinds soils. This is improvement of biodiversity on the particular cow pea growing area. The Excellent Manures are prepared by the senescence parts of long bean crop. The grains and leaves from cow pea food for all animals including humans and cattle, the cattle milk have rich nutrients as well as improve income of people. Below image describe the various uses of cowpea in daily life (**Figure 4**).

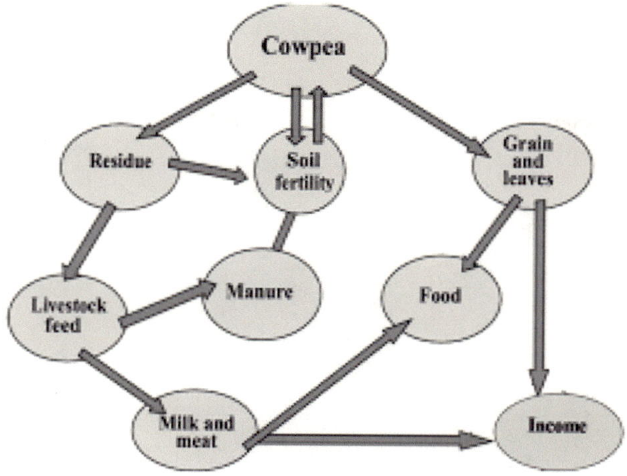

Figure 4.
Cow pea consider as source of income and multidimensional applications.

4. Nutritional aspects of cowpea

The seeds from cowpea seeds have high content of proteins, carbohydrates, low cholesterol, minerals and vitamins. The leaves of cowpea also have high proteins. The calorie values are higher in cowpea when compare to other crops. The nutrient value will be more and easily cultivable even in polluted or low mineral soils, so many countries consider it is important chief food source.

The homo-polysaccharide starch is present in all crops but in cowpea has to do delay in digestion than other crop starch, so it is good for all animals as well as

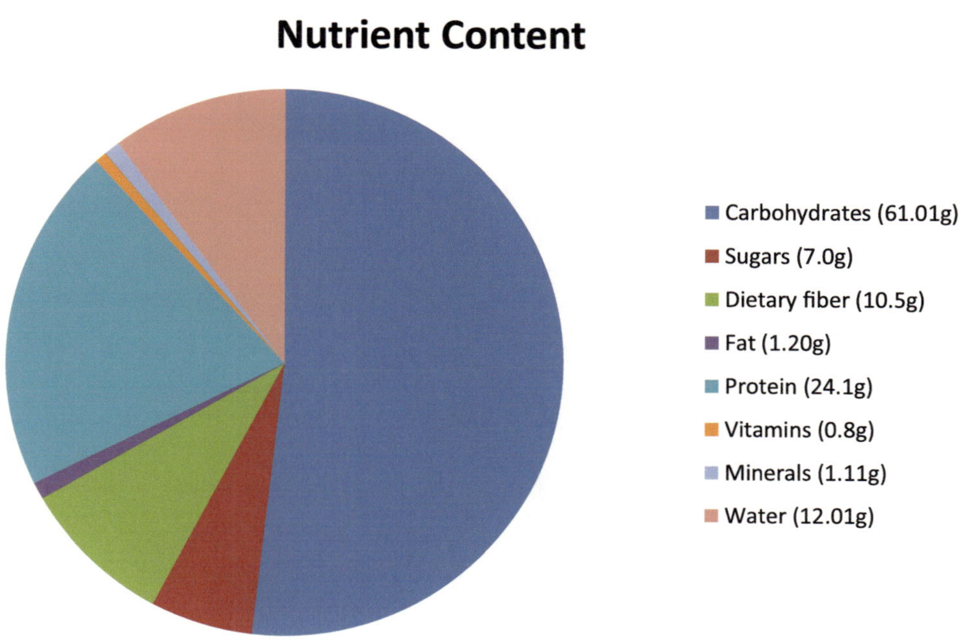

Figure 5.
Nutrient content of cow pea seeds (g/100 g).

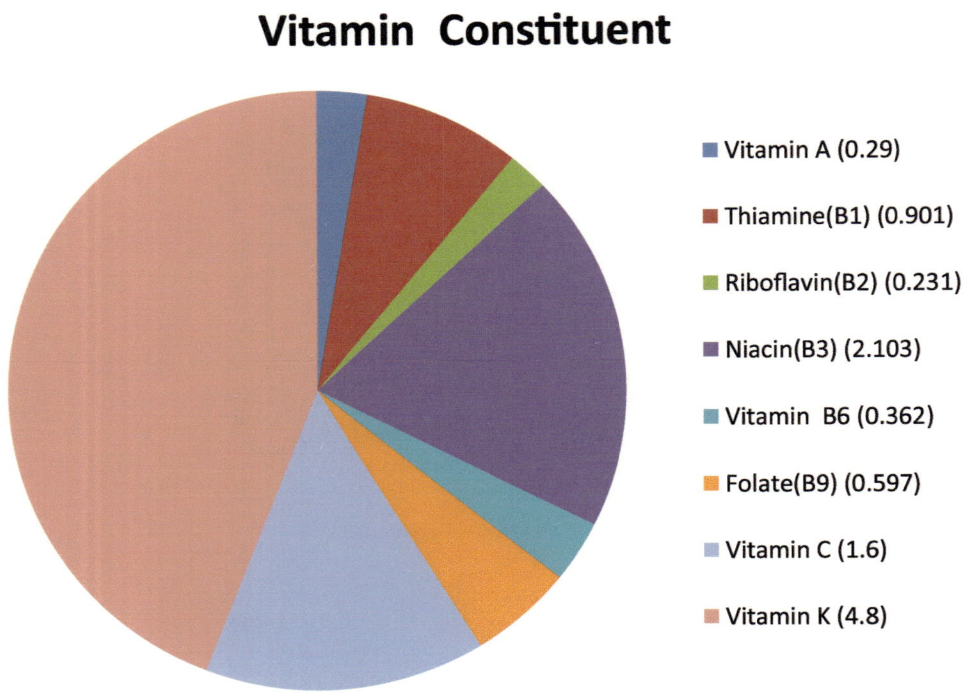

Figure 6.
Vitamin constituent of cow pea seeds (g/100 g).

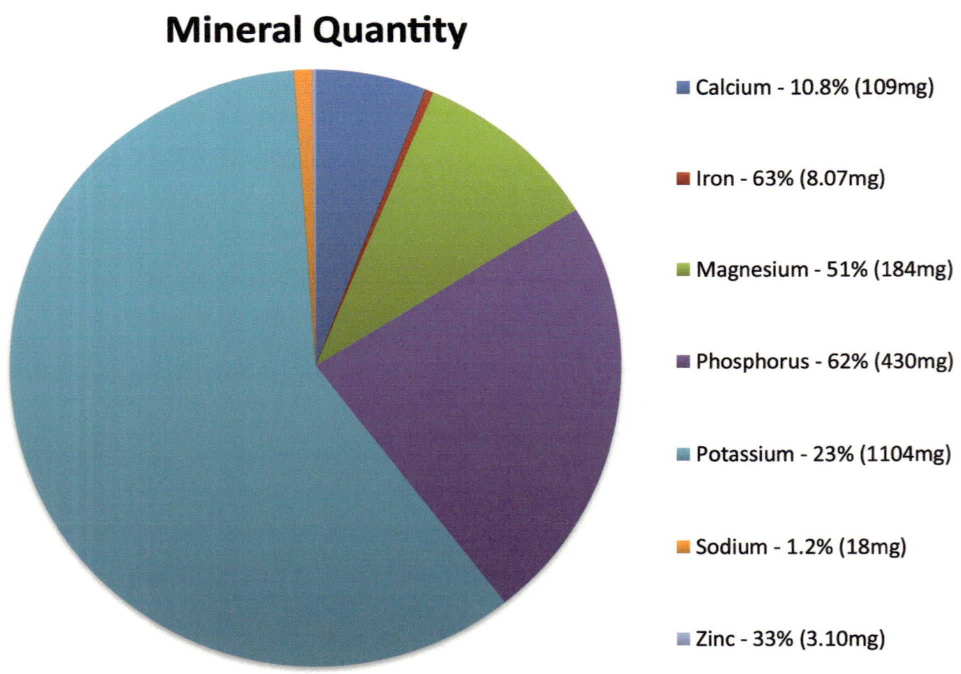

Figure 7.
Mineral quantity of cow pea seeds (g/100 g).

people consumption. The cowpeas seed have high content of vitamin B9 and helps in reduce or remove inborn error metabolism by means of efficiently develop neural tube in babies like fetus.

The cowpea cultivated by very poor people so it is called poor man crop and easy to grow under high drought conditions. These beans have phyticacids and some metabolic inhibitors which decrease the nutritional value. Some physiological methods like soaking, autoclaving fermentation, seed germination and stage of debranning used for the anti-nutritional properties of the cowpea by to is increased the cowpea nutrient bioavailability in soil.

Although the researchers has been identified importance of nutrients in many crop varieties. The anti-nutritional factors are high in dried seeds in some corps including cowpea than young or old leaves and pods. The below the images (**Figure 5**) describes the nutrient values, (**Figure 6**) describes the vitamin constituents and (**Figure 7**) describes the mineral quantity present in gram/100 gram of dried seeds or leaves.

5. Nutraceutical aspects of cowpea

Vegetable-based food systems are more sustainable than meat-based ones because it requires less energy, land, and water resources. The proteins from pulses is used for the balanced growth with have high nutrients to prevent human diseases and maintain good health [2].

The cowpea plants have highest medical constituents such as high fibers, important vitamins like vitamin B complex, folic acids and vitamin K for to promote good health. Cowpea has more potassium content, it lowers cholesterol, low fat, have high iron and antioxidants and do weight loss (**Figure 8**). The phenolics and carotenoids components and anthocyanins pigments are involved and maintain good anabolic and catabolic processes [3, 4].

The seed content of cowpea is used for decrease cholesterol formation so to decrease the body weight [5], cowpea seeds also increase blood cells in hematopoiesis to improve cardiac circulation and do efficient food digestion and also reduce digestive problems such as constipation, etc. [6]. The cow plants have lack of glycemic content of index is to delay the digestion of starch, so the cowpea seeds have high dietary fiber involved in insulin metabolism to decrease hunger for approximately in twelve hours [7].

Cowpea regular consumption to cause prevent or reduce many chronic communicable and non communicable diseases [8], the cow pea seeds have high nutrients to remove many gastric syndromes [6]. Cowpea beans also eradicate many heart diseases,

Figure 8.
V. unguiculata *(L.) Walp have approximately ten medical applications.*

decrease cholesterol diseases and improve weight loss [9]. Cowpea seeds or leaves have many medical properties, including reduce carcinoma, improve insulin, reduce lipid diseases, decrease pathogenic infections and remove certain allergies [10, 11].

The high content of carbohydrates, proteins as well as essential amino acids of cowpea seeds used for many health or therapeutic purposes. Cowpea proteins are highly valuable for both consumption and some culinary and some cowpea seed proteins also used for designing the texture of foods [12]. High regular intake of cowpea seeds may reduce blood plasma cholesterol like low density lipoprotein and also reduce non communicable diseases like many cancers and atherosclerosis or myogenic heart diseases [13].

6. Symbiotic relationship of cowpea for environmental protection

Legume plants are known to form symbioses with extremely broad range of beneficial soil microorganisms (BSM), the examples of almost all plant-microbe mutuality systems present in soil. The different legume plants have interaction with many beneficial microorganisms to increase nutritious, minerals and also do the intake of water from soil via prominent root hairs (**Figure 9**).

Cowpea plants promote the other plant development and also give protection to many bacterial of fungal plant pathogens and insect-pests. In the field of environmental sciences and agricultural aspects, the cowpea consider do most important beneficial legume to do the symbioses, they are bacterial specifically root nodule (RN) symbiosis and the fungi particularly arbuscular mycorrhiza (AM).

These symbioses explains the metabolic and genetically relations with host plants and have that plant growth promoting rhizosphere bacteria which are

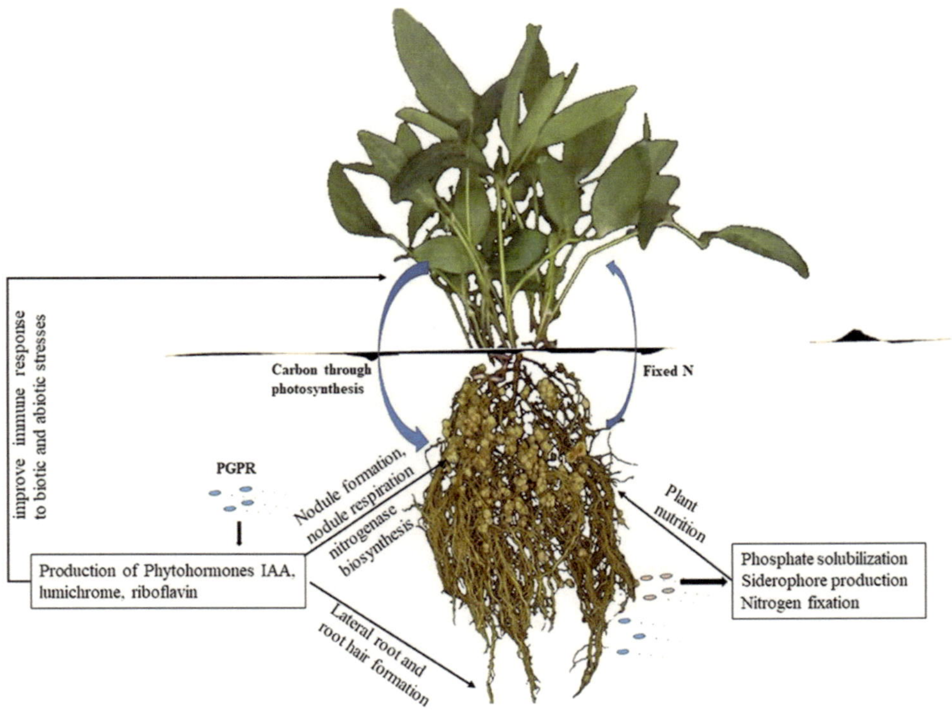

Figure 9.
Cowpea (Legumes) involved in N2 fixation and Carbon fixation. Courtesy: Sanjay K. Jaiswal, Frontiers in Sustainable Food Systems.

beneficial endophytic bacteria interact with legumes such as cowpea. The RNS requires a set of genes belonging to common symbiosis signaling pathway (CSSP), most of which are present in all land plants including legumes that can form a second symbiosis with, the fungi arbuscular mycorrhiza (AM) [14, 15].

Legumes are able to fix nitrogen from the air in symbiosis with Rhizobium bacteria [16]. The symbiotic relationship between plants with Rhizobium bacteria or any fungi to do fix the atmospheric nitrogen and compensate the ecological gas balance and also do the growth of crop plants to ultimately improve agriculture and maintain sustainable development of life.

The cowpea plants is eager to absorb high heavy metals such as copper from highly polluted soil or industrial effluents or etc., so it is otherwise called hyper-accumulator plants [17]. The cow pea consider as great pollution eradicator.

7. Conclusion

Cowpea have morphologically and taxonomically well adapted in tropical countries. The herbaceous nature of cowpea is easy for crop cultivation, crop rotation as well as intercrop. It have able to withstand drought or seasonal conditions. The nutrient aspects have more starchy content to proof that stable food for poor people. Cowpeas have other high protein less fat and high vitamins and minerals for maintain excellent metabolism for consuming organisms. Cowpea do fixing atmospheric nitrogen by to symbiotic relations with microorganisms with do recycling with biogeochemical cycles to enhance atmospheric gases.

The cowpea is one of the the largest cultivator crop among legume family and is commonly cultivated crop in most of the countries globally. The cow pea has highest nutrient values so used for maintain good nutrient calories required for our body. This crop plant is considered as good medical applications to cure or reduce the most non-communicable diseases in animals and or human systems. Cow also pea maintain ecological balances by fixing nitrogen, to chelate heavy metals for water and soil pollution removal and it used as bio-fuel production to improve our green environment and reduce fossil fuels to control other environmental pollutions.

8. Summary

V. unguiculata (L.) Walp is considered as green vegetable crop and the dried seeds were placed in pulses. Excellent to cultivation of the crop in any soil type and able to grow most arid climates and induce and develop more farm industries in global scale. Cowpea has poor digestion ability and considered as most stable food for very poor or developing countries. In Neolithic times the black eyed seed is most common starchy food. This crop have more chances to grow in tropical lands.

The morphological aspects ultimately deals with, the cowpea is recognized as an annual crop, herbaceous woody and greenish in nature. The seedlings were early to germinate in less moisture conditions. Cowpea leaves are rich in main green pigments such as chlorophyll *a* and chlorophyll *b* and also have reticulate venation made by vascular bundles. Inflorescence of the plant is raceme with five free petals with five fused sepals. The Androecium is nine plus one in number and is responsible for the development of male gametes. The gynoecium is ready to provide an egg. The double fertilization of the egg with pollen grain to fuse to form fertile black eyed seeds. The seeds nutritional information of cowpea deals with the presence high concentrations of carbohydrates, proteins, minerals and vitamins. The major ingredient is starch among the chemical components was highly recommended.

All vitamins regulate metabolism, specifically high concentration of niacin as coenzymes in cowpea and improve cellular growth. The irons and phosphorous are dominant minerals present in cowpea to regulate growth anabolism and catabolism.

The medical aspects enumerate the cowpea is no limit. The cowpea has high potassium to improve cell signaling in animals. The legume lowers the blood glucose level to reduce the diabetes among humans and it lowers blood cholesterol. The seeds have high in full of dietary fibers to regulate metabolism. The fiber content of cowpea is responsible to maintain very low fat in our body. Cowpeas have marvelous cell antioxidant mechanism to remove reactive oxygen species and reduce the effect certain toxic substances due to stress and radiation. The seeds rich in protein contents, the ingredients of cowpea show wonderful circulatory system and improve cardiac muscle function and lower the blood pressure. The black eyed seeds also induce gall bladder bile secretion for digestion purpose. Cowpea role in to decrease body weight, it means that improve weight loss.

The ecological review deals with by symbiotic relationship among various types of beneficial microorganisms. The root nodules of cowpea have to fix atmospheric nitrogen to ammonia with the help of *Vesicular Arbuscular Mycorrhiza* fungi and also certain bacteria like *Rhizobium* and other some organisms. The *Agrobacterium* also do fix atmospheric nitrogen to by *Nif* genes cowpea induce their flavonoids with acetosyringone to transfer DNA to host plant induce cancers or other mechanisms to improve growth of crop plants.

Author details

M. Duraipandian*, K.E. Poorani, H. Abirami and M.B. Anusha
Vivekanandha College of Arts and Sciences for Women (Autonomous), Elayampalayam, India

*Address all correspondence to: drduraipandian@vicas.org

References

[1] Xu P. "Genome wide linkage disequilibrium in Chinese asparagus bean (Vigna unguiculata ssp. sesquipedialis) germplasm: Implications for domestication history and genome wide association studies". Heredity. 2012;**109**(1):34-40 DOI: 10.1038/hdy.2012.8

[2] Martín-Cabrejas MA. Legumes: An overview. In: Martín-Cabrejas MA, editor. *Food Chemistry, Function and Analysis*. Vol. 8. 2019. pp. 3-18. DOI: 10.1007/s12117-001-1015-5,1

[3] Mudryj AN, Yu N, Hartman TJ, Mitchell DC, Lawrence FR, Aukema HM. Pulse consumption in Canadian adults influences nutrient intakes. British Journal of Nutrition. 2012;**108**(S1): S27-S36. DOI: 10.1017/S00071 14512000724

[4] Liyanage R, Perera OS, Weththasinghe P, Jayawardana BC, Vidanaarachchi JK, Sivakanesan R. Nutritional properties and antioxidant content of commonly consumed cowpea cultivars in Sri Lanka. Journal of Food Legumes. 2014;**27**(3):215-217

[5] Oboh HA, Agu K. The effects of various traditional processing methods on the glycemic index and glycemic load of cowpeas (*Vigina Unguiculata*). Journal of Food Biochemistry. 2010;**34**(6):1332-1342. DOI: 10.1111/j.17 454514.2010.00423.x

[6] Trehan I, Benzoni NS, Wang AZ. Common beans and cowpeas as complementary foods to reduce environmental enteric dysfunction and stunting in Malawian children: Study protocol for two randomized controlled trials. Trials. 2015;**16**(1):520. DOI: 10.1186/s13063-015-1027-0

[7] Pereira HVR, Saraiva KP, Carvalho LMJ, Andrade LR, Pedrosa C, Pierucci APTR. Legumes seeds protein isolates in the production of ascorbic acid microparticles. Food Research International. 2009;**42**(1):115-121. DOI: 10.1016/j.foodres.2008.10.008

[8] Frota K, dos Santos RD, Ribeiro VQ, Arêas JAG. Cowpea protein reduces LDL-cholesterol and apolipoprotein B concentrations, but does not improve biomarkers of inflammation or endothelial dysfunction in adults with moderate hypercholesterolemia. Embrapa Meio- Norte-Artigo em periódico indexado (ALICE). 2015

[9] Frota KMG, Mendonça S, Saldiva PHN, Cruz RJ, Arêas JAG. Cholesterol-lowering properties of whole cowpea seed and its protein isolate in hamsters. Journal of Food Science. 2008;**73**:235-240. DOI: 10.1111/j.1750-3841.2008.00953.x

[10] Jayathilake C, Visvanathan R, Deen A, Bangamuwage R, Jayawardana BC, Nammi S, et al. Cowpea: An overview on its nutritional facts and health benefits. Journal of the Science of Food and Agriculture. 2018;**98**(13):4793-4806. DOI: 10.1002/jsfa.9074

[11] Peyrano F, de Lamballerie M, Avanza MV, Speroni F. Calorimetric study of cowpea protein isolates. Effect of calcium and high hydrostatic pressure. Food Biophysics. 2017;**12**(3):374-382. DOI: 10.1007/s11483-017-9493-4

[12] Rudra SG, Sethi S, Jha SK, Kumar R. Physico-chemical and functional properties of cowpea protein isolate as affected by the dehydration technique. Legume Research-An International Journal. 2016;**39**(3):370-378. DOI: 10.18805/lr.v0iOF.9441

[13] Mune MAM, Minka SR, Mbome IL. Optimising functional properties during preparation of cowpea protein concentrate. Food Chemistry. 2014;**154**:32-37. DOI: 10.1016/j.foodchem.2013.12.108

[14] Favre P, Bapaume L, Bossolini E, Delorenzi M, Falquet L, Reinhardt D. A novel bioinformatics pipeline to discover genes related to arbuscular mycorrhizal symbiosis based on their evolutionary conservation pattern among higher plants. BMC Plant Biology. 2014;**14**:333

[15] Bravo A, York T, Pumplin N, Mueller LA, Harrison MJ. Genes conserved for arbuscular mycorrhizal symbiosis identified through phylo-genomics. Nature Plants. 2016;**2**:15208

[16] Dequiedt B, Moran D. The cost of emission mitigation by legume crops in French agriculture. Ecological Economics. 2015;**110**:51-60

[17] Duraipandian M, Sevugaperumal R, Ramasubramanian V, Ganesh D. Alleviation of the effects of copper chloride on *Vigna unguiculata* (L) using *Ulva lactuca*. Biochemistry & Physiology: Open Access. 2016;**5**:4. DOI: 10.4172/2168-9652.100021

Chapter 12

Health Risks Associated with the Consumption of Legumes Contaminated with Pesticides and Heavy Metals

Motunrayo Ganiyat Akande

Abstract

Legumes have high nutritional value and they are important sources of protein, carbohydrates, fats and dietary fiber. The contamination of legumes with pesticides and heavy metals has been reported in scientific literature. Human beings are mainly exposed to the residues of pesticides and heavy metals through the dietary route. The purpose of this review chapter is to highlight the acute and chronic health risks that human beings may be exposed to as a result of the ingestion of legumes polluted with pesticides and heavy metals. Additionally, the mechanisms through which pesticides and heavy metals engender different undesirable health outcomes in human beings were stated. Scientific literature were perused and the information contained in them were collated to derive this chapter. Pesticides cause short-term health effects including hypersensitivity and mortality, while heavy metals induce acute effects like seizures and death. Some chronic untoward effects of pesticides are congenital disabilities and neurological damage. Heavy metals elicit disorders like anemia, hypertension and cancer. It is envisaged that the findings documented in this review will create awareness of the health risks posed by the contamination of legumes with the residues of pesticides and heavy metals so that food safety measures can be enforced globally.

Keywords: Legumes, Health risks, Pesticides, Heavy metals, Contamination

1. Introduction

Legumes are plants that belong to the family, *Fabaceae* or *Leguminosae*, and they include chickpeas, cowpea, lentils, soy, etc. [1]. They are highly nutritious and can be consumed as food by human beings and animals [2].

Pesticides are chemical substances that are manufactured for the control of pests in domestic, agricultural and industrial settings, among others [3]. They are classified as herbicides, fungicides, bactericides and insecticides mainly for agricultural applications [4]. Insecticides comprise carbamates, organochlorines, organophosphates, neonicotinoids, pyrethroids; herbicides include benzothiazolyl urea, carbamic and sulfanilic acids, isoxazolyl urea, phenylpyrazole and pyridinium; while fungicides contain carboxamides, carbamates, dithiocarbamates, etc. [5, 6].

Heavy metals (HMs) are a group of metals that possess high atomic weights and densities beyond 5 g/cm^3 [7]. They are usually derived from agricultural, mining and industrial activities, as well as effluents [8]. It has been observed that HMs for instance, lead (Pb), cadmium (Cd), zinc (Zn), nickel (Ni) and copper (Cu), build up in the soil and in plant uptake structures and evoke injurious effects on the environment and the health status of human beings [7, 9, 10]. Besides, HMs have the propensity to be toxic when the populace are exposed to them or when they are consumed in proportions beyond the acceptable daily limits [11].

Different food stuffs such as legumes, cereals, fruits, vegetables, etc. have been contaminated with pesticides and HMs in various parts of the world. It is noteworthy that the common environmental pollutants, pesticides and HMs, increasingly cause numerous health hazards to the populace because of their permeation and upsurge in the food chain, as well as their persistence in the bionetwork [12].

In 2020, [13] identified the organophosphate pesticides (malathion, parathion, ethion and carbophenothion) in cowpea (an African legume) from Gwagwalada market in Abuja, Nigeria, through the use of Gas Chromatography–Mass Spectrometry. The levels of the pesticides found in the cowpea samples exceeded the maximum residue limits set by the European Union and the Agency for Toxic Substances and Disease Registry. Additionally, [14] confirmed the contamination of cowpea by residues of organochlorine (endosulfan and lindane), and organophosphorus (malathion) pesticides in high levels in Northern Cameroun.

Furthermore, heavy metals including nickel, cadmium, manganese and cobalt were detected in cowpea by [15]. In the research, the concentrations of nickel, cobalt and manganese discovered in the legumes were below the acceptable limits by FAO/WHO, while the cadmium concentration was beyond the FAO/WHO limit permitted. However, in a study conducted in Saudi Arabian markets, it was discovered that kidney beans and haricots contained high levels of Mn, while peas had elevated levels of zinc beyond the standard permissible levels [16]. Hence, regular monitoring of food stuffs for HM content was advocated.

It has been observed that fertilizers and pesticides are responsible for the increased level of soil pollution by non-essential micronutrients such as arsenic, cadmium, mercury, nickel and lead [17]. These non-essential micronutrients in the soil are conveyed through various plants as they accumulate in the edible portions [18]. Subsequently, the general population and animals may be exposed to the residues of these environmental contaminants (pesticides and HMs) when they consume plants that have been polluted with them.

The aim of this review chapter is to underscore the acute and chronic health risks that human beings may be exposed to during their life time as a result of the consumption of legumes contaminated with pesticides and HMs. In addition, the mechanisms through which pesticides and HMs engender different undesirable health outcomes in biological systems were highlighted.

It is envisaged that the information in this review chapter will stimulate and enhance concerted efforts towards the regular monitoring of legumes and other foodstuffs for pollutants in order to attain food safety.

2. Mechanisms through which pesticides induce adverse health effects

When human beings consume pesticides in legumes and other food sources through the dietary route, the pesticides may cause acute and chronic health risks through diverse mechanisms. For instance, organophosphate pesticides (chlorpyrifos, diazinon, dichlorvos, fenitrothion, malathion, parathion, etc.) bring about deleterious health effects by inhibiting the function of acetylcholinesterase, reducing the secretion of insulin, and perturbation of the metabolism of nutrients in living systems [19–21].

Moreover, organophosphate pesticides stimulate the release of reactive oxygen species and this phenomenon might cause oxidative stress [22, 23]. Oxidative stress refers to a disparity between the levels of prooxidants and antioxidants thereby culminating in damage to vital molecules including DNA, RNA, lipids and proteins in biological systems [24]. Oxidative stress has been identified as an important mechanism through which different kinds of pesticides cause biological injuries to human beings and animals.

The organochlorine pesticides are chlorinated hydrocarbons that persist in the environment and they include methoxychlor, dieldrin, chlordane, mirex and lindane, among others [25]. They cause the stimulation of the nervous system through the perturbation of action potentials, thereby leading to paralysis and death [26]. Dietary exposure brings about the bioaccumulation of organochlorine pesticides in the body, and this may terminate in derangements in human health [27].

Carbamates such as methiocarb, carbaryl, aldicarb, propoxur and carbofuran reversibly inhibit acetylcholinesterase in mammals and humans [19]. This group of pesticides are capable of causing apoptosis and necrotic changes in the cells of the immune system [28]. Consequently, the immune system becomes compromised and affected individuals become susceptible to various diseases when they are exposed to these pollutants in food and other sources.

Pyrethroids (e.g. allethrin, permethrin, cypermethrin and deltamethrin) are neurotoxicants and they disrupt the muscular structures and modify voltage-dependent sodium channels in the body [29]. They exhibit low acute toxicity to mammals and avian species. Conversely, at low levels, pyrethroids elicit acute toxicity to diverse aquatic species and arthropods [30].

3. Acute health risks associated with dietary exposure to pesticides

Acute health risks may become visible instantaneously or within 24 hours sequel to exposure to a pesticide [31]. Pesticides can have access to human bodies through the dermal [32], ocular [33], oral and respiratory [34] routes. According to [35], the exposure of the populace to pesticide residues through the oral route is about five orders of enormity compared to the other routes. It has been affirmed that pesticides can produce acute health disorders such as hypersensitivity, asthma and mortality in people [8, 36–38].

4. Chronic health risks related to dietary exposure to pesticides

It has been reported that chronic detrimental health risks do not manifest in human beings even within a day subsequent to pesticide exposure [31]. Chronic health risks evoked by exposure to pesticides in food such as legumes include congenital disabilities and decreased birth weight [36, 37]. Additionally, many organophosphate insecticides bring about declines in sperm counts, viability, density and motility; inhibition of spermatogenesis, reductions in testes weights, and sperm DNA damage in males [38]. Other chronic adverse effects of pesticides entail hindering the activities of hormones, their time of release, or mimicking these hormones, thereby culminating in reduced fertility and deformities in the male and female reproductive tracts [39]. These harmful effects may result in declines in the population of human beings affected. Also, pesticides perturb the immune system function and elicit carcinogenicity [40, 41].

Some investigators have asserted that organochlorine pesticides and their active metabolites are linked to neurological aberrations, cancers, hypertension, cardiovascular and dermatological disorders in humans [42–44]. Besides, the exposure of the populace to organophosphate, organochlorine and carbamate pesticides may evoke chronic neurological disorders such as Alzheimer's and Parkinson's disease [45]. According to [46], pesticides affect neuronal function negatively by hyperphosphorylation and the disruption of microtubules thereby ultimately causing Alzheimer's disease.

5. Mechanisms by which heavy metals induce health risks in the human population

Heavy metals can accumulate in the bones or adipose tissues of human beings through dietary intake (for example, the consumption of legumes and other crops), thereby leading to the diminution of critical nutrients and undermined immune defenses [8].

Lead induces oxidative stress in tissues and it is plausible that it brings about deleterious effects in humans that consume legumes and other plants contaminated with it through this mechanism [47]. Also, lead, a ubiquitous lethal HM, may activate the development of tumors through the production of reactive oxygen species, as well as the promotion of damage to DNA and its repair mechanisms in living organisms [48].

Furthermore, it has been documented that arsenic sets off neoplasms through the alteration of the genome of human beings, perturbation of DNA and induction of oxidative injury [49, 50]. Another common HM, mercury, generates carcinogenesis through oxidative damage and interference with the structure, mending and preservation of DNA in living organisms [51].

These myriads of mechanisms may be responsible for the untoward acute and chronic health hazards reported in human beings who have consumed foodstuffs such as legumes, cereals, fruits, vegetables, etc. that have been tainted with HMs over a period of time.

6. Acute health risks associated with dietary exposure of individuals to heavy metals

Heavy metal contamination is a health menace to both adults and children [52]. When HMs build up in human tissues internally, for instance, when legumes polluted with HMs are eaten for some time, they may harm the central nervous system, and produce seizures, headache and coma [8].

There is scientific evidence that lead may engender neurotoxicity, nephrotoxicity and impaired haeme synthesis [53]. When children have contact with cadmium through a range of routes, for example the oral route (by feeding on legumes such as chickpeas, cowpea, etc.), they become exceedingly prone to lead intoxication and permanent neurological abnormalities may ensue [54]. Other acute impacts of lead exposure documented in children are inattention, hyperactivity, increased dullness, irritability, headache, convulsion, coma and death [55, 56].

Copper is important for the normal functioning of the brain, but it can be noxious if the cellular concentration surpasses the metabolic requirement [57]. Elevated levels of copper can cause dysfunctions in the working memory of individuals [58], while short-term exposure to copper through dietary and other routes have been linked with stomach pain, haematemesis, melena, jaundice, anorexia and vomiting [59]. Moreover, rhabdomyolysis, cardiac and renal failure, hepatic necrosis, haemolysis, methemoglobinemia, encephalopathy, as well as mortality can occur in severe copper toxicity [60]. These harmful acute impacts of high copper levels may be attributed to its induction of oxidative stress, DNA damage and lessening of cell proliferation [61].

Moreover, the distortion of the concentrations of high-density lipoproteins and impairment of the immune system have been ascribed to excessive concentrations of zinc in the bodies of human beings [62]. People are usually exposed to zinc through the ingestion of contaminated food, for instance, legumes, pulses, among others [63].

7. Chronic health risks linked with the exposure of people to heavy metals through the oral route

Chronic health effects like diabetes, neurodegenerative diseases, renal damage, bone disorders and tumors of the breast, prostate and lungs have been elicited by Cd in people [64–66].

There are existing reports that indicate that the long-term effects of lead exposure through food and other routes are manifested as anemia, abdominal colic, miscarriages, male infertility, birth defects, renal diseases and behavioral dysfunctions in children [67, 68]. It has also been observed that lead may elicit decreased circulating maternal thyroid hormone thereby influencing growth patterns adversely [67, 68].

It is known that tremendous amounts of arsenic in the soil, food crops (including legumes, cereals, vegetables, etc) and groundwater can stimulate cancer and dermal aberrations, as well as disorders in the heart, stomach, intestines, liver, kidneys and brain, among others [62, 69–73].

Furthermore, excessive copper consumption can activate hepatic damage and other gastric-related problems in people [62, 74, 75]. Wilson's disease, a form of chronic Cu toxicity in human beings, is characterized by alterations in mental states, motor disorders, dysphagia, incoordination, haemolytic anemia, renal and hepatic dysfunctions [72]. It has been shown that human beings become susceptible to chronic copper poisoning when they ingest food items (for instance, legumes, pulses, cereals, etc.) contaminated with copper probably through polluted irrigation sources [76]. Additionally, persistent bronchitis, emphysema, pulmonary disorders and fibrosis were reported in people following long-term exposure to nickel through dietary sources [77].

8. Conclusion

This review chapter highlighted the mechanisms through which the widespread pollutants, pesticides and HMs, evoke acute and chronic health disorders in human beings. Pesticides are usually applied in agricultural settings for the control of vectors and the enhancement of crop yield, while most HMs are utilized for industrial purposes.

Human beings may be susceptible to various health risks engendered by pesticides and HMs through the consumption of food crops, for example, legumes, contaminated with them. This may eventually pose serious threats to the wellbeing and survival of the populace except if regular monitoring of food items for the residues of these ubiquitous contaminants are conducted globally by the appropriate agencies.

It is envisioned that the information contained in this review chapter will provide a springboard for scientists, researchers and agriculturists, among others, to create innovative techniques for the minimization of human exposures to the residues of pesticides and HMs in order to forestall their pernicious effects in the general population.

Acknowledgements

The author is grateful to the members of staff of the Faculty of Veterinary Medicine, University of Abuja, Federal Capital Territory, Nigeria, for their support.

Conflict of interest

The author declares that there is no conflict of interest.

Author details

Motunrayo Ganiyat Akande
Department of Veterinary Pharmacology and Toxicology, Faculty of Veterinary Medicine, University of Abuja, Federal Capital Territory, Nigeria

*Address all correspondence to: akande.ganiyat@uniabuja.edu.ng

References

[1] Stagnari F, Pisante M. Managing faba bean residues to enhance the fruit quality of the melon (*Cucumis melo* L.). Crop Sci Hort. 2010; 126: 317-323. https://doi.org/10.1016/j.scienta.2010.07.004.

[2] Adepoju FA, Adepoju AF, Fadina OO. Heavy metal levels in beans (*Vigna unguiculata*) in selected markets in Ibadan, Nigeria. J Exp Agric Int. 2018; 25: 1-8. DOI: 10.9734/jeai/2018/35900.

[3] Damalas CA, Eleftherohorinos IG. Pesticide exposure, safety issues, and risk assessment indicators. Int J Environ Res Public Health. 2011; 8: 1402-1419. DOI: 10.3390/ijerph8051402.

[4] Abdel Khalek ST, Mostafa ZK, Hassan HA, Abd El-Bar MM, Abu El-Hassan GMM. A new list to the entomofauna associated with faba bean, *Vicia faba* L. (*Fabales: Fabaceae*) grown in El-Kharga Oasis, New Valley Governorate, Egypt. Egypt. Acad J Biol Sci. 2018; 11: 95-100.

[5] Kaur R, Mavi GK, Raghav S, Khan I. Pesticides classification and its impact on environment. Int J Curr Microbiol Appl Sci. 2019; 8: 1889-1897. https://doi.org/10.20546/ijcmas.2019.803.224.

[6] Lippmann M, Leikauf G, editors. Environmental Toxicants Human Exposures and Their Health Effects. 4th ed. New Jersey; 2020. p. 1-40.

[7] Zhang X, Yan L, Liu J, Zhang Z, Tan C. Removal of different kinds of heavy metals by novel PPG-nZVI beads and their application in simulated stormwater infiltration facility. Appl Sci. 2019; 9: 4213. DOI:10.3390/app9204213.

[8] Alengebawy A, Abdelkhalek ST, Qureshi SR, Wang MQ. Heavy metals and pesticides toxicity in agricultural soil and plants: ecological risks and human health implications. Toxics. 2021; 9: 42. https://doi.org/10.3390/toxics9030042.

[9] Su C, Jiang L, Zhang W. A review on heavy metal contamination in the soil worldwide: situation, impact and remediation techniques. Environ Skept Critics. 2014; 3: 24-38.

[10] Tóth G, Hermann T, Da Silva MR, Montanarella L. Heavy metals in agricultural soils of the European Union with implications for food safety. Environ Int. 2016; 88: 299-309. https://doi.org/10.1016/j.envint.2015.12.017.

[11] Onakpa MM, Njan AA, Kalu OC. A review of heavy metal contamination of food crops in Nigeria. Annals of Global Health. 2018; 84: 488-494. DOI: https://doi.org/10.29024/aogh.2314.

[12] Verger PJP, Boobis AR. Reevaluate pesticides for food security and safety. Science. 2013; 341: 717-718. https://doi.org/10.1016/j.envint.2015.12.017.

[13] Akande MG, Sanni FS, Enefe NG. Human health risk evaluation of organophosphate insecticide residues in post-harvest cowpea in Gwagwalada, Abuja, Nigeria. J Health Pollution. 2020; 10: 201203. DOI: 10.5696/2156-9614-10.28.201203.

[14] Jean S, Benoit NM, Bosco TJ, Srivastava AK, Srivastava LP. Contamination of cowpea and by-products by organophosphorous pesticide residues in Ngaoundere markets: dietary risk estimation and degradation study. Afr J Food Sci. 2013; 7: 92-102. https://doi.org/10.5897/AJFS2013.1013.

[15] Akande MG, Sanni FS, Enefe NG. Assessment of the concentrations and health risk of some heavy metals in cowpea (*Vignus unguiculata*) in Gwagwalada, Nigeria. Drug and

Chemical Toxicology. 2019; doi: 10.1080/01480545.2019.1621334.

[16] Ali MHH, Al-Qahtani KM. Assessment of some heavy metals in vegetables, cereals and fruits in Saudi Arabian markets. Egyptian Journal of Aquatic Research. 2012; 38: 31-37. https://doi.org/10.1016/j.ejar.2012.08.002.

[17] Gunatilake SK, Samaratunga SS, Rubasinghe RT. Chronic kidney disease (CKD) in Sri Lanka - current research evidence justification: a review. Sabaragamuwa University Journal. 2014; 13: 31-58. DOI:10.4038/suslj.v13i2.7680.

[18] Edirisinghe EMRKB, Jinadasa BKKK. Arsenic and cadmium concentrations in legumes and cereals grown in the North Central Province, Sri Lanka and assessment of their health risk. International Journal of Food Contamination. 2019; 6:3. https://doi.org/10.1186/s40550-019-0073-x.

[19] Shah MAA, Srivastava AK Garg. Toxicology of Agrochemicals. In: Garg SK, editor. 1st ed. CBS Publishers & Distributors Pvt. Ltd: New Delhi; 2002. p. 146-179.

[20] Jaga K, Dharmani C. Sources of exposure to and public health implications of organophosphate pesticides. Rev Panam Salud Publica. 2003; 14:171-185. DOI: 10.1590/s1020-49892003000800004.

[21] Karami-Mohajeri S, Abdollahi M. Toxic influence of organophosphate, carbamate, and organochlorine pesticides on cellular metabolism of lipids, proteins, and carbohydrates: a systematic review. Hum Exp Toxicol. 2011; 30(9):1119-1140. DOI: 10.1177/0960327110388959.

[22] Abdollahi M, Ranjbar A, Shadnia S, Nikfar S, Rezaie A. Pesticides and oxidative stress: a review. Med Sci Monit. 2004; 10: 141-147. PMID: 15173684.

[23] Sharma D, Kaur GS, Khera KS. Triazophos-induced oxidative stress and histomorphological changes in ovary of female Wistar rats. Pestic Biochem Physiol. 2014; 117: 9-18. DOI: 10.1016/j.pestbp.2014.09.004.

[24] Conti V, Izzo V, Corbi G, Russomanno G, Manzo V, De Lise F, Di Donato A, Filippelli A. Antioxidant supplementation in the treatment of aging-associated diseases. Front Pharmacol. 2016; 7:24. DOI: 10.3389/fphar.2016.00024.

[25] Jones KC, de Voogt P. Persistent organic pollutants (POPs): state of the science. Environmental Pollution. 1999; 100: 209-221. https://doi.org/10.1016/S0269-7491(99)00098-6.

[26] Aliu YO. Veterinary Pharmacology. 1st ed. Wusasa-Zaria: Tamaza Publishing Company Limited; 2007. 466 p.

[27] Snedeker S. Pesticides and breast cancer risk: a review of DDT, DDE and dieldrin. Environ Health Perspect. 2001; 109 (suppl 1): 35-47. DOI: 10.1289/ehp.01109s135.

[28] Li Q, Kobayashi M, Kawada T. Ziram induces apoptosis and necrosis in human immune cells. Arch Toxicol. 2011; 85: 355-361. https://doi.org/10.1007/s00204-010-0586-9.

[29] Ray DE, Fry JR. A reassessment of the neurotoxicity of pyrethroid insecticides. Pharmacol Ther. 2006; 111: 174-193. DOI: 10.1016/j.pharmthera.2005.10.003.

[30] Pérez-Fernández V, Garcia MA, Marina ML. Characteristics and enantiomeric analysis of chiral pyrethroids. J Chromatogr A. 2010; 1217: 968-989. DOI: 10.1016/j.chroma.2009.10.069.

[31] Singh NS, Sharma R, Parween T, Patanjali PK. Pesticide contamination and human health risk factor. In: Oves M, et al. editors. Modern Age Environmental Problems and their Remediation. International Publishing AG: Springer; 2018. p. 50-68. DOI: 10.1007/978-3-319-64501-8_3.

[32] Anderson SE, Meade BJ. Potential health effects associated with dermal exposure to occupational chemicals. Environ Health Insights. 2014; 8s1, EHI. S15258. DOI: 10.4137/EHI.S15258.

[33] Fareed M, Kesavachandran CN, Pathak MK, Bihari V, Kuddus M, Srivastava AK. Visual disturbances with cholinesterase depletion due to exposure of agricultural pesticides among farm workers. Toxicol Environ Chem. 2012; 94: 1601-1609. https://doi.org/10.1080/02772248.2012.718780.

[34] Amaral AFS. Pesticides and asthma: challenges for epidemiology. Front Public Health. 2014; 2: 6. DOI: 10.3389/fpubh.2014.00006.

[35] Olutona GO, Aderemi MA. Organochlorine pesticide residue and heavy metals in leguminous food crops from selected markets in Ibadan, Nigeria. Legume Science. 2019;1-9. https://doi.org/10.1002/leg3.3.

[36] Sharon M, Bhawana M, Anita S, Gothecha VK. A short review on how pesticides affect human health. Int J Ayurvedic Herb Med. 2012; 2: 935-946.

[37] Wickerham EL, Lozoff B, Shao J, Kaciroti N, Xia Y, Meeker JD. Reduced birth weight in relation to pesticide mixtures detected in cord blood of full-term infants. Environ Int. 2012; 47: 80-85. DOI: 10.1016/j.envint.2012.06.007.

[38] Mehrpour O, Karrari P, Zamani N, Tsatsakis AM, Abdollahi M. Occupational exposure to pesticides and consequences on male semen and fertility: a review. Toxicol Lett. 2014; 230: 146-156. DOI: 10.1016/j.toxlet.2014.01.029.

[39] Kabir ER, Rahman MS, Rahman I. A review on endocrine disruptors and their possible impacts on human health. Environ Toxicol Pharmacol. 2015; 40: 241-258. DOI: 10.1016/j.etap.2015.06.009.

[40] Garry VF. Pesticides and children. Toxicol Appl Pharmacol. 2004; 198: 152-163. DOI: 10.1016/j.taap.2003.11.027.

[41] Özkara A, Akyıl D, Konuk M. Pesticides, environmental pollution and health. In: Larramendy ML, Soloneski S, editors. Environmental Health Risk-Hazardous Factors to Living Species: IntechOpen; 2016. DOI: 10.5772/63094.

[42] Walbel H, Henckes C, editors. Pesticides use and policies in Ghana: an economic analysis of current practice and factors influencing pesticide use. Pesticide Policy Project Publication Series No. 10. Hannover: Germany; 2001.

[43] Mansour SA. Pesticide exposure: Egyptian scene. Toxicology. 2004; 198: 91-115. DOI: 10.1016/j.tox.2004.01.036.

[44] Parbhu B, Rodgers G, Sullivan JE. Death in a toddler following endosufan ingestion. Clin Toxicol. 2009; 47: 899-901. DOI: 10.3109/15563650903328879.

[45] Keifer MC, Firestone J. Neurotoxicity of pesticides. J Agromed. 2007; 1:17-25. DOI: 10.1300/J096v12n01_03.

[46] Zaganas I, Kapetanaki S, Mastorodemos V, Kanavouras K, Colosio C, Wilks MF, Tsatsakis AM. Linking pesticide exposure and dementia: what is the evidence? Toxicology. 2013; 10: 3-11. DOI: 10.1016/j.tox.2013.02.002.

[47] Jaishankar M, Tseten T, Anbalagan N, Mathew BB, Beeregowda KN. Toxicity, mechanism and health effects of some heavy metals. Interdiscip Toxicol. 2014; 7: 60-72. DOI: 10.2478/intox-2014-0009.

[48] Silbergeld EK, Waalkes M, Rice JM. Lead as a carcinogen: Experimental evidence and mechanisms of action. Am J Ind Med. 2000; 38: 316-323. DOI: 10.1002/1097-0274(200009)38:3<316::aid ajim11>3.0.co;2-p.

[49] Martinez VD, Vucic EA, Becker-Santos DD, Gil L, Lam WL. Arsenic exposure and the induction of human cancers. J Toxicol. 2011; 2011: 1-13. DOI: 10.1155/2011/431287.

[50] Bjorklund G, Aaseth J, Chirumbolo S, Urbina MA, Uddin R. Effects of arsenic toxicity beyond epigenetic modifications. Environ Geochem Health. 2017; 39: 1-11. DOI: 10.1007/s10653-017-9967-9.

[51] Crespo-Lopez ME, Macedo GL, Pereira SI, Arrifano GP, Picanco-Diniz DL, do Nascimento JL, Herculano AM. Mercury and human genotoxicity: Critical considerations and possible molecular mechanisms. Pharmacological Research. 2009; 60(4):212-220. DOI: 10.1016/j.phrs.2009.02.011.

[52] Lu X, Zhang X, Li LY, Chen H. Assessment of metals pollution and health risk in dust from nursery schools in Xi'an, China. Environ Res. 2014; 128: 27-34. https://doi.org/10.1016/j.envres.2013.11.007.

[53] Malik RN, Zeb N. Assessment of environmental contamination using feathers of Bubulcus ibis L., as a biomonitor of heavy metal pollution, Pakistan. Ecotoxicology. 2009; 18: 522-536. DOI: 10.1007/s10646-009-0310-9.

[54] Maret W. The bioinorganic chemistry of lead in the context of its toxicity. In: Sigel A, Sigel H, Sigel RKO, editors. Lead: its effects on environment and health. Volume 17. Berlin: De Gruyter; 2017. p. 1-20.

[55] Pfadenhauer LM, Burns J, Rohwer A, Rehfuess EA. A protocol for a systematic review of the effectiveness of interventions to reduce exposure to lead through consumer products and drinking water. Syst Rev. 2014; 3: 36. DOI: 10.1186/2046-4053-3-36

[56] Evens A, Hryhorczuk D, Lanphear BP, Rankin KM, Lewis DA, Forst L, Rosenberg D. The impact of low-level lead toxicity on school performance among children in the Chicago Public Schools: A population-based retrospective cohort study. Environ Health. 2015; 14: 21. https://doi.org/10.1186/s12940-015-0008-9.

[57] Johnson WT. Nutritional neuroscience. In: Lieberman HR, Kanarek RB, Prasad C, editors. Nutritional Neuroscience. Boca Raton: CRC Press (Taylor and Francis Group); 2005. p. 289-305.

[58] Zhou G, Ji X, Cui N, Cao S, Liu C, Liu J. Association between serum copper status and working memory in schoolchildren. Nutrients. 2015; 7: 7185-7196. DOI: 10.3390/nu7095331.

[59] Gamakaranage CSSK, Rodrigo C, Weerasinghe S, Gnanathasan A, Puvanaraj V, Fernando H. Complications and management of acute copper sulphate poisoning: a case discussion. J Occup Med Toxicol. 2011; 6: 34. DOI: 10.1186/1745-6673-6-34.

[60] Harris ED. Cellular copper transport and metabolism. Annu Rev Nutr. 2000; 20: 291-310. DOI: 10.1146/annurev.nutr.20.1.291.

[61] Oe S, Miyagawa K, Honma Y, Harada M. Copper induces hepatocyte

injury due to the endoplasmic reticulum stress in cultured cells and patients with Wilson disease. Exp Cell Res. 2016; 347: 192-200. DOI: 10.1016/j.yexcr.2016.08.003.

[62] Zhou H, Yang WT, Zhou X, Liu L, Gu JF, Wang WL, Zou JL, Tian T, Peng, PQ, Liao BH. Accumulation of heavy metals in vegetable species planted in contaminated soils and the health risk assessment. Int J Environ Res Public Health. 2016; 13: 289. DOI: 10.3390/ijerph13030289.

[63] Hough RL, Breward N, Young SD, Crout NM, Tye AM, Moir AM, Thornton I. Assessing potential risk of heavy metal exposure from consumption of home produced vegetables by urban populations. Environ. Health Perspect. 2004; 112: 215-221. DOI: 10.1289/ehp.5589.

[64] Avenant-Oldewage A, Marx HM. Bioaccumulation of chromium, copper and iron in the organs and tissues of *Clarias gariepinus* in the Olifants River, Kruger National Park. Water SA. 2000; 26: 569-582.

[65] Jiang LF, Yao TM, Zhu ZL, Wang C, Ji LN. Impacts of Cd (II) on the conformation and self-aggregation of Alzheimer's tau fragment corresponding to the third repeat of microtubule-binding domain. Biochim Biophys Acta-Proteins Proteomics. 2007; 1774: 1414-1421. DOI: 10.1016/j.bbapap.2007.08.014.

[66] Chunhabundit, R. Cadmium exposure and potential health risk from foods in contaminated area, Thailand. Toxicol. Res. 2016; 32: 65-72. DOI: 10.5487/TR.2016.32.1.065.

[67] Gundacker C, Hengstschläger M. The role of the placenta in fetal exposure to heavy metals. Wien Med Wochenschr. 2012; 162: 201-206. DOI: 10.1007/s10354-012-0074-3.

[68] Wani AL, Ara A, Usmani JA. Lead toxicity: a review. Interdiscip Toxicol. 2015; 8: 55-64. DOI: 10.1515/intox-2015-0009.

[69] Chiou HY, Hsueh YM, Liaw KF, Horng SF, Chiang MH, Pu YS, Lin JS, Huang CH, Chen CJ. Incidence of internal cancers and ingested inorganic arsenic: a seven-year follow-up study in Taiwan. Cancer Res. 1995; 55: 1296-1300. PMID: 7882325.

[70] Kapaj S, Peterson H, Liber K, Bhattacharya P. Human health effects from chronic arsenic poisoning - a review. J Environ Sci Health A Tox Hazard Subst Environ Eng. 2006; 41: 2399-2428. DOI: 10.1080/10934520600873571.

[71] Lin HJ, Sunge T, Cheng CY, Guo HR. Arsenic levels in drinking water and mortality of liver cancer in Taiwan. J Hazard Mater. 2013; 262: 1132-1138. DOI: 10.1016/j.jhazmat.2012.12.049.

[72] Islam S, Rahman MM, Rahman MA, Naidu R. Inorganic arsenic in rice and rice-based diets: health risk assessment. Food Control. 2017; 82: 196-202. https://doi.org/10.1016/j.foodcont.2017.06.030.

[73] El-Kady AA, Abdel-Wahhab MA. Occurrence of trace metals in foodstuffs and their health impact. Trends Food Sci Technol. 2018; 75: 36-45. https://doi.org/10.1016/j.tifs.2018.03.001.

[74] Gaetke LM, Chow CK. Copper toxicity, oxidative stress, and antioxidant nutrients. Toxicology. 2003; 189: 147-163. DOI: 10.1016/s0300-483x(03)00159-8.

[75] Rahman MA, Rahman MM, Reichman SM, Lim RP, Naidu R. Heavy metals in Australian grown and imported rice and vegetables on sale in Australia: health hazard. Ecotoxicol Environ Saf. 2014; 100: 53-60. https://doi.org/10.1016/j.ecoenv.2013.11.024.

[76] El-Kady AA, Abdel-Wahhab MA. Occurrence of trace metals in foodstuffs and their health impact. Trends Food Sci. Technol. 2018; 75: 36-45. https://doi.org/10.1016/j.tifs.2018.03.001.

[77] United States Department of Health and Human Services, Agency for Toxic Substances and Disease Registry. 2005; 1-397.

Chapter 13

Fermentation as Strategy for Improving Nutritional, Functional, Technological, and Sensory Properties of Legumes

Michela Verni, Erica Pontonio, Marco Montemurro and Carlo Giuseppe Rizzello

Abstract

Compared with cereals and other plant-derived food matrices, legumes can be considered as valuable sources of proteins with high biological value, dietary fibers, minerals, oligosaccharides, and phenolic compounds. Nevertheless, the presence of different antinutritional factors (ANFs) limited the large-scale use of such ingredients by the food industry. The potential of several biotechnological processes and enzymatic treatments in decreasing ANF in legumes and legume-derived ingredients was investigated. Among these options, fermentation is traditionally recognized as suitable tool to improve the overall quality of legumes in different areas of the world. The scientific community demonstrated the effectiveness of the use of selected lactic acid bacteria and biotechnologies inspired to sourdough fermentation in ANF degradation, improving technological and sensory profile of legume grains and flours as well as contributing to their safety in terms of spoilage or pathogenic microorganisms and toxic compounds. Apart from their consumption as they are, legumes are the main ingredient of many traditional food products, and fermentation allows them to be used as ingredients in innovative formulations of staple foods, such as baked goods and pasta with high nutritional and functional profile.

Keywords: fermentation, legumes, lactic acid bacteria, antinutritional factor, biotechnologies, sourdough, bread, pasta

1. Introduction

The challenge of feeding the growing world population and the necessity to provide a nutritionally balanced diet while reducing greenhouse gas emissions, as well as a transition to a diet higher in plant- rather than animal-derived proteins, require relevant increases in vegetables production. In this context, the fortification of foods and beverages has been identified as an effective, sustainable, and promising intervention capable of modulating the diet toward healthier choices, addressing environmental concerns, and meeting nutritional deficiencies and recommendations. To date, several studies investigated the nutritional value of additional ingredients to be used as wheat alternatives in cereal-based products, such as bread and pasta.

Legumes are considered as good source of high biological value proteins and dietary fibers. Moreover, they are rich in phenols, minerals, vitamins, and oligo-saccharides. The optimal technological properties of the legume flours (e.g., high water-binding capacity and solubility) make them suitable ingredients for gluten-free foods.

Nevertheless, legumes contain part of their nutritional compounds under a nonbioavailable form and several antinutritional factors (ANFs) that may decrease digestibility of other nutrients or cause physiological discomfort or conditions. Furthermore, legumes have poor technological, rheology, and sensory attributes if compared with gluten-containing cereals. Hence, the full exploitation of such food matrices goes through the most suitable bioprocessing.

Lactic acid bacteria (LAB) are the group of microorganisms most largely used at food industrial level, having the status of Generally Recognized as Safe (GRAS). Used as natural (e.g., sourdough and spontaneous fermentation) or selected starters, LAB have the capability to conjugate desired functional activities, sensory properties, and microbiological safety.

Overall, bioprocessing including LAB fermentation is considered a safe, sustainable, and effective tool for improving the functional and nutritional features of many plant-derived matrices and to obtain suitable technological, sensory, and shelf-life characteristics of fermented foods and beverages (**Figure 1**). The positive effects of LAB fermentation are in part related to the acidification, although further effects can be observed, such as those related to the synthesis of metabolites and the activation of the flour endogenous enzymes. The properties of the fermented matrix are often profoundly different from the unfermented ingredients. Among the main nutritional advantages of the LAB fermentation, the increase of the protein digestibility and the decrease of the glycemic index have

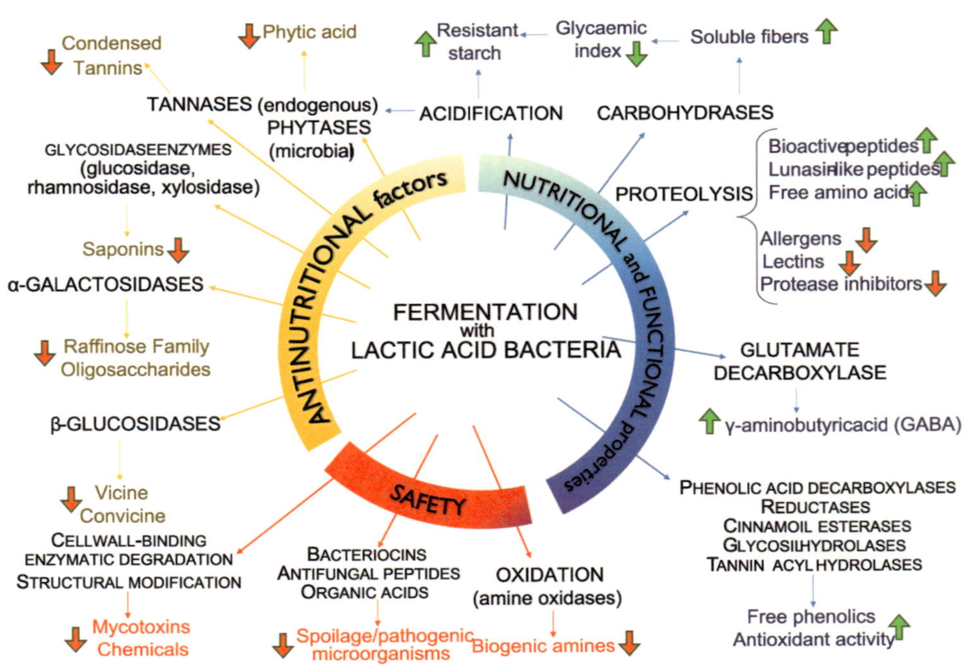

Figure 1.
Main nutritional, functional, and safety properties deriving from LAB fermentation.

been largely investigated. More recently, also the degradation of the antinutritional compounds (e.g., trypsin inhibitors, phytic acid, saponins, condensed tannins, and α-galactosides) and the synthesis of bioactive compounds have been described. Starting from the conventional application of the sourdough-inspired procedures, innovative biotechnological protocols, based on the use of selected starters, automatized bioreactors, and semiliquid formulations have been recently proposed to extend to a large-scale application the use of legumes in food industry.

Indeed, fermentation (both spontaneous or guided by selected LAB) has been recognized as the most suitable and sustainable process to exploit the potential of legumes to fortify staple foods such as baked goods, pasta, extruded snacks, and plant-based fermented beverages.

In this chapter, the scientific evidence confirming the nutritional, functional, rheology, sensory, and shelf-life improvements of fermented legumes and derived food products is described.

2. Nutritional insights

As recommended by global organizations, due to the growing concerns related to the environmental impact of animal breeding and the health risks associated with high meat intake, the decrease in animal-derived foods consumption led to the need for more plant-based foods in diet and more energy-efficient processing [1]. Simultaneously, the large market growth of foods designed for vegetarian, vegan, and gluten-free diets generated an increased consideration in improving the nutritional quality of grains-derived ingredients to be used in food preparation [2].

Leguminosae family, belonging to the Dicotyledonae group, includes 18,000 different species. After cereals, legumes are the most important group of crops, and their consumption is widely distributed all over the world.

A large variety of legumes used for human diet are cultivated extensively or locally [3, 4]. The economic importance of the Leguminosae family is related to the low input required for their cultivation, the positive impact on the soil fertility, and the great adaptability to underrestrictive pedoclimatic conditions [4]. Moreover, the advantages of cereal-legume intercropping, also providing an efficient exploitation of natural resources, have been abundantly demonstrated [5].

Legumes are excellent sources of proteins with high biological value, providing many essential amino acids, contain carbohydrates and dietary fibers, and supply relevant levels of vitamins, minerals, oligosaccharides, and phenolic compounds [6]. The frequent consumption of legumes is effective to prevent or decrease risks of cardiovascular disease (CVD) [7], type 2 diabetes [8], some types of cancer [9], and overweight and obesity [10].

When cereals and legumes are combined in food formulations, protein efficiency improved thanks to complementary essential amino acid profiles [11]. Overall, compared with cereal, legumes contain less starch, more protein, and more fiber, whereas lipid content is either equal or higher. Starch content in wheat varies between 60 and 80%, whereas it ranges from 40 to 65% for legumes except for lupin, having a markedly lower starch content [1]. Proteins in legume flours vary between 20 and 30% and can reach up to 40% in faba and lupin flours, against the 9–18% in wheat and other cereals [1]. Fiber content is circa 2% (on dry matter) in wheat flour and semolina, while it can reach 10% in pea and faba flours, and even 20–40% in chickpea, lentil, and lupin flours [1]; however, legume flours are often obtained from whole grains (not dehulled) resulting in a higher proportion of fiber. Ultimately, lipid content varies between 1 and 3% (on dry matter) in wheat and legume flours except for chickpea and lupin flours in which it can reach 10–13% [1].

Besides nutritional composition, the main proteins contained in cereals and legumes also present several differences in terms of type and functionality. In wheat, for example, gluten proteins (gliadins and glutenins) are the most abundant, accounting for 80% of total protein fraction [12]. In legumes, globulins are the dominant group, accounting for 50–70% of total proteins [13]. Wheat gliadins and glutenins contain higher concentration of sulfur amino acids compared with legume globulins, meaning they have more reactive cysteine residues [13, 14]. Moreover, low-molecular-mass albumins are present in both cereal and legume grains, reaching, respectively, 15 and 15–40% of the total proteins content [13]. Just as for proteins, starch granules in wheat and legumes show differences. They both contain linear amylose and branched amylopectin organized in semicrystalline and amorphous structures; however, they differ in shape and amylose/amylopectin ratio [15]. Legume starches have a higher proportion of amylose than wheat starch, ranging from 24/76 to 40/60 for pea and lentil starches and from 23/77 to 35/65 for chickpea starch [16].

3. Antinutritional factors and microbial degradation

Legumes contain several ANFs, such as raffinose, phytic acid, condensed tannins, saponins, alkaloids, lectins, pyrimidine glycosides, and protease inhibitors [17]. Overall, ANFs decrease the bioaccessibility and bioavailability of other nutrients, and, in some cases, are responsible for adverse reactions to the ingestion.

The content of raffinose-family oligosaccharides (RFOs, raffinose, verbascose, and stachyose) in legumes ranges from 1 to 6% with stachyose as the most abundant compound [18]. While in cereals, it is commonly lower than 1.5%, with raffinose as the sole or the most abundant compound [19, 20]. RFOs are nondigestible oligosaccharides that may result in adverse digestive symptoms when about 15 g/person per day are exceeded [21], a threshold that is readily reached in legume-based diets. Raffinose and RFO are indeed fermented by the intestinal microbiota with abundant gas production, causing discomfort and flatulence.

Phytic acid is the main storage compound for phosphorous and minerals in cereal and legume seeds. In legumes, its concentration can reach 20 g/kg [22, 23]. Phytic acid and divalent minerals (e.g., Ca^{2+}, Zn^{2+} and iron) form stable complexes (phytates) that are insoluble and not hydrolyzed in the gastrointestinal tract, thus reducing the bioavailability of minerals for the monogastrics. Ca^{2+} and Zn^{2+} deficiencies are commonly observed in developing countries, and complexation of dietary minerals by phytates in plant-derived foods contributes to the mineral deficiency [17]. Iron uptake from plant-derived foods is impeded not only by complexation with phytate but also by complexation with condensed tannins [24, 25].

Proanthocyanidins, gallotannins. and ellagitannins, commonly referred to as tannins, are phenolic compounds that occur in a wide variety of plant foods. Their presence in cereals and legumes is dependent on the plant species and the cultivar [26]. Tannins impart bitter taste, reduce protein and starch digestibility by inhibition of pancreatic enzymes, and reduce iron uptake [26, 27]. The presence of tannins reduces the caloric content and the glycemic index of foods [28], but the abundance in diet reduces the supply of macro- and micro-nutrients.

Lectins and specific inhibitors of digestive enzymes (proteases and amylases) further reduce the digestibility of starch and proteins in legumes [26, 29].

Some ANFs are heat-labile (e.g., protease inhibitors and lectins) and easily removed by thermal treatments. Nevertheless, phytic acid, raffinose, tannins, and saponins are rather thermostable. Dehulling, soaking, air classification, extrusion, steaming, and pregelatinization are the main technological options for decreasing the negative impact of ANF on legume consumption [30–32]. Nevertheless, biological methods such as germination, enzyme treatments, and especially, fermentation seem to be more efficient [30, 31, 33, 34].

Proteolysis, enzyme inhibition due to acidification, acid activation of flour endogenous enzymes (e.g., phytases) and/or microbial enzyme activities (e.g., α-galactosidase, β-glucosidase, phytases, tannases) are responsible for the inactivation of most ANFs.

Raffinose family oligosaccharides are hydrolyzed through the activity of α-galactosidases, levansucrase, and sucrose-phosphorylase activities of lactic acid bacteria [35, 36] or corresponding enzymes of fungal cultures; their removal in legume fermentations has been amply reported [37].

In cereal matrices [22], the phytase activity is often sufficient to degrade phytates, especially in acidic conditions [18, 38]. Therefore, phytate degradation in LAB-fermented matrices spontaneously occurs without microbial enzymes involvement [18]. The optimal pH for the activity of the cereal phytases corresponds to 5.5; nevertheless, phytases are still active at pH levels lower than those commonly reached by sourdough (3.8–4.2) [18]. Sourdough fermentation and other types of traditional bioprocesses involving LAB (e.g., fermentations for production of cereal porridges or beverages) allow the increase of the mineral bioavailability [39]. Compared with that found in cereals, the phytase activity in legumes is poor [22, 40]. Nevertheless, pretreatments and processing conditions including fractionation, germination, soaking, thermal treatments, and fermentation drastically decrease phytate levels in legumes [41]. In many spontaneously fermented legume products, substrate-derived phytases are inactivated, and phytate degradation is achieved by fermentation with bacilli or fungal cultures, for example, *Rhizopus stolonifer* or *Aspergillus oryzae*, which hydrolyze phytate with extracellular enzymes [42, 43].

Metabolism of tannins or other polyphenols by LAB was deeply characterized only in a few fermented plant-derived matrices [44, 45]. *Lactiplantibacillus plantarum*, *Lactiplantibacillus paraplantarum*, *Lactiplantibacillus pentosus* have been identified, among the LAB, as the species that could decrease the tannins concentration through their tannases (tannin acyl hydrolase, EC 3.1.1.20) [46–48]. However, characterization of fermented cassava allowed identifying uncommon tannase producers such as *Weissella cibaria* and *Leuconostoc mesenteroides* ssp. *mesenteroides* [49]. Most of tannase producers were found in fermented vegetables but also in human feces. In *L. plantarum*, tannase is very well characterized. Its activity was demonstrated and characterized by Rodríguez et al. [47], and genetic analysis showed that it constitutes a novel family of tannases [50]. LAB tannases are intracellular. Genes involved in tannins degradation are regulated in a coordinated way and are inducible by tannin and other phenolic compounds [51].

The lactic fermentation of grass pea (*Lathyrus sativus*) with *L. plantarum* lowered the levels of phytic acid and trypsin inhibitory activity [52]. Selected strains of *L. plantarum* and *Levilactobacillus brevis* decreased the content of raffinose up to circa 64% during sourdough fermentation of different legume flours. Sourdoughs made with different legume flours (bean, lentil, pea, grass-pea, chickpea) contained an increased phytase activity compared with the unfermented controls [34]. The combination of legume sprouting and sourdough fermentation decreased the content of phytic acid, condensed tannins and raffinose, and trypsin inhibitory activity [53, 54].

Besides the abovementioned ANFs, faba bean is rich in two glucosidic aminopyrimidine derivatives, vicine and convicine, which, upon hydrolysis of the β-glucosidic bond, generate the aglycones divicine (2,6-diamino-4,5-dihydroxypyrimidine) and isouramil (6-amino-2,4,5-trihydroxypyrimidine), respectively [55].

Divicine and isouramil trigger favism disease in susceptible individuals. Technological processes (air classification, roasting, and boiling) and selection of cultivars with low content of such compounds seemed to be only in part effective [55, 56]. On the contrary, β-glucosidase from LAB effectively degraded the pyrimidine glycosides from faba bean suspension and flour [30]. When used as starter to ferment fava bean flour, *L. plantarum* expressed β-glucosidase activity and decreased the content of vicine and convicine by more than 90%. The degradation was complete after 48 h of fermentation, and aglycone derivatives were not detectable [57]. Similar results were obtained when flours from different faba bean accessions collected from the Mediterranean area were subjected to the LAB fermentation [58]. *Ex-vivo* hemolysis assays on human blood confirmed the lack of toxicity of the fermented fava bean [57].

4. Decrease of allergens, biogenic amines, mycotoxins, and chemicals through fermentation

Different legume proteins act in susceptible individuals as allergens. Their complex structures are difficult to degrade. The selection of legumes' natural variants or the use of specific biotechnological processes has been exploited to solve this issue. However, some side effects such as an increase in the protein synthesis pathways of the seed and the synthesis of other proteins that might be allergenic have been also reported [59–62]. Overall, plant proteins exhibit low digestibility compared with animal proteins. Poor protein digestibility can cause gastrointestinal disorder, and the increase in protein digestibility could reduce the level of immunoreactive proteins in their active forms, thus reducing the risk of food allergies symptoms [63]. Several studies showed that LAB fermentation increases the digestibility of plant proteins through the combined activity of microbial and endogenous proteases and peptidases [64, 65]. The use of fermentation to reduce or eliminate allergenicity of soy products represents an interesting opportunity to produce hypoallergenic food products from legumes [66, 67]. It was indeed shown that fermentation of soybean meal with *L. plantarum* or *Bifidobacterium lactis* allowed a significant increase in the total amino acids and a low immunoreactivity.

Besides allergens, many undesirable substances, contaminating foods and feeds, are harmful to human and animal health. These include mycotoxins, which are widely present in food and feeds commodities. The role of different microorganisms including fungi, yeasts, and bacteria in mycotoxins degradation has been investigated. Several studies extensively reported that mycotoxin degradation mechanisms are different and include cell wall binding, enzyme degrading, or structure modification. However, the degradative mechanisms are strain-dependent [68–73].

For example, patulin is a mycotoxin synthesized by different fungi, such as *Penicillium expansum,* able to colonize different fruits and vegetables [74]. Its toxicity is due to the high reactivity with thiols [75], which leads to the decrease of cellular glutathione levels. The capability of some yeasts or heterofermentative lactobacilli to release thiols during fermentation allows the patulin inactivation. Patulin degradation can also occur thanks to the conversion in inactive forms by *L. plantarum* esterase and reductase activities [76]. It was also reported that fermentation of legumes and cereals allows the decrease of aflatoxin concentration [77]; however, the mechanisms have not been completely clarified [78]. The mycotoxins absorption by the bacterial biomass has also been hypothesized [79].

Fermented foods often contain biogenic amines, derived from microbial metabolisms, and characterized by a dose-dependent toxicity. Biogenic amines (BAs) are produced not only by Gram-positive and Gram-negative bacteria, but also by yeasts and molds [80]. Also LAB are considered as BAs producers in fermented foods and *Enterococcus*, species of the former *Lactobacillus* genus, *Streptococcus, Lactococcus, Oenococcus, Pediococcus, Weissella, Carnobacterium, Tetragenococcus, Leuconostoc, Sporolactobacillus* are the main genera showing this trait [81]. BAs production is a strain specific feature, and some studies revealed that the involved enzyme is encoded by unstable plasmids [82, 83]. Therefore, horizontal gene transfer is essential to disseminate this ability in LAB [82, 83].

Many intrinsic and extrinsic parameters affect the BAs production (e.g., pH, temperature, and water activity); nevertheless, their control is often difficult during food processes. The BAs production is strain-dependent; therefore, the starter selection is an efficient tool to decrease their accumulation in fermented foods. Another effective strategy includes the use of amine oxidizing selected starters [84].

Through their oxidases, such microorganisms catalyze the oxidative deamination of BAs and their conversion to aldehydes, hydrogen peroxide, and ammonia [85]. Kim et al. [86] isolated strains of *Bacillus subtilis* and *Bacillus amyloliquefaciens* from fermented soybean foods. They observed the ability of *B. subtilis* to degrade putrescine and cadaverine and of *B. amyloliquefaciens* to oxidize histamine and tyramine. Similarly, Kang et al. [87] showed the ability of *B. subtilis* and *B. amyloliquefaciens* strains to reduce tyramine in Cheonggukjang. Eom et al. [88] isolated from buckwheat sokseongjang, a Korean traditional fermented soybean food, three strains (belonging to *B. subtilis* and *Bacillus idriensis* species), which were able to degrade histamine and tyramine but also unable to produce them. Lee et al. [89] recently proposed the use of *L. plantarum* strains to reduce BAs content during Miso fermentation. The possibility to use amine oxidizing starter cultures is an effective tool to decrease the BAs concentration in fermented foods obtained with legumes, especially when traditional production methods are used.

Another growing concern for the consumer is represented by the potential presence of chemicals and pesticides in foods, especially if correlated to the global recommendation to increase the dietary uptake of fruit and vegetables. It has been reported, for example, that the cumulative intake of pesticides by high consumers of fruits and vegetables in Brasil exceeds the Acute Reference Dose [90]. There is a consensus that the level of residual pesticides in foods needs to be decreased. However, the replacement of conventional pesticides in agriculture is a slow and difficult process. Therefore, the possibility to degrade pesticides through fermentation has been investigated. Several chemicals can be converted by microorganisms, but many of the most effective species characterize the environmental microbiota and are not easily usable in food processing.

The conversion of pesticides during food fermentation has been investigated in correlation, for example, to the large diffusion of contaminated soy (genetically resistant to the herbicide glyphosate). The degradation of organophosphorus insecticides was observed during the fermentation of Kimchi by *Leuc. mesenteroides, Lv. brevis, L. plantarum*, and *Latilactobacillus sakei* strains [91].

Lv. brevis was also seen as an active catalyst against the same family of compounds during the fermentation of milk products [92]. The degradation of organochlorine pesticides has also been investigated in milk during yogurt and cheese production showing the effect of starters [93]. Other examples refer to the capability of *Micrococcus varians* to degrade DDT (dichlorodiphenyltrichloroethane) to DDD (ddichlorodiphenyldichloroethane) and lindane to 2,4-, 2,5-, 2,6-, and 3,4-dichlorophenol and of *Lactococcus lactis subsp. lactis* to degrade dinitrotoluene isomers [94, 95].

5. LAB as biopreservation agents against pathogenic and spoilage microorganisms

Besides decreasing antinutritional factors and allergy, LAB can fulfill a task of biopreservation [96]. This word can be defined as the extension of shelf-life and food safety by means of natural or controlled microbiota and/or their antimicrobial compounds [97]. Overall, LAB fermentation is one of the most common methods of food biopreservation.

In South-East Asia, specific biopreservation strategies to limit pathogens and spoilage microorganisms contamination in foods have been proposed. Overall, the most common contamination of legumes in the field is represented by sporulating bacteria; then, fungi can develop and produce mycotoxins. Finally, different pathogens can occasionally derive from cross-contamination with other foods.

Phan et al. [98] studied LAB strains isolated from fermented products from Vietnam, including dua gia (bean sprouts), identifying *L. plantarum*, *Limosilactobacillus fermentum,* and *Lactobacillus helveticus* strains as dominant. In legumes, such as in other products, it is important to use bacteria that can grow rapidly to become dominant compared with the endogenous microbial contaminants.

The biopreservation mechanisms by which LAB inhibit spoilage organisms include the destabilization of cell membrane and subsequent interference with the proton gradient, inhibition enzyme activity, and creation of reactive oxygen species [96]. Moreover, LAB strains are able to produce antimicrobial compounds such as low-molecular-weight metabolites (reuterin, reutericyclin, diacetyl, fatty acids), hydrogen peroxide, antifungal compounds (propionate, phenyllactate, hydroxyphenyl-lactate, and 3-hydroxy fatty acids), and bacteriocins that may be exploited in the biopreservation of foods [99]. There is a wide number of bacteriocins produced by LAB that are classified into three classes: Class I (Lantibiotics), class II (Non Lantibiotics), and class III (Big peptides) depending on their chemical and genetic characteristics. The antibacterial activity of nisin, the most studied lantibiotics, has been demonstrated against *Listeria* spp., *Micrococcus* spp., and sporulating bacteria such as *Bacillus* spp. and *Clostridium* spp. [100]. Nguyen et al. [101] isolated the LAB from nem chua and determined their antimicrobial activity against pathogenic and sporulating strains such as *Bacillus cereus*, *Listeria monocytogenes*, *Escherichia coli*, and *Salmonella typhimurium*. Five strains NH3.6, NT1.3, NT1.6, NT2.9, and NT3.20 showed a broadened antimicrobial activity against both pathogenic Gram-positive *B. cereus* and *Ls. monocytogenes* and Gram-negative *E. coli* [101]. *L. plantarum* HA2, HA3, HA5, HA8, and HA9 and *L. fermentum* HA6, HA7, and HA10 isolated from Vietnamese fermented vegetables showed an intense antifungal activity against different indicator molds and yeasts (*Aspergillus terreus*, *Aspergillus fumigatus*, *Aspergillus niger*, *Absidia corymbifera*, *Paecilomyces lilacinus*, *Geotrichum candidum*, *Fusarium sp.*, *Scopulariopsis brevicaulis*, *Curvularia lunata*, *Penicillium* spp., and *Candida albicans*) [102]. These LAB strains are currently investigated for the specific use in legume-fermented products [96].

Fungi are the most common spoilage microorganisms of baked goods and represent a huge economic problem in bakery sector. The use of chemical preservatives is currently the only effective tool to prolong the microbial shelf-life of baked goods [103, 104]. Nevertheless, the European directive on preservatives has recently decreased the allowed concentrations of preservatives, and consumers require clean label and preservative-free baked goods. Therefore, the scientific and industrial research is now oriented toward the search for new preservatives, derived from natural sources. Overall, plants produce proteins and peptides involved in fungal resistance mechanisms, and seeds of many different species of leguminous plants are rich in such active compounds [105]. It was reported that the water-soluble extract of *Phaseolus vulgaris* cv. Pinto showed inhibitory activity toward a large spectrum of fungal species isolated from bakeries. The antifungal proteins corresponded to phaseolin alpha-type precursor, phaseolin, and erythroagglutinating phytohemagglutinin precursor. Bread manufactured with the addition of this water-soluble extract (27%, v/w) did not show fungal contamination until at least 21 days of storage at room temperature, ensuring a level of protection comparable with that afforded by calcium propionate (0.3%, w/w) [106]. A pea (*Pisum sativum*) protein hydrolysate, obtained by a food-grade protease, showed high inhibitory activity toward several fungi isolated from bakeries. The antifungal activity was correlated to pea defensins 1 and 2, nonspecific lipid transfer protein (nsLTP), and a mixture of peptides, encrypted in leginsulin A, vicilin, provicilin, and nsLTP, and released by the enzymatic activity of the protease [107]. A mixture of legumes-derived protein hydrolysates inhibited *Aspergillus parasiticus*, *Penicillium carneum*, *Penicillium paneum*, and *Penicillium polonicum*. Several native proteins and a mixture of peptides, encrypted in legume vicilins, lectins, and chitinases, were identified as the compounds responsible for the antifungal activity [108].

More recently, a LAB-fermented chickpea flour was proposed as fresh pasta ingredients aiming at prolonging the shelf-life of the product, moreover, achieving different nutritional advantages [109].

6. Traditional and novel fermented legume products

6.1 Traditional foods

Legumes are used as food ingredients worldwide, but only in few geographical areas they are commonly used for the production of fermented foods (**Table 1**), such as Japanese natto, Nigerian dawadawa or iru, Nepalese kinema, and Thai thua nao. Fermented legumes are consumed directly or used as ingredients or flavoring agents [124]. Yukiwari-natto and hama-natto spontaneous microbiota are dominated by molds, while *Bacillus* spp. is commonly isolated in itohiki-natto. Molds are also responsible for meitauza, oncom, and sufu fermentation, while yeasts dominate the fermentation of the Indian papad/papadam. Tempe is characterized by a microbial consortium including molds and LAB. Also Indian idli, wadi, and dhokla are produced by the combined fermentation activities of LAB and yeasts. Complex microbiota including LAB, yeasts, and molds characterize the fermentation processes for obtaining inyu, kecap asin, kecap manis, meju, miso, soy sauce, and tauco [118, 125].

Fermentation has an important impact on the nutritional and sensory profile of legumes [2, 96]. However, production of traditional fermentation products is often managed empirically, with rudimentary equipment, and based on the activity of endogenous microorganisms [96]. The quality of raw materials as well as the biotechnologies is not standardized [96]. These products are characterized by the local cultural identity. Despite their important sensorial role in Asian food, bringing, for instance, the umami taste to the meals [126], the necessity to improve overall quality and to minimize food safety hazards has been recently highlighted [127].

LAB have an important role in some of the traditional fermented legume products (such as in vietnamese tuong and cambodian sieng), but many other microorganisms (bacteria, yeasts, and molds) are involved in spontaneous fermentation processes. Nevertheless, the advantages of legumes fermentation with LAB are gaining interest from the scientific and food industry community [2].

Product	Main ingredients	Microorganisms	Area	Reference
Adai	Legume seeds and cereal grains	Lactic acid bacteria (*Pediococcus* spp., *Streptococcus* spp., and *Leuconostoc* spp.)	South India	[110]
Afiyo (okpehe or kpaye)	Mesquite bean (*Prosopis* sp.)		Nigeria	[111]
Aisa	*Albizia saman* seeds	Bacilli (*Bacillus* spp.) and staphylococci (*Staphylococcus* spp.)	Nigeria	[112]
Amriti	Black gram dal (*Vigna mungo*)	Aerobic mesophilic bacteria	India	[113]
Bedvin roti	Black gram dal, opium seed or walnut flour		India	[114]
Chungkokjang (cheonggukjang or jeonkukjang)	Soybean (*Glycine max*)	Bacilli (*Bacillus* spp.)	Korea	[115, 116]
Dawadawa (soydawadawa)	Soybean		Nigeria	[117]
Dawadawa, kinda, iru, soumbala	Locust bean (*Parkia biglobosa*)		West and Central Africa	[117]
Dhokla	Rice grains and bengal gram dal (*Cicer arietinum*)		India	[118]
Dosa	Black gram dal and rice grains		India	[118]
Douchi	Black soybean		China	[117]
Gochujang/Kochujang	Soybean, red pepper, rice, barley malt powder	Bacilli (*Bacillus* spp.)	Korea	[115]
Idli	Black gram dal and rice grains		India, Sri Lanka	[118]
Kinema, hawaijar, tungrymbai, aakhone, bekang, peruyyan	Soybean		Darjeeling hills and North East of India, Bhutan, Nepal	[117]
Maseura (masyaura)	Black gram dal/ ricebean (*Vigna umbellate*)	Lactic acid bacteria, bacilli, and yeast	India, Nepal	[119]
Meitauza	Okara (soybean press cake)		China, Taiwan	[117]
Meju	Soybean	Fungi and bacilli (*Bacillus* spp.)	Korea	[115]
Natto	Soybean		Japan, Korea	[117]

Product	Main ingredients	Microorganisms	Area	Reference
Oncom: Hitam (black) and merah (red)	Peanut (*Arachis hypogaea*) press cake		Indonesia	[118]
Oso	*Cathormion altissimum* seeds		Nigeria	[117]
Otiru	African yam bean (*Sphenostylis stenocarpa*)		Nigeria	[117]
Owoh	African yam bean		Nigeria	[117]
Papad or papadam	Black gram, bengal gram, lentil (*Lens culinaris*), red gram (*Cajanus cajan*) or green gram (*Vigna radiata*) flour		India	[118]
Pitha (chakuli, enduri, munha, chhuchipatra, podo)	Black gram dal and rice grain	Lactic acid bacteria	India	[110]
Sepubari	Black gram dal		India	[120]
Soybean paste: Doenjang or jang, miso, tauco, tao chieo	Soybean, wheat or rice grains		China, Indonesia, Japan, Korea, Thailand	[117]
Soy sauce: Jiang you, shoyu or tamari shoyu, kanjang, kicap, kecap, taosi, ketjap, inyu	Soybean/black soybean and wheat grains		China, Japan, Korea, Malaysia, Indonesia, Philippines, Indonesia, Taiwan, Hong Kong	[117]
Sufu or furu	Soybean		China, Taiwan	[118, 121]
Tempeh	Soybean		East Java, Indonesia	[118, 122]
Thua nao	Soybean		Thailand	[117]
Tuong	Soybean	Bacilli (*Bacillus* spp.), enterobacteria	North and central Vietnam	[123]
Vada	Legume and cereal	Lactic acid bacteria (*Pediococcus* spp., *Streptococcus* spp., and *Leuconostoc* spp.)	India	[110]
Ugba/ukpaka	African oil bean (*Pentaclethra macrophylla*)		West and Central Africa	[117]

Product	Main ingredients	Microorganisms	Area	Reference
Wadi	Black gram dal		Northern India	[118]
Yandou	Soybean		China	[117]

Table 1.
Main traditional food products containing fermented legumes.

6.2 Sourdough-inspired fermentation, sprouted flours, and baked good fortification

Besides the direct consumption as conventional dishes, legumes have a great potential as ingredients in various baked goods and pasta. Their use as fortifiers should increase their consumption as strongly recommended in many dietary guidelines. With this goal in mind, in the past decades, many researchers focused on using legume flours (also sprouted), fermented or not, as part of food formulations. Fermentation of legumes mainly determines improvement of the protein digestibility and related nutritional values and the biological availability of fibers and total phenols (**Table 2**). However, unlike cereal flour sourdoughs, very little is known about the microbiota of sourdough-type propagation, when only legume flour is used. Coda et al. [136] explored this topic investigating, through 16S rRNA gene pyrosequencing and culture-dependent analysis, the microbial ecology of faba bean sourdoughs obtained from an Italian and a Finnish cultivar, belonging respectively to *Vicia faba major* and *V. faba minor* groups. Among the LAB isolates, *Pediococcus pentosaceus, Leuc. Mesenteroides, and Weissella koreensis* had the highest frequency of occurrence in both sourdoughs. The presence of hulls and the different microbial composition reflected on biochemical characteristics of Finnish sourdoughs, including acidification and phenolic compounds [136].

Traditional varieties and biotypes, often replaced by modern cultivars selected for improved agronomic and commercial traits, can also be rediscovered and valorized through fermentation [34, 58, 130, 133]. Nineteen Italian legume flours, fermented with selected strains of *L. plantarum* and *Lv. brevis*. and compared with doughs without bacterial inoculum, had higher concentrations of free amino acids, soluble fibers, and total phenols. During sourdough fermentation, the level of γ-aminobutyric acid (GABA) markedly increased reaching up to 624 mg/kg [34]. GABA-producing strains of *L. plantarum* and *Lc. lactis* subsp. *lactis* were employed as starters for sourdough fermentation of a blend of chickpea and pseudo-cereals resulting in sourdough bread with very high levels of free amino acids and GABA (up to 504 mg/kg) [131]. The pairing between sourdough fermentation and legumes to accumulate GABA was performed also using adzuki bean flour [128] and extracts from kidney beans subjected to liquid state fermentation [129]. Type I sourdough, containing wheat-legume flour mixtures, was also used (15%, w/w) in bread making. The fortification increased the antioxidant activity and the *in vitro* protein digestibility (IVPD). According to the levels of carbohydrates, dietary fibers, and resistant starch, the bread fortified with wheat-legume sourdough had a decreased value of starch hydrolysis index [32].

Nevertheless, either considering gluten-free products or wheat-based baked goods, the lack of gluten is one of the challenges deriving from the use of legumes. The addition of wheat-legume flours increases water absorption providing more water for dough starch gelatinization during baking and preventing stretching and tearing of gluten strands [53]. Substitution of wheat flour with legumes at levels higher than 20–30% causes detrimental effects on dough and bread properties, which results in sticky and excessively compact [53, 140]. Hence, maintaining good

Legume	Fermentation type	Effects	References
Bean (Adzuki bean)	*Lactococcus lactis* subsp. *lactis* and *Lacticaseibacillus rhamnosus* GG	Increase of GABA concentration	[128]
Bean	Spontaneous fermentation	Decrease of α-galactosides, phytic acid, trypsin inhibitors and condensed tannins concentrations	[33]
Bean (Kidney beans)	Spontaneous fermentation; inoculum with *L. plantarum* ATCC 14917	Increase of GABA concentration	[129]
Bean, chickpea, grass pea, lentil, pea (local cultivars)	*Lactiplantibacillus plantarum* C48 and *Levilactobacillus brevis* AM7	Increase of phytase and antioxidant activity; increase of free amino acids, γ-aminobutyric acid (GABA), soluble fibers, and total phenols concentrations. Decrease of raffinose and condensed tannins concentrations	[34]
Bean, chickpea, grass pea, lentil, pea (local cultivars)	*Lactiplantibacillus plantarum* C48 and *Levilactobacillus brevis* AM7	Release of lunasin-like polypeptides; inhibition of the proliferation of human adenocarcinoma Caco2 cells	[130]
Chickpea	*Lactiplantibacillus plantarum* C48 and *Lactococcus lactis* subsp. *lactis* PU1	Increase of free amino acid and GABA concentrations; decrease of the starch hydrolysis index (HI); increase of antioxidant activity; increased palatability and overall acceptability of bread	[131]
Chickpea	*Weissella confusa Ck15*	Synthesis of linear dextran from sucrose	[132]
Chickpea (black chickpea)	*Lactiplantibacillus plantarum* T0A10	Increase of free amino acids, resistant starch, and protein digestibility; release of bound phenolic compounds; decrease of raffinose, condensed tannins, trypsin inhibitors, and saponins. Decrease of HI, increase of antioxidant potential and overall acceptability of fortified pasta	[133]
Chickpea, lentil	*Furfurilactobacillus rossiae* LB5, *Lactiplantibacillus plantarum* 1A7 and *Fructilactobacillus sanfranciscensis* DE9	Increase in the concentrations of peptides, free amino acids and GABA, increase of protein digestibility and decrease of starch availability. Decrease of phytic acid, condensed tannins, raffinose concentrations and trypsin inhibitory activity	[54]
Cowpea	Spontaneous fermentation	Increase of lysine concentration and essential amino acids concentration	[53]
Cowpea, mottled cowpea, speckled kidney bean, small rice bean	Spontaneous fermentation; inoculum with *Lactiplantibacillus plantarum* (WCSF1 and ATCC 149170) and *Lactobacillus paracasei* ASCC 279	Increase of antioxidant activity	[134, 135]
Faba bean	*Lactiplantibacillus plantarum* DPPMAB24W	Decrease of vicine and convicine concentration, trypsin inhibitor activity, starch hydrolysis index. Increase of protein digestibility, and free amino acids and GABA concentrations	[57]
Faba bean (*Vicia faba* var. *major* and *minor*)	Type I sourdough	Increase of free amino acid content and antioxidant activity. Decrease of α-galactosides and condensed tannins concentrations	[136]

Legume	Fermentation type	Effects	References
Faba bean (Mediterranean accessions)	*Lactiplantibacillus plantarum* DPPMAB24W	Increase in the concentrations of peptides, free amino acids, and GABA, increase of protein digestibility; decrease of α-galactosides, trypsin inhibitors, condensed tannins, and vicine concentrations	[58]
Faba bean (high protein content)	*Leuconostoc citreum TR116*	Improved amino acid profile, increased nitrogen utilization rate and PER of bread; decrease of anti-nutritional compounds and increase antioxidant potential in bread	[137]
Faba bean	*Propionibacterium freudenreichii DSM 20271*	Synthesis of vitamin B12	[138]
Faba bean	*Weissella confusa* VTT E–143,403 (E3403) and *Leuconostoc pseudomesenteroides* DSM 20193	Increase of protein concentration. Increase of viscoelastic behavior, specific volume of bread. Decrease of crumb hardness of bread	[139]
Faba bean	*Pediococcus pentosaceus* I02	Increase of protein digestibility and protein biological indexes. Increase of volume and hardness of bread. Decrease of glycemic index	[140]
Faba bean	*Lactiplantibacillus plantarum* DPPMAB24W	Increase of protein digestibility, nutritional indexes, and resistant starch. No detrimental effect on pasta texture and cooking loss	[141]
Grass pea	*Lactiplantibacillus plantarum*	Decrease of phytic acid concentration and trypsin inhibitory activity	[52]
Lentil	*Lactiplantibacillus plantarum*	Release of potentially bioactive peptides having antioxidant and angiotensin I-converting enzyme (ACE) inhibitory activities	[142]
Lentil	*Lactiplantibacillus plantarum* CECT 748 and Savinase® 16L	Release of bioactive peptides showing ACE-inhibitory properties	[143]
Lentil (native and sprouted)	*Weissella confusa* SLA4	Synthesis of dextran from sucrose. Increase of total and soluble fiber content, specific volume and decrease of crumb hardness and staling rate in wheat bread supplemented with 30% of lentil sourdough	[144]
Lentil, bean, chickpea, and pea flours, raw and gelatinized	*Lactiplantibacillus plantarum* MRS1 and *Levilactobacillus brevis* MRS4	Increase of free amino acids and protein digestibility; degradation of phytic acid, condensed tannins and raffinose; decrease of trypsin inhibitory activity and starch hydrolysis index	[31]
Lupin	*Latilactobacillus sakei* KTU05–6, *Pediococcus acidilactici* KTU05-7 and *Pediococcus pentosaceus* KTU05–8	Increase of protein bioavailability and digestibility	[145]
Lupin	*Saccharomyces cerevisiae* and *Saccharomyces boulardii* w/o probiotic lactic acid bacteria	Increase of protein content; degradation of anti-nutritional factors (α-galactosides, phitic acid and alkaloids)	[146]

DOI: http://dx.doi.org/10.5772/intechopen.102523

Legume	Fermentation type	Effects	References
Mixture of soybean and African breadfruit	Spontaneous fermentation	Increase of protein digestibility and improvement of the sensory properties	[147]
Mixture of chickpea/ lentil/bean	Type I sourdough	Increase of free amino acid concentration; increase of antioxidant and phytase activities	[32]
Mixture of chickpea and pseudo-cereals	*Lactiplantibacillus plantarum* C48 and *Lactococcus lactis* subsp. *lactis* PU1	Increase of free amino acid and GABA concentrations; decrease of the starch hydrolysis index (HI); increase of antioxidant activity; increased palatability and overall acceptability of bread	[131]

Table 2.
Main advantages of the LAB fermentation on legume flours and legume-fortified bread.

technological properties is a key factor in the success of products that go beyond laboratory-scale levels. Sourdough fermentation of legume flours, mainly interfering with starch gelatinization, and fibers hydration lead to the improvement of the structural characteristics of the fortified bread [32, 128, 148].

Fermentation can further contribute to improving the structural properties of fortified baked goods if exopolysaccharides-producing LAB are selectively employed. Indeed, the replacement of wheat flour (up to 43%) with a faba bean sourdough fermented with *Weissella confusa* strains [132, 139] compensated the gluten dilution and improved bread volume and crumb softness. The gluten-dextran interactions might have strengthened gas cells and, hence, prevented their collapse during proofing and baking [143]. This, combined with water-binding capacity, led to higher loaf volume and softer crumb.

The increase of the antioxidant activity during fermentation was largely documented in legume flours most likely associated with the biotransformation between soluble phenols and the release of bound phenols [31, 34, 132–135, 143]. The bioconversion of phenolic compounds into more available and biologically active forms mainly relies upon acidification and microbial enzymes. In LAB phenolic compounds metabolism comes from the need to detoxify such compounds but also have a role in preserving the cellular energy balance [149–151]. Fermentation of black chickpea with *L. plantarum* T0A10 enabled the release of 20% of bound phenolic compounds and the conversion of free phenolic acids leading to high scavenging activity against 2,2-diphenyl-1-picrylhydrazyl (DPPH) and 2,2'-azino-bis(3-ethylbenzothiazoline-6-sulfonate) (ABTS) radicals and intense inhibition of linoleic acid peroxidation [133]. Caffeic, coumaric, ferulic, phenyllactic, and 4-hydroxibenzoic acids were found in high amount in faba bean flour subjected to air classification and fermented with *Leuconostoc citreum* TR116, resulting in a bread having better nutritional and technological performances compared with bread obtained with unfermented faba bean [137]. Indeed, phenolic acids are not only appreciated for their potential antioxidant activity after ingestion, they can also be advantageous with regard to the microbial shelf-life of food products [152].

Fermentation can also be used to enhance the content of compounds lacking in vegetable matrices such as vitamin B12. Species of the former *Lactobacillus* genus were found to produce pseudo-vitamin B12, an inactive form for humans, whereas *Propionibacterium freudenreichii* DSM 20271 was effectively used to singly ferment faba bean, soy bean, and lupin flours [138], increasing vitamin B12 content up to 400 ng/g.

The release of bioactive peptides showing *in vitro* activities toward cancer, cardiovascular diseases, oxidative damage, inflammation, hypertension, and high cholesterol [142, 153] is also an appealing trait of lactic acid fermentation. Lunasin is a bioactive peptide (43-amino acid residues) already characterized for anticancer, antioxidant, anti-inflammatory, and cholesterol-lowering functional activities. Lunasin is mainly recovered from soy and used as dietary supplements and for pharmaceutical formulations. The fermentation of different legumes with selected strains of *L. plantarum* and *Lv. brevis* allowed the enrichment of the matrices in lunasin-like polypeptides, released from native proteins in which they are encrypted in nonactive form. Extracts from these legume sourdoughs showed marked inhibition on the proliferation of human adenocarcinoma Caco-2 cells [130]. Fermentation with *L. plantarum* CECT 748 and treatment with a commercial protease of lentil flour led to the release of several antihypertensive peptides showing angiotensin-converting enzyme inhibitory activity (up to 85%) [143].

As an ancient practice, germination of legumes is becoming an emerging process because of the significant enhancement in bioactive components (e.g., vitamins, dietary fibers, peptides and amino acids, and phenols) and palatability. The fortification of baked goods with flours from sprouted legumes has been proposed recently

[154]. During germination, reserves within the storage tissues of the seed undergo hydrolysis in low-molecular-weight compounds and mobilize to support seedling growth [155]. Parameters such as temperature, humidity, steeping (soaking), and length of germination determine the degree of these changes [156]. Nevertheless, the combination of germination and sourdough fermentation seems to better exploit the nutritional modification of grains in terms of protein and starch hydrolysis and mineral solubility [157]. Sprouting and sourdough fermentation with *Furfurilactobacillus rossiae*, *L. plantarum*, and *Fructilactobacillus sanfranciscensis* enhanced the nutritional and functional features of chickpea and lentil by increasing the concentrations of peptides, free amino acids, and GABA.

Fermented sprouted flours were used to make breads with high protein digestibility and low starch availability and appreciable sensory attributes [54]. Germination followed by sourdough fermentation improved the IVPD and enhanced the sensory properties of soybean and African breadfruit seeds [147]. The same occurred for the germinated and fermented cowpea flour, which fortified the bread formula with high lysine content and optimal essential amino acid balance [53]. While more recently, sprouted lentil sourdough, added with 25% sucrose, and fermented with *W. confusa* SLA4, led to the synthesis of dextran up to 9.7% [144]. Wheat bread supplemented with 30% of this sourdough showed increased specific volume and decreased crumb hardness and staling rate, compared with the control wheat bread, as well as increased total and soluble fibers content [144]. Attempts to enhance the nutritional properties of legumes were also made combining gelatinization to fermentation with lactic acid bacteria. Fermentation of gelatinized flours (red and yellow lentils, white and black beans, chickpeas, and peas) with *L. plantarum* MRS1 and *Lv. brevis* MRS4 led to the further degradation of the antinutritional factors (condensed tannins, raffinose, phytic acid, and trypsin inhibitors), increased the protein digestibility, and reduced the starch hydrolysis index [31].

6.3 Use of fermented legumes in pasta making

Just like bread, pasta is considered a staple food worldwide with the potential to modulate the diet, and the addition of fermented legumes accounts for a further step toward this goal. Regardless, the biotechnology used for the production, higher content of proteins and fibers, and lower starch content characterize legume-containing pasta. Nonetheless, fermentation contributes to improving not only the nutritional profile, but also the technological features of fortified pasta [158].

Faba bean flour, either raw or fermented (spontaneously or with selected starters), used as dough or freeze-dried material, is among the most reported legume flours in pasta-making [141, 159–161]. The percentage of semolina replacement mostly ranges from 10 to 50% [141, 160, 161], reaching up to 100%, as in the case of gluten-free faba bean pasta described by Rosa-Sibakov and colleagues [159].

Besides the increase in proteins and dietary fibers content, which is directly proportional to the percentage of semolina replacement with both raw and fermented faba bean, as consequence of the proteolysis occurred during fermentation, a higher content of peptides and FAA was observed in pasta containing faba bean fermented by *L. plantarum* DPPMAB24W [141]. The proteolysis occurring during the LAB fermentation also allowed the increase of the protein digestibility. Moreover, essential amino acids (EAAI), biological value (BV), and protein efficiency ratio (PER) indexes increased when 30 and 50% of the semolina was replaced by fermented faba bean flour [141]. The Nutritional Index (NI) of the pasta fortified with 30% of fermented faba bean flour was twofold higher than that of the conventional semolina pasta. This parameter is commonly considered as a global predictor of the protein quality of foods, since qualitative and quantitative factors are included in its calculation [162]. Replacement level higher than 30% led to the decrease of the NI, as a consequence of a weakening of the gluten network, unable to retain the soluble protein fraction during cooking [141]. The use of fermented faba bean flour as ingredient allowed a marked reduction of the starch hydrolysis index (HI) and, consequently, of the glycemic index (GI) [30, 141, 160]. As previously demonstrated, [163], this decrease can be correlated to the high level of dietary fibers and resistant starch and also to the effect of biological acidification [163].

Experimental pasta was also produced using exclusively fermented faba bean flour [159]. Whereas protein and starch content were similar between fermented and unfermented faba bean pasta (circa 35% and 43%, respectively), RS was found progressively higher in fermented fava bean pasta suggesting the possibility to use fermentation as a mean to decrease GI of commercial gluten-free products [164], usually higher than that of conventional foods [165].

Similar effects to those obtained in pasta fortified with fermented faba bean were obtained when spontaneously fermented pigeon pea (*Cajanus cajan*) (presumably due to LAB growth) was also used in pasta making [166–168]. Compared with semolina pasta, true protein digestibility (TD) and PER markedly improved (6 and 73%, respectively) in pasta fortified with fermented pigeon pea as consequence of the complementarity of amino acids composition deriving from legumes and cereal proteins [167, 168].

A Mediterranean black chickpea flour was fermented with *L. plantarum* T0A10 in semiliquid conditions and used (15% replacement level) to fortify a semolina pasta (116). Fermentation with the selected starter enabled the release of 20% of bound phenolic compounds and the conversion of free compounds into more active forms (dihydrocaffeic and phloretic acid) in the dough. Moreover, fortified cooked pasta, showing scavenging activity against DPPH and ABTS radicals and intense inhibition of linoleic acid peroxidation, was appreciated for its peculiar organoleptic profile [133].

Despite all the nutritional advantages deriving from the use of fermented legumes in pasta making, good sensory and textural properties remain a necessary foundation to achieve products approved by consumers. Differences in sensorial attributes and textural properties between pasta fortified with prefermented ingredients and the conventional one are often perceived unpleasant by trained assessors especially when semolina replacement exceeds 50% [169]. Increased chewiness, sourness, flavor, and off-flavor intensity were observed when fermented faba bean was added to pasta [159], as well as the onset of the red color, as the consequence of Maillard reaction [170]. However, fermentation also showed an important role in the improvement of sensory and textural characteristics of legume flours since it allowed the elimination of beany flavor [171]. Since the balance between flavors and off-flavors often lies in the amount of fortifier added [167], the right compromise between higher nutritional and functional properties and acceptable sensory and rheological ones should be addressed.

7. Conclusion and future perspectives

The rising demand for healthier plant-based food lies in the increasing awareness of the adverse risks associated with the consumption of animal proteins as well as the environmental impact animal farming entails. In this evolving agricultural system, legumes play a fundamental role in regard to both the support of good and sustainable agronomical practices and the maintenance of healthier diets.

Apart from their consumption as they are, legumes are the main ingredient of many traditional food products. Nevertheless, their consumption is often limited by antinutritional compounds and poor sensory and technological properties. Recently, the effectiveness of sourdough fermentation-inspired biotechnologies has proved to be pivotal in improving legumes and legume-based foods acceptability and safety. Through the release of bioactive peptides, phenolic compounds, and soluble fibers or the degradation of antinutritional compounds, fermentation with selected starters proved to be able to improve the nutritional and functional properties of legumes. By synthesizing exopolysaccharides, better rheological properties can be obtained while microbiological safety can be achieved through the degradation of biogenic ammines, mycotoxins, or activity toward spoilage or pathogenic microorganisms.

Fermentation allows overcoming the issues that hold back legumes' potential and intensifies their use as ingredients in innovative formulations of staple foods, such as baked goods and pasta with a more balanced nutritional and functional profile.

The underlining idea behind functional foods is to reduce the prevalence of diet-related diseases by modulating the consumption of commonly eaten foods fortified with high-value ingredients. Fermented legumes fit the profile of such ingredients, but educating consumers on their health benefits, so that they can make an informed choice, is of paramount importance. It is necessary to get rid of the stigma of legumes as "poor man's meat" and recognize their value not only in agricultural practices but also their pivotal role in healthy and sustainable diets. Furthermore, there is growing recognition that changes in nutrition are critical to achieve several of the Sustainable Development Goals developed by the United Nations to promote prosperity while protecting the planet. In order to meet the global food demands, focus should be put into promoting the cultivation and utilization of local or underutilized legume crops often neglected and underexploited, which yet have a great impact on the biodiversity as well as in enhancing food and nutrition security. Whereas, from an academia point of view, those mechanisms, which are still unclear or need more exploiting, behind the advantages of fermentation in terms of biopreservation and safety in general, should be pursued as research topics, since they can further unleash legumes' potential.

Conflict of interest

The authors declare no conflict of interest.

Author details

Michela Verni[1], Erica Pontonio[1], Marco Montemurro[1] and Carlo Giuseppe Rizzello[2]*

1 Department of Soil, Plant, and Food Sciences, University of Bari Aldo Moro, Bari, Italy

2 Department of Environmental Biology, Sapienza University of Rome, Rome, Italy

*Address all correspondence to: carlogiuseppe.rizzello@uniroma1.it

References

[1] Monnet AF, Laleg K, Michon C, Micard V. Legume enriched cereal products: A generic approach derived from material science to predict their structuring by the process and their final properties. Trends in Food Science and Technoogy. 2019;**86**:131-143. DOI: 10.1016/j.tifs.2019.02.027

[2] Gobbetti M, De Angelis M, Di Cagno R, Polo A, Rizzello CG. The sourdough fermentation is the powerful process to exploit the potential of legumes, pseudo-cereals and milling by-products in baking industry. Critical Review in Food Science and Nutrition. 2020;**60**:2158-2173. DOI: 10.1080/10408398.2019.1631753

[3] Duranti M. Grain legume proteins and nutraceutical properties. Fitoterapia. 2006;**77**:67-82. DOI: 10.1016/j.fitote.2005.11.008

[4] Smartt J, Nwokolo E. Food and Feed from Legumes and Oilseeds. London, UK: Chapman and Hall; 1996

[5] Pelzer E, Bazot M, Makowski D, Corre-Hellou G, Naudin C, Al Rifaï M, et al. Pea-wheat intercrops in low-input conditions combine high economic performances and low environmental impacts. European Journal of Agronomy. 2012;**40**:39-53. DOI: 10.1016/j.eja.2012.01.010

[6] Roy F, Boye JI, Simpson BK. Bioactive proteins and peptides in pulse crops: Pea, chickpea and lentil. Food Research International. 2010;**43**:432-442. DOI: 10.1016/j.foodres.2009.09.002

[7] Widmer RJ, Flammer AJ, Lerman LO, Lerman A. The Mediterranean diet, its components, and cardiovascular disease. The American Journal of Medicine. 2015;**128**:229-238. DOI: 10.1016/j.amjmed.2014.10.014

[8] Jenkins DJ, Kendall CW, Augustin LS, Mitchell S, Sahye-Pudaruth S, Blanco Mejia S, et al. Effect of legume as part of low glycemic index diet on glycemic control and cardiovascular risk factors in type 2 diabetes mellitus: A randomized controlled trial. Archives of International Medicine. 2012;**172**: 1653-1660. DOI: 10.1001/2013.jamainternmed.70

[9] Feregrino-Perez AA, Berumen LC, Garcia-Alcocer G, Guevara-Gonzalez RG, Ramos-Gomez M, Reynoso-Camacho R, et al. Composition of chemopreventive effect of polysaccharides from common beans (*Phaseolus vulgaris* L.) on azoxymethane-induced colon cancer. Journal of Agricultural and Food Chemistry. 2008;**56**:8737-8744. DOI: 10.1021/jf8007162

[10] Mollard RC, Luhovyy BL, Panahi S, Nunez M, Hanley A, Andersona GH. Regular consumption of pulses for 8 weeks reduces metabolic syndrome risk factors in overweight and obese adults. British Journal of Nutrition. 2012; **108**:111-122. DOI: 10.1017/S0007114512000712

[11] Young VR, Pellett PL. Plant proteins in relation to human protein and amino acid nutrition. The American Journal of Clinical Nutrition. 1994;**59**:1203-1212. DOI: 10.1093/ajcn/59.5.1203S

[12] Shewry PR, Tatham AS, Forde J, Kreis M, Miflin BJ. The classification and nomenclature of wheat gluten proteins: A reassessment. Journal of Cereal Science. 1986;**4**:97-106. DOI: 10.1016/S0733-5210(86)80012-1

[13] Boye JI, Zare F, Pletch A. Pulse proteins: Processing, characterization, functional properties and applications in food and feed. Food Research International. 2010;**43**:414-431. DOI: 10.1016/j.foodres.2009.09.003

[14] Mann J, Schiedt B, Baumann A, Conde-Petit B, Vilgis TA. Effect of heat treatment on wheat dough rheology and wheat protein solubility. Food Science and Technology International. 2014;**20**:341-351. DOI: 10.1177/1082013213488381

[15] Buléon A, Colonna P, Planchot V, Ball S. Starch granules: Structure and biosynthesis. International Journal of Biological Macromolecules. 1998;**23**:85-112. DOI: 10.1016/S0141-8130(98)00040-3

[16] Hoover R, Hughes T, Chung HJ, Liu Q. Composition, molecular structure, properties, and modification of pulse starches: A review. Food Research International. 2010;**43**:399-394. DOI: 10.1016/j.foodres.2009.09.001

[17] Kumssa DB, Joy EJ, Ander EL, Watts MJ, Young SD, Walker S, et al. Dietary calcium and zinc deficiency risks are decreasing but remain prevalent. Scientific Reports. 2015;**5**:1-11. DOI: 10.1038/srep10974

[18] Gänzle MG. Food fermentations for improved digestibility of plant foods–an essential ex situ digestion step in agricultural societies? Current Opinion in Food Science. 2020;**32**:124-132. DOI: 10.1016/j.cofs.2020.04.002

[19] Kuo TM, VanMiddlesworth JF, Wolf WJ. Content of raffinose oligosaccharides and sucrose in various plant seeds. Journal of Agricultural and Food Chemistry. 1988;**36**:32-36

[20] Loponen J, Gänzle MG. Use of sourdough in low FODMAP baking. Food. 2018;**7**:96. DOI: 10.3390/foods7070096

[21] Yan YL, Hu Y, Gänzle MG. Prebiotics, FODMAPs and dietary fiber—Conflicting concepts in development of functional food products? Current Opinion in Food Science. 2018;**20**:30-37. DOI: org/10.1016/j.cofs.2018.02.009

[22] Viveros A, Centeno C, Brenes A, Canales R, Lozano A. Phytase and acid phosphatase activities in plant feedstuffs. Journal of Agricultural and Food Chemistry. 2000;**48**:4009-4013. DOI: 10.1021/jf991126m

[23] Reddy NR. Occurrence, distribution, content, and dietary intake of phytate. In: Reddy NR, Sathe SR, editors. Food Phytates. Boca Raton, Florida. United States: CRC Press; 2001. pp. 41-68

[24] Zimmermann MB, Hurrell RF. Nutritional iron deficiency. The Lancet. 2007;**370**:511-520. DOI: 10.1016/S0140-6736(07)61235-5

[25] Engels C, Gänzle MG, Schieber A. Fractionation of gallotannins from mango (*Mangifera indica* L.) kernels by high-speed counter-current chromatography and determination of their antibacterial activity. Journal of Agricultural and Food Chemistry. 2010;**58**:775-780. DOI: 10.1021/jf903252t

[26] Savelkoul FHMG, Van der Poel AFB, Tamminga S. The presence and inactivation of trypsin inhibitors, tannins, lectins and amylase inhibitors in legume seeds during germination. A review. Plant Foods for Human Nutrition. 1992;**42**:71-85. DOI: 10.1007/BF02196074

[27] Awika JM, Rooney LW. Sorghum phytochemicals and their potential impact on human health. Phytochemistry. 2004;**65**:1199-1221. DOI: 10.1016/j.phytochem.2004.04.001

[28] Ali Asgar MD. Anti-diabetic potential of phenolic compounds: A review. International Journal of Food Properties. 2013;**16**:91-103. DOI: 10.1080/10942912.2011.595864

[29] Woyengo TA, Beltranena E, Zijlstra RT. Effect of anti-nutritional

factors of oilseed co-products on feed intake of pigs and poultry. Animal Feed Science and Technology. 2017;**233**:76-86. DOI: 10.1016/j.anifeedsci.2016.05.006

[30] Coda R, Melama L, Rizzello CG, Curiel JA, Sibakov J, Holopainen U, et al. Effect of air classification and fermentation by *lactobacillus plantarum* VTT-133328 on faba bean (*Vicia Faba L.*) flour nutritional properties. International Journal of Food Microbiology. 2015;**193**:34-42. DOI: 10.1016/j.ijfoodmicro.2014.10.012

[31] De Pasquale I, Pontonio E, Gobbetti M, Rizzello CG. Nutritional and functional effects of the lactic acid bacteria fermentation on gelatinized legume flours. International Journal of Food Microbiology. 2020;**316**:108426. DOI: 10.1016/j.ijfoodmicro.2019.108426

[32] Rizzello CG, Calasso M, Campanella D, De Angelis M, Gobbetti M. Use of sourdough fermentation and mixture of wheat, chickpea, lentil and bean flours for enhancing the nutritional, texture and sensory characteristics of white bread. International Journal of Food Microbiology. 2014;**180**:78-87. DOI: 10.1016/j.ijfoodmicro.2014.04.005

[33] Granito M, Frias J, Doblado R, Guerra M, Champ M, Vidal-Valverde C. Nutritional improvement of beans (*Phaseolus vulgaris*) by natural fermentation. European Food Research and Technology. 2002;**214**:226-231. DOI: 10.1007/s00217-001-0450-5

[34] Curiel JA, Coda R, Centomani I, Summo C, Gobbetti M, Rizzello CG. Exploitation of the nutritional and functional characteristics of traditional Italian legumes: The potential of sourdough fermentation. International Journal of Food Microbiology. 2015;**196**:51-61. DOI: 10.1016/j.ijfoodmicro.2014.11.032

[35] Teixeira JS, McNeill V, Gänzle MG. Levansucrase and sucrose phoshorylase contribute to raffinose, stachyose, and verbascose metabolism by lactobacilli. Food Microbiology. 2012;**31**:278-284. DOI: 10.1016/j.fm.2012.03.003

[36] Gänzle M, Follador R. Metabolism of oligosaccharides and starch in lactobacilli: A review. Frontiers in Microbiology. 2012;**3**:340. DOI: 10.3389/fmicb.2012.00340

[37] Mital BK, Steinkraus KH. Fermentation of soy milk by lactic acid bacteria. A review. Journal of Food Protection. 1979;**42**:895-899. DOI: 10.4315/0362-028X-42.11.895

[38] Leenhardt F, Levrat-Verny MA, Chanliaud E, Rémésy C. Moderate decrease of pH by sourdough fermentation is sufficient to reduce phytate content of whole wheat flour through endogenous phytase activity. Journal of Agricultural and Food Chemistry. 2005;**53**:98-102. DOI: 10.1021/jf049193q

[39] Poutanen K, Flander L, Katina K. Sourdough and cereal fermentation in a nutritional perspective. Food Microbiology. 2009;**26**:693-699. DOI: 10.1016/j.fm.2009.07.011

[40] Frias J, Doblado R, Antezana JR, Vidal-Valverde C. Inositol phosphate degradation by the action of phytase enzyme in legume seeds. Food Chemistry. 2003;**81**:233-239. DOI: 10.1016/S0308-8146(02)00417-X

[41] Nout MR. Rich nutrition from the poorest–cereal fermentations in Africa and Asia. Food Microbiology. 2009;**26**:685-692. DOI: org/10.1016/j.fm.2009.07.002

[42] Nout MJR, Rombouts FM. Recent developments in Tempe research. Journal of Applied Bacteriology. 1990;**69**:609-633. DOI: 10.1111/j.1365-2672.1990.tb01555.x

[43] Tsuji S, Tanaka K, Takenaka S, Yoshida KI. Enhanced secretion of natto phytase by *Bacillus subtilis*. Bioscience, Biotechnology, and Biochemistry. 2015;**79**:1906-1914. DOI: 10.1080/09168451.2015.1046366

[44] Starzyńska-Janiszewska A, Stodolak B, Mickowska B. Effect of controlled lactic acid fermentation on selected bioactive and nutritional parameters of tempeh obtained from unhulled common bean (*Phaseolus vulgaris*) seeds. Journal of the Science of Food and Agriculture. 2014;**4**:359-366. DOI: 10.1002/jsfa.6385

[45] Svensson L, Sekwati-Monang B, Lutz DL, Schieber A, Ganzle MG. Phenolic acids and flavonoids in nonfermented and fermented red sorghum (*Sorghum bicolor* (L.) Moench). Journal of Agricultural and Food Chemistry. 2010;**58**:9214-9220. DOI: 10.1021/jf101504v

[46] Osawa RO, Kuroiso K, Goto S, Shimizu A. Isolation of tannin-degrading lactobacilli from humans and fermented foods. Applied and Environmental Microbiology. 2000;**66**:3093. DOI: 10.1128/aem.66.7.3093-3097.2000

[47] Rodríguez H, de las Rivas B, Gómez-Cordovés C, Muñoz R. Characterization of tannase activity in cell-free extracts of *lactobacillus plantarum* CECT 748T. International Journal of Food Microbiology. 2008;**121**:92-98. DOI: 10.1016/j.ijfoodmicro.2007.11.002

[48] Vaquero I, Marcobal Á, Muñoz R. Tannase activity by lactic acid bacteria isolated from grape must and wine. International Journal of Food Microbiology. 2004;**96**:199-204. DOI: 10.1016/j.ijfoodmicro.2004.04.004

[49] Kostinek M, Specht I, Edward VA, Pinto C, Egounlety M, Sossa C, et al. Characterisation and biochemical properties of predominant lactic acid bacteria from fermenting cassava for selection as starter cultures. International Journal of Food Microbiology. 2007;**114**:342-351. DOI: 10.1016/j.ijfoodmicro.2006.09.029

[50] Iwamoto K, Tsuruta H, Nishitaini Y, Osawa R. Identification and cloning of a gene encoding tannase (tannin acylhydrolase) from *lactobacillus plantarum* ATCC 14917T. Systematic and Applied Microbiology. 2008;**31**:269-277. DOI: 10.1016/j.syapm.2008.05.004

[51] Reverón I, Jiménez N, Curiel JA, Peñas E, de Felipe FL, de Las RB, et al. Differential gene expression by lactobacillus plantarum WCFS1 in response to phenolic compounds reveals new genes involved in tannin degradation. Applied and Environmental Microbiology. 2017;**83**:e03387-e03316. DOI: 10.1128/AEM.03387-16

[52] Starzyńska-Janiszewska A, Stodolak B. Effect of inoculated lactic acid fermentation on antinutritional and antiradical properties of grass pea (*Lathyrus sativus* 'Krab') flour. Polish Journal of Food Nutrition Sciences. 2011;**61**:245-249

[53] Hallén E, İbanoğlu Ş, Ainsworth P. Effect of fermented/germinated cowpea flour addition on the rheological and baking properties of wheat flour. Journal of Food Engineering. 2004;**63**:177-184. DOI: 10.1016/S0260-8774(03)00298-X

[54] Montemurro M, Pontonio E, Gobbetti M, Rizzello CG. Investigation of the nutritional, functional and technological effects of the sourdough fermentation of sprouted flours. International Journal of Food Microbiology. 2019;**302**:47-58. DOI: 10.1016/j.ijfoodmicro.2018.08.005

[55] Crépon K, Marget P, Peyronnet C, Carrouee B, Arese P, Duc G. Nutritional

value of faba bean (*Vicia faba* L.) seeds for feed and food. Field Crops Research. 2010;**115**:329-339. DOI: 10.1016/j.fcr.2009.09.016

[56] Pulkkinen M, Gautam M, Lampi AM, Ollilainen V, Stoddard F, Sontag-Strohm T, et al. Determination of vicine and convicine from faba bean with an optimized high-performance liquid chromatographic method. Food Research International. 2015;**76**:168-177. DOI: 10.1016/j.foodres.2015.05.031

[57] Rizzello CG, Losito I, Facchini L, Katina K, Palmisano F, Gobbetti M, et al. Degradation of vicine, convicine and their aglycones during fermentation of faba bean flour. Scientific Reports. 2016;**6**:32452. DOI: 10.1038/srep32452

[58] Verni M, De Mastro G, De Cillis F, Gobbetti M, Rizzello CG. Lactic acid bacteria fermentation to exploit the nutritional potential of Mediterranean faba bean local biotypes. Food Research International. 2019;**125**:108571. DOI: 10.1016/j.foodres.2019.108571

[59] Kroghsbo S, Rigby NM, Johnson PE, Adel-Patient K, Bøgh KL, Salt LJ, et al. Assessment of the sensitizing potential of processed peanut proteins in Brown Norway rats: Roasting does not enhance allergenicity. PLoS One. 2014;**9**:e96475. DOI: 10.1371/journal.pone.0096475

[60] Mills EC, Sancho AI, Rigby NM, Jenkins JA, Mackie AR. Impact of food processing on the structural and allergenic properties of food allergens. Molecular Nutrition & Food Research. 2009;**53**:963-969. DOI: 10.1002/mnfr.200800236

[61] Nowak-Wegrzyn A, Fiocchi A. Rare, medium, or well done? The effect of heating and food matrix on food protein allergenicity. Current Opinion in Allergy and Clinical Immunology. 2009;**9**:234-237. DOI: 10.1097/ACI.0b013e32832b88e7

[62] Rahaman T, Vasiljevic T, Ramchandran L. Effect of processing on conformational changes of food proteins related to allergenicity. Trends in Food Science & Technology. 2016;**49**:24-34. DOI: 10.1016/j.tifs.2016.01.001

[63] Untersmayr E, Jensen-Jarolim E. The role of protein digestibility and antacids on food allergy outcomes. Journal of Allergy and Clinical Immunology. 2008;**121**:1301-1308. DOI: 10.1016/j.jaci.2008.04.025

[64] El Hag ME, El Tinay AH, Yousif NE. Effect of fermentation and dehulling on starch, total polyphenols, phytic acid content and in vitro protein digestibility of pearl millet. Food Chemistry. 2002;**77**:193-196. DOI: 10.1016/S0308-8146(01)00336-3

[65] Pranoto Y, Anggrahini S, Efendi Z. Effect of natural and *lactobacillus plantarum* fermentation on in-vitro protein and starch digestibilities of sorghum flour. Food Bioscience. 2013;**2**:46-52. DOI: 10.1016/j.fbio.2013.04.001

[66] Song YS, Frías J, Martinez-Villaluenga C, Vidal-Valdeverde C, de Mejia EG. Immunoreactivity reduction of soybean meal by fermentation, effect on amino acid composition and antigenicity of commercial soy products. Food Chemistry. 2008;**108**:571-581. DOI: 10.1016/j.foodchem.2007.11.013

[67] Song YS, Martinez-Villaluenga C, De Mejia EG. Quantification of human IgE immunoreactive soybean proteins in commercial soy ingredients and products. Journal of Food Science. 2008;**73**:T90-T99. DOI: 10.1111/j.1750-3841.2008.00848.x

[68] El-Sharkawy SH, Abul-Hajj YJ. Microbial transformation of zearalenone. 2. Reduction, hydroxylation, and methylation products. The Journal of Organic Chemistry. 1988;**53**:515-519

[69] Kakeya H, Takahashi-Ando N, Kimura M, Onose R, Yamaguchi I, Osada H. Biotransformation of the mycotoxin, zearalenone, to a non-estrogenic compound by a fungal strain of *Clonostachys* sp. Bioscience, Biotechnology, and Biochemistry. 2002;**66**:2723-2726. DOI: 10.1271/bbb.66.2723

[70] Takahashi-Ando N, Ohsato S, Shibata T, Hamamoto H, Yamaguchi I, Kimura M. Metabolism of zearalenone by genetically modified organisms expressing the detoxification gene from *Clonostachys rosea*. Applied and Environmental Microbiology. 2004;**70**:3239-3245. DOI: 10.1128/AEM.70.6.3239-3245.2004

[71] Vekiru E, Hametner C, Mitterbauer R, Rechthaler J, Adam G, Schatzmayr G, et al. Cleavage of zearalenone by *Trichosporon mycotoxinivorans* to a novel nonestrogenic metabolite. Applied and Environmental Microbiology. 2010;**76**:2353-2359. DOI: 10.1128/AEM.01438-09

[72] El-Nezami H, Polychronaki N, Salminen S, Mykkänen H. Binding rather than metabolism may explain the interaction of two food-grade *lactobacillus* strains with zearalenone and its derivative ά-zearalenol. Applied and Environmental Microbiology. 2002;**68**:3545. DOI: 10.1128/AEM.68.7.3545-3549.2002

[73] Zhao Z, Liu N, Yang L, Wang J, Song S, Nie D, et al. Cross-linked chitosan polymers as generic adsorbents for simultaneous adsorption of multiple mycotoxins. Food Control. 2015;**57**:362-369. DOI: 10.1016/j.foodcont.2015.05.014

[74] de Souza Sant'Ana A, Rosenthal A, de Massaguer PR. The fate of patulin in apple juice processing: A review. Food Research International. 2008;**41**:441-453. DOI: 10.1016/j.foodres.2008.03.001

[75] Fliege R, Metzler M. Electrophilic properties of patulin. N-acetylcysteine and glutathione adducts. Chemical Research in Toxicology. 2000;**13**:373-381. DOI: 10.1021/tx9901480

[76] Hawar S, Vevers W, Karieb S, Ali BK, Billington R, Beal J. Biotransformation of patulin to hydroascladiol by *lactobacillus plantarum*. Food Control. 2013;**34**:502-508. DOI: 10.1016/j.foodcont.2013.05.023

[77] Wacoo AP, Mukisa IM, Meeme R, Byakika S, Wendiro D, Sybesma W, et al. Probiotic enrichment and reduction of aflatoxins in a traditional African maize-based fermented food. Nutrients. 2019;**11**:265. DOI: 10.3390/nu11020265

[78] Loi M, Fanelli F, Liuzzi VC, Logrieco AF, Mulè G. Mycotoxin biotransformation by native and commercial enzymes: Present and future perspectives. Toxins. 2017;**9**:111. DOI: 10.3390/toxins9040111

[79] Ahlberg SH, Joutsjoki V, Korhonen HJ. Potential of lactic acid bacteria in aflatoxin risk mitigation. International Journal of Food Microbiology. 2015;**207**:87-102. DOI: 10.1016/j.ijfoodmicro.2015.04.042

[80] Gardini F, Özogul Y, Suzzi G, Tabanelli G, Özogul F. Technological factors affecting biogenic amine content in foods: A review. Frontiers in Microbiology. 2016;**7**:1218. DOI: 10.3389/fmicb.2016.01218

[81] Barbieri F, Montanari C, Gardini F, Tabanelli G. Biogenic amine production by lactic acid bacteria: A review. Food. 2019;**8**:17. DOI: 10.3390/foods8010017

[82] Lucas PM, Wolken WA, Claisse O, Lolkema JS, Lonvaud-Funel A. Histamine-producing pathway encoded on an unstable plasmid in *lactobacillus hilgardii* 0006. Applied and Environmental Microbiology.

2015;**71**:1417-1424. DOI: 10.1128/AEM. 71.3.1417-1424.2005

[83] Satomi M, Furushita M, Oikawa H, Yoshikawa-Takahashi M, Yano Y. Analysis of a 30 kbp plasmid encoding histidine decarboxylase gene in *Tetragenococcus halophilus* isolated from fish sauce. International Journal of Food Microbiology. 2008;**126**:202-209. DOI: 10.1016/j.ijfoodmicro.2008. 05.025

[84] Naila A, Flint S, Fletcher G, Bremer P, Meerdink G. Control of biogenic amines in food—Existing and emerging approaches. Journal of Food Science. 2010;**75**:R139-R150. DOI: 10.1111/j.1750-3841.2010.01774.x

[85] Guarcello R, De Angelis M, Settanni L, Formiglio S, Gaglio R, Minervini F, et al. Selection of amine-oxidizing dairy lactic acid bacteria and identification of the enzyme and gene involved in the decrease of biogenic amines. Applied and Environmental Microbiology. 2016;**82**:6870-6880. DOI: 10.1128/AEM.01051-16

[86] Kim YS, Cho SH, Jeong DY, Uhm TB. Isolation of biogenic amines-degrading strains of Bacillus subtilis and *bacillus amyloliquefaciens* from traditionally fermented soybean products. Korean Journal of Microbiology. 2012;**48**:220-224. DOI: 10.7845/kjm.2012.042

[87] Kang HR, Lee YL, Hwang HJ. Potential for application as a starter culture of tyramine-reducing strain. Journal of the Korean Society of Food Science and Nutrition. 2017

[88] Eom JS, Seo BY, Choi HS. Biogenic amine degradation by *bacillus* species isolated from traditional fermented soybean food and detection of decarboxylase-related genes. Journal of Microbiology and Biotechnology. 2015;**25**:1519-1527. DOI: 10.4014/ jmb.1506.06006

[89] Lee YC, Kung HF, Huang YL, Wu CH, Huang YR, Tsai YH. Reduction of biogenic amines during miso fermentation by *lactobacillus plantarum* as a starter culture. Journal of Food Protection. 2016;**79**:1556-1561. DOI: 10.4315/0362-028X.JFP-16-060

[90] Caldas ED, Jardim ANO. Exposure to toxic chemicals in the diet: Is the Brazilian population at risk? Journal of Exposure Science & Environmental Epidemiology. 2012;**22**:1-15. DOI: 10.1038/jes.2011.35

[91] Cho KM, Math RK, Islam SMA, Lim WJ, Hong SY, Kim JM, et al. Biodegradation of chlorpyrifos by lactic acid bacteria during kimchi fermentation. Journal of Agricultural and Food Chemistry. 2009;**46**:1561-1567. DOI: 10.1021/jf803649z

[92] Zhang YH, Xu D, Liu JQ, Zhao XH. Enhanced degradation of five organophosphorus pesticides in skimmed milk by lactic acid bacteria and its potential relationship with phosphatase production. Food Chemistry. 2014;**164**:173-178. DOI: 10.1016/j.foodchem.2014.05.059

[93] Duan J, Cheng Z, Bi J, Xu Y. Residue behavior of organochlorine pesticides during the production process of yogurt and cheese. Food Chemistry. 2018;**245**:119-124. DOI: 10.1016/j. foodchem.2017.10.017

[94] Zhan H, Feng Y, Fan X, Chen S. Recent advances in glyphosate biodegradation. Applied Microbiology and Biotechnology. 2018;**102**:5033-5043. DOI: 10.1007/s00253-018-9035-0

[95] Shin KH, Lim Y, Ahn JH, Khil J, Cha CJ, Hur HG. Anaerobic biotransformation of dinitrotoluene isomers by *Lactococcus lactis* subsp. *lactis* strain 27 isolated from earthworm intestine. Chemosphere. 2005;**61**:30-39. DOI: 10.1016/j.chemosphere.2005. 03.020

[96] Licandro H, Ho PH, Nguyen TKC, Petchkongkaew A, Van Nguyen H, Chu-Ky S, et al. How fermentation by lactic acid bacteria can address safety issues in legumes food products? Food Control. 2020;**110**:106957. DOI: 10.1016/j.foodcont.2019.106957

[97] Ananou S, Maqueda M, Martínez-Bueno M, Gálvez A, Valdivia E. Bactericidal synergism through enterocin AS-48 and chemical preservatives against *Staphylococcus aureus*. Letters in Applied Microbiology. 2007;**45**:19-23. DOI: 10.1111/j.1472-765X.2007.02155.x

[98] Phan YTN, Tang MT, Tran TTM, Nguyen VH, Nguyen TH, Tsuruta T, et al. Diversity of lactic acid bacteria in vegetable-based and meat-based fermented foods produced in the central region of Vietnam. AIMS Microbiology. 2017;**3**:61. DOI: 10.3934/microbiol.2017.1.61

[99] Siedler S, Balti R, Neves AR. Bioprotective mechanisms of lactic acid bacteria against fungal spoilage of food. Current Opinion in Biotechnology. 2019;**56**:138-146. DOI: 10.1016/j.copbio.2018.11.015

[100] Parada JL, Caron CR, Medeiros ABP, Soccol CR. Bacteriocins from lactic acid bacteria: Purification, properties and use as biopreservatives. Brazilian Archives of Biology and Technology. 2007;**50**:512-542. DOI: 10.1590/S1516-89132007000300018

[101] Nguyen H, Elegado F, Librojo-Basilio N, Mabesa R, Dizon E. Isolation and characterisation of selected lactic acid bacteria for improved processing of Nem chua, a traditional fermented meat from Vietnam. Beneficial Microbes. 2010;**1**:67-74. DOI: 10.3920/BM2009.0001

[102] Ho PH, Luo JB, Adams MC. Lactobacilli and dairy propionibacterium with potential as biopreservatives against food fungi and yeast contamination. Applied Biochemistry and Microbiology. 2009;**45**:414-418. DOI: 10.1134/S0003683809040115

[103] Dal Bello F, Clarke CI, Ryan LAM, Ulmera H, Schobera TJ, Strom K, et al. Improvement of the quality and shelf life of wheat bread by fermentation with the antifungal strain *lactobacillus plantarum* FST 1.7. Journal of Cereal Science. 2007;**45**:309-318. DOI: 10.1016/j.jcs.2006.09.004

[104] Ponte JG, Tsen CC. Bakery products. In: Beuchat LR, editor. Food and Beverage Mycology. 2nd ed. New York, NY: AVI Van Nostrand Reinhold; 1987. pp. 233-267

[105] Xia L, Ng TB. An antifungal protein from flageolet beans. Peptides. 2005;**26**:2397-2403. DOI: 10.1016/j.peptides.2005.06.003

[106] Coda R, Rizzello CG, Nigro F, De Angelis M, Arnault P, Gobbetti M. Long-term fungal inhibitory activity of water-soluble extracts of *Phaseolus vulgaris* cv. Pinto and sourdough lactic acid bacteria during bread storage. Applied and Environmental Microbiology. 2008;**74**:7391-7398. DOI: 10.1128/AEM.01420-08

[107] Rizzello CG, Lavecchia A, Gramaglia V, Gobbetti M. Long-term fungal inhibition by *Pisum sativum* flour hydrolysate during storage of wheat flour bread. Applied and Environmental Microbiology. 2015;**81**:4195-4206. DOI: 10.1128/AEM.04088-14

[108] Rizzello CG, Verni M, Bordignon S, Gramaglia V, Gobbetti M. Hydrolysate from a mixture of legume flours with antifungal activity as an ingredient for prolonging the shelf-life of wheat bread. Food Microbiology. 2017;**64**:72-82. DOI: 10.1016/j.fm.2016.12.003

[109] Schettino R, Pontonio E, Gobbetti M, Rizzello CG. Extension of the shelf-life of fresh pasta using chickpea flour fermented with selected lactic acid bacteria. Microorganisms. 2020;**8**:1322. DOI: 10.3390/microorganisms8091322

[110] Ray M, Ghosh K, Singh S, Mondal KC. Folk to functional: An explorative overview of rice-based fermented foods and beverages in India. Journal of Ethnic Foods. 2016;**3**:5-18. DOI: 10.1016/j.jef.2016.02.002

[111] Egwim Evans AM, Abubakar Y, Mainuna B. Nigerian indigenous fermented foods: Processes and prospects. Mycotoxin and Food Safety in Developing Countries. 2013;**153**:153-180. DOI: 10.5772/52877

[112] Ogunshe AA, Ayodele AE, Okonko IO. Microbial studies on Aisa: A potential indigenous laboratory fermented food condiment from *Albizia saman* (Jacq.) F. Mull. Pakistan Journal of Nutrition. 2006;**5**:51-58. ISSN 1680-5194

[113] Roy A, Moktan B, Sarkar PK. Microbiological quality of legume-based traditional fermented foods marketed in West Bengal, India. Food Control. 2007;**18**:1405-1411. DOI: 10.1016/j.foodcont.2006.10.001

[114] Thakur N, Bhalla TC. Characterization of Some Traditional Fermented Foods and Beverages of Himachal Pradesh. New Delhi, India: CSIR; 2004

[115] Shin D, Jeong D. Korean traditional fermented soybean products: Jang. Journal of Ethnic Foods. 2015;**2**:2-7. DOI: 10.1016/j.jef.2015.02.002

[116] Su-Yeon K, Hyeong-Eun K, Yong-Suk K. The potentials of bacillus licheniformis strains for inhibition of *B. cereus* growth and reduction of biogenic amines in cheonggukjang (Korean fermented unsalted soybean paste). Food Control. 2017;**79**:87-93. DOI: 10.1016/j.foodcont.2017.03.028

[117] Sarkar PK, Nout MR, editors. Handbook of Indigenous Foods Involving Alkaline Fermentation. Boca Raton, Florida. United States: CRC Press; 2014

[118] Nout MR, Sarkar PK, Beuchat LR. Indigenous fermented foods. In: Food Microbiology: Fundamentals and Frontiers. 3rd ed. Washington, D.C. United States: American Society of Microbiology; 2007. pp. 817-835. DOI: 10.1128/9781555815912.ch38

[119] Chettri R, Tamang JP. Microbiological evaluation of maseura, an ethnic fermented legume-based condiment of Sikkim. Journal of Hill Research. 2008;**21**:1-7

[120] Bhalla TC. Traditional Foods and Beverages of Himachal Pradesh. New Delhi, India: CSIR; 2007

[121] Guan RF, Liu ZF, Zhang JJ, Wei YX, Wahab S, Liu DH, et al. Investigation of biogenic amines in sufu (furu): A Chinese traditional fermented soybean food product. Food Control. 2013;**31**:345-352. DOI: 10.1016/j.foodcont.2012.10.033

[122] Nout MJR, Ruikes MMW, Bouwmeester HM, Beljaars PR. Effect of processing conditions on the formation of biogenic amines and ethyl carbamate in soybean Tempe. Journal of Food Safety. 1993;**13**:293-303. DOI: 10.1111/j.1745-4565.1993.tb00114.x

[123] Nguyen LA. Health-promoting microbes in traditional Vietnamese fermented foods: A review. Food Science and Human Wellness. 2015;**4**:147-161. DOI: 10.1016/j.fshw.2015.08.004

[124] Sharma A. A review on traditional technology and safety challenges with regard to antinutrients in legume foods. Journal of Food Science and Technology.

2020;**1**:21. DOI: 10.1007/s13197-020-04883-8

[125] Aidoo KE, Rob Nout MJ, Sarkar PK. Occurrence and function of yeasts in Asian indigenous fermented foods. FEMS Yeast Research. 2006;**6**:30-39. DOI: 10.1111/j.1567-1364.2005.00015.x

[126] Hajeb P, Jinap S. Umami taste components and their sources in Asian foods. Critical Reviews in Food Science and Nutrition. 2015;**55**:778-791. DOI: 10.1080/10408398.2012.678422

[127] Anal AK, Perpetuini G, Petchkongkaew A, Tan R, Avallone S, Tofalo R, et al. Food safety risks in traditional fermented food from South-East Asia. Food Control. 2020;**109**:106922. DOI: 10.1016/j.foodcont.2019.106922

[128] Liao WC, Wang CY, Shyu YT, Yu RC, Ho KC. Influence of preprocessing methods and fermentation of adzuki beans on γ-aminobutyric acid (GABA) accumulation by lactic acid bacteria. Journal of Functional Foods. 2013;**5**:1108-1115. DOI: 10.1016/j.jff.2013.03.006

[129] Limón RI, Peñas E, Torino MI, Martínez-Villaluenga C, Dueñas M, Frias J. Fermentation enhances the content of bioactive compounds in kidney bean extracts. Food Chemistry. 2015;**172**:343-352. DOI: 10.1016/j.foodchem.2014.09.084

[130] Rizzello CG, Hernández-Ledesma B, Fernández-Tomé S, Curiel JA, Pinto D, Marzani B, et al. Italian legumes: Effect of sourdough fermentation on lunasin-like polypeptides. Microbial Cell Factory. 2015;**14**:168. DOI: 10.1186/s12934-015-0358-6

[131] Coda R, Rizzello CG, Gobbetti M. Use of sourdough fermentation and pseudo-cereals and leguminous flours for the making of a functional bread enriched of γ-aminobutyric acid (GABA). International Journal of Food Microbiology. 2010;**137**:236-245. DOI: 10.1016/j.ijfoodmicro.2009.12.010

[132] Galli V, Venturi M, Coda R, Maina NH, Granchi L. Isolation and characterization of indigenous *Weissella confusa* for in situ bacterial exopolysaccharides (EPS) production in chickpea sourdough. Food Research International. 2020;**138**:109785. DOI: 10.1016/j.foodres.2020.109785

[133] De Pasquale I, Verni M, Verardo V, Gómez-Caravaca AM, Rizzello CG. Nutritional and functional advantages of the use of fermented black chickpea flour for semolina-pasta fortification. Food. 2021;**10**:182. DOI: 10.3390/foods10010182

[134] Dueñas M, Fernández D, Hernández T, Estrella I, Muñoz R. Bioactive phenolic compounds of cowpeas (*Vigna sinensis* L). Modifications by fermentation with natural microflora and with *lactobacillus plantarum* ATCC 14917. Journal of Science and Food Agriculture. 2005;**85**:297-304. DOI: 10.1002/jsfa.1924

[135] Gan RY, Shah NP, Wang MF, Lui WY, Corke H. Fermentation alters antioxidant capacity and polyphenol distribution in selected edible legumes. International Journal of Food Science & Technology. 2016;**51**:875-884. DOI: 10.1111/ijfs.13062

[136] Coda R, Kianjam M, Pontonio E, Verni M, Di Cagno R, Katina K, et al. Sourdough-type propagation of faba bean flour: Dynamics of microbial consortia and biochemical implications. International Journal of Food Microbiology. 2017;**248**:10-21. DOI: 10.1016/j.ijfoodmicro.2017.02.009

[137] Hoehnel A, Bez J, Sahin AW, Coffey A, Arendt EK, Zannini E. *Leuconostoc citreum* TR116 as a microbial cell factory to functionalise high-protein

faba bean ingredients for bakery applications. Food. 2020;**9**:1706. DOI: 10.3390/foods9111706

[138] Xie C, Coda R, Chamlagain B, Edelmann M, Varmanen P, Piironen V, et al. Fermentation of cereal, pseudo-cereal and legume materials with *Propionibacterium freudenreichii* and *Levilactobacillus brevis* for vitamin B12 fortification. LWT-Food Science and Technology. 2021;**137**:110431. DOI: 10.1016/j.lwt.2020.110431

[139] Wang Y, Sorvali P, Laitila A, Maina NH, Coda R, Katina K. Dextran produced *in situ* as a tool to improve the quality of wheat-faba bean composite bread. Food Hydrocolloids. 2018;**84**:396-405. DOI: 10.1016/j.foodhyd.2018.05.042

[140] Coda R, Varis J, Verni M, Rizzello CG, Katina K. Improvement of the protein quality of wheat bread through faba bean sourdough addition. LWT-Food Science and Technology. 2017;**82**:296-302. DOI: 10.1016/j.lwt.2017.04.062

[141] Rizzello CG, Verni M, Koivula H, Montemurro M, Seppa L, Kemell M, et al. Influence of fermented faba bean flour on the nutritional, technological and sensory quality of fortified pasta. Food and Function. 2017;**8**:860-871. DOI: 10.1039/C6FO01808D

[142] Torino MI, Limón RI, Martínez-Villaluenga C, Mäkinen S, Pihlanto A, Vidal-Valverde C, et al. Antioxidant and antihypertensive properties of liquid and solid state fermented lentils. Food Chemistry. 2013;**136**:1030-1037. DOI: 10.1016/j.foodchem.2012.09.015

[143] Bautista-Expósito S, Peñas E, Dueñas M, Silván JM, Frias J, Martínez-Villaluenga C. Individual contributions of Savinase and *lactobacillus plantarum* to lentil functionalization during alkaline pH-controlled fermentation. Food Chemistry. 2018;**257**:341-349. DOI: 10.1016/j.foodchem.2018.03.044

[144] Perri G, Coda R, Rizzello CG, Celano G, Ampollini M, Gobbetti M, et al. Sourdough fermentation of whole and sprouted lentil flours: In situ formation of dextran and effects on the nutritional, texture and sensory characteristics of white bread. Food Chemistry. 2021;**129638**. DOI: 10.1016/j.foodchem.2021.129638

[145] Bartkiene E, Krungleviciute V, Juodeikiene G, Vidmantiene D, Maknickiene Z. Solid state fermentation with lactic acid bacteria to improve the nutritional quality of lupin and soya bean. Journal of the Science of Food and Agriculture. 2015;**95**:1336-1342. DOI: 10.1002/jsfa.6827

[146] Romero-Espinoza AM, Vintimilla-Alvarez MC, Briones-García M, Lazo-Vélez MA. Effects of fermentation with probiotics on anti-nutritional factors and proximate composition of lupin (*Lupinus mutabilis* sweet). LWT-Food Science and Technology. 2020;**130**: 109658. DOI: 10.1016/j.lwt.2020.109658

[147] Ariahu CC, Ukpabi U, Mbajunwa KO. Production of African bread-fruit (*Treculia africana*) and soybean (*Glycine max*) seed based food formulations, 1: Effects of germination and fermentation on nutritional and organoleptic quality. Plant Foods for Human Nutrition. 1999;**54**:193-206. DOI: 10.1023/A:1008153620287

[148] Naqash F, Gani A, Gani A, Masoodi FA. Gluten-free baking: Combating the challenges—A review. Trends in Food Science and Technology. 2017;**66**:98-107. DOI: 10.1016/j.tifs.2017.06.004

[149] Filannino P, Di Cagno R, Gobbetti M. Metabolic and functional paths of lactic acid bacteria in plant

foods: Get out of the labyrinth. Current Opinion in Biotechnology. 2018;**49**:64-72. DOI: 10.1016/j.copbio.2017.07.016

[150] Sanchez-Maldonado AF, Schieber A, Ganzle MG. Structure-function relationships of the antibacterial activity of phenolic acids and their metabolism by lactic acid bacteria. Journal of Applied Microbiology. 2011;**111**:1176-1184. DOI: 10.1111/j.1365-2672.2011.05141.x

[151] Reveron I, Rivas B, Munoz R, Felipe F. Genome-wide transcriptomic responses of a human isolate of *lactobacillus plantarum* exposed to *p*-coumaric acid stress. Molecular Nutrition & Food Research. 2012;**56**:1848-1859. DOI: 10.1002/mnfr.201200384

[152] Axel C, Zannini E, Arendt EK. Mold spoilage of bread and its biopreservation: A review of current strategies for bread shelf life extension. Critical Reviews in Food Science and Nutrition. 2017;**57**:3528-3542. DOI: 10.1080/10408398.2016.1147417

[153] López-Barrios L, Gutiérrez-Uribe JA, Serna-Saldívar SO. Bioactive peptides and hydrolysates from pulses and their potential use as functional ingredients. Journal of Food Science. 2014;**79**:273-283. DOI: 10.1111/1750-3841.12365

[154] Mäkinen OE, Arendt EK. Nonbrewing applications of malted cereals, pseudocereals, and legumes: A review. Journal of the American Society of Brewing Chemists. 2015;**73**:223-227. DOI: 10.1094/ASBCJ-2015-0515-01

[155] Bewley JD. Seed germination and reserve mobilization. In: Encyclopedia of Life Sciences. London, U.K.: Nature Publishing Group; 2001

[156] Koehler P, Hartmann G, Wieser H, Rychlik M. Changes of folates, dietary fiber, and proteins in wheat as affected by germination. Journal of Agricultural and Food Chemistry. 2007;**55**:4678-4683. DOI: 10.1021/jf0633037

[157] Katina K, Liukkonen KH, Kaukovirtanorja A, Adlercreutz H, Heinonen SM, Lampi AM, et al. Fermentation-induced changes in the nutritional value of native or germinated rye. Journal of Cereal Science. 2007;**46**:348-355. DOI: 10.1016/j.jcs.2007.07.006

[158] Montemurro M, Coda R, Rizzello CG. Recent advances in the use of sourdough biotechnology in pasta making. Food. 2019;**8**:129. DOI: 10.3390/foods8040129

[159] Rosa-Sibakov N, Heiniö RL, Cassan D, Holopainen-Mantila U, Micard V, Lantto R, et al. Effect of bioprocessing and fractionation on the structural, textural and sensory properties of gluten-free faba bean pasta. LWT-Food Science and Technology. 2016;**67**:27-36. DOI: 10.1016/j.lwt.2015.11.032

[160] Petitot M, Boyer L, Minier C, Micard V. Fortification of pasta with split pea and faba bean flours: Pasta processing and quality evaluation. Food Research International. 2010;**43**:634-641. DOI: 10.1016/j.foodres.2009.07.020

[161] Petitot M, Barron C, Morel MH, Micard V. Impact of legume flour addition on pasta structure: Consequences on its in vitro starch digestibility. Food Biophysics. 2010;**5**:284-299. DOI: 10.1007/s11483-010-9170-3

[162] Crisan EV, Sands A. Edible Mushrooms, Nutritional Value. Dalam. In: Chang ST, Hayes WA, editors. The Biology and Cultivation of Edible Mushrooms. New York: Hangeri Academic Press; 1978

[163] De Angelis M, Damiano N, Rizzello CG, Cassone A, Di Cagno R,

Gobbetti M. Sourdough fermentation as a tool for the manufacture of low-glycemic index white wheat bread enriched in dietary fibre. European Food Research and Technology. 2009;**229**:593-601. DOI: 10.1007/s00217-009-1085-1

[164] Elli L, Branchi F, Tomba C, Villalta D, Norsa L, Ferretti F, et al. Diagnosis of gluten related disorders: Celiac disease, wheat allergy and non-celiac gluten sensitivity. World Journal of Gastroenterology. 2015;**21**:7110. DOI: 10.3748/wjg.v21.i23.7110

[165] Atkinson FS, Foster-Powell K, Brand-Miller JC. International tables of glycemic index and glycemic load values: 2008. Diabetes Care. 2008;**31**:2281-2283. DOI: 10.2337/dc08-1239

[166] Torres A, Frias J, Granito M, Guerra M, Vidal-Valverde C. Chemical, biological and sensory evaluation of pasta products supplemented with agalactoside-free lupin flours. Journal of the Science of Food and Agriculture. 2007;**87**:74-81. DOI: 10.1002/jsfa.2673

[167] Torres A, Frias J, Granito M, Vidal-Valverde C. Fermented pigeon pea (*Cajanus cajan*) ingredients in pasta products. Journal of Agricultural Food Chemistry. 2006;**54**:6685-6691. DOI: 10.1021/jf0606095

[168] Martínez-Villaluenga C, Torres A, Frias J, Vidal-Valverde C. Semolina supplementation with processed lupin and pigeon pea flours improve protein quality of pasta. LWT-Food Science and Technology. 2010;**43**:617-622. DOI: 10.1016/j.lwt.2009.11.001

[169] Curiel JA, Coda R, Limitone A, Katina K, Raulio M, Giuliani G, et al. Manufacture and characterization of pasta made with wheat flour rendered gluten-free using fungal proteases and selected sourdough lactic acid bacteria. Journal of Cereal Science. 2014;**59**:79-87. DOI: 10.1016/j.jcs.2013.09.011

[170] Schoenlechner R, Drausinger J, Ottenschlaeger V, Jurackova K, Berghofer E. Functional properties of gluten-free pasta produced from amaranth, quinoa and buckwheat. Plant Foods for Human Nutrition. 2010;**65**:339-349. DOI: 10.1007/s11130-010-0194-0

[171] Frías J, Granito M, Vidal-Valverde C. Fermentation as a process to improve the nutritional quality of grain legumes. In: Proceedings of the 5th European Conference on Grain Legumes. Dijon, France: Legumes for the Benefit of Agriculture, Nutrition and the Environment: Their Genomics, Their Products and Their Improvement; June, 2004

Chapter 14

Enzymatic Processing of Pigeon Pea Seed Increased Their Techno-Functional Properties

Zainab Muhammad Bello, Sanusi Muhammad, Adamu Aliyu Aliero and Ibrahim Aliyu Dabai

Abstract

Neglected and underutilized crops (NUS) are those crops that are entirely ignored or little attention is paid to them by agriculture researchers, plant breeders and policy-makers. There has been renewed interest in NUS as many of these varieties and species, along with a wealth of traditional knowledge are being lost at an alarming rate. This chapter provides an overview of underutilized legumes in Sub-Saharan Africa (SSA). There is a recognized need to explore the diversity of indigenous micro symbionts associated with underutilized legumes. The biochemical mechanism in legumes remains elusive to date as evidence is mounting for allelopathic inhibition of nitrifying microorganisms by root exudation of phenolic compounds. A cross-sectional study was undertaken to explore the potential relationship between enzymatic processes of certain legumes and high tolerance to drought stresses, high biomass productivity, erosion control and dune stabilization and general soil health. Pigeon pea among other legumes have a huge untapped potential for improvement of both in quantity and quality of production in Africa.

Keywords: NUS, PGPR, PGPF, African legumes, pigeon pea, food security

1. Introduction

Cajanus cajan (L.) Millsp. (*C. cajan*) (Family: Fabaceae) also known as pigeon pea, is a famous food and cover/forage crop bearing a high amount of key amino acids like methionine, lysine and tryptophan [1]. It is one of the key promising sources of antioxidants and key enzyme inhibitors that can be exploited for future bioproduct development in tropical and subtropical regions in the world. This legume is one of the underutilized species in sub-Saharan Africa. Other economically important legumes include sword beans (*Canavalia ensiformis*), Bambara groundnut (*Vigna subterranea*), and Lima beans (*Phaseolus lunatus*) among several others [2, 3]. Neglected and underutilized species (NUS) has been a question of great interest in a wide range of fields.

The role of NUS in fighting poverty, hunger and malnutrition has received increased attention across several disciplines in recent years. There has been a growing recognition of the vital links between NUS, nutritional diversity, food sustainability and wealth creation. One well-known study that is often cited in research on NUS is that of Padulosi [2], who defined NUS as those species which are entirely ignored by agricultural researchers, plant breeders and policymakers. It is a widely held view that NUS is not traded as commodities. Agricultural species that are not among the major staple crops often come under the heading of 'neglected and underutilized species' (NUS) and are sometimes called 'orphan' crops.

If [2]'s findings are accurate, many of these varieties and species, along with a wealth of traditional knowledge about their cultivation and use, are being lost at an alarming rate. Evidence suggest that nearly 30,000 edible plant species have been identified, of which over 7000 plant species have been used in the history of civilization to meet food requirements [4]. Recent trends in NUS have led to a proliferation of studies as researchers have shown an increased interest in these species. It was established that barely 150 species are commercially cultivated and, of these, just 103 crops provide up to 90% of the calories in the human diet.

Several divergent accounts of NUS have been proposed, creating numerous controversies. The relationship between crop diversity, dietary diversity, nutrition and health has attracted conflicting interpretations from the scientific community. NUS offers opportunities to enrich diets with healthier food (particularly legumes, fruits and vegetables). The gradual shift from diets based on local foods to a 'Western-style' diet, high in fats, salt, sugar and processed foods, increases the incidence of non-communicable diseases, such as diabetes, obesity, heart disease and certain types of cancer.

African legumes can contribute significantly to the dietary supply of nutrients especially protein, essential amino acids, dietary fiber, vitamins, and minerals in the diet. However, compared to the well-known pulses, like the common bean, oilseed legumes, soybean, the African legumes are greatly underutilized and under-researched [5].

Numerous studies have attempted to explain the consistent contribution of NUS in fighting poverty, hunger and malnutrition and in generating income in both domestic and international markets [6–10]. This chapter examines the relationship between the biochemical mechanism through the enzymatic processes resulting to allelopathic inhibition of nitrifying microorganisms by root exudation of phenolic compounds in pigeon peas and related legumes. These could be exploited for future bio-product development to bring about sustainable crop production and to also ameliorate the effect of both biotic abiotic stresses to boost yield and restore depleted soil nutrients.

1.1 Neglected and underutilized legumes in Sub-Saharan Africa

Previous studies of NUS have established that in some countries like the Côte d'Ivoire, a considerable variety of minor leafy vegetables, legumes and vegetable crops are utilized and appreciated for their rich nutritional value. In the same way, minor forest and timber forest species are collected from the wild in Ghana which is utilized in developmental projects in the country such as housing construction, furniture and other arts and crafts. In the same vein, sword beans (*C. ensiformis*) are common food crops in the region. Similarly, in East Africa however, demand for leafy green vegetables is ever more recognized due to their nutritive value. In Uganda on the other hand, despite disregard on the part of stakeholders more especially the agricultural research and development programs, poor communities continue to grow and market miscellaneous underutilized species which are locally processed in form of artifacts and are mainly exported around the globe. NUS also serves as an alternative food source and is equally exploited for ornamental and medicinal value, generating significant income for some populations in West Africa. The use of some NUS is also significant in Malawi, particularly about Bambara groundnut (*V. subterranea*), sorghum, finger millets and pearl millets. In Nigeria,

NUS such as fonio (*Digitaria exilis*, *Digitaria iburua*), Ethiopian tef (*Eragrotis tef*), Lima beans (*P. lunatus*), Pigeon peas (*C. cajan*) and *V. subterranea*, *Crotalaria* spp. and *Solanum nigrum*. Zimbabwe reports significant use of NUS, especially finger and pearl millets [2, 3].

*1.1.1 Sword beans (*C. ensiformis*)*

C. ensiformis (Sword beans) is a leguminous crop that could potentially be used as a biocontrol [11]. The article provides valuable insight into this leguminous vine. It is native to South and Central America however; it has adapted well to thrive in depleted soils of the tropics and subtropics worldwide. *C. ensiformis* is beneficial in curbing erosion, sandbank stabilization, soil rejuvenation, conservation, improvement and general soil health. It is grown for food, fodder, as a cover crop or green manure. *C. ensiformis* is resistant to drought and pests but does not grow well in excessively wet soil and has certain compounds such as Canavaine that increase its resistance to pest and herbivory due to the presence Concanavalin A—a lectin that guards against diseases caused by microorganism infections Canavanine is lethal as, when consumed, it could interchange L-arginine during protein synthesis and cause proteins to become malformed [11]. The beetle *Caryedes brasiliensis* thrives on *C. ensiformis* seeds as it contains an arginyl-T-RNA synthase that can discriminate between canavanine and L-arginine, countering the toxic effects.

*1.1.2 Bambara groundnut (*V. subterranean*)*

Bambara groundnut (*V. subterranea* (L.) Verdc.) is an indigenous legume crop, cultivated by subsistence farmers throughout sub-Saharan countries [12]. Neglected and underutilized legumes have for long been a question of great interest in a wide range of fields. In recent years, researchers have shown an increased interest in this neglected legume and considerable literature has grown up around the theme. It serves as the main source of income generation to subsistence farmers in Zimbabwe, Swaziland, Uganda and Zambia with a suitable land area (84, 100, 98 and 95%) respectively for cultivating this legume. A pilot study was initiated in seven rural districts of Zimbabwe aimed at improving nutrition, food security and maximizing its production. Follow-up studies according to Mubaiwa [12] report revealed a decline in production due to increasing climate and weather variability which in turns negatively affected food security. In an effort to keep this in check, agriculture and policy-makers in Zimbabwe have recognized the hidden potentials of NUS in improving food security by adapting to current climate change.

One major finding on Bambara groundnut in Nigeria is by Onuche et al. [13] who referred to it as "special" legume as it can adapt to high temperature and resistant to drought and suitable also for marginal soils where other leguminous crops are unlikely to thrive as it requires little resource from the soil in general. It is regarded also as special because it is not prone to total harvest failure even in low and uncertain rainfall regions. It increases soil fertility and boost yield of other crops cultivated around it. Therefore, it could potentially serve as great source of bio-fertilizer which is safe and free from high cost of synthetic chemicals commonly used as fertilizers which are indeed detrimental to health.

Bambara groundnut (*V. subterranea*) is a grain legume indigenous to sub-Saharan Africa where it is widely cultivated by subsistence farmers. The center of origin is most likely North-Eastern Nigeria and Northern Cameroon, in West Africa. The species is also grown to a lesser extent in some Asian countries such as India, Malaysia, Philippines and Thailand [4]. It is cultivated for its subterranean pods, is extremely hardy and produces reasonable yields even under conditions of drought and low soil fertility. There is notable paucity of empirical research focusing specifically on neglected underutilized legumes in relation to soil microbial interaction.

To date, very little attention has been paid to the role of NUS such as Bambara nut in sustainable crop production. Previous studies have failed to find any consistent association between neglected underutilized legumes and rhizosphere and rhizoplane microbial relations. Research studies in these areas should contribute to an increase in the utilization of African legumes. It has been proposed that African legumes are drought-tolerant crops, thus excellent candidates utilized as climate friendly food crops. Global warming and climate change with the resultant effects of low agricultural productivity and food insecurity are important global issues. African legumes therefore have great global potential as sustainable food sources [3].

Miscellaneous underutilized grain legumes play significant roles in the food cultures around the world as absolute sources of quality protein, natural medicine, animal fodder, natural fertilizers, and environmental restoration products, alongside the well-established soil enrichment property of symbiosis with nitrogen-fixing bacteria also contain high-quality proteins and micronutrients which are comparable to those found in conventional legumes. They are also indispensable in crop rotation strategies to fertilize agricultural soils [5].

*1.1.3 Lima beans (*P. lunatus*)*

P. lunatus L. was similarly reported to have tolerance and resistant traits against insects' attack [14]. Lima Beans roles in sustainable crop production have been investigated by Ruiz-Santiago et al. identifying plant-insect interactions as a determining factor for sustainable food and nutritional security. This could be utilized to influence the impact of herbivorous insects and their expression under varying climatic condition to achieve sustainable environmentally safe crop production. *P. lunatus* was similarly reported is an important legume for the poor population of the Brazilian northeast region [15]. The legume is able to take advantage of the nitrogen fixation process, but research to date has not been able to explore the diversity of indigenous microsymbionts.

*1.1.4 Pigeon peas (*C. cajan*)*

The study by Ayenan et al. [16] offers probably the most comprehensive empirical analysis. It was established from the report findings that in terms of production and quantity, pigeon pea is the 5th legume after cowpea (*Vigna ungulata*), Bambara groundnut (*V. subterranean*), Soy bean (*Glycine max*) and groundnut (*Arachis hypoaea*). Pigeon pea is a perennial shrub that has many advantages over annual legumes as it ensures several harvests and capable of enhancing soil fertility. The legume can withstand harsh environmental stress and has high drought tolerance which are the major challenges in this present-day climatic variability constraints.

Sinan et al. [1] in a recent study carried out the enzymatic assays, revealed that *C. cajan* has high amount of essential amino acids such as methionine, lysine and tryptophan. The study similarly was able to determine the total phenolic (TPC) content and total antioxidant capacity of the legume. Their study therefore serves as a baseline which gives opportunity for further investigation as the methanolic extract among other extract (hexane, ethyl acetate, aqueous) of prepared from *C. cajan* stem bark shows prominent antioxidant ability. This suggest that *C. cajan* could serve as a new source of antioxidant and key enzymes inhibitor (α-amylase and α-glucosidase enzymes) among others which could be exploited for future bioproduct development.

From the archives of the ECHO's Seed Bank [16], in a paper titled "factors to consider when selecting a pigeon pea variety" highlighted how pigeon pea requires lesser amount of nitrogen (N_2) into a form useful to the plant. Interestingly, *C. cajan* are reported to require lesser amount of N_2 compared to other crops like maize. It obtains N_2 from the atmosphere, assimilate it and are released to the soil as the plant die provided that this legume residues are kept in the field. Motis [17] also highlighted that pigeon pea has extensive roots which could grow through hard layers of soil aggregates, increasing soil porosity and aeration which in turn improves soil fertility. This report therefore is to infer that mixed cropping could be beneficial to other crops when pigeon pea is incorporated.

*1.1.5 Common bean (*Phaseolus vulgaris *L.)*

Common bean is one of the most familiarized beans worldwide. The developing countries especially in the sub-saharan regions mainly rely on common bean as a protein source; an alternative to animal protein, which are highly costly and beyond the reach of many low-income earners especially among the rural population.

Ismail et al. [18] recently identified six bacterial and four fungal strains which were isolated from the common bean (*P. vulgaris*) root plant for their growth promoting properties using molecular techniques and all the microbial isolates showed varying activities to produce indole-3-acetic acid and different hydrolytic enzymes such as amylase, cellulase, protease, pectinase and xylase. All the six bacterial endophytes isolates were reported to display phosphate solubilizing capacity. To increase the reliability of measures, each bacterial and fungal were tested twice by conducting a field experiment to evaluate the promotion activity of the metabolites of the most potent endophytic bacterial and fungal strains, in comparison to two exogenous applied hormones; IAA and benzyl adenine (BA) on the growth and biochemical characteristics of the *P. vulgaris*. Base on the results findings of Ismail et al. [18] it was established that both bacterial and fungal endophytic metabolites surpassed the exogenously applied hormone in increasing the plant biomass, photosynthetic pigments, carbohydrates, and protein contents, antioxidants, enzyme activity, endogenous hormones and yield traits. Therefore, it could be deducted that endophytic bacteria *Brevibacillus agri* (PB5) shows promising traits as a stimulator for growth and productivity of the common bean.

2. Role of plant growth promoting microbes in sustainable crop production

PGPR play a pivotal role in promoting plant growth either directly (by enabling resource acquisition of essential nutrients and elements like nitrogen, phosphorus), moderating plant hormone levels, or indirectly serving as biocontrol agents by decreasing the inhibitory effects of various pathogens on plant growth and development [19, 20].

According to Finkel et al. [21], growing crops continuously in agricultural soils can result in pathogen build up as well as bring about the emergence of disease suppressive soils. These soils in the long run convey resistance to plant pathogens and contain biocontrol agents within their resident bacterial community. Hassan et al. [22] points out that plant growth promoting rhizobacteria may affect plant growth, development and disease suppression by one or more direct or indirect mechanisms.

Ahemad and Kibret [23], in their study on the current prospect of mechanisms and application of plant growth promoting rhizobacteria point out that 80% of microorganisms isolated from the rhizosphere of various crops possesses the ability to synthesis and release of auxins as secondary metabolites. Phytohormones like auxins, gibberellins, cytokinin, ethylene and abscisic acid facilitate plant cell enlargement and extension in symbiotic and non-symbiotic roots [24]. It is now well established from a variety of studies that indole acetic acid (IAA) is involved in multiple processes in plant growth and development such as aiding cell division, differentiation and vascular formation which are the three most essential processes involved in nodule formation. Therefore, it was suggested that auxins levels in host legume plants are necessary for nodule formation [23–30]. Supporting this view is the study by Camerini et al. [31] that introducing IAA biosynthesis pathway in the inoculation with *Rhizobium leguminosarum* bv Viciae produced potential nitrogen fixing root nodules containing up to 60-fold more IAA than nodules formed by wild-type counterpart in *Vicia hirsute*. Therefore, drawing from the extensive range of sources, *Rhizobium* sp. have shown to produce IAA.

Gouda et al. [32] claims that plants have always been in symbiotic relationship with soil microbes (bacteria and fungi). The leguminous crops and beneficial microbes create a mutual relationship in the rhizosphere to fix the atmospheric nitrogen into plant available form. Mehmood's et al. [33] work on plant growth promoting rhizobacteria is complemented by Hassan's [22] study of the interactions of rhizo deposits with plant growth-promoting rhizobacteria in the rhizosphere, holds the view that plant release certain exudates such as sugars, sterols, growth factors, etc. which stimulate the movement of beneficial microbes towards plant roots. On reaching the root, the beneficial microbes are known to form nodules.

It is documented that 80% of plant available nitrogen comes from biological nitrogen fixation (BNF), while remaining 20% is contributed by other non-symbiotic organisms [34]. Nitrogen fixing organisms are generally categorized as symbiotic N_2 fixing bacteria including members of the family rhizobiaceae which forms symbiosis with leguminous plants (e.g. *rhizobia*) [28]. Non-leguminous trees (e.g. Frankia) and non-symbiotic (free living, associative and endophytes) nitrogen fixing forms such as cyanobacteria (*Anabaena*, *Nostoc*), *Azospirillum*, *Azotobacter*, and symbiotic nitrogen fixing bacteria provide only a small amount of the fixed nitrogen that the bacterially-associated host plant requires [29]. On a global scale, BNF provides the largest input of nitrogen to agricultural soils. Inoculation of these efficient plant beneficial rhizo-microbe species usually increases plant's productivity. If *Rhizobium* is inoculated as biofertilizer in the crops such as groundnut, pigeon pea, soybean, etc., it can supply ~19–22 kg ha^{-1} which can raise the production by ~17–33% [34]. Gouda et al. [32] in a similar study on 'revitalization of plant growth promoting rhizobacteria for sustainable development in agriculture' reported that plant growth promoting rhizobacteria fix atmospheric nitrogen in the soil including the strains of *Azoarus* sp., *Beijinrinka* sp., *Pantoea agglomerans* and *Klebsiella pneumoniae*. Symbiotic nitrogen fixing rhizobia within the rhizobiaceae family (a-proteobacteria) infect and establish symbiotic relationship with the roots of leguminous plants. On the other hand, arbuscular mycorrhizal fungi (AMF) were similarly reported to improve growth, nodulation and nitrogen fixation in legume-Rhizobium symbiosis [26].

2.1 Phytohormone production

Plant hormones or plant growth regulators are a wide range of microorganisms within the rhizosphere involved in secretions of substances known as exudates that regulate growth and development of organic materials [32]. They are also referred to as organic compounds that influence physiological changes in the plant at low concentrations [35]. Like Shah et al. [35] and Gouda et al. [32] hold the view that at a minimum concentration (<1 mM), growth hormones/regulators can promote, inhibit, or modify growth and development of plants. Similarly, Kumar et al. [36] maintain that hormones are the basic chemical signals that influence plant's ability to respond to the environment even at a low concentration of these active organic compounds [33]. Surprisingly, these growth regulators (also known as exogenous plant hormones) can be induced or activated by certain microbes, such as PGPR in plants [22, 37].

Organic plants hormones are classified into five (5) groups; auxins, ethylene, cytokinins, gibberellins and abscisic acid [23, 32, 33, 35, 36]. Microbial synthesis of the phytohormone; auxin (indole-3-acetic acid/indole acetic acid (IAA)) has a long history. According to Ahemad and Kibret [23], 80% of microorganisms isolated from various crop rhizospheres have the ability to synthesis and release auxins as secondary metabolites. Auxins are a group of plant hormone relevant in the advancement of root formation which in turns leads to increase in absorption of essential nutrients by the roots of plants [34]. IAA has been implicated in almost all aspects of plant growth and development, likewise in defense responses.

Generally, IAA is a commonly produced phytohormone associated with cell division, cell enlargement, cell extension, cell differentiation; stimulates seeds and tuber germination; increases the rate of xylem and root development and adventitious root formation and initiation; controls vegetative growth process; mediates responses to light, gravity and fluorescence; affect photosynthesis, pigment formation, biosynthesis of various metabolites and resistance to stressful conditions. It was also reported that rhizobacterial IAA could increase root exudation by loosening the plant root cell walls, which in turn helps the rhizobacterial colonization and growth [38]. IAA produced by PGPR is also reported to modifying plant morphological functions to uptake more nutrients from the soil [23, 33–35].

Gibberellins can also alter the plant morphology by elongation of stem tissues [34–36]. Ethylene generally is an essential metabolite for the development of plants [23].

According to Sarkar et al. [34]; cytokinin has similar functions as gibberellins in stimulating lateral shoot development, leading to advance plant development. In addition to that, cytokinin encourages tissue expansion, cell division and cell enlargement, enhanced root hair formation, inhibition of root elongation, shoot initiation, or specific physiological responses in plants [35, 39]. Ethylene is the only gaseous hormone [36]. Kumar et al. [36] also referred to it as "wounding hormone" because its production can be induced by physical or chemical disturbance of plant tissues. Ethylene has many effects on plant growth and development, such as rapid seedling death and deprived growth when secreted in excess [33].

Ethylene production in excess can inhibit root growth. Mode of action of some PGPR involves the production of 1-aminocyclopropane-1-carboxylate (ACC), a deaminase enzyme which significantly improves growth parameters. In the same view, Dutta et al. [38] report that ACC deaminase production of PGPR brings about growth and advancement by decreasing the levels of ethylene, inducing salt tolerance [23] and reducing drought stress in plants. ACC is a predecessor of ethylene in the biosynthesis pathway of ethylene in plants [36].

Therefore, it was reported that although ethylene being an essential metabolite for plant growth and development, under different stress conditions such as drought or salinity stress, the ethylene level drastically increases, this could negatively influence plant growth [23, 35]. In a nutshell, several forms of stress are relieved by decreasing ethylene production [40]. Through ACC deaminase producers such as effects of phytopathogenic microorganisms (viruses, bacteria, fungi, etc.), and resistance to stress by polyaromatic hydrocarbons, heavy metals, radiation, wounding, insect predation, high salt concentration, drought, extremes of temperature (hot/cold), high light intensity and flooding.

The major noticeable effects of seed/root inoculation with ACC deaminase-producing microbe are root elongation, promotion of shoot growth and enhancement in rhizobial nodulation and N, P and K uptake as well as mycorrhizal colonization in various crops [23, 33, 38].

Kachroo and Kachroo [41] make an interesting contribution with regards to the impact of salicylic acid (SA) and jasmonic acid (JA) in plant defense mechanism. It is a widely held view that many phytohormones are now well known to mediate induced defense signaling, SA and JA are two major players that have been traditionally attributed roles in regulating plant defenses. SA and JA activate specific signaling pathways, which can act individually, synergistically or antagonistically, depending upon the pathogen involved. In addition to plant defense, SA and JA also regulate various developmental processes including flowering, root growth, floral nector secretion, senescence, development, cell growth, trichome development and thermogenesis. JA is an important regulator of plant responses to both biotic and abiotic stresses and is particularly well known for its role in plant defense against insects and herbivores. Febble's [11] work on *C. ensiformis*, complemented by Ruiz-Santiago's [14] study of *P. lunatus* taken together these studies provide important insights into the role of legumes in defense against herbivorous insects due to the activities of the vital phytohormones. **Table 1**

Phytohormones	Crops	PBRMs
Gibberellin	Alder	*Bacillus* sp.
	Soybean	
	Soybean	
Cytokinin	Soybean	*Pseudomonas fluorescens*
	Lettuce and rapeseed	*Rhizobium leguminosarum*
	Wheat	*Paenibacillus polymyxa*
Indole acetic acid (IAA)	Soybean	*Bradyrhizobium japonicum*
	Soybean	*Bradyrhizobium amyloliquefaciens*
	Wheat	*Azospirillum brasilense*
	Radish	*Bradyrhizobium* sp.
	Rice	*Enterobacter cloacae*
	Rice	*Aeromonas veronii*
	Lettuce	*Agrobacterium* sp.
	Lettuce	*Alcaligenes piechaudii*
	Lettuce	*Comamonas acidovorans*
Jasmonic acid (JA)	Soybean	*B. japonicum*
	Soybean	*Bradyrhizobium amyloliquefaciens*
Salicylic acid (SA)	Soybean	*B. japonicum*
	Soybean	*Bradyrhizobium amyloliquefaciens*

Adopted from [34].

Table 1.
Plant beneficial rhizosphere microorganisms (PBRMs) as an efficient phytohormone producer in various plants.

summarizes the essential phytohormones in certain crop species and the associated plant growth promoting rhizomicrobes.

2.2 Role of plant growth-promoting bacteria (PGPR) and fungi (PGPF) in sustainable agriculture

In a comprehensive literature review on role of Plant Growth-Promoting microorganisms in Sustainable Agriculture and Environmental Remediation [42], highlighted the most influential accounts of soil microorganisms, particularly the rhizobacteria that have the potential to influence growth, yield and nutrient uptake by crops either directly or indirectly as well as maintain soil fertility and health.

PGPR can enhance the plant tolerance by promoting the plant growth, even in poor growth conditions and increase agricultural produce of different crops under stressful environment [43–45]. Apart from the above-mentioned facts, recent reports suggest that application of PGPRs also improves nutritional quality and antioxidant status of the crops [46, 47]. Harnessing the above-mentioned plant-microbe interactions can also help in reclamation of degraded lands, reduction in usage of chemical fertilizers and agrochemicals [48]. A prominent agricultural symbiotic association exists between the rhizosphere bacteria and roots of the legumes by the formation of root nodules. Previous studies showed that plant-fungal associations are much older than the rhizobia-legume interaction. In various plant-fungal interactions fungi help in phosphate acquisition and make it available to plants [49]. Some reports also indicate that the DMI-2 protein is required for the initiation of the plant-arbuscular mycorrhizae interaction, which helps in phosphate solubilization. Although, the underlying mechanism of the PGPR and PGPF interactions with the plants are quite different, some studies showed a similarity between them. In *Medicago truncatula*, a type of arbuscular mycorrhizal fungi (AMF), the interaction releases some small diffusible factors to activate the similar genes by the rhizobacterial nod factor [50]. This confirms the analogy of the PGPR and PGPF to some extent and needs further clarification from cutting-edge research on the topic. Supporting this view [51], stated that plants use receptor-like kinases to monitor environmental changes and transduce signals into plant cells. The study further added that *M. trancula* DOES NOT MAKE INFECTION 2 (DMI2) protein functions as a co-receptor of rhizobial signals to initiate nodule development and rhizobial infection during nitrogen-fixing symbiosis. However, the mechanisms regulating DMI2 protein level and folding associated with arbuscular mycorrhizae are still unknown.

Mutualistic association (co-inoculation) of PGPR and PGPF (arbuscular mycorrhizae fungi) increases the growth, nutrient uptake potential, and yield of the plants [52]. PGPF can directly enhance the nutrient uptake (P, Zn) and water use efficiency of the inhabiting plant by increasing the root surface with hyphal network. With increased water use efficiency, AMF also controls the N_2O emission, which is a potent greenhouse gas emitted from agricultural fields [53]. AMF alone or in combination with certain PGPR enhances plant growth indirectly by inhibiting growth of root pathogens and optimizing soil structures [54]. Apart from this PGPF can also regulate soil health and fertility by improving the overall soil nutrient dynamics [55].

The negative effect of climate change is also mitigated by AMF through maintenance of proper soil aggregation and thereby providing another major advantage to agricultural crop production. More and more studies show that the mycorrhizae can play an essential role in plant growth by enhancing plant vigor in poorly performing soils, and through their ability to store large amounts of carbon, which in turn may improve some of the effects of climate change. In conclusion, we can say that application of PGPR and PGPF in combination or alone can negate the hazardous effect of chemical fertilizers, improve soil health, reduce environmental stresses and promote sustainable agriculture [56–58]. The interaction of AMF and rhizobacteria thus can promote plant growth by improving soil structural properties as well as the enhanced availability of nutrients and reduce disease progression in a sustainable manner [43].

3. Soil health, rhizosphere microbes and sustainable crop growth

The concept of soil health has a long history. Attention was much given to soil health in well developed countries initially. This may probably be due to increasing concern of global food security in terms demand for high quality food and feed as a result of exponential growth in population and to keep in check the state of the soil health in general.

According to some studies, some authors are of the opinion that SH has to have inherent components before it can be termed as healthy soil, considering the biological physical and chemical qualities. Health of the soil is a function of its environmental sustainability [59, 60]. There are certain properties to look out for to determine 'healthy soil'. One of the inherent elements of a healthy soil is the ability to be adapted to diverse ecological resources in an optimally balanced condition. It also has to be self-purifying from soil pollutants through elution or biotransformation and thirdly, it has to be suppressive of harmful biota (phytophytogenic) maintaining beneficial soil microbes [60].

While a variety of definitions to the term 'soil health' have been suggested, this chapter will use the definition suggested by Semenov et al. [60] that, soil fertility was traditionally defined as the ability of soil to provide plants with mineral elements, water, air and overall ability to live in favorable physical and chemical environment. Organic fertilizers when used solely, serves as biological control against harmful organisms. Similarly, exclusion of chemical pesticides and using genetically modified organisms in organic farming system offers improved physical and chemical soil properties. The report established that organic farming increases soil health and microbial biodiversity by limiting mass development of pathogenic microbes.

According to Soil Science Society of America (SSSA), soil quality is the ability of the soil to function within the limits of the ecosystem to preserve its biological productivity and quality, ensuring healthy plants and animals. It was further established that methods and parameters have been developed, including portable devices and sets of tools (Environmental Box Kit Soil quality) for the determination of the quality of the soil. These kits are essential in determining soil conductivity, its pH, resistance during plowing, density, organic carbon, carbon microbial biomass, humidity, moisture capacity, nitrate content, soil respiration rates, etc.

Increasing agricultural production to feed the snowballing population in sub-Saharan Africa (SSA) is very challenging [59]. These results due to poor soil quality resulting from soil degradation, unsustainable management practices, and available arable land for mass production to feed the ever-increasing population of subsistence farmers and non-agricultural population. Han et al.'s [61] report on organic and inorganic/chemical fertilizers established that both have their positives and negatives on plant growth and soil in general. Chemical fertilizers/inorganic fertilizers according to their study are reasonably affordable, very rich nutrient contents, and could easily be assimilated by plants. Despite these beneficial effects, chemical fertilizers when used excessively could affect both the plant and the soil community through nutrient loss, surface water and groundwater contamination, soil acidification or basification, reductions in useful microbial communities, and increased sensitivity to harmful insects.

On the other hand, organic manure similarly has several positive effects owing to the stable supply of both essential nutrients and micronutrients, improved microbial activity due to amplified nutrient availability resulting from breakdown of harmful elements, improved soil structure and root development and sufficient soil water for physiological activities [61]. Organic manure is said to be produced from the decomposition of animal byproducts which are utilized to overcome environmental degradation and low plant productivity resulting from excessive utilization of chemical fertilizers. Organic manure is cost-effective and environmentally friendly particularly when waste from livestock industries is recycled which in turn promote agricultural productivity.

In a recent development, Fasusi et al. [45] have shown that nodule formation is enhanced by the low availability of nitrogen, but microorganisms that produce an enzyme 1-aminocyclopropane-1-carboxylate (ACC) deaminase, have the potential to degrade 1-aminocyclopropane-1-carboxylate before its conversion to ethylene and may also enhance the formation of a nodule. Such formation is part of a common strategy developed by leguminous plants and Rhizobiaceae bacteria to decrease the concentration of oxygen to which the nitrogenase is exposed due to the inhibitory effect of oxygen on nitrogenase activity.

4. Conclusion

This study set out to review in detail the available information on enzymatic processes in pigeon peas and related legumes like the allelopathic inhibition of nitrifying microorganisms by root exudation of phenolic compounds, legumes' tolerance to drought stresses, high biomass productivity, relationship between plant growth promoting bacteria (PGPB) and plant growth promoting rhizofungi (PGPF) and general soil health towards sustainable crop production. This study has examined the peer reviewed literature on pigeon pea among others legumes as having huge untapped potential for improvement of both in quantity and in quality of production in Africa. The most obvious finding to emerge from this study is that pigeon pea requires lesser amount of nitrogen (N_2) compared to other crops like maize. Secondly, pigeon pea has extensive roots which could grow through hard layers of soil aggregates, increasing soil porosity and aeration which in turn improves soil fertility. Taken together, these findings suggest a role for pigeon pea in promoting crop yield. The present study is important in furthering our understanding of the role pigeon pea and related legumes play in boosting yield when mixed cropping is practiced. This is the first study to report an association between underutilized African legumes, enzymatic processes involved in mutualistic association of PGPR (rhizobacteria) and PGPF (arbuscular mycorrhizae fungi) and nodule formation to bring about increases in growth, nutrient uptake potential, and yield of the plants. The study is limited by the lack of information on pigeon pea and accurate state-of the art about the enzymatic processing of pigeon pea seed in promoting its techno-functional potentials. The strengths of the study included the in-depth analysis of Role of Plant Growth-Promoting Bacteria (PGPR) and Fungi (PGPF) in Sustainable Agriculture, Neglected and Underutilized Legumes in Sub-Saharan Africa and general Soil Health for Sustainable Crop Growth. Further research should be undertaken to explore how the mechanism of enzymatic processing of pigeon pea works for sustainable food security.

5. Challenges and future research perspectives

Role of Rhizosphere Microbes in Soil Fertility and Sustainability 'Zero Hunger Challenge' is the major goal of human efforts. This goal can be achieved by sustainable increment in the arena of agricultural productivity to satisfy the demands of abruptly spreading human population. Growing more and more foods from less and less farm holdings is a demand feature of our economy. The time of action is now. There is growing realization that agriculture must diversify. NUS more especially legumes have an important role to play in advancing agricultural development beyond the Green Revolution model of improving and raising the yields of staple crops.

NUS are local and traditional but are globally significant and thus require scientific and political attention beyond the local and national levels. Some of the major challenges to the sustainable use of NUS are neglected by agronomic researchers and policy-makers, genetic erosion, loss of local knowledge, marketing and climate change. The main purpose for the review is to encourage scientists and policy-makers to optimize and promote the benefit of NUS and to create links between scientific and traditional knowledge by establishing interdisciplinary research networks to identify suitable material for NUS breeding by means of strengthening cooperation among stakeholders and creating synergies at local, regional, national and international levels.

Acknowledgements

We acknowledge the support of Tetfund National Research under the theme of Science, Engineering, Technology and Innovation.

Conflict of interest

The authors declare that they have no known competing financial or personal relationships that could have appeared to influence the work reported in this paper.

Author details

Zainab Muhammad Bello[1*], Sanusi Muhammad[1], Adamu Aliyu Aliero[1] and Ibrahim Aliyu Dabai[2]

1 Department of Biological Sciences, Usmanu Danfodiyo University, Sokoto, Sokoto, Nigeria

2 Department of Microbiology, Usmanu Danfodiyo University, Sokoto, Sokoto, Nigeria

*Address all correspondence to: bello.zainab@udusok.edu.ng

References

[1] Sinan KI, Mahomoodally MF, Eyupoglu EO, Etienne OK, Sadeer NB, Gunes AK, et al. HPLC-FRAP methodology and biological activities of different stem bark extracts of *Cajanus cajan* (L.) Millsp. Journal of Pharmaceutical and Biomedical Analysis. 2021;**192**:113678. DOI: 10.1016/j.jpba.2020.113678

[2] Padulosi S, Thompson J, Rudebjer P. Fighting Poverty, Hunger and Malnutrition with Neglected and Underutilized Species (NUS): Needs, Challenges and the Way Forward. Rome: Biodiversity International; 2013

[3] Popoola J, Omena O, Conrad O, Adegoke A. Neglected and underutilized legume crops: Improvement and future prospects. In: Farooq S, Zafar K, Amjad I, Metin T, Murat O, editors. Recent Advances in Grain Crops Research. IntechOpen; 2019

[4] Food and Agriculture Organization of the UN (FAO). The State of the World's Plant Genetic Resources for Food and Agriculture. Rome: Commission on Genetic Resources for Food and Agriculture, FAO; 1997

[5] Duodu KG, Franklin BA. African legumes: Nutritional and health-promoting attributes. In: Gluten-Free Ancient Grains: Woodhead Publishing Series in Food Science, Technology and Nutrition. 2017. pp. 223-269. Available from: http://dx.doi.org/10.1016/B978-0-08-100866-9.00009-1 [Accessed: 30 July 2017]

[6] Food and Agriculture Organization of the UN (FAO). Traditional Crops: Bambara Groundnuts. Rome: FAO; 2021

[7] Schippers RR, Besong MT. *Gnetum africanum* Welw [internet] record from PROTA4U. In: GJH G, Denton OA, editors. PROTA (Plant Resources of Tropical Africa/Ressources végétales de l'Afrique tropicale). Wageningen, Netherlands; 2004. Available from: http://www.prota4u.org/search.asp [Accessed: 17 October 2021]

[8] Mwangi S, Kimathi M. African leafy vegetables evolve from underutilized species to commercial cash crops. In: Research Workshop on Collective Action and Market Access for Smallholders. Cali, Colombia; 2-5 October 2006

[9] Chadha ML, Oluoch MO. Healthy diet gardening kit for better health and income. Acta Horticulturae. 2007; **752**:581-583

[10] Jordaan D, du PS, Akinnifesi FK, Ham C, Ajayi OC. The feasibility of small-scale indigenous fruit processing enterprises in Southern Africa. In: Akinnifesi FK, RRB L, Ajayi OC, Sileshi G, Tchoundjeu Z, Matakala P, Kwesiga FR, editors. Indigenous Fruit Trees in the Tropics: Domestication, Utilization and Commercialization. Nairobi, Kenya and Wallingford, UK: World Agroforestry Centre and CAB International Publishing; 2008. p. 273

[11] Febles EAV. *Canavalia ensiformis* (Jack Bean). Invasive Species Compendium. Wallingford, UK: CABI; 2017. DOI: 10.1079/ISC.14486.20203482789

[12] Mubaiwa J, Fogliano V, Chidewe C, Bakker EJ, Linnemann AR. Utilization of bambara groundnut (*Vigna subterranea* (L.) Verdc.) for sustainable food and nutrition security in semi-arid regions of Zimbabwe. PLoS One. 2018;**13**(10):e0204817. DOI: 10.1371/journal.pone.0204817

[13] Onuche U, Jimoh IS, Anthony T. Profitability and efficiency of Bambara groundnut production in Nigeria: A case study. Review of Agricultural and

Applied Economics. 2020;**23**(2):92-101. DOI: 10.15414/raae.2020.23.02.92-101

[14] Ruiz-Santiago RR, Ballina-Gómez HS, Ruiz-Sánchez E, Martínez-Castillo J, Garruña-Hernández R, Andueza-Noh RH. Determining relevant traits for selecting landrace accessions of *Phaseolus lunatus* L. for insect resistance. PeerJ. 2021;**9**:e12088. DOI: 10.7717/peerj.12088

[15] Chibeba AM, Pereira CS, Antunes JEL, et al. Polyphasic characterization of nitrogen-fixing and co-resident bacteria in nodules of *Phaseolus lunatus* inoculated with soils from Piauí State, Northeast Brazil. Symbiosis. 2020;**80**:279-292. DOI: 10.1007/s13199-020-00672-1

[16] Ayenan MAT, Ofori K, Ahoton LE, Danquah A. Pigeonpea [(*Cajanus cajan* (L.) Millsp.)] production system, farmers' preferred traits and implications for variety development and introduction in Benin. Agriculture & Food Security. 2020;**6**:48. DOI: 10.1186/s40066-017-0129-1

[17] Motis T. Factors to Consider when Selecting a Pigeon Pea Variety. ECHO's Seed Bank. ECHO Development Notes (EDN)|EDN. 2020 Issue #150. Available from: http://edn.link/gzxhf3

[18] Ismail MA, Amin MA, Eid AM, Hassan SED, Mahgoub HAM, Lashin I, et al. Comparative study between exogenously applied plant growth hormones versus metabolites of microbial endophytes as plant growth-promoting for *Phaseolus vulgaris* L. Cell. 2021;**10**:1059. DOI: 10.3390/cells100 51059

[19] Rojas W, Soto JL, Pinto M, Jäger M, Padulosi S, editors. Granos Andinos Avances, logros y experiencias desarrolladas en quinua, cañahua y amaranto en Bolivia. Rome: Bioversity International; 2010

[20] Glick BR. Plant growth-promoting bacteria: Mechanisms and applications. Scientifica. 2012;**2012**:1-15

[21] Finkel OM, Castrillo G, Paredes SH, Gonzalez IS, Dangl J. Understanding and exploiting plant beneficial microbes. Current Opinion in Plant Biology. 2017;**38**:155-163

[22] Hassan MK, McInroy JA, Kloepper JW. The interactions of rhizodeposits with plant growth-promoting rhizobacteria in the rhizosphere: A review. Agriculture. 2019;**9**:1-13

[23] Ahemad M, Kibret M. Mechanisms and applications of plant growth promoting rhizobacteria: Current perspective. Journal of King Saud University - Science. 2014;**26**:1-20

[24] Glick BR. Bacteria with ACC deaminase can promote plant growth and help to feed the world. Microbiological Research. 2014; **169**:30-39

[25] Ahemad M, Khan MS. Ecotoxicological assessment of pesticides towards the plant growth promoting activities of Lentil (*Lens esculentus*)-specific *Rhizobium* sp. strain MRL3. Ecotoxicology. 2011;**20**:661-669

[26] Ahemad M, Khan MS. Plant growth promoting fungicide tolerant *Rhizobium* improves growth and symbiotic characteristics of lentil (*Lens esculentus*) in fungicide-applied soil. Journal of Plant Growth Regulation. 2011; **30**:334-342

[27] Ahemad M, Khan MS. Ecological assessment of biotoxicity of pesticides towards plant growth promoting activities of pea (*Pisum sativum*)-specific *Rhizobium* sp. strain MRP1. Emirates Journal of Food and Agriculture. 2012;**24**:334-343

[28] Ahemad M, Khan MS. Effects of pesticides on plant growth promoting

traits of Mesorhizobium strain MRC4. Journal of the Saudi Society of Agricultural Sciences. 2012;**11**:63-71

[29] Ahemad M, Khan MS. Productivity of green gram in tebuconazole-stressed soil, by using a tolerant and plant growth promoting *Bradyrhizobium* sp. MRM6 strain. Acta Physiologiae Plantarum. 2012;**34**:245-254

[30] Goswami D, Thakker JN, Dhandhukia PC. Portraying mechanics of plant growth promoting rhizobacteria (PGPR): A review. Cogent Food & Agriculture. 2016;**2**:1-19

[31] Camerini S, Senatore B, Lonardo E, Imperlini E, Bianco C, Moschetti G, et al. Introduction of a novel pathway for IAA biosynthesis to rhizobia alters vetch root nodule development. Archives of Microbiology. 2008;**190**:67-77

[32] Gouda S, Rout GK, Gitishree D, Spiros P, Han-seung SJ. Revitalization of plant growth promoting rhizobacteria for sustainable development in agriculture. Microbiological Research. 2018;**206**:131-140

[33] Mehmood U, Inam-ul-Haq M, Saeed M, Altaf A, Azam F. A brief review on plant growth promoting rhizobacteria (PGPR): A key role in plant growth promotion. Plant Protection. 2018;**2**:77-82

[34] Sarkar A, Saha M, Meena V. Plant beneficial rhizospheric microbes (PBRMs): Prospects for increasing productivity and sustaining the resilience of soil fertility. In: Meena VS, editor. AgriSculturally Important Microbes for Sustainable Agriculture. Odisha, India: © Springer Nature Singapore Pte Ltd.; 2017. pp. 3-29

[35] Shah SV, Ramanan AS, Anil KS. Potential and prospect of plant growth promoting rhizobacteria in Lentil. In: Bhakta N, Singh AK, Sangale UR, Manibhusran SPK, Kumar S, Yasin JK, editors. Scientific Lentil Production. 2018. pp. 431-451

[36] Kumar A, Maurya BR, Raghuwanshi R, Meena VS, Singh DK, Dixit J. Does a plant growth promoting rhizobacteria enhance agricultural sustainability? Journal of Pure and Applied Microbiology. 2015;**9**:715-724

[37] Saha ND. Application of Plant Growth Promoting Rhizobacteria (PGPR) for Enhanced Root Growth. New Delhi, India: Pusa; 2015. Available from: www.krishsewa.com [Accessed: 1 December 2015]

[38] Dutta J, Handique PJ, Thakur D. Assessment of culturable tea rhizobacteria isolated from tea estates of Assam, India for growth promotion in commercial tea cultivars. Frontiers in Microbiology. 2015;**6**:1-13

[39] Goswami D, Pithwa S, Dhandhukia P, Thakker JN. Delineating *Kocuria turferensis* 2M4 as a credible PGPR: A novel IAA-producing bacteria isolated from Saline Desert. Journal of Plant Interactions. 2014;**9**:566-576

[40] Shafi J, Tian H, Ji M. Bacillus species as versatile weapons for plant pathogens: A review. Biotechnology and Biotechnological Equipment. 2017;**31**:446-459

[41] Kachroo P, Kachroo A. Chapter 7: The Roles of Salicylic Acid and Jasmonic Acid in Plant Immunity. Lexington, KY, USA: Department of Plant Pathology, University of Kentucky; 2019. pp. 1-74. Available from: https://www.researchgate.net/publication/277705981

[42] Aggarwal A, Kadian N, Tanwar A, Yadav A, Gupta KK. Role of arbuscular mycorrhizal fungi (AMF) in global sustainable development. Journal of Applied and Natural Science. 2011;**3**(2):340-351

[43] Dubey RK, Tripathi V, Edrisi AS, Bakshi M, Pradeep KD, Singh A, et al.

Role of plant growth-promoting microorganisms in sustainable agriculture and environmental remediation. In: Singh HB, Kumar Sarma B, Keswani C, editors. Advances in PGPR Research. CAB International; 2017. pp. 75-125. DOI: 10.1079/9781786390325.0075

[44] Singh A, Jain A, Sarma BK, Upadhyay RS, Singh HB. Beneficial compatible microbes enhance antioxidants in chickpea edible parts through synergistic interactions. LWT- Food Science and Technology. 2014;**56**:390-397

[45] Nadeem SM, Ahmad M, Zahir ZA, Javaid A, Ashraf M. The role of mycorrhizae and plant growth promoting rhizobacteria (PGPR) in improving crop productivity under stressful environments. Biotechnology Advances. 2014;**32**:429-448

[46] Bisen K, Keswani C, Mishra S, Saxena A, Rakshit A, Singh HB. Unrealized potential of seed biopriming for versatile agriculture. In: Rakshit A, Singh HB, Sen A, editors. Nutrient Use Efficiency: From Basics to Advances. New Delhi: Springer; 2015. pp. 193-206

[47] Jain A, Singh A, Chaudhary A, Singh S, Singh HB. Modulation of nutritional and antioxidant potential of seeds and pericarp of pea pods treated with microbial consortium. Food Research International. 2014;**64**:275-282

[48] Singh DP, Prabha R, Yandigeri MS, Arora DK. Cyanobacteria-mediated phenylpropanoids and phytohormones in rice (*Oryza sativa*) enhance plant growth and stress tolerance. Antonie Van Leeuwenhoek. 2011;**100**:557-568

[49] Mishra S, Singh A, Keswani C, Saxena A, Sarma BK, Singh HB. Harnessing plant–microbe interactions for enhanced protection against phytopathogens. In: Arora NK, editor. Plant Microbes Symbiosis: Applied Facets. New Delhi, India: Springer; 2015. pp. 111-125

[50] Marx J. The roots of plant–microbe collaborations. Science. 2004;**304**: 234-236

[51] Pan H, Stonoha-Arther C, Wang D. Medicago plants control nodulation by regulating proteolysis of the receptor-like kinase DMI2. Plant Physiology. 2018;**177**:792-802

[52] Rathi MS, Paul S, Thakur JK. Response of wheat to inoculation with mycorrhizae alone and combined with selected rhizobacteria including *Flavobacterium* sp. as a potential bioinoculant. Journal of Plant Nutrition. 2014;**37**:76-86

[53] Lazcano C, Barrios-Masias FH, Jackson LE. Arbuscular mycorrhizal effects on plant water relations and soil greenhouse gas emissions under changing moisture regimes. Soil Biology and Biochemistry. 2014;**74**:184-192

[54] Smith SE, Read DJ. Mycorrhizal Symbiosis. London, UK: Academic Press; 2008

[55] Berta G, Copetta A, Gamalero E, Bona E, Cesaro P, Scarafoni A, et al. Maize development and grain quality are differentially affected by mycorrhizal fungi and a growth-promoting pseudomonad in the field. Mycorrhiza. 2014;**24**:161-170

[56] Abd-Alla MH, El-Enany AWE, Nafady NA, Khalaf DM, Morsy FM. Synergistic interaction of *Rhizobium leguminosarum* bv. viciae and arbuscular mycorrhizal fungi as a plant growth promoting biofertilizers for Faba bean (*Vicia faba* L.) in alkaline soil. Microbiological Research. 2014; **169**:49-58

[57] Keswani C. Proteomics studies of thermo-tolerant strain of *Trichoderma* spp. [Ph.D. thesis]. Varanasi, India: Banaras Hindu University; 2015. p. 126

[58] Takoutsing B, Weber J, Aynekulu E, Martín JAR, Shepherd K, Sila A, et al. Assessment of soil health indicators for sustainable production of maize in smallholder farming systems in the highlands of Cameroon. Geoderma. 2016;**276**:64-73. DOI: 10.1016/j.geoderma.2016.04.027

[59] Semenov AM, Sokolov MS, Glinushkin AP, Glazko VI. The health of soil ecosystem as self-maintenance and sustainable bioproductivity review article. Acta Phytopathologica et Entomologica Hungarica. 2017;**52**(1):69-82. DOI: 10.1556/038.52.2017.004

[60] Fasusi OA, Cruz C, Babalola OO. Agricultural sustainability: Microbial biofertilizers in rhizosphere management. Agriculture. 2021;**11**:163. DOI: 10.3390/agriculture11020163

[61] Han SH, Young AJ, Hwang J, Kim SB, Byung BP. The effects of organic manure and chemical fertilizer on the growth and nutrient concentrations of yellow poplar (*Liriodendron tulipifera* Lin.) in a nursery system. Forest Science and Technology. 2016;**12**(3):137-143. DOI: 10.1080/21580103.2015.1135827

Chapter 15

Function of Urease in Plants with Reference to Legumes: A Review

Peter S. Joseph, Dickson A. Musa, Evans C. Egwim and A. Uthman

Abstract

Urease (urea amidohydrolase, EC 3.5.1.5) is a nickel-containing enzyme produced by plants, fungi, and bacteria that catalyzes the hydrolysis of urea into ammonia and carbamate. Plant (especially legumes) ureases hold a special place in science history, participating on some important landmarks of biochemistry as it was the first enzyme ever to be crystallized in 1926. Finding nickel in urease's active site in 1975 was the first indication of a biological role for this metal. Despite the abundance of urease in tissues and seeds of some members of Legumes families, and its ubiquity in virtually all plants little has been revealed of the roles of urease. This review will explore many faces of these ureases from legumes and other plants, their roles, nutritional relationship between plants and the commensal bacteria with which they associate. In addition, we will explore the possibility that bacteria participate in turnover of the "plant" urea pool. Plant ureases possess insecticidal and fungitoxic properties independent of its ureolytic activity. Altogether, with this review we wanted to invite the readers to take a second look at ureases from versatile plants especially legumes for various biotechnological applications.

Keywords: legumes, plant, urease, urea

1. Introduction

Ureases (urea amidohydrolase, EC 3.5.1.5) are ubiquitous enzymes produced by plants, bacteria and fungi, animals do not produce these metalloenzymes. They are found to be the most proficient known enzymes to date, the enzyme catalyzes the hydrolysis of urea to form carbamate and ammonia; the carbamate then decomposes to form carbon dioxide and another molecule of ammonia, enabling the reaction rate to be faster by at least a factor of 10^{14} when it is compared to the decomposition of urea by elimination reaction [1–4]. The proficiency of urease Computational modeling brought about a proposal of a value equivalent to 10^{32} multiplied by the theoretical rate of uncatalyzed hydrolysis of urea [5]. But in solution, it can be debated upon that the value obtained is not visible based on some limitation imposed due to the substrate diffusion in water. Ureases from Plants maintain a special position in the history of science, involving in some relevant events in biochemistry. For example, Urease contributed about three landmarks in the history of Biochemistry. One, the *Canavalia ensiformis* (jack bean) seeds urease isolated and crystallized by a scientist called James B. Sumner, who in 1926 demonstrated the enzymes' proteinaceous nature [6], this findings in 1946 was laureated with the Nobel Prize in Chemistry. Two, the biological importance of nickel (N^{2+}) was in 1975 recognized as obligatory for urease' catalytic activity after studies conducted by Zerner's and colleagues who reveal the existence of nickel ions in jack bean urease' active site [7]. Three, the identification of a toxin in plant as a urease in the year 2001 may be regarded as another breakthrough that has to do with ureases, which brought about the discovery of properties of these enzymes that are non-catalytic [8]. This discovery increased our knowledge on the functions carried out by these proteins, asides their function in metabolism of nitrogen [9]. This urease enzymes belongs to the super family of phosphotriesterases and amidohydrolases, that possesses in their active sight two catalytic Nickel metal(s) with a few reported exceptions [8, 10]. Takeuchi's findings as stated in Real-Guerra et al. make all this possible after he observed that *Glycine max* (soybean) seeds' crude extracts shows high amounts of urease activity [11]. As at that time, research on urease only focus on microorganisms and in algae. Takeuchi's discovery was the first report that shows the existence of ureases in higher plants. The advantage of this finding is that it gave researchers knowledge on the large availability of urease globally, thereafter, so many other researches on ureases were carried out, taking advantage of the leguminous plant urease having the aim to understand functions of the enzyme. Urease from *G. max* was among the main focus on enzymology development, involving several researches associated with this enzyme which brought about hypothesis that were important to Michaelis and Menten's observation on the reaction rate of enzymes and the substrates they catalyzed [12]. Till date, after those first experiments more than a century ago, urease from legumes continued to generate attention by researchers globally, in various fields such as biochemistry, physiology and genetic.

A leguminous plant belongs to the family Leguminosae otherwise known as Fabaceae. They produce their seeds around a pod [13, 14]. This plant family is large with more than 18,000 species shrubs, climbers, trees and herbs whereby only few has been studied for urease extraction and utilization. Common legumes that have been used for the extraction of urease include mung bean, *Pisumsativum* seeds, peas, some beans species, peanuts, lentils, soybeans, upins, lotus, green beans and sprouts are known as food or grain legumes. Different legumes are shown in **Figure 1**.

Figure 1.
Some species of legumes.

In this review, we shall focus on ureases from legumes, providing information generated over time and exposes some areas that need to be focused by researchers.

2. Metabolic origins of urea in plants

Mammals synthesize urea via the Krebs-Henseleit cycle (also called the arginine, ornithine, or urea cycle) as nontoxic form of jettisoned ammonia [15]. Plants usually have the opposite problem, i.e., how to conserve nitrogen, which after carbon, is the most limiting element in plant nutrition [16]. This contention is consistent with the presence of urease in plants and in most bacteria and fungi [17] and its absence in mammals. Whereas in the latter, urea is a nontoxic "waste" form of ammonia in excess, in the former, ureolytic activity is necessary to recycle urea nitrogen (urea is 47% nitrogen). We discuss in this section the metabolic and tissue origins of plant urea.

2.1 The precursor of urea

Arginine, by the action of arginase, is the immediate precursor of urea in the mammalian arginine (urea) cycle. Although an active functional arginine cycle in plants has been debated [18], it is clear that arginase (EC 3.5.3.1) is widely spread in the plant world. Plant arginases resemble the animal enzyme in their high pH optima (approximately 9.7) and Mn^{2+} requirement [18–22]. As arginase is widely spread, so is its substrate abundant: arginine is a main nitrogen transport and plants' storage compound. It is the nitrogen main transport compound of deciduous [23] and coniferous [24] trees and a major component of underground storage organs (bulbs, roots, tubers [25]). It was shown to be among the amino acids in the seeds of 379 angiosperms that is predominant [26]. It was recalculated (in mol %) the reported average seed amino acid composition of these 379 species. Arginine accounted for 7.7% of seed amino acids and its "N-weighted" contribution was 21.1% of total amino acid nitrogen, the highest contribution of any amino acid, with glutamine a close second (18.6% based on the assumption that half of the glutamate in protein hydrolyses came from glutamine) [26]. In Glycine max, arginine contains 18% of seed protein-bound nitrogen [27]. At least 50% of free amino acid N in developing seeds of pea [28] and soybean [27] is in the arginine pool. In addition, many legume seeds are exceedingly rich in free canavanine [29], an arginine analog. Half of arginine nitrogen (and two-thirds of canavanine nitrogen), in the guanidino moiety, is convertible to urea by arginase kanavinase action.

3. Repercussion of urease activity elimination in plants

In higher plants it appears that urea can be assimilated only by urease action. Urease-negative plants and cultures, induced genetically [30] with urease inhibitor [31, 32] or by nickel deprivation [33–36], have been observed either to accumulate urea or to be blocked in the ability to employ urea as a nitrogen source. All plant [37] and bacterial [3] ureases are probably nickel metalloenzymes. Seed ureases from jackbean [7, 38] and soybean [39] have been shown to contain nickel. Duckweed plants [40] and callus of soybean, rice, and tobacco [34, 37] are dependent on nickel for maximal growth with urea as sole nitrogen source. Urease appears to be the only nickel-requiring enzyme in plants since, as indicated below, nickel-deprived soybean plants have the same phenotype as those genetically blocked in urease synthesis [41]. Thus, higher plants appear to lack the ATP-dependent urea amidohydrolase reported in algae and yeast [42–45]. This biotin-containing carboxylase/hydrolase appears not to be a nickel metalloenzyme and has a urea-assimilatory function (e.g., [46] in these urease-negative lower eukaryotes. In an interesting example of the potential of urease to provide nitrogen for the plant, [47] developed transgenic tobacco plants engineered for resistance to cyanamide. The resistance gene, from soil fungus *Myrothecium verrucaria,* codes a cyanamide hydratase that converts cyanamide to urea. Although urea levels are raised in such plants, the endogenous urease apparently hydrolyzes much of the liberated urea.

3.1 Effect urease on protein deposition and embryo development

Given that urease is the plant's only means of assimilating urea, the next question is the metabolic version of "If you are so smart how come you are not rich?" When applied to urease it would read, "If you're so important how come the plant survives without you?" Indeed, completely urease negative mutant soybean plants develop to maturity and produce a relatively good yield of seeds that germinate at normal frequency to propagate another generation. However, if the role of urease is to recycle urea nitrogen generated from arginine (and possibly ureide) degradation, then a protein-rich plant such as soybean may provide a "suppressing" background for a urease defect. Soybean has indeed been intensively bred for large and protein-rich seeds. However, even in soybean we question the dispensability of urease. Urease-negative mutant plants [41] and nickel-deprived wild type [35, 36] exhibit necrotic leaf tips, apparently due tourea "burn." Similar observations were made in nickel-deprived tomato [48, 49]. More important, perhaps, than the deleterious effects of leaf burn is the loss of significant quantities of nitrogen in a urea dead-end, lost nitrogen that could have a significant negative impact on seed protein deposition during pod fill. The assessment of the agronomic impact of a urease-negative phenotype on soybean performance requires extensive field testing of isogenic or nearly isogenic paired urease-positive and urease-negative lines, preferably in multiple environments. To obtain these lines we are currently engaged in the long process of backcrossing, generation advance, selection for uniformity in maturity group and plant architecture, etc., and amplification of seed stocks [50]. Our prediction that total seed protein of plant will decrease as suggested by [36], who reported a correlation between seed yield and nickel content per seed. However, there was too much variation in their material to obtain a statistically significant difference. It was observed that at the time of flowering completely urease-negative soybean mutants accumulated approximately 100 dry wt of total leaf (green plus necrotic tip) [41]. Assuming that 10% of the leaf dry weight is protein (16% nitrogen), much of which is destined to provide amino acids during pod fill, accumulated urea (47% nitrogen) represents 18% of the nitrogen in leaf protein. The developing soybean embryo does not generate urea [41]. Thus, other than recycling maternally derived urea, urease appears to play no direct role in embryo metabolism. However, it is possible that in monocots a urease role is more critical in embryo development. It has been reported that nickel is essential for development of viable barley embryos [51]. Post dormancy grains could not be rescued by nickel, whereas developing grains could be rescued by the feeding of nickel to the maternal plant. The role of nickel in barley embryo development is not known and may be unrelated to urease. (Urease is the only nickel metalloenzyme yet identified in plants). However, it is possible that a loss of urease activity under nickel-deprivation conditions leads either to urea poisoning or to nitrogen starvation of the embyro. It would be informative to study barley embryo development *in vitro,* where both nickel availability and nitrogen source can be manipulated.

3.2 Effects of urease on germination

What is the consequence of blocking urease action during germination when large of amounts of arginine and ureides are mobilized and generate urea? In soybean we observed a 7–8 h delay in germination of its protein-rich seeds imbibed in the potent urease inhibitor [52–54] phenylphosphorodiamidate (PPD) [55]. Protein-poor *Arabidopsis thaliana* seeds imbibed in water with 50 PPD did not germinate at all. That PPD acts specifically on urease and brings about inhibition of germination by nitrogen limitation is indicated by parallel dose-response curves for germination and urease inhibition, and by reversal of inhibition of germination with added nitrogen sources, NH, NO, (5 mM), or casein hydrolysate (1 mg/ml) [55]. Similar respose was observed for a cyclotriphosphazatriene urease inhibitor [56]. Within *A. thaliana*, large seeds germinated in distilled water tend to produce seedlings that survive longer than those from small seeds [57]. Phenyl phosphorodiamidate may effectively reduce seed size by depriving the seedling of that portion of its nitrogenous reserves catabolized to urea. Hydrolyzed *A. thaliana* seed meal contains 5.5 g arginine per 16 g total nitrogen [26]. Thus, 1.76 g or 11% of seed nitrogen (free plus that bound in protein and nucleic acids) is in arginine and, potentially, 5.5% (or more, if arginine is proportionally higher in "reserves") seed nitrogen can be converted to urea by arginase. Seed nitrogen in nucleic acids is another generator of urea; greater than 40% of urea generated in 6-day *Arabidopsis* seedlings is eliminated by allopurinol [55]. In other species, e.g., *Lupinus texensis* [58] and wild radish [59], larger seeds tend to germinate at a higher frequency than small seeds. We have observed a crude correlation, across species, between seed size and resistance to PPD inhibition of germination and will extend studies of relative PPD sensitivity to large seeded soybeans vs. selected small-seeded sister lines [60], when these lines are made available to the public.

3.3 Loss of chemical protection

As described in the next section, *G. max* contains two different isozymes of urease: the enzyme which is ubiquitous is made in all the examined tissues, more so, embryo-specific synthesis of urease is dependent to the embryo that is developing and it is conserved in the seed that is mature where its specific activity is about 1000-fold higher than that of the ubiquitous urease in any tissue [61, 62]. Since mutants that lacks the embryo-specific urease do not display any of the abnormalities related with loss of the ubiquitous enzyme necrotic leaf tips, accumulation of urea in leaves or seeds, retarded germination [41] it was concluded that this enzyme has no important physiological use. *In vitro* culture of developing cotyledons of pea [63] and *G. max* [41, 64] indicates that ureases play little or no role in embryo nutrition since urea was an extremely poor source of nitrogen. Indeed, urea is not normally generated within the developing *glycine max* cotyledon *in vivo* [41], in agreement with the lack of ureide delivery to the legume embryo from maternal tissues [65, 66]. The obvious question from the observations of the previous paragraph is why would the developing soybean embryo and those of jackbean, watermelon, and many other members of Leguminosae and Cucurbitaceae [67] invest in a very active ureolytic activity when it never "sees" urea. Although much urea is generated upon germination [41] and although much of the embryo-specific urease is retained in seedling cotyledons and roots [68], the loss of the embryo-specific urease causes no discernible increase in seedling urea levels over those of wild type [41]. It is suggested, therefore, that the embryo-specific urease plays no urea assimilatory role but rather that of chemical defense in seeds. To draw two parallels, at least, with the pathogenic effects of bacterial urease on vertebrates, active seed urease could cause either hepatic coma by subversion of the urea cycle or peptic ulceration by localized increases in NH^+ and OH-ions (urea + $3H_2O$ + $2NH_4^+$ + HCO^- + OH^-). A micro aerophilic, bacterium, *Helicobacter pylori,* can colonize gastric epithelium because its active urease creates a more basic micro environment in this acidic milieu [69, 70]. Arguments summarized by [71] link urease from bacteria to ulceration of the gastric mucosa either by cytotoxicity of ammonia directly or by its prevention of proton flux from gastric glands to the gastric lumen leading in a back-diffusion of protons. Hepatic coma results when intestinally derived nitrogenous compounds, e.g., ammonia, bypass the liver and get to the brain. Administration of urease inhibitors has proven effective in reducing hyperammonemia [72, 73]. It is easy to visualize an active seed urease mimicking these bacterial effects (the bacterial and plant seed ureases have >50% amino acid identity) [74], especially urease aided by other cytotoxic components in the seed i.e. protease inhibitors, lectins that disrupt intestinal brush borders [75], etc. Another postulated chemical defense for seed urease is its induction of a hostile environment upon microbial and perhaps insect attack. With this second model, wounding or infection of the immature embryo will lead to the release of arginase from mitochondria that is ruptured. Cytoplasmic arginase would generate urea from the large pool of arginine which is at least 50% of free amino acid nitrogen [27, 28] and cytoplasmic urease would rapidly convert urea to ammonia. It was observed that cultured cotyledons containing the embryo specific urease commit suicide (probably by the combined effects of ammonia toxicity and medium alkalinization) in the presence of urea (20 mM), whereas those containing only the ubiquitous urease isozyme survive and utilize this urea, albeit poorly [41]. The assessment of pest resistance and herbivore avoidance of soybean cultivars lacking the embryo-specific urease requires extensive field testing of isogenic lines exposed to a variety of pathogens and pests. In addition to the abundant storage proteins, seeds contain several moderate- to high-abundance proteins with enzymatic or other biological activity. It is easy to invoke a plant protection role for many of these proteins: phytohemagglutinins (lectins) [75], lipoxygenases [76], ribosomal inactivators [77] and inhibitors of amylase [78], and animal proteases [79]. The lack of an essential physiological role for many of these proteins in *plant is* suggested by the relatively high frequency of cultivars and varieties lacking one of the urease [61], lipoxygenases [80–82], etc. Lack of an essential physiological role for the seed (embryo specific) urease is further suggested by the high prevalence of seed urease nulls in Japanese populations of *Glycine soja* [83], a sexually compatible close relative of cultivated soybean (*Glycine max*).

4. Physiology of leguminous soybean ureases and their functions

Carbon followed by nitrogen is the major limiting element for the performance of plant [84], and a regular demand for utilization of nitrogen targeting the development of mechanism that are efficient for the uptake of nitrogen and metabolic pathways for remobilization if nitrogen in plants [85, 86]. A pressure like this eventually led to nitrogen content reduction of proteins from plants [87]. Since urea is known to be a major source of nitrogen in plants, arginase' activity is found to be the only known pathway involved in the *in vivo* generation of urea; it may be generated also due to ureides and purines degradation [88], even though this pathway happens to attract so much debate. The only way urea can be assimilated is after it has been hydrolyzed by urease into carbon dioxide and ammonia [89], this happens to be the major physiological function attributed to plant ureases [85]. Re-assimilation of ammonia will then be performed by an enzyme called glutamine synthetase utilizing glutamate as substrate [64]. The activity of urease is found in almost all species of plant and is found to be ubiquitous in all tissues of plants [62, 68, 90], this shows the great advantage of its physiological function to the entire plant. For ages, the importance of urease-mediated metabolism of urea in plant was not regarded because, there was an assumption that plants have difficulty in taking up urea, otherwise they are broken down by soil microorganisms leading to the absorption of nitrate and ammonia. Presently it has been established that plant can absorb urea from the soil actively, via the activity of a committed transporters urea [91] and can also process urea imported from the soil efficiently, even though the concentrations is high. These discoveries focuses at urease as an area for research on improving plant nitrogen metabolism urea that is urea based, the widely utilized fertilizer globally [92]. *G. max* happens to be an exciting model for research based on the physiological function of urease in import plants, because it has so far become the only plant that the genome have been sequenced which shows above one isoform of the enzyme. Soy bean urease ub-SBU has been identified to be the isoform accountable for recycling every urea that is derived metabolically [93–95]. This has been shown because mutants with noes-SBU never assemble urea in any tissue and do not have any deterioration on urea utilization as the major source of nitrogen, although ub-SBU is available at levels only 0.1–0.3% that of es-SBU [61, 93]. *G. max* mutants that have no activity of ub-SBU show a distinguishing feature phenotype consisting of necrosis of the tips of leaf, because of urea "burn", and accumulation of urea in many tissues [41, 96]. Urease is the only Nickel dependent enzyme discovered in plants and the same phenotype as mentioned earlier, has been studied for plants grown under the deprivation of Nickel [35].

It is interesting to note that there is no physiological function, either assimilatory or that of any other characteristics, could possibly be seen due to bountiful es-SBU. In reality, types of cultured cotyledons that are wild were not able to grow due to urea presence, because of a sudden increase in pH as a result released ammonia that was not controlled. A similar effect was not seen on mutants having ub-SBU only [41]. It was concluded that es-SBU could possibly be associated in protection against predators in plant. For the case of microbes or attack of insect, protection of chemical was postulated. Using a model like this, immature embryo that is infected would bring about the release of arginase release from mitochondria that is ruptured which will generate urea in abundance from the pool of arginine and urease from the cytoplasm would transform urea to ammonia rapidly [96]. This hypothesis still waiting for demonstration to be carried out, meanwhile a report has been published stating that mutants that lack the activity of urease were highly susceptible to be infected by microbes [97].

5. Biotechnological potential functions of urease from legumes

Many questions on why ureases are ubiquitous and multimeric has been asked, a possible answer is that the "earliest" enzyme may have gotten other form of "traits" during the evolutionary pressure of a biosphere that is complex which led to increase in competition [97]. Due to these "extra traits" findings on ureases, some applications of biotechnology can be suggested. Legumes such as *G. max* may be targeted and attacked by several organisms, which may include virus, fungi, nematodes, and insects. These pests and pathogens may cause serious damage in pods, seeds, stems, roots and leaves, and most times are specific to tissues [98]. Even though measures have been taken for control, pests decrease the *G. max* production globally by about 28% [99], there is need for urgent development of new technology for the control of these pests; also exploring natural compounds of plant is an important strategy. There are so many biotechnological potentials in respect to plant ureases and their derived peptides. Many sources that are edible are found to be in abundant in ureases, which include legumes and potatoes, they are even eaten in their raw form e.g. cucumbers, or some fruits like watermelon and melon [97]. Hence, issues of bio-safety could possibly be managed better since es-SBU, JBU and Jaburetox' derived peptide seems to be non-toxic to mammalians [100, 101], the fungi toxic and entomotoxic characteristics of these molecules are very important when putting into consideration the strategies biotechnology which aims to provide security to crops that are relevant commercially against natural enemies. The proof of an *in vivo* effect of *G. max* urease in protecting the plant against fungi [102] are creating excitement. The chances of choosing *G. max* cultivars having higher content of urease or to increase the production of these proteins found in the plant via genetic manipulations, so as to improve the resistance to fungi and insects, is encouraging. In addition, the premise of utilizing naturally occurring proteins in plant to increase resistance seems to be more appealing to the public more than the option of foreign genes insertion (from animals or microorganisms) into crops. As stated some time back that *G. max* having a high ureases content could be also agronomically valuable, not regarding the defense function, for allowing more thorough absorption of fertilizer containing urea by the plant [33, 39]. More so, considering the fact that *G. max* meal is widely used as feed for animals and the potential of becoming a source of protein to humans, a higher content of urease in *G. max* could be interesting for increasing nutritional quality of *G. max*, after processing it appropriately, because urease has higher methionine content more than many other *G. max* seed proteins. *G. max* has a very low amount of amino acids containing sulfur, about half of which are regarded as best as feed for animal. Even though this issue can be dealt with by supplementing the feed with free methionine, the supplementation are associated with problems, including leaching of the methionine when processing the degradation of bacteria leading to formation of volatile sulfides that are not desirable [103]. Improving methionine content in *G. max* by increasing the biosynthesis of endogenous proteins, like ureases, is an approach that is so interesting.

6. Some effects of legumes containing urease used as animal feed

It has been estimated that *G. max* meal accounts for about 67% of the total sources of protein utilized in feeds of animal around the globe [104], because of majorly its high concentration of protein which is around 44–48% [105]. Nonetheless, *G. max* possesses an unusual high amounts of bioactive compounds having antinutritional or/and properties that are toxic, that possesses on the body of animals' metabolism a negative effect [106]. The content of urease was not different evidently among 11 tested soybean cultivars [107]. Contrarily, the content of urease was found to vary among several other soybean cultivars [108–110], and the urease levels positively correlated with antinutritional effects in rats [109]. The negative effects of utilizing meals containing urease as feed for animals are reported. Urea is regularly mixed with the feed of animals and, when *G. max* that are not processed are mixed with urea, the activity of urease will enabled the release of ammonia, which is an effect that is not desired in a feed that are mixed [111]. In ruminant animals, ammonia quickly enters the blood which can cause affects that is adverse ranging from depressed feed intake and performance of the animal, to Mortality due to toxicity of ammonia [112]. In dairy cows, the liver whose responsibility is remove potentially toxic ammonia from circulation, was able to remove ammonia added to portal blood until the supply was up to 182 mg/min but, at infusion rates that is high, peripheral blood concentrations of ammonia increased, supporting the assessment that a rapid hydrolysis of dietary urea can rise above the capacity of the liver to remove it [113]. In chickens, it was shown that meals of *G. max* from a particular source regularly produced a high incidence of tibialdyschondroplasia (TD) and the major striking difference between the meals was the high antitrypsin and values of urease in those that induced the disease [114]. TD incidence was previously shown in broilers to increase when 1.5 or 30% ammonium chloride was added to the feed [115], meanwhile it was not shown when calcium chloride was added [116]. These could indicate that the ammonia released by urease may a function in TD incidence on meals of *G. max* fed on chickens. For addition of nitrogen supplement to be allowed in the feed of animals, so as to protect the animals from the production of high ammonia level which is toxic, there is need for pre-treatment of the *G. max* meal. The major method used is the treatment by heat to do away or reduce the anti-nutritional effects and/or factors that are toxic in *G. max*, which include urease [103, 117], meanwhile these treatments should be kept at a low level, because it may be possible to destroy some important constituents of the seed [106]. To do away with the activity of urease, many effective treatments are required, which includes steam-heating at 120°C for 7.5 min or at102°C for 40 min [103], boiling for 60 min. at 92°C [109], and dry-heating at 100°C and 2 kgf/cm^2 for 60 min [117]. All these treatments do away with the activity of urease, coupled with a reduction of many anti-nutritional factors. The best method to evaluate the processing adequacy and final quality of the *G. max* meals is by conducting biological tests. Nevertheless, the required time, complexity and cost of these tests have negative effect their use. From 1940s, the use of urease test has been in existence as an indirect method of evaluating the adequacy processing *G. max* using heat because it is fast, require minimum skill and minimum amount of laboratory equipment. A discovery revealed a correlation that is high among trypsin inhibitors activities, lectins and urease, which indicate that the *G. max* processing can be estimated adequately by these analytical criteria to a considerable extent. In the past years, several guidelines were developed to facilitate the measurement of urease activity [103]. These guidelines are used to quantify directly or indirectly the amount of ammonia released. The one that was first developed is the method by Caskey-Knapp [118] in a buffer solution the meal is incubated with urea and then the addition of phenol red. Meals that are not properly processed will lead to an increase in the pH of the solution after incubation, which will be observed by a change in color of the solution from red-orange to pink, but meals that are well processed will show a little or no color change. A study suggested another alternative method, having the potential to distinguish meals with low levels of the activity urease, based on the meal incubation with urea in a buffered solution and using colorimeter to determine the urea residual with *p*dimethylaminobenzaldehyde [119]. Also, a method was propose d for direct titration of ammonia as a measure of urease activity [120] and adapted [121], whereby urea incubation with the meal is carried out, maintaining the pH of the solution by adding HCl slowly. The system is therefore titrated with NaOH. The difference in titration between a urease inactivated (control) and the sample is considered as the urease activity of the meal. Other two methods were developed for determination of ammonia directly, which is based on phenol-hypochlorite reaction [122, 123]. So many adaptations and modifications of these methods were developed as time passes by. Meanwhile, regardless of the method used, the activity of urease is a very good indicator of *G. max* meals that are not properly processed. This method could also be applied to other meals suspected to contain urease. However, this activity is not a good indicator *G. max* meals that are over processed.

7. Mechanism of action for insecticidal property in leguminous plant

Ureases from plants such as *G. max* (soy bean) and jack bean (JBU and *canatoxin*) are discovered to exhibit insecticidal activity in insects with cathepsin B-based (cysteine protease) and cathepsin D-based (aspartyl protease) digestive system. But no effect was observed in insects that relay on digestive enzymes which are trypsin-like [124, 125]. It is interesting to note that, no change was seen in the insecticidal effects of *canatoxin* and *G. max* ureases was after the enzyme was treated with an irreversible inhibitor of ureolytic activity, which indicate that ureolytic and insecticidal activities are not related [100]. In canatoxin, it was shown that its entomotoxic effect depends on an internal 10-kDa peptide (called pepcanatox), which is released by canatoxinhydrolysis by cathepsins in the digestive system of insects that are susceptible. Nonetheless, a 13-kDa recombinant peptide named jaburetox-2Ec, analog to pepcanatox, was cloned in *Escherichia coli* and was discovered to have insecticidal activity [126]. But with ureases from plant, there was no insecticidal activity seen in bacterial (*Bacillus pasteurii*) urease [100] and it was proposed that the entomotoxic peptide released from canatoxin by insect cathepsins is not present in ureases from microbes due to the subunit structure which is made up of two or three chains [100] and peptides that links and connect the different chains exhibit the insecticidal activity. Canatoxins has been studied extensively in several applications including industrial, agricultural, and medical [127]. Insect pest are the main cause of damage in agriculture having negative effect on commercially important crops globally. Having transgenic crops with intrinsic resistance to pest propose an alternative that is promising for chemical pesticides to prevent the crop losses [128]. Since plant ureases exhibit insecticidal property, to know their three-dimensional structure and the structural basis of their mechanism of its endomotoxic peptide could be used effectively for the development of transgenic plants that are resistant to insects. From the industrial point of view, a large number of research papers have been published four decades ago in legumes urease immobilization studies. Legumes urease structural integrity and important resistance to chemical and thermal deactivation have been studied extensively for immobilization studies. Based on these facts, the present study reveals some details of leguminous canatoxin urease, which may be utilized by researchers to design better carriers for enzyme immobilization and areas by utilizing these plants.

8. Summary and prospects for research on urease from legumes

In summary there are several indications that urease is important for efficient nitrogen assimilation. The urease substrate urea is derived from ureides and arginine. Arginine, is the richest nitrogen repository among the amino acids of legumes and plant seeds storage proteins. Urease are significant during the fixation of nitrogen in "tropical" legumes for example soybean and other plants. Urease-negative plants accumulate substantial, non-utilizable urea in both maternal and embryonic tissue. During germination of urease-negative seeds, further urea accumulates as a dead end in nitrogen metabolism. Although this accumulation may not be a lethal defect for large protein-rich legume seeds like soybean, small or protein-poor seeds, such as *Arubidopsis*, may be severely retarded or blocked in germination by the lack of an active urease. The better known, abundant legume seed ureases, for example, canatoxin urease, may perform a chemical defense function, similar to those postulated for some ureases of pathogenic bacteria. The ureases characterized to date, both from plants and bacteria, resemble each other in primary structure and in their requirement for accessory genes. Although the functions of the accessory genes have not yet been fully characterized they are undoubtedly involved in insertion of a nickel cofactor at the urease active site. A study of urease has revealed what may be a universal phenomenon in plant biochemistry: that enzyme profiles may be a composite of both plant and microbial activities. Indeed, the microbial contributions may be physiologically significant to the plant; in legume such as soybean it has been observed that bacteria are responsible for generating urea from ureides and mitigating urea accumulation in some urease negative mutants. In the future, there is a real need to quantify nitrogen fluxes through urea and to assess in the field and greenhouse the effects of urease inactivation on germination, seedling vigor, protein deposition, etc. Such studies are best performed in isogenic plant lines differing in the presence or absence of urease. Leguminous plants such as Soybean is so far the best experimental subject in light of its battery of urease mutants, its importance as a protein crop, and knowledge of its physiology. The ability to eliminate the abundant embryo specific urease, exclusively, will allow us to focus on its possible defense roles, especially when the seed urease-negative trait is combined with those eliminating lectins [82], protease inhibitors [129], and other putative defense factors. Obviously, improved understanding of the extent of plant-bacterium interdependence requires that we successfully cure plants of the bacterium. It is not clear whether this has been achieved, even in plants regenerated from cell culture. Finally, we need the knowledge and understanding of the nature of ureases' control and also to find out if urease is participate in the ensemble of N assimilatory enzymes whose expression is coordinately controlled.

9. Conclusion

Plant ureases especially those from legumes are without any iota of doubt important events in the history of science, which has been an area of research as early as 1900s. Nevertheless, despite several researches, there are still lots of work that need to be carried out to completely understanding this complex molecule. The several characteristics exhibited by these enzyme shows that ureases are not just enzymes for the hydrolysis of urea, it also presents a wide array of biotechnological applications which are so interesting.

10. Future perspective of urease from legumes

Intensive study of the toxic properties of ureases from plants can be of great interest in developing an alternative strategy in agriculture as biosecurity which will be very important to crops against so many natural enemies. Urease from legumes can also be isolated, purified, characterized and immobilized for the diagnosis of urea level of patients in the medical laboratories.

Author details

Peter S. Joseph[1*], Dickson A. Musa[2], Evans C. Egwim[3] and A. Uthman[2]

1 Department of Chemistry, Air Force Institute of Technology, Kaduna, Nigeria

2 Department of Biochemistry, Ibrahim Badamasi Babangida University, Lapai, Nigeria

3 Department Biochemistry, Federal University of Technology Minna, Nigeria

*Address all correspondence to: jpshabs@afit.edu.ng; jpshabz@gmail.com

References

[1] Real-Guerra R, Staniscuaski F, Carlini CR. Soybean urease: Over a hundred years of knowledge. In: Board JE, editor. A Comprehensive Survey of International Soybean Research – Genetics, Physiology, Agronomy and Nitrogen Relationships. Rijeka, Croatia: Intech; 2013. pp. 317-339

[2] Maroney MJ, Ciurli S. Nonredox nickel enzymes. Chemical Reviews. 2014;**114**(8):4206-4228

[3] Mobley HLT, Hausinger RP. Microbial ureases—Significance, regulation, and molecular characterization. Microbiological Reviews. 1989;**53**(1):85-108

[4] Krajewska B, Ureases I. Functional, catalytic and kinetic properties: A review. Journal of Molecular Catalysis B: Enzymatic. 2009;**59**(1-3):9-21

[5] Callahan BP, Yuan Y, Wolfenden R. The burden borne by urease. Journal of the American Chemical Society. 2005;**127**(31):10828-10829

[6] Sumner JB. The isolation and crystallization of the enzyme urease. Journal of Biological Chemistry. 1926;**69**:435-441

[7] Dixon NE, Gazzola TC, Blakeley RL, Zermer B. Letter: Jack bean urease (EC 3.5.1.5). A metalloenzyme. A simple biological role for nickel? Journal of the American Chemical Society. 1975;**97**(14):4131-4133

[8] Follmer C, Barcellos GB, Zingali RB, Machado OL, Alves EW, Barja-Fidalgo C, et al. Canatoxin, a toxic protein from Jack beans (*Canavalia ensiformis*), is a variant form of urease (Ec 3.5.1.5): Biological effects of urease independent of its Ureolytic activity. Biochemical Journal. 2001;(Pt 1):217

[9] Carlini CR, Ligabue-Braun R. Ureases as multifunctional toxic proteins: A review. Toxicon. 2016;**110**: 90-109

[10] Carter EL, Tronrud DE, Taber SR, Karplus PA, Hausinger RP. Iron-containing urease in a pathogenic bacterium. Proceedings of the National Academy of Sciences of the United States of America. 2011;**108**(32):13095-13099

[11] Real-Guerra R, Staniscuaski F, Carlini CR. Soybean Urease: Over a Hundred Years of Knowledge. Intech; 2013. pp. 317-319

[12] Van Slyke DD, Cullen GE. The mode of action of urease and enzymes in general. Journal of Biological Chemistry. 1914;**19**(2):19-141

[13] Blakeley RL, Hinds JA, Kunze HE, Webb EC, Zerner B. Jack bean urease (Ec 3.5.1.5). Demonstration of a carbamoyl-transfer reaction and inhibition by hydroxamic acids. Biochemistry. 1969;**8**(5):1991-2000

[14] Carter EL, Flugga N, Boer JL, Mulrooney SB, Hausinger RP. Interplay of metal ions and urease. Metallomics. 2009;**1**(3):207-221

[15] Krebs HA, Henseleit K. Hoppe Seyler's Z. Physiological Chemistry. 1932;**210**:33-66

[16] Schubert KR. Annual Review of Plant Physiology. 1986;**37**:539-574

[17] Mobley HLT, Hausinger RP. Microbial urease significance, regulation and molecular characterization. Microbiological Reviews. 1989;**53**: 85-108

[18] Thompson JF. The Biochemistry of Plants. In: Miflin BJ, editor. Vol. 5. New York: Academic Press; 1980. pp. 375-402

[19] Kang JH, Cho YD. Purification and properties of Arginase from Soybean,

Glycine max, Axes. Plant Physiology. 1990;**93**:1230-1234

[20] Splittstoesser WE. The appearance of arginine and arginase in pumpkin cotyledons. Characterization of arginase. Phytochemistry. 1969;**8**:753-758

[21] Cheema PS, Padmanaban G, Sarma PS. Arginase from *Lathyrus sativus*. Phytochemistry. 1969;**8**:409-411

[22] Boutin JP. Purification, properties and subunit structure of Arginase from iris bulbs. European Journal of Biochemistry. 1982;**127**:237-243

[23] Tromp J, Ovaa JC. Uptake and distribution of nitrogen in young Apple tree after application of nitrate or ammonium, with special reference to asparagine and arginine. Plant Physiology. 1979;**45**:23-28

[24] Bidwell RGS, Durzan DJ. In: Davies PJ, editor. Historical and Current Aspects of Plant Physiology. Ithaca, NY: Cornell Univ. Press; 1975. pp. 152-225

[25] Schubert KR, Boland M. In: Miflin BJ, Lea PJ, editors. The Biochemistry of Plants. Vol. 16. San Diego: Academic Press; 1990. pp. 197-282

[26] VanEtten CH, Kwolek WF, Peters JE, Barclay AS. Plant seeds as protein sources of food or feed. Evaluation based on amino acid composition of 379 species. Journal of Agricultural and Food Chemistry. 1967;**15**:1077-1085

[27] Micallef BJ, Shelp BJ. Arginine metabolism in developing soybean cotyledons: I. Relationship to nitrogen nutrition. Plant Physiology. 1989;**90**: 624-630

[28] de Ruiter H, Kolloffel C. Arginine catabolism in the cotyledons of developing and germinating pea seeds. Plant Physiology. 1983;**73**:525-528

[29] VanEtten CH, Miller RW, Wolff IA, Jones Q. Nutrients in seeds, amino acid composition of seeds from 200 angiospermous plant species. Journal of Agricultural and Food Chemistry. 1963

[30] Meyer-Bothling LE, Cianzio SR, Polacco JC. Mutational analysis of the embryo-specific urease locus of soybean. Molecular & General Genetics. 1987;**209**: 439-444

[31] Krogmeier MJ, McCarty GW, Bremner JM. Phytotoxicity of foliar-applied urea. Proceedings of the National Academy of Sciences of the United States of America. 1989;**86**:8189-8191

[32] Stahlhut RW, Widholm JM. Ureide catabolism by soybean [Glycine-Max (L) Merrill] cell-suspension cultures. I. Urea is not an intermediate in allantoin degradation. Journal of Plant Physiology. 1989;**134**:85-89

[33] Polacco JC. Nitrogen metabolism in soybean tissue culture: I. Assimilation of urea. Plant Physiology. 1976;**58**(3): 350-357

[34] Polacco JC. Nitrogen metabolism in soybean tissue culture: II. Urea utilization and urease synthesis require Ni^{2+}. Plant Physiology. 1977;**59**:827-830

[35] Eskew DL, Welch RM, Cary EE. Nickel: An essential micronutrient for legumes and possibly all higher plants. Science. 1983;**222**(4624):621-623

[36] Eskew DL, Welch RM, Norvell WA. Nickel in higher plants: Further evidence for an essential role. Plant Physiology. 1984;**76**:691-693

[37] Polacco JC. Is nickel a universal component of plant ureases? Plant Science Letters. 1977;**10**:249-255

[38] Fishbein WN, Smith MJ, Nagarajan K, Scurzi W. Urease catalysis and structure. X. Alternate bonding-site isozymes of jackbean urease. Federation of American Societies for Experimental Biology. 1976;**25**:680

[39] Polacco JC, Havir EA. Comparisons of soybean urease isolated from seed and tissue culture. Journal of Biological Chemistry. 1979;**254**:1707-1715

[40] Gordon WR, Schwemmer SS, Hillman WS. Nickel and the metabolism of urea by Lemna paucicostata Hegelm. Planta. 1978;**140**:265-268

[41] Stebbins N, Holland MA, Cianzio SR, Polacco JC. Plant Physiology. 1991;**97**: 1004-1010

[42] Roon RJ, Levenberg BJ. Journal of Biological Chemistry. 1968;**243**:5213-5215

[43] Roon RJ, Levenberg BJ. Journal of Biological Chemistry. 1970;**245**:4593-4595

[44] Thompson JF, Muenster A-M. Biochemical and Biophysical Research Communications. 1971;**43**:1049-1055

[45] Whitney PA, Cooper TG. Journal of Biological Chemistry. 1972;**247**:1349-1353

[46] Whitney PA, Cooper TG, Magasanik B. Journal of Biological Chemistry. 1973;**248**:6203-6209

[47] Maier-Greiner UH, Klaus CBA, Estermaier LM, Hartmann GR. Herbicide resistance in transgenic plants through degredation of the phytotoxin to urea. Angewandte Chemie International Edition in English. 1991;**30**:1314-1315

[48] Shimada N, Ando T. Role of nickel in plant nutrition. II. Effect of nickel on the assimilation of urea by plants. Biochemistry of Nutrition I. Vol. 51. Baltimore, MD: University Park Press; 1980. pp. 493-496

[49] Shimada N, Ando T, Tomiyama M, Kaku H. Role of nickel in plant nutrition. I. Effects of nickel on growth of tomato and soybean. Nippon Dojo Hiryogaku Zasshi. 1980;**51**:487-492

[50] De Cianzio SR. Strategies in population development for improvement of Fe efficiency in soybean. Genetic Aspect of Plant Mineral Nutrition. 1993;387-397

[51] Brown PH, Welch RM, Carey EE. Nickel: A micronutrient essential for plant. Plant Physiology. 1987;**85**:801-803

[52] Liao CFH, Raines SG. Inhibition of soil urease activity by amido derivatives of phosphoric and thiophosphoric acids. Plant and Soil. 1985;**85**:149-152

[53] Kobashi K, Takebe S, Numata A. Specific inhibition of urease by N-Acylphosphoric Triamides. Journal of Biochemistry. 1985;**98**:1681-1688

[54] Zerner B. Recent advances in the chemistry of an old enzyme, urease. Bioorganic Chemistry. 1991;**19**:116-131

[55] Zonia LE. [Ph.D. dissertation]. Columbia: University of Missouri; 1992

[56] Savant NK, Jasmes AF, Peters GE, Medina R. Evaluation of phenoxy-aminocyclotriphosphazatrienes as sustained-action soil urease inhibitors. Journal of Agricultural and Food Chemistry. 1988;**36**:390-392

[57] Krannitz PG, Aarssen LW, Dow JM. The effect of genetically based differences in seed size on seedling survival in *Arabidopsis thaliana* (Brassicaceae). American Journal of Botany. 1991;**78**:446-450

[58] Schaal BA. Reproductive capacity and seed size in *Lupinus texensis*. American Journal of Botany. 1980;**67**:703-709

[59] Stanton ML. Seed variation in wild radish: Effect of size on component of seedling and adult fitness. Ecology. 1984;**65**:1105-1112

[60] LeRoy AR, Fehr WR, Cianzio SR. Direct and indirect selection for small seed of soybean in temperate and tropical environment. Crop Science. 1991;**31**: 693-697

[61] Polacco JC, Sparks RB. Patterns of urease synthesis in developing soybeans. Plant Physiology. 1982;**70**(1):189-194

[62] Polacco JC, Winkler RG. Soybean leaf urease: A seed enzyme? Plant Physiology. 1984;**74**(4):800-803

[63] Lea PJ, Hughes JS, Miflin BJ. Glutamine- and Aspara journal of gine-dependent protein synthesis in maturing legume cotyledons cultured in vitro. Journal of Experimental Botany. 1979;**30**:529-437

[64] Thompson JF, Madison JT, Muenster AE. In vitro culture of immature cotyledons of soya bean (*Glycine max* L. Merr.). Annals of Botany. 1977;**41**:29-39

[65] Rain Bird RM, Thorne JH, Hardy RWF. Role of amides, amino acids, and ureides in the nutrition of developing soybean seeds. Plant Physiology. 1984;**74**:329-334

[66] Peoples MB, Atkins CA, Page JS, Murray DR. Nitrogen nutrition and metabolic interconversions of nitrogenouse solute in developing cowpea fruits. Plant Physiology. 1985;**77**:382-388

[67] Bailey CJ, Boulter D. In: Harborne JB, Boulter D, Turner BL, editors. The Chemotaxonomy of the Leguminosae. New York: Academic Press; 1971. pp. 485-502

[68] Torisky RS, Polacco JC. Soybean roots retain the seed urease isozyme synthesized during embryo development. Plant Physiology. 1990;**94**(2):681-689

[69] Eaton KA, Brooks CL, Morgan DR, Krakowka S. Essential role of urease in pathogenesis of gastritis induced by *Helicobacter pylori* in gnotobiotic piglet. Infection and Immunity. 1991;**59**: 2470-2475

[70] Ferrero RL, Lee A. Genetic tests of the roles of the embryonic ureases of soybean. Microbial Ecology in Health and Disease. 1991;**4**:121-134

[71] Cussac V, Ferrero RL, Labigne A. An adenosine triphosphate-dependent, avidin-sensitive enzymatic cleavage of urea in yeast and green algae. Journal of Bacteriology. 1992;**174**:2466-2473

[72] Summerskill WHJ, Thorsell F, Feinberg JH, Adrete J. CO_2 fixation and the involvement of allophanate in the biotin-enzyme-catalyzed cleavage of urea. Gastroenterology. 1967;**54**:20-26

[73] Fishbein WN, Daly JE. Separation of the Chlorella ATP: Urea amido-lyase into two components. Proceedings of the Society for Experimental Biology and Medicine. 1970;**134**:1083-1090

[74] Mulrooney SB, Hausinger RP. Urea carboxylase and allophanate hydrolase: Two components of a multienzyme complex in *Saccharomyces cerevisiae*. Journal of Bacteriology. 1990;**172**: 5837-5843

[75] Liener IE. In: Neuberger IA, Jukes TH, editors. International Review of Biochemistry, Biochemistry of Nutrition I. Baltimore, MD: University Park Press; 1979. pp. 97-122

[76] Gardner HW. Journal of Agricultural and Food Chemistry. 1979;**27**:345-350

[77] Gatehouse AMR, Barbieri L, Stirpe F, Croy RRD. Effect of ribosome on inactivating proteins on insect development-differences between Lepidoptera and coleoptera. Entomologia Experimentalis et Applicata. 1990;**54**: 43-51

[78] Weselake RJ, MacGregor AW, Hill RD, Duckworth HW. Purification and characterisation of an endogenous alpha-amylase inhibitor from barley kernels. Plant Physiology. 1983;**72**: 809-812

[79] Gatehouse AMR, Boulter D. Assessment of the antimetabolic effects

of trypsin inhibitors from cowpea (*Vigina unguiculata*) and other legumes on development of the bruchid beetle callosobruchus maculatus. Journal of the Science of Food and Agriculture. 1983;**34**:220-239

[80] Davies CS, Nielsen NC. Genetic analysis of a null-allele for lipoxygenase-2 in soybean. Crop Science. 1986;**26**: 460-463

[81] Kitamura K, Davies CS, Kaizuma N, Nielsen NC. Genetic analysis of a null-allele for lipoxygenase-3 in soybean seeds. Crop Science. 1983;**23**:924-926

[82] Pull SP, Pueppke SG, Hymowitz T, Orf JH. Soybean lines lacking the 120,000-dalton seed lectin. Science. 1978;**200**:1277-1279

[83] Mervosh TL, Hymowitz T. DA *Glycine soja* collection for urease variants. Screening of the US. Soybean General News. 1987;**14**:265-267

[84] Vitousek PM, Howarth RW. Nitrogen limitation on land and in the sea- how can it occur. Biogeochemistry. 1991;**13**(2): 87-115

[85] Witte CP. Urea metabolism in plants. Plant Science. 2011;**180**(3):431-438

[86] Cao FQ, Werner AK, Dahncke K, Romeis T, Liu LH, Witte CP. Identification and characterization of proteins involved in rice urea and arginine catabolism. Plant Physiology. 2010;**154**(1):98-108

[87] Elser JJ, Fagan WF, Subramanian S, Kumar S. Signatures of ecological resource availability in the animal and plant proteomes. Molecular Biology and Evolution. 2006;**23**(10):1946-1951

[88] Polacco JC, Holland MA. Genetic control of plant ureases. Genetic Engineering; Principles and Methods. 1994;(16):33-48

[89] Sirko A, Brodzik R. Plant ureases: Roles and regulation. Acta Biochimica Polonica. 2000;**47**(4):1189-1195

[90] Witte CP, Medina-Escobar N. In-gel detection of urease with nitroblue tetrazolium and quantification of the enzyme from different crop plants using the indophenol reaction. Analytical Biochemistry. 2001;**290**(1):102-107

[91] Liu LH, Ludewig U, Frommer WB, von Wiren N. Atdur3 encodes a new type of high-affinity urea/H+ symporter in Arabidopsis. The Plant Cell. 2003; **15**(3):790-800

[92] Food and Agriculture Organization of the United States. 2011. Faostat. Available from: http://faostat.fao.org

[93] Polacco JC, Krueger RW, Winkler RG. Structure and possible Ureide degrading function of the ubiquitous urease of soybean. Plant Physiology. 1985;**79**(3):794-800

[94] Stebbins NE, Polacco JC. Urease is not essential for Ureide degradation in soybean. Plant Physiology. 1995;**109**(1): 169-175

[95] Witte CP, Tiller SA, Taylor MA, Davies HV. Leaf urea metabolism in potato. Urease activity profile and patterns of recovery and distribution of (15) N after foliar urea application in wild-type and urease-antisense transgenics. Plant Physiology. 2002;**128**(3):1129-1136

[96] Polacco JC, Holland MA. Roles of urease in plant-cells. International Review of Cytology—A Survey of Cell Biology. 1993;**145**:65-103

[97] Carlini CR, Polacco JC. Toxic properties of urease. Crop Science. 2008;**48**(5):1665-1672

[98] Grossi-de-Sa MF, Pelegrini PB, Fragoso RR. Genetically modified

soybean for insect-pests and disease control. In: Sudaric A, editor. Soybean-Molecular Aspects of Breeding. Rijeka: InTech; 2011. pp. 429-452

[99] Oerke EC, Dehne HW. Safeguarding production-losses in major crops and the role of crop protection. Crop Protection. 2004;**23**(4):275-285

[100] Follmer C, Real-Guerra R, Wasserman GE, Olivera-Severo D, Carlini CR. Jackbean, soybean and *Bacillus Pasteurii* ureases. Biological effects unrelated to ureolytic activity. European Journal of Biochemistry. 2004;**271**(7):1357-1363

[101] Mulinari F, Staniscuaski F, Bertholdo-Vargas LR, Postal M, Oliveira-Neto OB, Rigden DJ, et al. Jaburetox-2ec: An insecticidal peptide derived from an isoform of urease from the plant *Canavalia ensiformis*. Peptides. 2007;**28**(10):2042-2050

[102] Wiebke-Strohm B, Pasquali G, Margis-Pinheiro M, Bencke M, Buecker-Neto L, Becker-Ritt AB, et al. Ubiquitous urease affects soybean susceptibility to fungi. Plant Molecular Biology. 2012;**79**(1-2):75-87

[103] Qin G, Elstter ER, Bosch MW, Poelvander AFB. Thermal processing of whole soya beans: Studies on the inactivation of antinutritional factors and effects on ileal digestibility in piglets. Animal Feed Science and Technology. 1996;**57**(4):313-324

[104] American Soybean Association. Soystats. Soybean Urease: Over a Hundred Years of Knowledge. 2011. DOI: 10.5772/52106337. Available from: http://www.soystats.com/

[105] Cromwell GL. Soybean meal-The "gold standard". The Farmer's Pride, KPPA News. 1999;**11**(20):1-15

[106] Liener IE. Implications of antinutritional components in soybean foods. Critical Reviews in Food Science and Nutrition. 1994;**34**(1):31-67

[107] Gandhi AP, Nenwani MM, Ali N. Investigations on the trypsin-inhibitor, urease and cooking behavior of soybean glycine-max Merr. Food Chemistry. 1984;**15**(3):215-218

[108] Vasconcelos IM, Siebra EA, Maia AAB, Moreira RA, Neto AF, Campelo GJA, et al. Composition, toxic and antinutritional factors of newly developed cultivars of Brazilian soybean (*Glycine max*). Journal of the Science of Food and Agriculture. 1997;**75**(4):419-426

[109] Vasconcelos IM, Maia AA, Siebra EA, Oliveira JT, Carvalho AF, Melo VM, et al. Nutritional study of two Brazilian soybean (*Glycine max*) cultivars differing in the contents of antinutritional and toxic proteins. The Journal of Nutritional Biochemistry. 2001;**12**(1):55-62

[110] Becker-Ritt AB, Mulinari F, Vasconcelos IM, Carlini CR. Antinutritional and/or toxic factors in soybean (*Glycine max* (L) Merril) seeds: Comparison of different cultivars adapted to the southern region of Brazil. Journal of the Science of Food and Agriculture. 2004;**84**(3):263-270

[111] Food and Agriculture Organization of the United States. FAO; 2011. Available from: http://www.fao.org/

[112] Huntington GB, Harmon DL, Kristensen NB, Hanson KC, Spears JW. Effects of a slow-release urea source on absorption of ammonia and endogenous production of urea by cattle. Animal Feed Science and Technology. 2006;**130**(3-4): 225-241

[113] Symonds HW, Mather DL, Collis KA. The maximum capacity of the liver of the adult dairy-cow to metabolize ammonia. British Journal of Nutrition. 1981;**46**(3):481-486

[114] Edwards HM Jr. Effects of different soybean meals on the incidence of tibial dyschondroplasia in the chicken. Journal of Nutrition. 1985;**115**(8):1005-1015

[115] Veltmann JR, Jensen LS. Tibial dyschondroplasia in broilers-comparison of dietary additives and strains. Poultry Science. 1981;**60**(7):1473-1478

[116] Edwards HM. Studies on the etiology of tibial dyschondroplasia in chickens. Journal of Nutrition. 1984;**114**(6): 1001-1013

[117] Vasconcelos IM, Brasil IC, Oliveira JT, Campello CC, Maia FM, Campello MV, et al. Combination of chemical analyses and animal feeding trials as reliable procedures to assess the safety of heat processed soybean seeds. Journal of Agriculture and Food Science. 2009;**57**(11):4668-4673

[118] Caskey CD, Knapp FC. Method for detecting inadequately heated soybean oil meal. Industrial and Engineering Chemistry. 1944;**16**(10):640-641

[119] Schramm G, Aines PD. Colorimetric determination of urease activity in soybean meals. Journal of the American Oil Chemists Society. 1959;**36**(1):1-3

[120] Croston CB, Smith AK, Cowan JC. Measurement of urease activity in soybean oil meal. Journal of the American Oil Chemists Society. 1955;**32**(5):279-282

[121] Smith AK, Belter PA, Anderson RL. Urease activity in soybean meal products. Journal of the American Oil Chemists Society. 1956;**33**(8):360-363

[122] Weather burn, M. W. Phenol-hypochlorite reaction for determination of ammonia. Analytical Chemistry. 1967;**39**(8):971

[123] Kaplan A. The determination of urea, ammonia, and urease. Methods of Biochemical Analysis. 1969;**17**:311-324

[124] Ferreira-Dasilva CT, Gombarovits ME, Masuda H, Oliveira CM, Carlini CR. Proteolytic-activation of canatoxin, a plant toxic protein, by insect cathepsin-like enzymes. Archives of Insect Biochemistry and Physiology. 2000;**44**:162-171

[125] Carlini CR, Grossi-de-Sa MF. Plant toxic proteins with insecticidal properties. A review on their potentialities as bio insecticides. Toxicon. 2002;**40**:1515-1539

[126] Mulinari F, Staniscuaski F, Bertholdo-Vargas LR, Postal M, Oliveira-Neto OB, Rigden DJ, et al. Jaburetox-2Ec: An insecticidal peptide derived from an isoform of urease from the plant Canavaliaensi form is. Peptides. 2007;**28**:2042-2050

[127] Qin Y, Cabral JMS. Properties and applications of urease. Biocatalysis and Biotransformation. 2002;**20**:1-14

[128] Mohan Babu R, Sajeena A, Seetharaman K, Reddy MS. Advances in genetically engineered (transgenic) plants in pest management—An overview. Crop Protection. 2003;**22**:1071-1086

[129] Orf JH, Hymowitz T. Inheritance of the absence of the kunitz trypsin inhibitor in seed protein of Soybeans. Crop Science. 1979;**19**:107-109

Chapter 16

Legume Protein: Properties and Extraction for Food Applications

Elisha Onyango

Abstract

Grain legumes are important sources of protein for nutritional and techno-functional applications. Their protein content is 18-50% protein on dry matter basis. Most of the protein is of the storage type, of which 70% are globulins. The globulin proteins are mainly legumins and vicilins, which are also known as 7S and 11S globulins, respectively. Several methods comprising wet and dry processes are used to extract protein from legumes. Choice of extraction method mainly depends on legume type and desired purity and functionality of extracted protein. Dry processing is suitable for starch-rich legumes, and involves fine milling and air classification. Wet processing uses solubility differences to extract and separate protein from non-protein components. The major extracted protein products are protein concentrate and isolate. Functional properties of protein depend on its amino acid profile, protein structure, hydrophobic, and hydrophilic effects. The major functional properties for food applications are solubility, water absorption capacity, oil absorption capacity, gelling, texturization, emulsification and foaming. They indicate ability of a protein to impart desired physico-chemical characteristics to food during processing, storage and consumption. The food products where isolated legume protein can be used include bakery products, plant based dairy alternative products, beverages and meat analogues.

Keywords: grain legumes, protein content, protein extraction, functional properties, food applications

1. Introduction

Grain legumes are important sources of protein for food at household level and also for applications during food processing. Legume protein has nutritional and techno-functional benefits and uses. In terms of nutrition, the protein is a rich and low-cost source of some of the essential amino acids, which has relevance for food, nutrition security and health. As the world population continues to increase, they are sustainable and affordable alternatives to animal protein. They are also increasingly being used in processed food as functional ingredients to impart desirable characteristics such as texture. This emanates from techno-functional properties of protein such as solubility, emulsification, water holding capacity, foaming, gelation, and oil holding capacity. These properties are important in formulation of food products such as plant based milk products, bakery products, beverages, meat analogues, and other categories of food products which require incorporation of protein as functional ingredient to achieve a optimal product. The functional properties of legume protein varies, and is influenced by type of legume, cultivar, molecular weight, amino acid composition, and charge distribution, and processing [1, 2].

Grains legumes are contained in pods at harvest, and are used as food either as dry or immature seeds. They are in the class *Leguminosae*. Their major common characteristic is ability to fix nitrogen from the atmosphere as a result of nitrogen-fixing microorganisms that is present in their nodules. They have been important sources of food for humans for over a thousand years in most parts of the world. Grain legumes are generally classified as either pulses or oilseed. Pulses include dry beans (*Phaseolus* spp.), dry peas (*Pisum sativum*), dry broad beans (*Vicia faba*), lentil (*Lens culinaris*), chickpea (*Cicer arietinum*), lupins (*Lupinus* spp), dry cow pea (*Vigna unguiculata*), pigeon pea (*Cajanus cajan*), bambara groundnuts (*Vigna subterranea*), vetch (*Vicia sativa*), and other minor pulses. Oilseeds are soybean (*Glycine max*), peanut (*Arachis hypogaea*), rapeseed (*Brassica napus*), sunflower (*Helianthus annuus*), sesame (*Sesamum indicum*) and other minor oilseed crops. They form key part of human food, as they are generally good sources of protein, energy, and micronutrients [3, 4]. Grain legumes generally have a common characteristic of being rich in protein. Their protein content range from about 18 to 50% on dry matter basis [5]. About 40 species of grain legumes exist in the world [6]. Besides their use as food, protein from legumes is also added to food during processing to achieve desired functional properties in the end product. They are therefore good sources of protein for food applications. Soybean and pea are among the major legumes that have been most explored and used in food applications. The other legumes have generally found lesser applications in industrial use. Their extraction and characterization continue to generate more knowledge on their properties and food applications. This chapter discusses protein from grain legumes, their characteristics, functional properties, and extraction for food applications.

2. Grain legume composition and protein characteristics

Grain legumes content of protein, carbohydrates, lipid, and micronutrients vary with type of legume, environment and agronomic factors. The macro-nutrient content of major grain legumes is approximately 18–50%, 0.8–21%. 60–69% and 0.9–7.2%, representing protein, oil, carbohydrates and fiber, respectively [6, 7]. Composition of grain legumes is presented in **Table 1**.

Legume protein are generally classified into three groups, namely storage, biologically active, and structural. Storage protein are the most abundant in legumes. They are a store of nitrogen for germination. Biologically active protein include lectins, enzymes and enzyme inhibitors, while structural proteins are ribosomal, chromosomal and membrane proteins. The storage proteins can further be classified into albumins, globulins, prolamins and glutelins. At the basic level, they can be differentiated by their solubility in various solvents. Albumins are soluble in water, globulins in dilute salt solution, prolamins in 75% ethanol, and glutelins show difficulty to solubilize but are soluble in dilute alkali. They can also be differentiated by their occurrence in monocotyledonous, cereals; and dicotyledonous plants, legumes. Albumins and globulins are the major storage protein in legumes, whereas prolamins and glutelins are the major protein in cereals. Globulin and albumin fractions account for about 70% and 10–20% of total protein in legumes, respectively [34]. The molecular weight of albumins is about 5–80 kDa, and is therefore generally low. Trypsin inhibitors, lectins, and amylase inhibitors may in some cases be part of albumins. As nitrogen source for germination, the storage proteins are rich in nitrogen containing the amino acids asparagine, arginine and glutamine [35].

Legume	Composition (g/100 g)					Ref.
	Protein	Carbohydrate	Crude fiber	Fat	Ash	
Dry bean (*Phaseolus* spp)	16–25	39–56	6.3–18.9	1.5	2.9–4.3	[8]
Black gram (*Vigna mungo*)	24	51–59	5.5	1.44	1.4	[9]
Green gram (*Vigna radiata*)	26–27	62–70	4.6	1.6–1.8	1–3.7	[10, 11]
Dry peas (*Pisum sativum*)	20–30	41–46	3.3–6.5	1–4	2–3	[12–14]
Dry broad beans (*Vicia faba*)	26–38	55–60	6.4–8.4	0.7–1.1	2.7–3.5	[15, 16]
Lentils (*Lens culinaris*)	23–29	60	13–14	0.8	2.4–2.8	[17–20]
Chickpea (*Cicer arietnum*)	17–22	56–63	4.8–8	4.5–5.7	1.9–3.2	[21, 22]
Cowpea (*Vigna unguiculata*)	27	69.8–76	1.4–14	5	3.2–4.1	[23–25]
Pigeon peas (*Cajanus cajan*)	19–21.7	54	9.8–13	1.2–1.3	3.9–4.3	[26, 27]
Lupin (*Lupinus* spp)	40	47.7	10.8	8.9	3.15	[28]
Soybean (*Glycine max*)	40–44	30–34	4.5	18–19	3–4.8	[29–31]
Peanut (*Arachis hypogea*)	24–28	38	4.1–4.8	44–49.5	2.3–2.5	[32, 33]

Table 1.
Composition of selected legume sources of protein.

The major classes of globulin proteins in legumes are legumins and vicilins. They are also referred to as 7S and 11S globulins respectively, based on their Svedberg sedimentation values (S) [36]. This is similar for peas, soybean and other grain legumes. But they occur in different ratios in legumes, but both are oligomers. The 11S type is a hexamer made up of six subunits of molecular weight of about 340–360 kDa; whereas the 7S fraction has molecular weight of about 145–180 kDa made up of trimers of molecular weight of about 60–70 kDa [34, 37, 38]. Vicilins generally vary among the legumes in terms of molecular weight. Legume protein is low in cysteine, cysteine, methionine and tryptophan, which are sulfur containing amino acids. Only soybean contains all the essential amino acids required for human nutrition. Its protein also has good digestibility. Thus it has high quality protein almost comparable to protein from animal sources in terms of nutritional quality. But legumes in general have high nutritional value due to their high protein content, and content of other macro- and micro-nutrients.

Protein occur in structures described as primary, secondary, tertiary and quaternary. The primary structure denotes the linear sequence of the amino acid of the protein but providing little information that explains protein functionality. The higher structures are determined by conformational fold and helps in understanding protein molecular characteristics such as net charge, size, shape and other properties.

3. Technologies for extracting legume protein

Several technologies exist for extraction of protein from legumes. Some have found success in commercial processing, while others have not mainly due to economics of the technologies. The methods include several wet extraction and dry processes. Extraction of protein from legume to an appropriate level of purity is necessary for its use in food formulations. Choice of extraction process depends on amount of protein required in the protein extract, legume type, functionality desired in the protein and nature of previous processing [35]. The processing methods can generally be classified into two categories, namely aqueous and dry processing. The various legumes differ in composition as some are lipid rich, while other are starch rich. Each is suitable for protein extraction using a particular method. Dry processing which involves fine milling and air classification is suitable for starch-rich legumes as pea and faba bean, but generally not used to process oil-rich legumes such as soybean, peanut and similar seeds [5, 39].

3.1 Aqueous extraction

Aqueous processing involves using water or alkaline solution to solubilize protein from legume flour or flakes in the initial processing step. It uses solubility differences to separate protein from carbohydrates, fiber and other non-protein components. This is then followed by purification and drying. This method has been used for several decades in soy protein extraction. Extraction efficiency depends on pH, temperature, ionic environment and solvent to flour ratio. It has variations depending on raw material which is being used for protein extraction. It is desired to achieve highest protein recovery without the process being detrimental to protein functionality. The major extracted protein products in terms of application in food are protein concentrate and isolate.

3.1.1 Protein concentrate

Processing of grain legumes into protein concentrate follows the following general steps, namely cleaning, de-hulling, flaking, defatting, protein extraction, neutralization and drying. Production of protein concentrate from legumes is illustrated in **Figure 1**. De-fatted soy flakes or flour is the starting raw material. Protein extraction can be carried out using three methods. The methods are washing with (1) aqueous alcohol (60–90%), (2) acid leaching at pH 4.5 and (3) moist heat water leaching. If extraction was done using alcohol, solvent is removed in a process step referred to as desolventization. The protein may then be neutralized to pH 7 then dried. Drying options include drum drying and spray drying, with the latter being preferred and most commonly used. Protein concentrate products contain about 70% protein.

3.1.2 Protein isolate

Protein isolate is more refined than concentrate. Its processing generally involves alkaline extraction of protein followed by protein recovery at their isoelectric pH of about 4.0–5.0, depending on the legume. The general principles of this extraction approach can be applied, with appropriate modifications, to extract protein from different legumes. Process options exist for unit operations such as preparation, defatting, pH of carrying out protein recovery and drying. The general process and its possible modifications are described here. **Figure 2** illustrates processing steps involved in production of protein isolate from legumes.

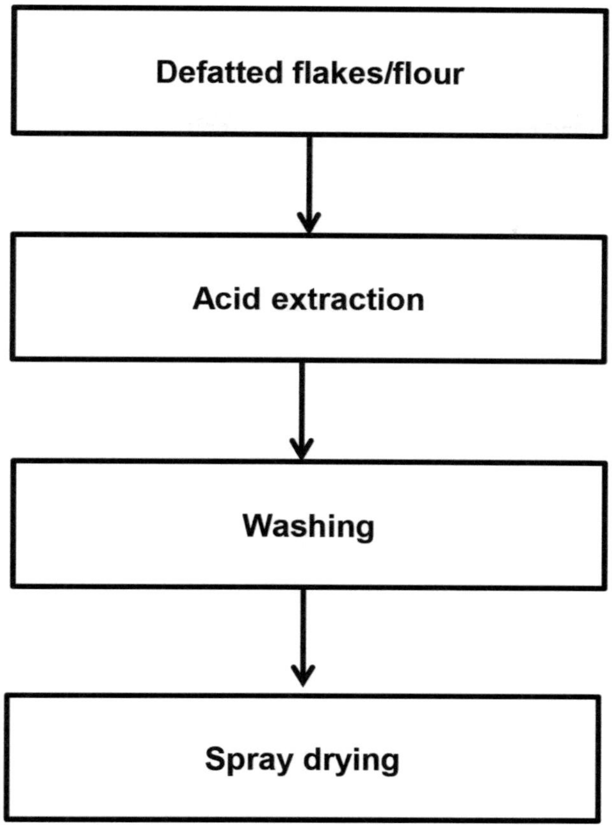

Figure 1.
Process flow diagram for production of protein concentrate.

Processing starts with cleaning of legume seeds to remove impurities such as stones, mold infected seeds, discolored seeds, sand, soil and other foreign matter. Grains such as soybean, chickpea, and pigeon pea are du-hulled to remove the fiber-rich seed coat. The seed coat is also referred to as hull, husk, testa, or husk depending on grain legume. The grain is then prepared into flakes or flour. Then defatting is carried out. The most commonly used method to remove fat is by hexane extraction. It can be carried at either low or high temperature of around 60–80°C. The high temperature also inactivates lipoxygenase, which contributes to causing off-flavors in the resultant extracted protein. Defatting using hexane leaves residual oil of about 0.5–1% (w/w) of the extracted protein. The residual lipid is composed of phospholipids and polar lipids. They cannot be extracted with hexane. They are sources of off-flavors as they participate in auto-oxidation and lipoxygenase mediated reactions. Following defatting, the solvent is the recovered from the meal and oil. Alternative defatting processes include supercritical carbon dioxide extraction, enzymatically aided extraction, organic solvent extraction, and aqueous extraction [40]. They have various degrees of efficiency and cost implications. They have however not been used for commercial processing due to cost and efficiency.

The intermediate product obtained at the above step can be toasted and cooled to make defatted soy flour. Toasting is applied to inactivate trypsin inhibitors. The product obtained from removal of solvent is also called flakes, which can also be milled into flour, as starting materials for aqueous extraction into protein concentrate and isolate.

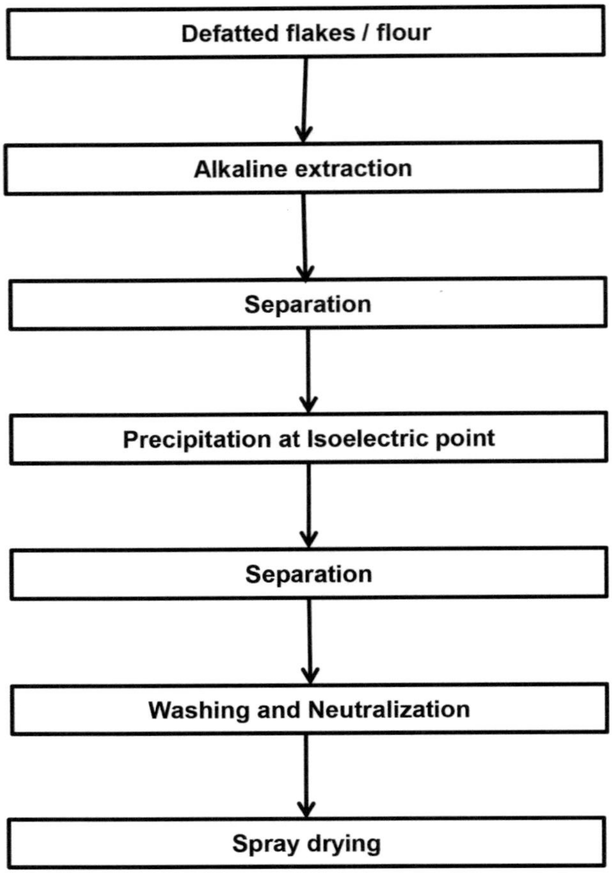

Figure 2.
Process flow diagram for production of protein isolate.

To extract proteins, the flakes is milled into flour and mixed with water at a ratio of 1:6. The pH is adjusted to 9 using lye so that protein goes into solution. Legume proteins are generally soluble in aqueous media. In solutions where pH is less or greater than isoelectric pH, a net positive or negative charge occurs resulting in repulsion and protein staying in solution [41]. Clarification is then carried out to remove non-protein insoluble materials such as carbohydrates and insoluble fiber. Clarification can be done using centrifugation, filtration or membrane process. Protein is then recovered from the solution. Recovery can be done using ultra-filtration or precipitation. Precipitation can be achieved using isoelectric precipitation, salting out, salting in, heating or organic solvents. Isoelectric precipitation is the most commonly used method during commercial scale processing. It is generally not detrimental to protein functionality. But it may cause protein to aggregate and also result in changes in solubility as a result of non-covalent interactions. Isoelectric precipitation is carried out by adjusting the acidity of the protein solution using dilute acid to pH 4.0–5.0. Separation is then done by centrifugation or decanting. The resultant protein is washed, then may be neutralized to pH 7. Final processing step involves drying by spray or drum drying. In the former method, a thin layer is applied to a heated drum to evaporate water. The preferred method is however spray drying because it gives protein that has less heat damage. Some aggregation may however occur. Protein isolate contain at least 90% protein.

3.2 Dry extraction

Dry fractionation can produce protein-enriched products which possess native functional properties. This is mainly because the process is mild [5]. Dry processing involves physical separation of starch and protein. It relies on the principle that milling can separate protein bodies from other seed components to give flour streams that is fractionated into different components. This has applications in starch rich legumes such as pea and faba bean. The method is however not applicable to grains that are rich in oil. In general, size of starch granules partly determines suitability of a particular legume for dry fractionation [5]. Milling is considered effective when it removes the protein bodies from starch granules, the latter being bigger in size. For pea and faba bean, starch and protein can be separated by milling followed by air classification. During air classification, protein separate as fine particles while starch are the coarse particles. Protein are separated as the light fraction, and starch as the heavy fraction based on their different shape, density, and size. This results in protein and starch enriched fractions.

Dry process involves milling and air classification. Starch granules are separated from protein bodies during fine milling usually by pin milling. This enables their separation during fractionation by air classification. Two processes are therefore involved in dry processing, namely, reducing particle size by milling and separation of the particles using their shape, density, size, and electrostatic properties [42]. The starch rich fraction is re-milled and fractionated. Particle size and shape are the main properties which are manipulated during dry extraction. The particle size can be varied by size reduction in the range, coarse, >500 μm; fine, 50–500 μm and ultrafine, <50 μm. Particles sizes selection enhance protein and starch enrichment and also reduce content of undesirable components such as anti-nutrients. Impact mill can produce the above range of particle size, while hammer milling can be used to produce only coarse and fine particle size. Ultrafine milling is used for protein and starch enrichment with pulses such as peas and faba beans. Air classification is applied after fine milling to obtain protein enriched fraction. The resulting end product has its protein content doubled compared to the raw material. Dry milling of faba bean and Lima bean have been reported to give protein yields of 63–75% and 43–50%, respectively [5, 42]. Air classification has not been found to be successful with oilseeds due to generation of free lipids [5].

4. Functional properties and food applications of legume proteins

Legume proteins have diverse application in foods due to their functional properties. They contribute to food having desirable textural characteristics during processing, storage and consumption. The major functional properties that are relevant for food applications are solubility, water absorption capacity, oil absorption capacity, gelling property, texturization, flavor binding, emulsification and foaming. These properties depend on amino acid profile, protein structure, hydrophobic, and hydrophilic effects. The functional properties that are key to their application in food are reviewed below.

4.1 Functional properties

4.1.1 Solubility

Solubility of legume protein is important in formulation of products such as plant based dairy alternative products and beverages among many other similar products. Protein possess minimum solubility at pH corresponding to their p1. This is because of a zero net charge at their pI which results in aggregation of proteins. In general, legume proteins have minimum solubility in the range pH 4–4.5, and solubility maxima at above pH 8 and below pH 2.5. There solubility is high at low and conditions, and high alkaline conditions. Solubility is lowest at the isoelectric points of proteins. This is taken advantage of during protein extraction when purifying proteins. Solubility of proteins is affected during processing depending on level of heat treatment. Protein of high solubility is achieved by mild processing conditions mainly in terms of heat treatment. Solubility of legume proteins generally vary with heat treatment and pH [43].

4.1.2 Water absorption capacity

Water absorption capacity (WHC) refers to the amount of water that protein can absorb. It is also be referred to water binding capacity. Commercially prepared soy protein isolate can absorb water at 4–5.5 times its weight, and 2.4–3.4 for concentrates [4]. In fava bean, lentils, cow peas, chickpeas, soybean, beans and peas WHC was reported to be in the range 2.39–6.78 g of water per gram of protein concentrate [44]. This generally indicates wide variation in WHC, and suggests that some of the protein isolates are better suited than others for use where WHC is intended to be achieved. Soybean and pea protein isolates have been reported to improve viscosity and prevent syneresis in cultured dairy products [45]. These are partly functions of their WHC. In another study [46], WHC was found to be 4.09–6.13 g/g for pinto bean, lima bean, red bean, kidney bean, black bean, navy bean, red bean, mung bean, lentil and chickpea. WHC of flour produced from 21 legume samples which included green gram, chickpea, lentil, soybean and several bean varieties was found in another study to be 1.32–3.14 g/g [47]. Protein products from legumes are in the form of flour, protein concentrates and isolates. They differ in level of refining, with flour being less refined and isolate being highly refined source of protein. The various forms of protein have different uses. Refining whole flour such as by defatting and removal of other non-protein components as much as possible improve functional properties such as WHC and oil holding capacity [48]. Compared to other legumes, soybean and pea have generally been extensively characterized for functional properties and also used in food applications. WHC of 7 protein isolates from 7 pea cultivars were reported to range be in the range 1.88–2.37 g/g [49].

4.1.3 Oil absorption capacity

Oil absorption capacity (OAC) refers to the weight of oil absorbed per unit weight of protein. Legume proteins have been reported to have OAC in the range 1–4 g/g protein [50, 51]. OAC has been reported to be in the range 3.46–6.37 grams oil per gram of protein concentrate in fava bean, lentils, cow peas, chickpeas, soybean, beans and peas [44]. Like WHC, OAC also varies among legumes, thus they vary in their effectiveness in achieving OAC. Lentil and chickpea and several types of common beans was reported to have OAC in the range 0.93–1.38 g/g [46]. In another study, flour prepared from 21 legume samples comprising green gram, chickpea, lentil, soybean and several bean varieties had OHC of 0.62–2.57 g/g [47]. OHC as well as other functional properties will vary with type of legume as well as form of protein. In a study, OHC of protein isolates from 7 pea cultivars ranged from 1.07 to 1.40 g/g [49].

4.1.4 Emulsification

Plant protein exhibit emulsifying properties. This relates to their ability to stabilize oil in water and water in oil emulsions thereby reducing interfacial tension and phase separation [15, 52, 53]. Thus they play a role as emulsifiers due to hydrophobic and hydrophilic balance at the interface of oil and water. Proteins align at interface so that the hydrophobic part face the oil phase, while its hydrophilic part face the water phase. Emulsification properties can be measured using methods such as turbidimetric method and droplet size measurement. The former comprise emulsifying activity index (EAI) and emulsifying stability index (ESI). Emulsification properties of legume protein affected by pH, temperature, protein concentration, and ionic environment [46]. Bean and pea protein isolates were reported in a study to have similar emulsifying capacity of about 27%, but emulsion stability was higher for pea protein isolate [54]. A study reported emulsifying activity index and emulsifying capacity index of protein isolates from 7 cultivars of pea to be 31.09–39.05 m^2/g and 10.97–11.26 min, respectively [49]. Pea and soybean protein products are one of the most characterized and used in food applications.

4.1.5 Foam formation and stability

Legume protein are also important in foam formation and stability. When protein unfold to form a interfacial skin that hold air, this is referred to as foam. Foaming property of a given legume protein can be measured by homogenizing known concentration of protein dispersion to form foam. Foam capacity (FC), also known as form expansion (FE), is calculated as percent increase in volume from whipping, and foam stability (FS) measures change in volume over time [39]. Foaming is utilized in food applications such as whipped toppings, cakes, meringues, leavened breads and beverages. In a study, protein from fava beans had high foaming capacity than protein from lentils, cow peas, chickpeas, soybean, beans and peas [44].

4.1.6 Gelation

Legume proteins, as other proteins, can also form gels upon heating. Heat induced gelation involves unfolding of proteins, and exposure of reactive groups leading to aggregation into a gel. Gelation imparts characteristics such as water holding capacity to the food. It also holds ingredients. This has applications in foods such as meat and desserts. Gelation affected by pH and ionic environment. Gelling effectiveness is evaluated using the parameter, least gelling concentration (LGC) which is the lowest concentration of protein required to form a stable gel [19]. Legume protein can provide suitable gels for food applications [55–57].

4.2 Food applications

Legume proteins applications in food formulation are in the areas of nutritional enhancement, technological and functional properties. The advantages of using plant proteins include their being abundant, low cost, and healthful compared to alternatives such as chemical based ingredients. Legume proteins can be used in the form of flours, concentrates and isolates depending on the particular food application. Soybean is currently the most widely used legume protein. The areas where the proteins can be used include bakery products, plant protein based dairy type products, meat analogues, and emulsifiers [5, 58–60]. Bambara groundnuts can have applications in acidified high acid beverages owing to properties of vicilin fraction of its protein [61].

5. Conclusions

Grain legumes are important sources of protein and other macro- and micronutrients. Besides having a significant role in nutrition and food security where they are grown, they are also source of proteins whose functional properties have diverse food applications. As the world population continues to increase, legume protein have potential to play a significant role to fill the gap for demand for more protein. Methods for extracting legume protein for food applications can generally be classified into dry and aqueous methods. The various legume proteins still need to be characterized for their ability to impart desired physico-chemical properties in food systems during processing, storage and consumption.

Conflict of interest

The authors declare no conflict of interest.

Author details

Elisha Onyango
Kenya Industrial Research and Development Institute, Nairobi, Kenya

*Address all correspondence to: elisha.onyango@kirdi.go.ke

References

[1] Makri E, Papalamprou E, Doxastakis G. Study of functional properties of seed storage proteins from indigenous European legume crops (lupin, pea, broad bean) in admixture with polysaccharides. Food Hydrocolloids. 2005;**19**(3):583-594. DOI: 10.1016/j.foodhyd.2004.10.028

[2] Zha F, Rao J, Chen B. Modification of pulse proteins for improved functionality and flavor profile: A comprehensive review. Comprehensive Reviews in Food Science and Food Safety. 2021;**20**:1-25. DOI: 10.1111/1541-4337.12736

[3] Asif M, Rooney LW, Ali R, Riaz MN. Applications and opportunities of pulses in food system. A review. Critical Reviews in Food Science and Nutrition. 2013;**53**(11):1168-1179. DOI: 10.1080/10408398.2011.574804

[4] Gueguen J, Cerletti P. Proteins of some legume seeds: Soybean, pea, faba bean and lupin. In: Hudson BJF, editor. New and Developing Sources of Food Proteins. Dordrecht: Springer Science+Business Media, B.V.; 1994. pp. 145-183. DOI: 10.1007/978-1-4615-2652-0_6

[5] Schutyser MAI, Pelgrom PJM, Van der Goot AJ, Boom RM. Dry fractionation for sustainable production of functional legume protein concentrates. Trends in Food Science and Technology. 2015;**45**(2):1-11. DOI: 10.1016/j.tifs.2015.04.013

[6] Vollmann J. Soybean versus other food grain legumes: A critical appraisal of the United Nations International Year of Pulses 2016. Journal of Land Management, Food and Environment. 2016;**67**(1):17-24. DOI: 10.1515/boku-2016-0002

[7] González-Pérez S, Arellano JB. Vegetable protein isolates. In: Phillips GO, Williams PA, editors. Handbook of Hydrocolloids. 2nd ed. Oxford: Woodhead Publishing; 2009. pp. 383-419

[8] Celmeli T, Sari H, Canci H, Sari D, Adak A, Eker T, et al. The nutritional content of common bean (*Phaseolus vulgaris* L.) landraces in comparison to modern varieties. Agronomy. 2018;**8**(9):166. DOI: 10.3390/agronomy8090166

[9] Girish TK, Pratape V, Rao UP. Nutrient distribution, phenolic acid composition, antioxidant and alpha-glucosidase inhibitory potentials of black gram (*Vigna mungo* L.) and its milled by-products. Food Research International. 2012;**46**:370-377. DOI: 10.1016/j.foodres.2011.12.026

[10] Mubarak AE. Nutritional composition and antinutritional factors of Mung bean seeds (*Phaseolus aureus*) as affected by some home traditional processes. Food Chemistry. 2005;**89**:489-495. DOI: 10.1016/j.foodchem.2004.01.007

[11] Hung PV, Nguyen THY, Nguyen TLP, Nguyen PHT, Nguyen THT. Nutritional composition, enzyme activities and bioactive compounds of mung bean (*Vigna radiata* L.) germinated under dark and light conditions. LWT—Food Science and Technology. 2020;**133**:110100. DOI: 10.1016/j.lwt.2020.110100

[12] Gebreegziabher BG, Tsegay BA. Proximate and mineral composition of Ethiopian pea (*Pisum sativum* var. *abyssinicum* A. Braun) landraces vary across altitudinal ecosystems. Cogent Food & Agriculture. 2020;**6**(1). DOI: 10.1080/23311932.2020.1789421

[13] Nikolopoulou D, Grigorakis K, Stasini M, Alexis MN, Iliadis K. Differences in chemical composition of field pea (*Pisum sativum*) cultivars:

Effects of cultivation area and year. Food Chemistry. 2007;**103**:847-852. DOI: 10.1016/j.foodchem.2006.09.035

[14] Gueguen J, Barbot J. Quantitative and qualitative variability of pea (*Pisum sativum* L.) protein composition. Journal of the Science of Food and Agriculture. 1988;**42**:209-224. DOI: 10.1002/jsfa.2740420304

[15] Barratt DHP. Chemical composition of mature seeds from different cultivars and lines of *Vicia faba* L. Journal of the Science of Food and Agriculture. 1982;**33**:603-608. DOI: 10.1002/jsfa.2740330703

[16] Bhatty RS. Chemical composition of some faba bean cultivars. Canadian Journal of Plant Science. 1974;**54**:413-421. DOI: 10.4141/cjps74-063

[17] Tahir M, Lindeboom N, Båga M, Vandenberg A, Chibbar R. Composition and correlation between major seed constituents in selected lentil (*Lens culinaris* Medik) genotypes. Canadian Journal of Plant Science. 2011;**91**(5):825-835. DOI: 10.4141/cjps2011-010

[18] Wang N, Hatcher D, Toews R, Gawalko E. Influence of cooking and dehulling on nutrional compostion of several varieties of lentils (*Lens culinaris*). LWT—Food Science and Technology. 2009;**42**:842-848. DOI: 10.1016/j.lwt.2008.10.007

[19] Hefnawy TH. Effect of processing methods on nutritional composition and anti-nutritional factors in lentils (*Lens culinaris*). Annals of Agricultural Sciences. 2011;**56**:57-61. DOI: 10.1016/j.aoas.2011.07.001

[20] Gharibzahedi S, Mousavi M, Jafari S, Faraji K. Proximate composition, mineral content, and fatty acids profile of two varieties of lentil seeds cultivated in Iran. Chemistry of Natural Compounds. 2012;**47**:976-978. DOI: 10.1007/s10600-012-0119-2

[21] Frimpong A, Sinha A, Tar'an B, Warkentin TD, Gossen BD, Chibbar RN. Genotype and growing environment influence chickpea (*Cicer arietinum* L.) seed composition. Journal of the Science of Food and Agriculture. 2009;**89**:2052-2063. DOI: 10.1002/jsfa.3690

[22] Wang N, Hatcher DW, Tyler RT, Toews R, Gawalko EJ. Effect of cooking on the composition of beans (*Phaseolus vulgaris* L.) and chickpea (*Cicer arietinum* L.). Food Research International. 2010;**43**:589-594. DOI: 10.1016/j.foodres.2009.07.012

[23] Preet K, Punia D. Proximate composition, phytic acid, polyphenols and digestibility (*in vitro*) of four brown cowpea varieties. International Journal of Food Sciences and Nutrition. 2000;**51**(3):189-193. DOI: 10.1080/09637480050029692

[24] Maia FM, Oliveira JT, Matos MR, Moreira RA, Vasconcelos IM. Proximate composition, amino acid content and haemagglutinating and trypsin-inhibiting activities of some Brazilian *Vigna unguiculata* (L) Walp cultivars. Journal of the Science of Food and Agriculture. 2000;**80**:453-458. DOI: 10.1002/(SICI)1097-0010(200003)80:4<453::AID-JSFA548>3.0.CO;2-X

[25] Farinu GO, Ingrao G. Gross composition, amino acid, phytic acid and trace element contents of thirteen cowpea cultivars and their nutritional significance. Journal of the Science of Food and Agriculture. 1991;**55**:401-410. DOI: 10.1002/jsfa.2740550308

[26] Amarteifio JO, Munthali DC, Karikari SK, Morake TK. The composition of pigeon peas (*Cajanus cajan* (L. Millsp.)) grown in Botswana. Plant Foods for Human Nutrition. 2002;**57**:173-177. DOI: 10.1023/A:1015248326920

[27] Onimawo I, Akpojovwo AE. Toasting (dry heat) and nutrient

composition, functional properties and antinutritional factors of pigeon pea (*Cajanus cajan*) flour. Journal of Food Processing and Preservation. 2006;**30**:742-753. DOI: 10.1111/j.1745-4549.2006.00093.x

[28] Tizazu H, Emire S. Chemical composition, physicochemical and functional properties of lupin (*Lupinus albus*) seeds grown in Ethiopia. African Journal of Food, Agriculture, Nutrition and Development. 2010;**10**:3029-3046. DOI: 10.4314/ajfand.v10i8.60895

[29] Da Silva J, Carrão-Panizzi M, Prudencio S. Chemical and physical composition of grain-type and food-type soybean for food processing. Food Technology Pesquisa Agropecuaria Brasileira—PAB. 2009;**44**(7). DOI: 10.1590/S0100-204X2009000700019

[30] Giami SY. Chemical composition and nutritional attributes of selected newly developed lines of soybean (*Glycine max* (L) Merr). Journal of the Science of Food and Agriculture. 2002;**82**:1735-1739. DOI: 10.1002/jsfa.1239

[31] Redondo-Cuenca A, Villanueva-Suárez M, Rodriguez-Sevilla MD, Mateos-Aparicio I. Chemical composition and dietary fibre of yellow and green commercial soybeans (*Glycine max*). Food Chemistry. 2007;**101**:1216-1222. DOI: 10.1016/j.foodchem.2006.03.025

[32] Khalil JK, Chughtai MID. Chemical composition and nutritional quality of five peanut cultivars grown in Pakistan. Plant Foods for Human Nutrition. 1983;**33**:63-70. DOI: 10.1007/BF01093738

[33] Tharanathan RN, Wankhede DB, Rao M, Rao RR. Carbohydrate composition of groundnuts (*Arachis hypogea*). Journal of the Science of Food and Agriculture. 1975;**26**(6):749-754. DOI: 10.1002/jsfa.2740260605

[34] Shevkani K, Singh N, Chen Y, Kaur A, Yu L. Pulse proteins: Secondary structure, functionality and applications. Journal of Food Science and Technology. 2019;**56**:2787-2798. DOI: 10.1007/s13197-019-03723-8

[35] Wright DJ, Bumstead MR. Legume proteins in food technology. Philosophical Transactions of the Royal Society of London, Series B, Biological Sciences. 1984;**304**:381-393. DOI: 10.1098/rstb.1984.0033

[36] Sironi E, Sessa F, Duranti M. A simple procedure of lupin seed protein fractionation for selective food applications. European Food Research and Technology. 2005;**221**:145-150. DOI: 10.1007/s00217-005-1151-2

[37] Boyce J, Zare F, Pletch A. Pulse proteins: Processing, characterization, functional properties and applications in food and feed. Food Research International. 2010;**43**:414-431. DOI: 10.1016/j.foodres.2009.09.003

[38] Arise AK, Nwachukwu ID, Aluko RE, Amonsou EO. Structure, composition and functional properties of storage proteins extracted from bambara groundnut (*Vigna subterranea*) landraces. International Journal of Food Science and Technology. 2017;**52**:1211-1220. DOI: 10.1111/ijfs.13386

[39] Gueguen J. Legume seed protein extraction, processing, and end product characteristics. Plant Foods for Human Nutrition. 1983;**32**:267-303. DOI: 10.1007/BF01091191

[40] Russin TA, Boye JI, Arcand Y, Rajamohamed SH. Alternative techniques for defatting soy: A practical review. Food and Bioprocess Technology. 2011;**4**:200-223. DOI: 10.1007/s11947-010-0367-8

[41] Singhal A, Karaca AC, Tyler R, Nickerson M. Pulse proteins: From processing to structure-function

relationships. In: Goyal A, editor. Grain Legumes. London: IntechOpen Limited; 2016. pp. 55-78 DOI: 10.5772/ 64020

[42] Tyler RT, Youngs CG, Sosulski FW. Air classification of legumes. I. Separation efficiency, yield, and composition of the starch and protein fractions. Cereal Chemistry. 1981;**58**: 144-148

[43] Barac MB, Pesic MB, Stanojevic SP, Kostic AZ, Bivolarevic V. Comparative study of the functional properties of three legume seed isolates: Adzuki, pea and soy bean. Journal of Food Science and Technology. 2015;**52**(5):2779-2787. DOI: 10.1007/s13197-014-1298-6

[44] Lafarga T, Álvarez C, Villaró S, Bobo G, Aguiló-Aguayo I. Potential of pulse-derived proteins for developing novel vegan edible foams and emulsions. International Journal of Food Science and Technology. 2020;**55**:475-481. DOI: 10.1111/ijfs.14286

[45] Akin Z, Ozcan T. Functional properties of fermented milk produced with plant proteins. LWT—Food Science and Technology. 2017;**86**:25-30. DOI: 10.1016/j.lwt.2017.07.025

[46] Shuang-Kui D, Hongxin J, Xiuzhu Y, Jay-Lin J. Physicochemical and functional properties of whole legume flour. LWT—Food Science and Technology. 2014;**55**:308-313. DOI: 10.1016/j.lwt.2013.06.001

[47] Gupta S, Chhabra GS, Liu C, Bakshi JS, Sathe SK. Functional properties of select dry bean seeds and flours. Journal of Food Science. 2018;**83**:2052-2061. DOI: 10.1111/ 1750-3841.14213

[48] Aditya J, Changqi L, Shridhar S. Functional properties of select seed flours. LWT—Food Science and Technology. 2015;**60**:325-331. DOI: 10.1016/j.lwt.2014.08.038

[49] Stone AK, Avarmenko NA, Warkentin TD, Nickerson MT. Functional properties of protein isolates from different pea cultivars. Food Science and Biotechnology. 2015;**24**:827-833. DOI: 10.1007/s10068-015-0107-y

[50] Kaur M, Singh N. Characterization of protein isolates from different Indian chickpea (*Cicer arietinum* L.) cultivars. Food Chemistry. 2007;**102**:366-374. DOI: 10.1016/j.foodchem.2006.05.029

[51] Fernandez-Quintella A, Macarulla MT, Del Barrio AS, Martinez JA. Composition and functional properties of protein isolates obtained from commercial legumes grown in Northern Spain. Plant Foods for Human Nutrition. 1997;**51**:331-342. DOI: 10.1023/A:1007936930354

[52] Tsoukala A, Papalamprou E, Makri E, Doxastakis G, Braudo EE. Adsorption at the air-water interface and emulsification properties of grain legume protein derivatives from pea and broad bean. Colloids and Surfaces B: Biointerfaces. 2006;**53**(2):203-208. DOI: 10.1016/j.colsurfb.2006.08.019

[53] Sharif HR, Williams PA, Sharif MK, Abbas S, Majeed M, Masamba KG, et al. Current progress in the utilization of native and modified legume proteins as emulsifiers and encapsulants—A review. Food Hydrocolloids. 2018;**76**:2-16. DOI: 10.1016/j.foodhyd.2017.01.002

[54] Pedrosa MM, Varela A, Domínguez-Timón F, Tovar CA, Moreno HM, Borderías AJ, et al. Comparison of bioactive compounds content and techno-functional properties of pea and bean flours and their protein isolates. Plant Foods for Human Nutrition. 2020;**75**:642-650. DOI: 10.1007/s11130-020-00866-4

[55] Yeon-Ji J, Weijuan H, Lingyun C. Fabrication and characterization of lentil protein gels from fibrillar aggregates and the gelling mechanism

study. Food & Function. 2020;**11**:10114-10125. DOI: 10.1039/D0FO02089C

[56] Diedericks CF, Shek C, Jideani VA, Venema P, van der Linden E. Physicochemical properties and gelling behaviour of Bambara groundnut protein isolates and protein-enriched fractions. Food Research International (Ottawa, Ont.). 2020;**138**(Pt B):109773. DOI: 10.1016/j.foodres.2020.109773

[57] Guldiken B, Stobbs J, Nickerson M. Heat induced gelation of pulse protein networks. Food Chemistry. 2021;**350**:129158. DOI: 10.1016/j.foodchem.2021.129158

[58] Vogelsang-O'Dwyer M, Zannini E, Arendt EK. Production of pulse protein ingredients and their application in plant-based milk alternatives. Trends in Food Science and Technology. 2021;**110**:364-374. DOI: 10.1016/j.tifs.2021.01.090

[59] Chandra S, Singh S, Kumari D. Evaluation of functional properties of composite flours and sensorial attributes of composite flour biscuits. Journal of Food Science and Technology. 2015;**52**(6):3681-3688. DOI: 10.1007/s13197-014-1427-2

[60] Foschia M, Horstmann SW, Arendt EK, Zannini E. Legumes as functional ingredients in gluten-free bakery and pasta products. Annual Review of Food Science and Technology. 2017;**8**(1):75-96. DOI: 10.1146/annurev-food-030216-030045

[61] Alabi O, Ali N, Nwachukwu ID, Aluko R, Amonsou E. Composition and some functional properties of Bambara groundnuts vicilin fraction. LWT—Food Science and Technology. 2020;**125**:109256. DOI: 10.1016/j.lwt.2020.109256

Chapter 17

A Review on the Cooking Attributes of African Yam Bean (*Sphenostylis stenocarpa*)

Ndenum Suzzy Shitta, Alex Chukwudi Edemodu, Wosene Gebreselassie Abtew and Abush Abebe Tesfaye

Abstract

African yam bean, an underutilized legume usually cultivated for its edible tubers and seeds, is known for its nutrition-rich qualities; however, the crop's level of consumption is low. The underutilization of the crop could be attributed to several constraints, including long cooking hours of up to 24 hours. Cooking time is an important food trait; it affects consumers' choices, nutrients content, and anti-nutrient conditions. Additionally, foods requiring long cooking hours are non-economical in terms of energy usage and preparation time. The prolonged cooking time associated with AYB places enormous limitations on the invaluable food security potentials of the crop. Therefore, the availability of AYB grains with a short cooking time could lift the crop from its present underused status. To efficiently develop AYB grains with reduced cooking time, information on the crop's cooking variables is a prerequisite. This review presents available information on variations in cooking time, cooking methods, and processing steps used in improving cooking time and nutrient qualities in AYB. Likewise, the review brings to knowledge standard procedures that could be explored in evaluating AYB's cooking time. This document also emphasizes the molecular perspectives that could pilot the development of AYB cultivars with reduced cooking time.

Keywords: Seed hardness, Mattson Bean Cooker, GWAS, QTL, Cooking Time

1. Introduction

Food and nutrition security which is part of livelihood, is notably attracting the attention of stakeholders, spanning across nations, research organizations, the general public, academic institutions, and policymakers. At present, the world population is estimated at 7 billion; however, by 2050, the population is expected to reach 9.3 billion. As of 2017, the number of food-insecure people worldwide was estimated at 690 million [1]; however, by 2050, a 70–85% increase in food production will be needed to feed the projected 9.3 billion people [2, 3]. Notwithstanding, upscaling the adoption and utilization of sustainable crops offers considerable potentials in boosting food production amidst the prevailing challenges.

Grain-Legumes are sustainable, capable of surviving under harsh climate conditions. The grain legumes require minimal fertilizer inputs because of their ability to fix atmospheric nitrogen through symbiosis with soil *Rhizobia*.

Also, intercropping legumes with other crops has improved soil fertility and crop productivity [4–6]. Importantly, legumes are a good source of food and feed for humans and animals, respectively; crops within the legume category are nutritionally rich; most significantly, they provide affordable sources of protein [7, 8]. The contribution of legumes as food and feed differs across types, while some legumes are known worldwide and considerably utilized (soy bean) (*Glycine max L*), common bean (*Phaseolus vulgaris* L), cowpea (*Vigna unguiculata* L.)) others are less known and underutilized (African yam bean (*Sphenostylis stenocarpa* Harms), lablab beans, (*Lablab purpureus* L) wing bean (*Psophocarpus tetragonolabus* L.). Adopting and accepting underutilized legumes such as African yam bean as a food crop is vital for their survival; nevertheless, AYB's adoption and utilization is intertwined with several factors, including cooking time, nutrient potentials, palatability, and value-added products.

1.1 African yam bean

African yam bean which, is commonly referred to as AYB, is one among the underutilized grain legumes of tropical Africa. The crop is grown for its edible seeds and tuberous roots. **Figure 1** presents AYB seeds harvested from a field evaluation in 2020. AYB seeds are enclosed in pods measuring about 3–15 cm long, such that a single pod can accommodate up to 30 seeds. The crop is a climber usually grown in mixed cropping with major crops [10–12]. AYB is locally adopted and has wide adaptability across diverse environmental conditions [13, 14]. Even though the crop is usually cultivated as an annual crop [15–17], some schools of thought consider it as perennial [18–20]. The cultivation of AYB majors among smallholder farmers across sub-Saharan Africa, of which Nigeria is one country prominent on the list [21]. The consumption of AYB is known to contribute to daily nutrition, food availability, and diet diversification to communities utilizing it; this date back to the Nigerian civil war of 1967–1970, where the crop's food and nutritional potentials were efficiently utilized in fighting malnutrition and hunger [15, 22–24].

The seeds of AYB provide an affordable source of protein when compared with other plant sources and animal extract. Aside from its rich protein content, its high carbohydrate content [25, 26] is comparable to the amount reported in grain cereals.

Figure 1.
Dried AYB seeds. (A) Non variegated seeds (B) Variegated seeds. Source: field evaluation (Shitta et al. [9]).

AYB's amino acid (histidine, isoleucine, lysine, methionine) profile is more in quantity than the amount observed in soybean [27–29]. Likewise, several authors have reported the presence of essential nutrients in AYB's seeds [25, 26, 30–36]. AYB tubers (**Figure 2**) contain considerable amount of magnesium (167 mg/100 g), potassium (1010 mg/100 g), protein (15–16%), and carbohydrate (67–68%) [34]. In addition to the crop's nutritional qualities, the crop is flexible for use in various diets; it can be utilized as a condiment, or as a whole meal, or as a snack. The contribution of AYB in feeds enrichment is an added advantage of the crop's food and nutrition attributes [37, 38].

Considering the enormous potential of AYB and its role in some African traditions [39–41]; the efficient utilization of AYB can reduce hunger and nutritional challenges in sub-Saharan Africa. Nevertheless, the food potential of the crop remains widely untapped, which can be attributed to several constraints such as long cooking hours of up to 24 hours [41–44], a long-maturity cycle of 9–10 months [16, 17, 45], and the abundance of anti-nutrition factors [35, 46–49]. However, the genetic variability reported in the crop [9, 50–53] provides a foundation for breeders to develop improved cultivars. In particular, the availability of AYB cultivars with reduced cooking time could boost the cultivation and consumption of the crop. Up-to-date information on cooking-related attributes is a prerequisite for improving cooking time trait. Keeping the above in view, the present review brings to knowledge cooking variables reported in AYB. Also, the review proposes the application of standard procedures and molecular technology for advanced studies. Furthermore, the present document is intended to stimulate more research interest towards improving cooking time in the crop.

Figure 2.
AYB tubers. Source: field evaluation (Shitta et al. [9]).

Seed thickness (mm)	Seed length (mm)	Seed width (mm)	100 seed weight (g)	References
—	8.90	6.90	24.70	[23]
6.99	8.69	6.80	29.79	[56]
6.66	8.57	6.77	28.80	[51]
6.30	8.21	—	20.83	[48]
6.22	8.06	6.31	22.42	[53]
6.20	8.40	6.35	25.30	[9]

Table 1.
Means of some physical properties reported in AYB seeds.

1.2 Structure of African yam bean seeds

Past research investigations have explained the relationship between seed properties, variety type, seed storage conditions, and cooking time [54, 55]. **Table 1** presents the physical properties reported in AYB seeds. AYB seeds are, dicot in nature and they can measure up to 10 mm in length and 7 mm in width and thickness[9, 50–53, 56]. The seeds of AYB differ in texture across germplasm; they could be rough, wrinkled, or smooth. The electron microstructure study of seeds revealed the presence of smooth starch granules exhibiting different sizes and shapes [57]. The cells were bounded by cell walls same as observed in other legumes [58, 59]. Likewise, the round undulating surface observed in the cotyledon is similar in structure to that of cowpea [59, 60]. For seeds subjected to milling, the cotyledon and cell components showed structural change. Equally, cell wall materials and protein matrix were reduced to flakes and particles; however, the structure of starch granules remained unchanged. The micrographs of cotyledon, flour, and starch showed the size of starch granules within the range of 4–40 μm for lengths and 4–25 μm for diameter [57].

2. Cooking quality in African yam bean

Preparing and cooking food is an integral part of daily living [61, 62]. For example most grain legumes are subjected to cooking before consumed; the cooking process converts raw food into a ready-to-eat product. Also, cooking facilitates the destruction of foodborne pathogens, thereby eliminating microbial hazards and achieving quality [63]. Moreover, the physical and chemical changes that occur during cooking increases the digestibility and availability of nutrient for use and storage in the body [64]; through processes including inactivation of antinutrient, starch gelatinization, proteins denaturation, leaching of polyphenols and solubilization of polysaccharides among other factors [59, 65, 66]. Despite the importance of cooking in food and nutrition the cooking culture is dwindling, especially in industrialized societies where individuals are exposed to a busy lifestyle with little time at their deposal. To cope with busy schedules, consumptions are choosing convenience food that requires less cooking time. Also, reports have shown that consumers are ready to pay more in exchange for long cooking hours [67, 68].

Cooking time, an attribute of cooking quality is defined as the time from the beginning of cooking up to when the food becomes tender and suitable to eat [66, 69]. AYB, the same as most legumes is characterized by seed hardness, requiring long

Source	Cooking method	Cooking time (mins)	End product	References
Whole AYB seeds	Boiling	480	Flour	[70]
Whole AYB seeds	Boiling	228	Paste	[47]
Whole AYB seeds	Boiling	60	Porridge	[71]
Whole AYB seeds	Roasting	60	Flour	[47]
Whole AYB seeds	Roasting	10	Flour	[43]
Whole AYB seeds	Boiling	155	—	[43]
Dehulled AYB and maize flour	Frying	10	Kokoro	[72]
Dehulled AYB wet flour	Frying		Cheese	[73]
Dehulled AYB-wheat flour	Baking	20	Cookies	[74]
Dehulled AYB cowpea flour	Steaming	50	Moi-moi	[75]
Dehulled AYB wet flour	Steaming	60	Moi-moi	[71, 76]
AYB-maize-coconut fiber	Roasting	5	Flour blend	[77]
Whole grain	Roasting	45	Flour	[78]
Whole grain	Roasting	300	Flour	[79]

Table 2.
Source, cooking method, cooking time, end product, and references reported in AYB cooking experiments.

cooking hours of up to 24 hours (**Table 2**) in some scenarios [80]. Seed hardness has been identified as a heritable trait but also affected by seed composition, production, and, storage environment [54, 81, 82]. The mechanism by which seeds become hard-to-cook is categorized as a very complex phenomenon; it includes processes such as changes in the intracellular cell wall, middle lamella, polysaccharides, and other components. The hard-to-cook mechanism in seeds has been extensively reviewed by authors [83–85]. According to a particular study, an increase in calcium ion concentration led to a subsequent increase in seed hardness and a decrease in phytate concentration. It was also reported that a higher rate of leaching in phytate and peptic acid occurred in cooked and soaked hard-to-cook seeds than in fast-to-cook seeds [85].

Generally, grains with short cooking time are more preferred by consumers; because less time is invested in their preparation, and importantly less energy is spent when compared to energy requirements for grains with long cooking time. In addition, several studies have shown that nutrients such as minerals and proteins are conserved when grains are cooked over a short period. In contrast grains requiring long cooking hours usually lose a significant amount of nutrients [55, 86]. Cooking methods reported in AYB include boiling, steaming, roasting, and frying. However, advanced procedures including, sensory analysis: involving sensory panel [87, 88]; tactile method: [89] a method of compressing seeds within the thumb; texture analysis: [87] a method that measures the resistance of seed compression using a texture analyzer [90] have been investigated in major legumes.

2.1 Cooking method reported in AYB

2.1.1 Boiling

Boiling cooking method is a moist approach whereby the target food is submerged into a liquid. Cooking is achieved through the transfer of heat from the cooking equipment to the liquid in contact with the food. The food surface absorbs the heat and through conduction, the heat passes through to cook the food. The boiling method was experimented with selected AYB grains. The steps included boiling the grains in water for 480 minutes (**Table 2**) and thereafter oven drying for 24 hours before milling into flour [70]. In another report, AYB grains were boiled for 228 minutes. The analysis of the boiled seeds showed a reduction in phytate content and an increase in moisture content [47]. In addition, the boiling cooking method was reportedly used in preparing porridge. The procedure included presoaking seeds overnight and boiling them for 60 minutes. The porridge analysis showed an increase in carbohydrate, gross energy, fiber, lipid, water absorption capacity, oil absorption, bulk density, and gelation capacity however a decrease in protein and moisture content was observed [71].

2.1.2 Roasting

The roasting method is commonly used in preparing "roasted AYB grain," a popular snack consumed in combination with other food in South-East Nigeria [19, 40, 43]. Roasting was effective in increasing the level of phosphorus and in-vitro protein digestibility of grains. An increase in phytic acid was also reported; however, the tannin level was shown to be at the barest minimum [43]. In the preparation of breakfast cereal from AYB grains in combination with maize and coconut fiber, the blends were roasted for 5 minutes at 280^0c temperature. The formulated blends revealed a protein content of 18.26%, moisture content of 4.20%, ash content of 7.36%, and energy content of 339.47% [77]. The roasting approach was likewise used in preparing AYB flour. The grains were subjected to roasting for 45 minutes (**Table 2**) using firewood as the energy source. Then, the roasted grains were dehulled and milled. The analysis of the roasted flour showed a decrease of about 0.27 mg/100 g in the level of the tannin content [78]. In a separate study, AYB grains were roasted in an oven at 120^0c for 300 minutes; and the roasted grains were dehulled and milled. The analysis of the dehulled flour showed a reduction in the emulsifying capacity, foam capacity, and stability of the flour, also the samples presented a high water and oil absorption capacity [79]. In a further experiment, researchers investigated the effect of roasting on the proximate, mineral, and anti-nutrient content of AYB grains. The study preceded the roasting of grains over firewood for 1 hour at 300^0c temperature condition. An increase was reported in the levels of calcium, potassium, copper, iron, manganese, magnesium, phosphorus, and sodium, and a drastic reduction in the percentage level of phytate, oxalate, tannins, hydrogen cyanide, and trypsin inhibitor was reported. On the contrary, there was no significant increase in the nutrient content [47].

2.1.3 Steaming

The steaming approach involves the use of steam as the cooking medium; the steam is mostly generated from vigorously boiling water. Unlike reported in boiling method, the steaming procedure does not require submerging the food directly into the water; in steaming, the target food gets cooked as the result of the steam or vapors generated from the boiling water. Steam is considered a good heat conductor, nevertheless, the temperature release from steam does not exceed that of boiling water except in the pressure system [91]. Steaming was reported to have minimal effects on chlorophyll, soluble protein, sugar, vitamin c, and glucosinolates [92]. The steaming process helped preserve antioxidant properties and maintained the lowest biogenic amine content in bean varieties [93]. In AYB, the steaming approach was reportedly used in preparing a traditional snack called "Moi-Moi". The procedure involved dehulling and wet milling of the grains accompanied by spicing. For the Moi-Moi to get cooked, it was steamed for about 60 minutes [71, 76]. The analysis of the AYB Moi-Moi showed a lower gelation capacity, higher water absorption capacity, lower oil absorption capacity when compared to Moi-Moi made from cowpea. The sensory analysis of AYB Moi-Moi showed no significant difference in color and flavor from Moi-Moi made from cowpea (cowpea is the most common grain for preparing Moi-Moi). Additionally, the acceptance level of the AYB Moi-Moi was similar to Moi-Moi constituted from cowpea [71]. Some researchers utilized the steaming cooking method in making Moi-Moi from AYB and cowpea blends, they reported a total steaming time of about 50 minutes [75].

2.1.4 Frying

Frying is one of the ancient and well-known cooking methods used for food preparation; the procedure is known for its ease, speed, and unique flavor and taste [94]; in addition, frying gives an attractive color, texture to food. The frying process involves the use of fat or oil which serves as the medium of direct heat transfer with the food [63, 95]. The transfer of heat, oil, and air during the frying process brings about changes like loss of moisture, oil uptake, starch gelatinization, aromatization, denaturation of protein, and changes in the color of the food. The changes in food and oil are largely dependent on the food property, the quality of oil, heating process, length of immersion, the rate at which air mixes with the oil, temperature, and the quality of the frying medium [96]. Frying could lead to the release of toxic products through oxidation, which usually occurs when oil is continuously used under high temperatures and atmospheric air [97]. The frying method of cooking was reportedly used in the preparation of traditional snacks commonly known as "akara" or "beans ball", a snack widely eaten in Nigeria. The grains were soaked overnight and dehulled before wet milling (paste) and spicing. The frying medium (groundnut oil) was heated to 185-190^{0}c, and the total frying time was about 5 minutes (**Table 2**). The end product (akara) showed an increase in carbohydrate, gross energy, water absorption capacity, oil absorption capacity, bulk density, and gelation capacity. Meanwhile, no significant difference was reported in accepting the AYB akara from the usual cowpea akara [71]. In like manner, the frying method was used in preparing Kokoro a popular snack in South-West Nigeria. The Kokoro process involved deep-frying the paste constituted from the AYB-Maize blend for about 10 minutes. The proximate analysis conducted on the Kokoro showed an increase in protein, sugar, ash, moisture, potassium, and calcium as the proportion of AYB flour increases. On the contrary, a decrease in fat and starch was observed with an increase in AYB flour [72]. Furthermore, the frying process was used to produce AYB cheese, using palm oil as the frying medium. The sensory evaluation indicated a general acceptance of the AYB cheese [73].

2.1.5 Baking

The baking process is a method whereby the raw dough is transformed into crumb and crust texture, under the influence of heat. During baking, the changes that occur include the crust formation, yeast inactivation, coagulation of protein, volume expansion, starch gelatinization, and moisture loss [98–100]. The baking approach was used in producing cookies from AYB-wheat composite flour. The cookies were baked for 20 minutes using an oven mark of 180^{0}c. The nutritional analysis of the cookies showed an increase in protein content from 8.59 to 9.35% fat from 3.84 to 4.63%, ash from 4.84 to 5.21%, and crude fiber from 3.84 to 4.22%.

An increase in mineral content corresponding to a percentage increase in the level of AYB flour was also observed [74].

2.2 Technological gap in the evaluation of AYB cooking time

In AYB, the majority of the cooking time investigations were conducted using basic approaches like firewood, gas, and kerosene stove. No information is documented on the use of standard equipment such as texture analyzer and Mattson bean cooker; however, the use of Matson bean cooker and texture have been reported in several legumes.

2.2.1 Mattson bean cooker

One standard method of measuring cooking time in pulses is to evaluate using a Mattson bean cooker [101]. The equipment is easy to use, cost-effective, and generates unbiased data compared to other methods [90]. The use of Mattson cooker is recommended in grain genetic improvement for evaluating new varieties [66]. Mattson first developed the Mattson bean cooker, having 100 plungers [102], but was later redesigned to have 25 plungers [103]. The usage of the equipment involves placing individual presoaked seeds on each of the saddle on the rack such that the tip of each plunger comes in contact with the surface of the seed. The weight of each plunger can be optimized to suit the size of the target grain by adjusting the number of lead buckshot inside each plunger. To initiate the cooking test, the lower part of the cooking rack is immersed in a boiling water bath up to half of its height. When a seed reaches tenderness, the plunger penetrates that particular seed and drops a short distance through the hole in the saddle. The top of a plunger that has dropped (penetrated a seed) will be lower than the top of the plungers which are yet to drop. The scenario makes it visibly easy to identify the plunger that has penetrated its seed [66, 90]. The cooking time for a set of seeds (25) has been explained differently by researchers; the cooking time was defined as the time required for 100% of the seeds to get penetrated [104]. In an additional study, the cooking time was recorded as the time 92% of seeds got penetrated [105]. Operating the Mattson cooker requires the uninterrupted attention of the user; the user manually records the time each plunger penetrates a seed the situation becomes more critical when multiple plungers penetrate at the same time. To overcome the bottleneck of manual recording several researchers have reported the use of an automated Mattson cooker where the cooking time is automatically recorded [66, 90, 106].

2.2.2 Texture analysis

The texture is an important trait of food characterized by its mechanical, geometrical, surface, and body attributes detected by senses of vision, hearing, touch, and kinesthetics [107, 108]. The mechanical attributes have to do with the qualities of the food under stress conditions; like hardness, cohesiveness, elasticity, and adhesiveness. In contrast, the geometrical attributes are related to the size, shape, and structural arrangement of the product. The surface attribute has to do with the sensations produced (in the mouth) around or in the surface of the product by moisture and fat or either of the two; similarly, the body attributes are related to the feelings produced in the mouth and how the moisture and fat or both are released [109]. Of recent, instrumental texture analysis has proven to be efficient in evaluating the mechanical and physical qualities of the raw and finished product, of which the application of texture analyzer is a well-established protocol. A texture analyzer is used for evaluating the hardness, fragility, adhesiveness, springiness, cohesiveness, gumminess, chewiness, and resilience of food [54, 88]. The instrument is easy to operate; it eliminates subjective judgment, as may be found in sensory evaluations [109]. The selection of a probe for use during analysis is dependent on the type of test, which could be a compression test, penetration (puncture) test, traction (tension). The different texture analysis test types were previously reviewed [109]. The texture analyzer has been applied in several texture studies in legumes, fruits, vegetables, meat, milk, among others [109, 110].

3. Methods of reducing cooking time in AYB

Several studies have reported a significant decrease in cooking time after seeds were subjected to processing methods like presoaking, dehulling, frying, steaming, and blanching [43, 71, 111, 112].

3.1 Presoaking of seeds

Presoaking is a long-age traditional practice used in homes to reduce cooking time, especially in grain legumes. The approach is flexible, simple, and common both at the domestic and industrial levels. The process involves the imbibition of water through the outer cuticle, the seed coat, and then into the cotyledons; [69, 113]. The first step in imbibition is the penetration of water by the seed, and the process can be through the seed coats since the seed coat has high fiber content and thus high-water holding capacity. Water inhibition can also occur through the micropyle or hilum; when the water reaches the cotyledons, the seed starts to absorb water and swell until the seeds attain their maximum water uptake capacity. Presoaking of seed before cooking enables the easy identification of unhydrated seeds, which can be discarded to achieve uniform cooking time. The procedure reduces cooking time because the hydrated seeds acquire a soft texture and thereby speeding up the cooking process and shortening the cooking time [114]. Also, soaking aid the easy identification of hydratable seeds and improves the nutrient quality of foods since the soaked content is usually discarded. Soaking grains before cooking is a good practice used traditionally in increasing food safety especially in situations when consumers have no idea of the storage preservatives used for the target grain.

The effect of presoaking in shortening the cooking time of AYB's seed was reported by several authors. Presoaking AYB seeds in distilled water over a varying time of 6, 12, 18, and 24 hours reduced cooking time by 50%. The process also reduced the level of tannin and phytate, in addition to improving in-vitro protein digestibility. Soaking for 12 hours was the most effective in reducing cooking time, tannin, phytate, and in-vitro protein digestibility; however, soaking for 24 hours before dehulling was observed to significantly increase crude protein level by 16% [43]. In a similar study, AYB seeds were presoaked each in 0.20%, 0.40%, 0.60%, 0.80% and 1.00% of akanwu (sodium sesquicarbonate), and common salt (sodium chloride) and water for a duration of 6, 12, 18, 24, 30, 36 hours. Seeds soaked for 6 hours in 0.060% akanwu and 1.00% common salt showed a 50% decrease in cooking time, while seeds soaked in tap water achieved a 50% reduction in cooking time after 24 hours of presoaking. Meanwhile, seeds presoaked in tap water took about 180 minutes to get tender [112]. According to a study, a 50% reduction in cooking time was achieved when seeds were presoaked for 12 hours in either 1% potash or 4% common salt. Seeds soaked for 12 hours in 4% common salt reached tenderness after 45 minutes of cooking however seeds that were not soaked remained hard even after 60 minutes of cooking [111]. In a similar experiment, presoaking seeds in a different medium (water, alkali, brine, alkaline-brine) reduced the cooking time to a considerable level; the most effective medium was alkaline-brine, with a maximum cooking time of 100 minutes as against 210 minutes reported for cooking dry raw seeds [115]. In a separate study, AYB grains soaked overnight reached tenderness after 60 minutes of cooking [71]. Notably, aside from reducing cooking time, presoaking is also effective in investigating nutrient and anti-nutrient content [15, 43, 116–119].

3.2 Dehulling

Dehulling is a procedure through which seed coats or testa are removed either mechanically or using a machine. In most traditional setting, the process is carried out using either mortar and pestle or grinding stone, depending on the available option. Dehulled seeds have a good appearance in texture, cooking quality, palatability, and ease in digestibility. The approach reduces cooking time in grains legumes because during the dehulling process impermeable seed coats which usually prevent water uptake are removed [120]. Dehulled AYB grains showed the shortest cooking time of 35 minutes as against 80 and 150 minutes reported for whole seeds and soaked seeds, respectively [121]. The dehulling approach was observed to have a significant effect on the functional properties of AYB flour; a higher bulk density (0.93 g/cm^3) was reported as against the bulk density (0.59 g/cm^3) in cowpea and pigeon pea (0.70 g/cm^3). Similarly, the swelling index (5.9 g/cm^3) of dehulled AYB flour is more than the observed value in cowpea (3.7 g/cm^3) and pigeon (4.1 g/cm^3). The water absorption capacity (2.8 ml/h_20/g) in dehulled AYB flour was also higher than the observed in cowpea (1.2 ml/h_20/g) and pigeon pea (2.4 ml/h_20/g) [122]. In a further experiment, a higher water capacity of 71 ml/g was observed for dehulled AYB than the value of 60 ml/g reported for raw samples [71].

About 80–90% of the total amount of potential anti-nutrient factors (polyphenols) in grain legumes are found in the seed coats, and thus dehulling has proven to be effective in reducing anti-nutrient contents especially those found in the seed coats [123, 124]. Authors reported a drastic reduction in oxalate, phytate, saponin, trypsin inhibitor, and tannin content of dehulled AYB flour [122]. Similarly, an increase in protein but a decrease in calcium and iron was reported for dehulled AYB flour [43]. In a separate study, the proximate analysis of dehulled AYB flour showed high protein content, high carbohydrate concentration, and sufficient level of amino acid [125].

3.3 Other processing methods in AYB

3.3.1 Fermentation

Fermentation increases the bioaccessibility and bioavailability of nutrients and sensory quality in addition to shelf life [126, 127]. The process involves the biochemical modification of food by microorganisms and their enzymes [128]; the process is capable of disrupting the activities of pathogens [126, 129]. The fermentation process was explored for the preparation of "tempeh" from AYB grains; tempeh is a traditional food usually made from fermented soybean or soybean already broken down by microorganisms. The procedures for making AYB tempeh included: cooking presoaked grains for 45 minutes at 100^{0}c and inoculating the cooked grains with spore suspension to initiate fermentation. The inoculated grains were allowed to ferment over 42 hours. The final product showed significant changes in crude protein and carbohydrate. An increase in protein and amino nitrogen content was reported whereas a decrease in carbohydrates was observed. The quality of the AYB tempeh was acceptable to a large number of sensory panelists [130]. Meanwhile, some authors reported the minimal effect of fermentation on calcium, iron, magnesium, and zinc contents. However, they reported about a 34% reduction in phytate level and only tannin traces were detected [43]. Further research investigated the solid (3 days) and liquid (62 days) state fermentation approaches in making sauce from AYB grains. The prepared sauce revealed an increase of 11.94%, 4.85%, and 16.75% in ash, protein, and carbohydrate contents respectively. The sensory evaluation showed the acceptability of the AYB sauce was not significantly different from the level of acceptance of the commercial soy sauce in terms of color, aroma, and flavor [131].

Other studies used the fermentation process to formulate a yogurt-like product from dehulled and whole AYB grains. The process involved: the extraction of milk from grains which was followed by inoculation with a starter culture. For fermentation to occur, the inoculated milk was kept undisturbed over a time frame of 12 hours. The analysis of the formulated AYB yogurt presented a high total viable and *Lactobacilli* counts. As storage time increases, a decrease in the microbial load of the yogurt was observed [132]. In a similar experiment, raw AYB grains fermented for 48 hours showed an increase in protein and oil content [70]. "Dawa-Dawa" a traditional condiment was reportedly prepared through fermentation. The grains were boiled in water laced with "potash", the boiled grains were later dehulled and allowed to ferment at room temperature for 72 hours. The proximate analysis of the "Dawa-Dawa" showed an increase in crude protein from 22.00 to 32.80% and crude fibers from 5.70 to 7.77%, ash content increased from 3.20 to 4.60%, and lipid from 1.20 to 1.38%. Nevertheless, a decrease in carbohydrates from 74.20 to 57.21% was observed in the product [133].

3.3.2 Germination

Germination is a complex process that involves a mature seed to make an immediate change from maturation to the germination-driven stage and prepare for seedling growth [134]. The stages of germination include uptake of water by the seeds (imbibition) and the second phase is the reinitiating of metabolic processes followed by the emergence of the radicle through the seed envelopes. The germination process was used to prepare flour from AYB grains. The grains were soaked in water at room temperature for 48 hours. After soaking, the grains were allowed to sprout for 96 hours and subjected to oven drying. The dried grains were further dehulled and milled into flour. The germinated AYB-wheat composite flour showed an increase in protein; for every increase in the percentage of AYB flour [74].

4. Molecular perspectives for shortening cooking time in AYB

4.1 Seed hardness attribute

Seed hardness is an important quality of grain legumes; the trait acts as a barrier against seed coat pathogens and seed damage. Likewise, it affects germination, seed processing, and cooking time [82, 135]. Seed hardness is heritable but can also be influenced by environmental conditions at production and storage time [81, 82]. The genetic factors responsible for seed hardness are not well understood; however, the roles of a few genes have been documented [82]. The influence of the environment on seed hardness is reflected in the hard-to-cook phenomenon, which is not also independent of genetic influence [82, 84]. Understanding the genetic basis of cooking time in AYB is a necessity for improving the trait. It is noteworthy that genetic architecture in cooking time is yet to be reported in AYB; thus, no molecular approach has been documented in studying AYB's cooking time. Molecular techniques like GWAS and QTL could locate loci that controlled cooking time and thereby facilitate the identification of fast cooking lines. Likewise, new breeding techniques, including ZFNs, TALENS, and CRISPR/Cas9, have provided researchers the flexibility to insert desired traits precisely and quickly.

4.2 DNA technology

Previously, it would require about 7–10 years to transfer a target trait from a species to an adapted cultivar. The conventional process requires, handling a large number of progenies and several cycles of field evaluation. However, with molecular biology, a gene can be transferred in a single experiment, and within 5–6 years the new cultivar could exhibit a stable gene expression [136]. Presently, advances in plant molecular biology have provided processes and platforms through which the genetic architecture of traits can be well understood, manipulated, and transferred from different backgrounds [136, 137]. In addition, through DNA technology, gene sequences and functions can be accessed. Similarly, specific region (s) on the chromosome can be identified, molecular markers can be developed and genetic maps can be constructed, among many other possibilities. Genetic manipulation using physical, chemical, and biological mutagenesis presents added advantages with an enormous contribution to crop improvement. Among the widely used DNA technology reported in crop improvement programs are Genome-Wide Association Study (GWAS), Quantitative Trait Loci (QTL) Mapping, and Genome Editing.

4.2.1 GWAS

Over the years, GWAS has been implemented across a wide variety of crops such as soybean, maize, common bean, sorghum, and rice [55, 138–141]. GWAS identifies genetic variants across the genome and associates the variants with the target phenotype. The commonly used GWAS approach involves identifying single nucleotide polymorphism (SNPs) markers and testing each marker for evidence of an association between the marker and the trait of interest. The marker-trait association approach relies on linkage disequilibrium (LD) between markers and causal polymorphisms [142, 143]. To minimize false genotype–phenotype association that may arise from population structure, a linear mixed model analysis option is usually implemented. The application of GWAS has contributed significantly to identifying candidate genes; identified markers can be mapped to reference genomes, and thereafter candidate genes can be identified [143]. Once genomic regions of a target trait and the corresponding alleles at each locus are identified, the allele can be incorporated into another variety through crosses. The resultant progenies with the desired allele combination can be subjected to marker-assisted selection. GWAS in combination with marker-assisted breeding offers great gains for improving quantitative traits with low heritability [136].

4.2.2 QTL mapping

QTLs are phenotypically defined regions on the chromosome that contribute to allelic variation for a biological trait [144]. QTL technique has become a popular approach [144, 145] used to study complex traits [146, 147]. The application of QTL analysis in crop improvement was reported by several authors [82, 148]. Regions on the chromosomes that significantly affect variations of quantitative traits are identifiable through QTL mapping. The ability to locate chromosomal region (s) is important in identifying target genes and in understanding the genetic mechanism of genetic variation. Majorly, QTL mapping reveals information on QTL's having a significant effect on trait variation, and also answers the question to what extent is the variation due to additive, dominant, and epistasis effects of the QTL? The mapping of QTL also shows the genetic correlation of different traits and also answers the question does the QTL interact with the environment? [149]. The ability of QTL mapping to unravel and, at the same time provide answers to genetic questions makes it a powerful technique in crop improvement.

4.2.3 Genome editing

The discovery of genome editing technologies has revolutionized plant and animal research. Through genome editing, researchers can introduce sequence-specific modifications into the genome of different cell types and organisms. The site-specific nucleases (SSNs) have successfully been used in precise gene editing. The SSNs create double-stranded breaks (DSB) in the target DNA. The DSB is repaired through non-homologous end joining (NHEJ) or homolog-directed recombination (HDR) pathways resulting in insertion/deletion (INDELS) and substitution mutations in the target region (s), respectively [150, 151]. The technology produces defined mutant; also, the edited crops typically carry the desired trait [152]. Gene editing has been reported in plants including Arabidopsis [153], rice [154], and other crops, The genome editing techniques include meganucleases, zinc finger nucleases (ZFNs), transcription activator-like effector nucleases (TALENs), clustered regularly interspaced palindromic repeats (CRISPR/Cas9). These techniques have been extensively reviewed [151, 155].

5. Conclusion

Despite the unique attribute of AYB as a seed and tuber producing crop, the crop is underutilized due to identified limitations, including long cooking hours and the abundance of anti-nutrition. Different cooking hours have previously been reported for AYB grains; the lengthiest cooking duration was 24 hours. The cooking hours were observed to be dependent on the cooking methods used, the energy source, and the germplasm considered. The boiling cooking method presented the most prolonged cooking hours (24) while roasting gave rise to the least cooking time of 5 minutes. The diverse cooking methods experimented within AYB effectively reduced the level of anti-nutrient content in the grains. Nevertheless, processing methods such as presoaking and dehulling were observed as the most effective in improving both cooking time and nutritional contents. Fermentation and germination likewise showed positive effects in enhancing the nutrient quality of AYB food products.

Furthermore, the application of recommended equipment like the Mattson bean cooker and texture analyzer could efficiently evaluate cooking time and seed hardness across AYB germplasm. The adequate phenotyping of cooking traits using basic and standard equipment will provide definite baseline information that breeders could use to select parental materials for hybridization and genetic improvement of cooking traits. Additionally, DNA technology which has proven to be effective in providing solutions to complex problems could be exploited through GWAS, QTL mapping, and genome editing for the improvement of AYB's cooking attributes. Conclusively, the present review is targeted at stimulating researchers' interest in developing AYB cultivars with reduced cooking time.

Conflict of interest

The authors declare no conflict of interest.

Author details

Ndenum Suzzy Shitta[1*], Alex Chukwudi Edemodu[2], Wosene Gebreselassie Abtew[1] and Abush Abebe Tesfaye[2]

1 Jimma University, Jimma, Ethiopia

2 International Institute of Tropical Agriculture, Ibadan, Nigeria

*Address all correspondence to: suzzynde@yahoo.com

References

[1] Unicef. 2020. Available from: https://www.unicef.org/reports/state-of-food-security-and-nutrition-2020 [Accessed 2021-06-04]

[2] FOA. 2017. The future of food and agriculture—Trends and challenges. Rome: Food and Agriculture Organization of the United Nations (2017). Available from: http://www.fao/3/i6583e.pdf [Accessed 2021-06-23]

[3] Ray DK, Mueller ND, West PC, Foley JA. Yield trends are insufficient to double global crop production by 2050. PLoS one. 2013;8.e66428. DOI: 10.1371/journal.pone.0066428

[4] Cazzato E, Laudadio V, Stellacci AM, Ceci E, Tufarelli V. Influence of sulphur applicationon protein quality, fatty acid composition and nitrogen fixation of white lupin (*Lupinus albus* L.). European Food Research Technology. 2012;235:963-969. DOI: 10.1007/s00217-012-1817-5

[5] Ntatsi G, Karkanis A, Yfantopoulos D, Olle M, Travlos I, Thanopoulos R, Bilalis D, Bebeli P, Savvas D. Impact of variety and farming practices on growth, yield, weed flora and symbiotic nitrogen fixation in faba bean cultivated for fresh seed production. Acta Agriculturae Scandinavica Section B-Soil and Plant Science. 2018; 38:619-630. DOI: 10.1080/09064710.2018.14522286

[6] Maitra S, Shankar T, Banerjee P. Potential and advantages of maize-legume intercropping system. In: Hossain A, editor. Maize Production and Use. London: IntechOpen; 2020. p. 1-14. DOI: 10.5772/intechopen.91722

[7] Cakir O, Ucarli C, Tarhan C, Pekmez M, Turgut-kara N. Nutritional and health benefits of legumes and their distinctive genomic properties. 2019;39:1-12. Food Science and Technology. DOI: 10.1590/fst.42117

[8] Alemneh AA, Zhou Y, Ryder MH, Denton MD. Mechanism in plant growth-promoting rhizobacter that enhance legume-rhizobial symbioses. Journal of Applied Microbiology. 2020;129:1133-1156. DOI: 10.1111/jam.14754

[9] Shitta NS, Abtew WG, Ndlovu N, Oselebe HO, Edemodu AC, Abebe AT. Morphological characterization and genotypic identity of African yam bean (*Sphenostylis stenocarpa* Hochst ex.A. Rich. Harms) germplasm from diverse ecological zones. Plant Genet. Resources Characterization and Utilization. DOI: 10.1017/s1479262121000095

[10] Amoatey HM, Klu GYP, Bansa D, Kumaga FK, Aboagye LM, Benett-larley SO, Gamedoagbao DK (2000) African yam bean (*Sphenostylis stenocarpa*) A neglected crop in Ghana. West African Journal of Applied Ecology. 2000; 1:53-60. DOI: 10.4314/wajae.v1i1.40570

[11] Saka JO, Adeniyan ON, Akande SR, Balogun MO. An economic evaluation of intercropping African yam bean, Kenaf and Maize in the rain forest zone of Nigeria. Middle East Journal of Scientific Research 2007;2:1-8

[12] Adewale BD, Odoh NC. A review on genetic resources, diversity, and agronomy of African yam bean (*Sphenostylis stenocarpa* (Hochst. ex A. Rich.) Harms): A potential future food crop. Sustainable Agriculture Research. 2013;2:32-43. DOI: 10.5539/sar.v2n1p32

[13] Aremu CO, Ariyo OJ, Adewale BD. Assessment of selection techniques in genotypes × environment interaction in cowpea *Vigna unguiculata* (L) walp. African Journal of Agricultural Research. 2007;2:352-355

[14] Malumba P, Denis BM, Joseph KK, Doran L, Danthine S, Bera F. Structural and physicochemical characterization of

Sphenostylis stenocarpa (Hochst. ex A. Rich.) Harms tuber starch. Food Chemistry. 2016;212:305-312. DOI: 1016/j.foodchem.2016.05.181

[15] Nwokolo E. African yam bean (*Sphenostylis stenocarpa* (Hoechst ex. A. Rich) Harms.) In: Nwokolo E, Smart J, editors. Food and Feed from Legumes and Oilseed, Boston: Springer; 1996. p. 182-188. DOI: 10.1007/978-1-4613-0433-3_18

[16] Afolabi CG, Ogunsanya OM, Lawal OI. Evaluation of some African yam bean (*Sphenostylis stenocarpa* (Hochst. Ex A. Rich) accessions for resistance to flower bud and pod rot diseases. Current Plant Biology. 2019;20:1-5. DOI: 10.1016/j.cpb.2019.100126

[17] Ecocrop.Ecocrop database, FAO.2009. Available from: http://ecocrop.fao.org/ecocrop/srv/en/home. [Accessed: 2020-08-23]

[18] Verdcourt B, Doygaard S. *Sphenostylis* E. Mey. In: Pope GV, Polhill RM, editors. Flora Zambesiaca. London: Kew; 2001.p. 68-73

[19] Nwosu JN. Evaluation of the proximate composition and antinutritional properties of African yam bean (*Sphenostylis sternocarpa*) using malting treatment. International Journal of Basic and Applied Sciences. 2013;2:157-169

[20] Nnamani CV, Ajayi SA, Oselebe HO, Atkinson CJ, Igboabuchi AN, Ezigbo EC. *Sphenostylis stenocarpa* (ex. A. Rich.) Harms a fading genetic resource in a changing climate: prerequisite for conservation and sustainability. Plants. 2017;6:1-16. DOI: 10.3390/plants6030030

[21] Adewale BD, Kehinde OB, Odu BO, Dumet DJ. The potentials of African yam bean (*Sphenostylis stenocarpa* Hochst. ex. A. Rich) Harms in Nigeria: character distribution and genetic diversity. In: Smart J, Haq N Editors. New Crops and Uses: their role in a rapidly changing world. Chichester: RPM; 2008. p. 265-276

[22] Ukom AN, Ndudim S, Nwanagba LN. Effects of fermentation periods on the nutrient quality and sensory acceptability of African yam bean (*Sphenostylis stenocarpa*) porridge. Nigerian Journal of Biotechnology. 2019;36:9-16. DOI: 10.4314/njb.v36i1.2

[23] Adewale BD. Genetic diversity, stability and reproductive biology of African yam bean *Sphenostylis stenocarpa* (Hochst. ex A.Rich) Harms [thesis]. Abeokuta: University of Agriculture; 2011.

[24] Soetan KO, Olaiya CO, Karigidi KO. Comparative invitro antioxidant activities of six accessions of African yam beans (*Sphenostylis stenocarpa*) Annals. Food Science and Technology. 2018;19:455-446

[25] Baiyeri SO, Uguru MI, Ogbonna PE, Samuel-Baiyeri CCA, Okechukwu R, Kumaga FK, Amoatey C. Evaluation of the nutritional composition of the seeds of some selected African yam bean (*Sphenostylis stenocarpa* Hochst ex. A. Rich (Harms) accessions. Journal of Tropical Agriculture, Food, Environment and Extension. 2018;17:37-44. DOI: 10.4314/as.v17i2.5

[26] Adegboyega TT, Abberton MT, Abdelgadir AH, Dianda M, Maziya-Dixon B, Oyatomi OA, Ofodile S, Babalola OO. Evaluation of nutritional and anti-nutritional properties of African yam bean (*Sphenostylis stenocarpa* (Hochst ex.A.Rich.) Harms.) seeds. Journal of Food Quality. 2020;20:1-11. DOI:10.1155/2020/6569420

[27] Abbey BW, Berezi PE. Influence of processing on the digestibility of African yam bean *Sphenostylis stenocarpa*

(Hochst Ex A. Rich) Harms. Nutrition Reports International. 1988;37:819-827

[28] National Research Council (NRC). Lost Crops of Africa: Volume II: Vegetables, Development, Security, and Cooperation. Washington D.C; National Academy of Science. 322-344 p. DOI. 2007;**10**:17226/11763

[29] Ohanmu EO, Ikhajiagbe B, Edegbai BO. Nitrogen distribution pattern of African yam bean (*Sphenostylis stenocarpa*) exposed to cadminum stress. Journal of Applied Sciences and Environmental Management. 2018;22:1053-1057. DOI: 10.4314/jasem.v22i7.10

[30] Dukes JA. Handbook of legumes of world economic importance. New York: Plenum Press; 1981. 1-354 p. DOI: 10.1007/978-1-4684-8151-8

[31] Uguru MI, Madukaife SO. Studies on the variability in agronomic and nutritive characteristics of African yam bean (*Sphenostylis stenocarpa* Hochst. Ex. A. Rich Harms). Plant Production and Research Journal. 2001;6:10-19

[32] Akande SR, Olakojo SA, Ajayi SA, Owolade OF, Adetumbi JA, Adeniyan ON, Ogunbodede BA. 2012. Planting date effects cowpea seed yield and quality at southern guinea savanna of Nigeria. Seed Technology. 2012;34:51-60. Available from: http://jstor.org/stable/23433635 [Accessed: 2021-03-12]

[33] Oagile O, Mmolotsi R, Segwagwe A, Babili TP. African yam bean (*Sphenostylis stenocarpa*) nodulates promiscuously with rhizobium indigenous to soils of Botswana. Journal of Plant Studies. 2012;1:109-113. DOI:10.5539/jps.vIn2p109

[34] Ojuederie OB, Balogun MO. Genetic variation in nutritional properties of African yam bean (*Sphenostylis stenocarpa* Hochst ex. A. Rich. Harms) accessions. Nigerian Journal of Agriculture Food and Environment. 2017;13:180-187

[35] Anya MI, Ozung PO. Proximate, mineral and anti-nutritional composition of raw and processed African yam bean (*Sphenostylis stenocarpa*) seeds in Cross River state, Nigeria. Global Journal of Agricultural Sciences 2019;18:19-29. DOI: 10.4314/gjass.v18i1.3

[36] Konyeme TE, Nyananyo BL, Tanee FBG. Diversity in proximate analysis of tubers of some African yam bean (*Sphenostylis stenocarpa*) (Hochst Ex. A. Rich.) Harms (Fabaceae) accessions. Journal of Applied Sciences and Environmental Management. 2020;24:1787-1793. DOI: 10.4314/jasem.v24i10.12

[37] Onuoha CH, Harry BJ, Eze SO. Evaluation of nutrients and anti-nutritional factors of different species of African Yam Bean (*Sphenostylis stenocarpa*). European Journal of Basic and Applied Sciences. 2017;4:1-8

[38] Ojuederie OB, Ajiboye JA, Babalola OO. Biochemical and histopathological studies of key tissues in healthy male wistar rats fed on African yam bean seed and tuber meals. Journal of Food Qulaity. 2020;20:1-10. DOI: 10.1155/2020/8892618

[39] Potter D. Economic Botany of *Sphenostylis* (Leguminosae). Economic Botany 1992;46: 262-275. DOI: 10.1007/bf02866625

[40] Nnamani CV, Atkinson CJ, Nwite JE. Etymology of folk nomenclatures for *Sphenostylis stenocarpa* (Hochst ex A. Rich) Harms. Preprints. 2019;1-19. DOI: 10.20944/preprints201903.0044.V1.

[41] KIu GYP, Amoatey HM, Bansa D, Kumaga FK. Cultivation and use of African yam bean (*Sphenostylis stenocarpa*) in the Volta Region of

Ghana. Journal of Food Technology in Africa. 2001;6:74-77. DOI: 10.4314/jfta.v6i3.19292

[42] Okigbo BN. Introducing the yam bean (*Sphenostylis sternocarpa*) (Hochst. ex A. Rich) Harms. In: Proceedings of the first IITA Grain Legume Improvement Workshop 29 October-2 November 1973; Nigeria. IITA: Ibadan; 1973. p. 224-238

[43] Ene-Obong HN, Obizoba IC. Effect of domestic processing on the cooking time, nutrients, anti-nutrients and in vitro protein digestibility of the African yam bean (*Sphenostylis stenocarpa*). Plant Foods for Human Nutrition. 1996;49:43-52. DOI: 10.1007/BF01092521

[44] Njoku HO, Ofuya CO. Effect of pretreatment on the cooking time of the African yam bean (*Sphenostylis stenocarpa*) Journal of Food Science. 1989;54:758-759. DOI: 10.1111/j.1365-2621.1989.tb04700.x

[45] Ezueh MI. African yam bean as a crop in Nigeria. World Crops. 1984;36:199-200

[46] Ajibade SR, Balogun MO, Afolabi OO, Ajomale KO, Fasoyiro SB. Genetic variation in nutritive and antinutritive content of African yam bean. Tropical Science 2005;5:144-148. DOI: 10.1002/ts.14

[47] Ndidi US, Ndidi CU, Abbas O, Aliyu M, Francis GB, Oche O. Proximate, antinutrients and mineral composition of raw and processed (Boiled and Roasted) *Sphenostylis stenocarpa* seeds from Southern Kaduna, Northwest Nigeria. International Scholarly Research Notices. 2014;2014:1-9. DOI: 10.1155/2014/280837

[48] Ajibola GO, Olapade AA. Physical, proximate and anti-nutritional composition of African yam bean (*Sphenostylis stenocarpa*) seeds varieties. Journal of Food Research. 2016;5:67-72. DOI: 10.55539/jfr.v5n2p67

[49] Sam SM. Nutrient and anti-nutrient constituents in seeds of *Sphenostylis stenocarpa* (Hochst. Ex. A. Rich.) Harms. African Journal of Plant Science. 2019;13:107-112. DOI: 10.5897/AJS2019.1763

[50] Adewale BD, Vroh-Bi I, Dumet DJ, Nnadi S, Kehinde OB, Ojo DK, Adegbite AE, Franco J. Genetic Diversity in African yam bean accessions based on AFLP markers: towards a platform for germplasm improvement and utilization. Plant Genetic Resources Characterization and Utilization.2015;13:111-118. DOI: 10.1017/S1479262114000707

[51] Ojuederie OB, Balogun MO, Akande SR, Korie S, Omodele T. Intraspecific variability in agro-morphological traits of African yam bean *Sphenostylis stenocarpa* (Hochst ex. A. Rich) Harms. Journal of Crop Science and Biotechnology. 2015;18:53-62. DOI: 10.1007/s12892-014-0109-y

[52] Agbolade JO, Popoola JO, Kioko JI, Adewale BD, Ajiboye AA, Ajewole TO, David OA, Komolafe RJ. Comparative genetic variability and traits heritability in vegetative and floral characters in accessions of two minor legumes. Indian Journal of Agricultural Research. 2018;53:78-183

[53] Aina IA, Ilori CO, Ekanem UO, Oyatomi O, Potter D, Abberton MT. Morphological Characterization and Variability Analysis of African Yam Bean (*Sphenostylis stenocarpa* Hochst. ex. A. Rich) Harms. International Journal of Plant Research. 2020;3:45-52. DOI: 10.5923/j.plant.20201003.01

[54] Wani IA, Sogi DS, Wani AA, Gill BS. Physical and cooking characteristics of some Indian bean (*Phaseolus vulgaris* L.) cultivars. Journal of the Saudi Society of Agricultural

Sciences. 2017;16:7-15. DOI: 10.1016/j.jssas.2014.12.002

[55] Diaz S, Ariza-Suarez D, Ramdeen R, Aparicio J, Arunachalam N, Hernandez C, Diaz H, Ruiz H, Piopho HP, Raatz B. Genetic architecture and genomic prediction of cooking time in common bean (*Phaseolus vulgaris* L.). Frontiers in Plant Science. 2021;11:622213. DOI: 10.3389/fpls.2020.622213

[56] Popoola JO, Adegbite AE, Adewale BD, Odu BO. Morphological intraspecific variabilities in African yam bean (*Sphenostylis stenocarpa* Hochst. Ex. A. Rich) Harms. Scientific Research Essays. 2011;6:507-515

[57] Enwere NJ, Hung YC, Ngoddy PO. Texture and microstructure of African yam bean (*Sphenostylis sternocarpa*) products. Journal of Texture Studies 1990;21:377-394. DOI: 10.1111/j.1745-4603-1990.TB00489.X

[58] Silva HC, Luh BS. Scanning electron microscope studies on starch granules of red kidney beans and bean sprouts. Journal of Food Science. 1978;43:1405-1408. DOI: 10.1111/j.1365-2621.1978.tb02503.x

[59] Sefa-Dedeh S, Stanley DW. Textural implications of the microstructure of legumes. Food Technol. 1979;33:77-83

[60] Avanza MV, Chaves MG, Acevedo BA, Anon MC. Functional properties and microstructure of cowpea cultivated in north-east Argentina. LWT-Food Science and Technology. 2012;49:123-130. DOI: 10.1016/j.lwt.2012.04.015

[61] Mechling LC, Gast DL, Fields EA. Evaluation of a portable DVD player and system of least prompts to self-prompt cooking task completion by young adults with moderate intellectual disabilities. The Journal of Special Education. 2008;42: 179-190. DOI: 10.1177/0022466907313348

[62] McGowan L, Caraher M, Raats M, Lavelle F, Hollywood L, McDowell D, Spence M, McCloat A, Mooney E, Dean M. Domestic cooking and food skills: a review. Critical Reviews in Food Science and Nutrition. 2017;57:2412-2243. DOI: 10.1080/10408398.2015.1072495

[63] Pathare PB, Roskilly AP. Quality and energy evaluation in meat cooking. Food Engineering Reviews. 2016;8:435-447. DOI: 10.1007/s12393-016-9143-5

[64] Gibson R. The role of diet and host related factors in nutrient bioavailability. Food and Nutrition Bulletin. 2007;28:S77-100. 10.1177/15648265070281S108

[65] Siddiq M, Uebersax MA editors. Dry beans and pulses: Production, processing and nutrition. Hoboken: Wiley; 2012. 379 p. DOI: 10.1002/9781118448298

[66] Sabadoti VD, Miano AC, Augusto PED. Automation of Mattson bean cooker: A simple and a low-cost approach. Journal of food processing and preservation. 2020;44:e14769:1-10. DOI: 101111/jfpp.14769

[67] Faye M, Fulton J, Ibro G, Dushwaha S, Lowenberg-DeBoer J. Developing cowpea market opportunities in West Africa. Bean/Cowpea CRSP, Regional project research and training annual technical progress reports. Bean/Cowpea CRSP 2004, pp. 5-13

[68] Yeung H, Ehlers JD, Waniska RD, Alviola JN, Rooney LW. Rapid screening methods to evaluate cowpea cooking characteristics. Field Crops Research. 2009;112:245-252. DOI; 10.1016/j.fcr.2009.03.010

[69] Wood JA. Evaluation of cooking time in pulses: A review. Cereal

Chemistry Journal. 2016;94:32-48. DOI: 10.1094/CCHEM-05-16-0127-FI

[70] Obatolu VA, Fasoyiro SB, Ogunsunmi L. Processing and functional properties of yam bean (*Sphenostylis stenocarpa*). Journal of Food Processing and Preservation. 2007;31:240-249. DOI: 10.1111j.1745-4549.2007.00112.x

[71] Achinewhu, S., Akah, G (2003) Chemical, functional and sensory properties of processed African yam beans (*Sphenostylis stenocarpa*) and cowpeas (*Vigna unguiculata*). Plant Foods for Human Nutrition. 2003; 58:1-6. DOI: 10.1023/B:QUAL. 0000040364.74521.d8

[72] Idowu AO. Nutrient composition and sensory properties of kokoro (a Nigerian snack) made from maize and African yam bean flour blends. International Food Research Journal. 2015;22:739-744

[73] Yusufu MI, Ikya JK, Gwaza TT. Production and quality evaluation of African yam bean cheese and cow milk cheese in different palm oil samples. International Journal of Innovative Food Science and Technology. 2018;1:20-27. DOI: 10.25218/IJIFST.2018.01.001.03

[74] Okoye JI, Obi CD (2017) Chemical composition and sensory properties of wheat-African yam bean composite flour cookies. Discourse Journal of Agriculture and Food Sciences 5:21-27

[75] Nwosu JN, Onuegbu NC, Ogueke CC, Kabuo NO, Omeire GC. Acceptability of moin-moin produced from blends of African yam bean (*Sphenostylis stenocarpa*) and cowpea (*Vigna unguiculata*). International Journal of Current Microbiology and Applied Science. 2014; 5:996-1004

[76] Frank-Peterside, N., Dosumu, DO, Njoku HO. Sensory evaluation and proximate analysis of African yam bean (*Sphenostylis stenocarpa* Harms) moimoi. Journal of Applied Sciences and Environmental Management. 2002;6:43-48. DOI: 10.4314/jasem. v6i2.17175

[77] Okafor GI, Usman GO. Production and evaluation of breakfast cereals from blends of African yam bean (*Sphenostylis stenocarpa*), maize (*Zea mays*) and defatted coconut (*cocus nucifera*). Journal of food processing and preservation. 2013;38: 1037-1043. DOI: 10.1111/jfpp.12060

[78] Obasi NE, Uchechukwu N, Eke-Obia E. Production and evaluation of biscuits from Africa yam bean (*Sphenostylis stenocarpa*) and wheat (*Triticum aestivum*) flours. Food science and quality management. 2012;7:5-12

[79] Eke OS, Akobundu ENT. Functional properties of African yam bean (*Sphenostylis stenocarpa*) seed flour as affected by processing. Food Chemistry. 1993;48:337-340. DOI: 10.1016/0308-8146(93)90314-6

[80] Ene-Obong HN. Nutritional evaluation, consumption pattern and processing of the African yam bean (*Sphenostylis stenocarpa*) [Thesis]. Nsukka: University of Nigeria;1992.

[81] Argel PJ, Paton CJ. Overcoming legume hard seededness. In: Loch DS, Ferguson JE, editors. Forage seed production: Tropical and Subtropical Species. Wallingford, CAB International; 1999. p. 247-265

[82] Sandhu KS, You FM, Conner RL, Balasubramanian PM, Hou A (2018) Genetic analysis and QTL mapping of the seed hardness trait in a black common bean (Phaseolus vulgaris) recombinant inbred line (RIL) population. Molecular Breeding 2018;38:34:1-13. DOI: 10.1007/ s11032-018-0789-y

[83] Medoua GN, Mbone IL, Agbor-Egbe T, Mbofung CMF. Study of

the hard-to-cook property of stored yam tubers (*Dioscorea dumetorum*) and some determining biochemical factors. Food Research International. 2005;38:143-149. DOI: 10.1016/j.foodres.2004.09.005

[84] Shiga TM, Lajolo FM, Filisetti TMCC. Changes in the cell wall polysaccharides during storage and hardening of beans. Food Chemistry. 2004;84:53-64. DOI: 10.1016/s0308-8146(03)00166-3

[85] Galiotu-Panayotou M, NB Kyriakidis, Margaris I. Phytase-phytate-pectin hypothesis and quality of legumes cooked in calcium solutions. Journal of the Science of Food and Agriculture. 2008;88:355-361. DOI: 10.1002/jsfa.2973

[86] Chinedum E, Sanni S, Theressa N, Ebere A. Effect of domestic cooking on the starch digestibility, predicted glycemic indices, polyphenol contents and alpha amylase inhibitory properties of beans (*Phaseolis Vulgaris*) and breadfruit (*Treculia Africana*). International Journal of Biological Macromolecules. 2018;106, 200-206. DOI: 10.1016/j,ijbionmac.2017.08.005

[87] Pevicharova G, Sofkova-Bobcheva S, Zsivanovits G. Sensory and instrumental texture of snap bean (*Phaseolus vulgaris* L.). International Journal of Food Properties. 2015;18:1169-1180. DOI: 10.1080/10942912.2014.891610

[88] Erkan SB, Gurler HN, Bilgin DG, Germec M, Turhan I. Production and characterization of tempehs from different sources of legume by Rhizopus oligosporus. LWT-Food Science and Technology. 2020;119:108880. DOI: 10.1016/j.IWT.2019.108880

[89] Kinyanjui P, Njoroge D,Makokha A, Christiaens S, Sila D, Hendrickx M. Quantifying the effects of postharvest storage and soaking pre-treatments on the cooking quality of common beans (*Phaseolus Vulgaris*). Journal of Food Processing and Preservation. 2017;41;1-10. DOI: 10.1111/jfpp.13036

[90] Wang N, Daun JK. Determination of cooking times of pulses using an automated Mattson cooker apparatus. Journal of the Science of Food and Agriculture. 2005;85:1631-1635. DOI: 10.1002/jsfa.2134

[91] Schock DR, Harrison DL, Anderson LL. Effect of dry and moist heat treatments on selected beef quality factors. Journal of Food Science. 1970;35:195-198. DOI: 10.1111/j.1365-2621.1970.tb12137.x

[92] Yuan GF, Sun B, Yuan J, Wang OM. Effects of different cooking methods on health-promoting compounds of broccoli Journal of Zhejiang University Science. 2009;10:580-588. Doi 10.1631/jzus.B0920051

[93] Preti R, Rapa M, Vinci G (2017) Effect of steaming and boiling on the antioxidant properties and biogenic amines content in green bean (*Phaseolus vulgaris*) varieties of different colours. Journal of Food Quality. 2017;2017. DOI: 10.1155/2017/5329070

[94] Ngadi M, Xue J, Food Frying: Modifying the Functional Properties of Batters. In: Ahmed J, Ramaswamy HA, Kasapis S, Boye JI, editors. Novel Food Processing: Effects on Rheological and Functional Properties. Boca Raton: CRC Press; 2009. p. 437-457

[95] Singh RP. Energy in food processing. Drying Technology. New York: Elsevier; 1989. 839 p. DOI: 10.1080/07373938908916635

[96] Bordin K, Kunitake MT, Aracava KK, Trindade CSF. Changes in food caused by deep fat frying-A review. Archivos Latinoamericanos De Nutricion. 2013;63:5-13

[97] Del Ré PV, Jorge N. Behavior of vegetable oils for frying discontinuous frozen pre-fried products. Food Science and Technology.2006;26: 56-53

[98] Pyler EJ. Baking Science and Technology. 3rd ed. Kansas: Sosland Company; 1988. 1359 p

[99] Eliasson AC, Larsson K. Cereals in Breadmaking: A Molecular Colloidal Approach. Ist ed. CRC; 1993. 392 p.

[100] Therdthai N, Zhou W. Recent advances in the studies of bread baking process and their impacts on the bread baking technology. Food Science Technology Research. 2003;9:219-226. Available from: http://scholarbank,nus.edu.sg/handle/10635/77644 [Accessed: 2021-04-25]

[101] Carvalho B, Patto, M, Vieira I Barbosa A. New strategy for evaluating grain cooking quality of progenies in dry bean breeding programs. Crop Breeding and Applied Biotechnology. 2017;17:115-123. DOI; 10.1590/1984-70332017v17n2a18

[102] Mattson S. The cookability of yellow peas. A colloid-chemical and biochemical study. Acta Agriculturae Suecana. 1946;2:185-231

[103] Jackson GM, Varriano-Marston E. Hard-to-cook phenomenon in beans: effect of accelerated storage on water absorption and cooking time. Journal of Food Science. 1981;46:799-803. DOI: 10.1111/j.1365-2621.1981.tb15351.x

[104] Hsieh HM, Pomeranz Y, Swanson BG. Composition, cooking time, and maturation of azuki (*Vigna angularis*) and common beans (*Phaseolus vulgaris*). Cereal Chemistry. 1992;69:244-248

[105] Proctor J, Watts B. Development of a modified Mattson bean cooker procedure based on sensory panel cookability evaluation. Canadian Institute of Food Science and Technology Journal. 1987;20:9-14. DOI:10.1016/S0315-5463(87)70662-2

[106] De Oliveira JP, Inês, T, De Souza M, Pereira S, Chauca MNC (2015). Cozedor de mattson adaptado monitorado pela Plataforma arduino. In *Congresso Técnico Científico da Engenharia e da Agronomia*, Fortaleza, Brasil.

[107] Szezesniak AS. Texture is a sensory property. Food Quality Preference. 2002;13:215-225. DOI:10.1016/S0950-3293(01)00039-8

[108] IOS (International Organization for Standardization). Sensory Analysis–Vocabulary. 2nd ed. 2008. 107. p

[109] Rolle L, Siret R, Segade SR, Maury C, Gerbi V, Jourjon F. Instrumental texture analysis parameters as markers of Table-grape and wine-grape quality: A Review. American Journal for Enology and Viticulture. 2012;63:11-28. DOI: 10,5344/ajev.2011.11059

[110] Shimelis EA. Nutrient dense dairy product diversification and quality evaluation. East African Journal of Sciences. 2019;13:195-206

[111] Njoku HO, Ofuya CO. Effect of pretreatment on the cooking time of the African yam bean (*Sphenostylis stenocarpa*). Journal of Food Science. 1989;54:758-759. DOI: 10.1111/j.1365-2621.1989.tb04700.x

[112] Onyeike EN, Uzogara SG. Effects of soaking in salt solutions on water absorption, PH and cooking time of African yam bean seeds (*Sphenostylis stenocarpa* hochst ex. A, rich harms). Global Journal of Pure and Applied Sciences. 2000;6:67-73. DOI: 10.4314/gjpas.v6i1.16079

[113] Wood JA, Harden S. A method to estimate the hydration and swelling properties of chickpeas (Cicer arietinum

L.). Journal of Food Science. 2006;71:E190-E195. DOI: 10.1111/j.1750-3841.2006.00009.x

[114] Vasanthakaalam H, Karayire A, Nyagahungu I. Hydration and culinary profile of improved common bean (*Phaseolus vulgaris* L) cultivars. Rwanda Journal of Agricultural Sciences. 2020;2:68-75

[115] Agunbiade SO. Cookability of African yam bean, *Sphenostylis stenocarpa* (Hochst ex. A. Rich) Harms. Nahrung. 1996;40:37-40

[116] Aminigo ER, Metzger LE. Pretreatment of African yam bean (*Sphenostylis stenocarpa*): effect of soaking and blanching on the quality of African yam bean. Plant Foods for Human Nutrition. 2005;60:165-171. DOI: 10.1007/s11130-005-9551-4

[117] Adeye EI, Agesin OO. Dehulling the African yam bean (*Sphenostylis stenocarpa* Hochst. Ex.A, Rich seeds: Any nutritional importance? Note I. Bangladesh Journal of Scientific and Industrial Research. 2007;42:163-174. DOI: 10.3329/bjsir.v42i2.469

[118] Oke MO, Sobowale SS, Ogunlakin GO. Evaluation of the Effect of Processing Methods on the Nutritional and Anti-nutritional Compositions of Two Under-utilized Nigerian Grain Legumes. Pakistan Journal of Biological Sciences. 2013;16:2015-2020. DOI: 10.3923/pjbs.2013.2015.2020

[119] Ihemeje A, Nwanekezi EC, Odimegwu EN, Ekwe CC. Effect of processing methods of toasting, soaking, boiling, sprouting on dietary fiber and anti-nutrient contents of African yam bean and red kidney bean flour. European Journal of Food Science and Technology. 2018;6:40-48

[120] Williams PC, Nakoul H, Singh KB. Relationship between cooking time and some physical characteristics in chickpeas (*Cicer arietinum* L.). Journal of the Science of Food and Agriculture. 1983;34:492-496. DOI: 10.1002/jsfa.2740340510

[121] Obizoba IC, Souzey J. The nutritive value of African yam bean (*Sphenostylis stenocarpa*): nitrogen and mineral utilization. Ecology of food and nutrition. 1989;22:297-305. DOI: 10.1080/03670244.1989.9991078

[122] Abiodun OA, Adeleke RO. Effect of dehulling/soaking on the chemical, functional and anti-nutritional contents of three varieties of beans. Nutrition and Food Science. 2011;41:117-122. DOI: 10.1108/00346651111117382

[123] Rao W, Dcosthale YG. Tannin contents of pulses: Varietal differences and effects of cooking and germination. Journal of the Science of Food and Agriculture.1982;33:1013-1016. DOI: 10.1002/jsfa.2740331012

[124] Singh U. Methods for dehulling of pulses: A critical appraisal. Journal of Food Science Technology. 1995;32:81-93

[125] Adeyeye EI. Amino acid composition of six varieties of dehulled African yam bean (*Sphenostylis stenocarpa*) flour. International Journal of Food Sciences and Nutrition. 1997;48:345-351. DOI: 10.3109/09637489709028581

[126] Li S, Tayie FAK, Young MF, Rocheford T, White WS. Retention of provitamin A carotenoids in high β-carotene maize (*Zea mays*) during traditional African household processing. Journal of Agricultural and Food Chemistry. 2007;55: 10744-10750. 10.1021/jf071815v

[127] Chaves-Lopez C, Serio A, Grande-Tovar CD, Cuervo-Mulet R, Delgado-Ospina J, Paparella A. Traditional fermented foods and beverages from a microbiological and

nutritional perspective: The Colombian Heritage. Comprehensive Reviews in Food Science and Food Safety. 2014;13:1031-1048. DOI: 10.1111/1541-4337.12098

[128] Kahajdova Z, Karovicova J. Fermentation of cereals for specific purpose. Journal of Food and Nutrition Research. 2007;46:51-57

[129] Nkhata SG, Ayua E, Kamau EH, Shingiro JB. Fermentation and germination improve nutritional value of cereals and legumes through activation of endogenous enzymes. Food Science and Nutrition. 2018;6:2446-2458. DOI:10.1002/fsn3.846

[130] Njoku HO, Ofuya CO, Ogbulie JN. Production of tempeh from the African yam bean (*Sphenostylis stenocarpa* Hams). Food Microbiology. 1991;8:209-214

[131] Arisa NU, Ogbuele OC. Production quality assessment and acceptability of African yam bean (*Sphenostylis stenocarpa*) sauce. Journal of food processing and preservation. 2007;31:771-778. DOI: 10.1111/j.1745-4549.2007.00152.x

[132] Amakoromo ER, Innocent-Adiele HC, Njoku HO (2012) Microbiological quality of yogurt-like product from African yam bean. Nature Science. 2012;10:6-9

[133] Wokoma EC, Aziagba GC (2001) Microbiological, physical and nutritive changes occurring during the natural fermentation of African yam bean into dawa-dawa.Global Journal of Pure and Applied Sciences. 2001;7:219-224. DOI: 10.4314/gjpas.v7i2.16233

[134] Nonogaki H, Bassel GW, Bewley JD. Germination—still a mystery. Plant Science. 2010;179:574-581. DOI: 10.1016/j.plantsci.2010.02.010

[135] Zhang X, Zhao J, Bu Y, Xue D, Liu Z, Li X, Guo N, Wang H, Xing H, Qiu L. Genome -wide association studies of soybean seed hardness in the Chinese mini core collection. Plant Molecular Biology Reporter. 2018;36, 605-617. DOI: 10.1007/s11105-018-1102-2

[136] Sharma HC, Crouch JH, Sharma KK, Seetharama N, Hash CT. Applications of biotechnology for crop improvement: prospects and constraints. Plant Science. 2002;163: 381-395. DOI: 10.1016/50168-9452(02)00133-4

[137] Karp A, Edwards KJ, Bruford M, Funk S, Vosman B, Morgante M, Seberg O, Kremer A, Boursot P, Arctander P, Tautz D, Hewitt GM. Molecular technology for biodiversity evaluation: Opportunities and challenges. Nature Biotechnology. 1997;15: 625-628. DOI: 10.1038/nbt0797-625

[138] Fang C, Ma Y, Wu S, Liu Z, Wang Z, Yang R, Hu G, Zhou Z, Yu H, Zhang M, Pan Y, Zhou G, Ren H, Du W, Yan H, Wang Y, Han D, Shen Y, Liu S, Liu T, Zhang J, Qin H, Yuan J, Yuan X, Kong F, Liu B, Li J, Zhang Z, Wang G, Zhu B, Tian Z. Genome-wide association studies dissect the genetic networks underlying agronomical traits in soybean. Genome Biology. 2017;18:1-13. DOI: 10.1186/s13059-017-1289-9

[139] Jiang S, Zhang H, Ni P, Yu S, Dong H, Zhang A, Cao H, Zhang L, Ruan Y, Cui Z. Genome-wide association study dissects the genetic architecture of maize husk tightness. Frontiers in Plant Science. 2020;11:1-12. DOI: 10.3389/fpls.2020.00861

[140] Kimani W, Zhang LM, Wu XY, Hao HQ, Jing HC. Genome-wide association study reveals that different pathways contribute to grain variation in sorghum (*Sorghum bicolor*). BMC

Genomics. 2020; 21:1-19. DOI: 10.1186/s12864-020-6538-8

[141] Volante A, Tondelli A, Desiderio F, Abbruscato P, Barbara M, Biselli C, Cassella L, Singh N, McCouch SR, Tharreau D, Zampieri E, Cattivelli L, Vale G. Genome wide association studies for japonica rice resistance to blast in field and controlled conditions. Rice. 2020;13:1-17. DOI: 10.1186/s12284-020-00431-2

[142] Khan MA, Korban SS. Association mapping in forest trees and fruit crops. Journal of Experimental Botany. 2012;63:4045-4060. DOI: 10.1093/jxb/ers105

[143] O'Connor K, Hayes B, Hardner C, Nock C, Baten A, Alam M, Henry R, Topp B. Genome-wide association studies for yield component traits in a macadamia breeding population. BMC Genomics. 2020;21:1-12. DOI: 10.1186/s12864-020-6575-3

[144] Kumar S, Hash CT, Nepolean T, Satyavathi T, Singh G, Mahendrakar MD, Yadav RS, Srivastava RK. Mapping QTLs controlling flowering time and important agronomic traits in pearl millet. Frontiers in Plant Science. 2017;8:1-13. DOI: 10.3389/fpls.2017.11731

[145] Holland JB. Genetic architecture of complex traits in plants. Current Opinion in Plant Biology. 2007;10:156-161. DO: 1016/j.pbi.2007.01.003

[146] Doerge RW. Mapping and analysis of quantitative trait loci in experimental populations. Nature Reviews Genetics. 2002;3:43-52. DOI: 10.1038/nrg703

[147] Jansen RC, Jannink JL, Beavis WD. Mapping quantitative trait loci in plant breeding populations: use of parental haplotype sharing. Crop Science 2003;43:829-834

[148] Zhang B, Tamura M, Berger-Doyle J, Chen P. Comparison of instrumental methods for measuring seed hardness o food-grade soybean. Journal Texture Studies. 2008;39:28-39. DOI: 10.1111/j.1745-4603.2007.00128.x

[149] Zeng ZB. QTL mapping. In: Maloy S, Hughes K editors. Brenner's Encyclopaedia of Genetics. 2nd ed. Philadelphia: Elsevier; 2001. p. 587-1593

[150] Jinek M, Chylinski K, Fonfara I, Hauer M, Doudna JA, Charpentier E. A programmable dual-RNA–guided DNA endonuclease in adaptive bacterial immunity. Science. 2012;337:816-821. DOI: 101126/science.1225829

[151] Jaganathan D, Ramasamy K, Sellamuthu G, Jayabalan S, Venkataraman CRISPR for crop improvement: An update review. Frontiers in Plant Science. 2018;9:1-7. DOI: 103389/fpls.2018.00985

[152] Malzahn A, Lowder L, Qi Y. Plant genome editing with TALEN and CRISPR. Cell and Bioscience. 2017;7:1-18. DOI: 10.1186/s13578-017-0148-4

[153] Cermak T, Doyle EL, Christian M, Wang L, Zhang Y, Schmidt C, Baller JA, Somia NV, Bogdanove AJ, Voytas DF. Efficient design and assembly of custom TALEN and other TAL-effector based constructs for DNA targeting. Nucleic Acids Research. 2011;39:e82. DOI: 10.1093/nar/gkr218

[154] Li T, Liu B, Spalding MH, Weeks DP, Yang B. High-efficiency TALEN-based gene editing produces disease-resistant rice. Nature Biotechnology. 2012;30:390-392. DOI: 10.1038/nbt.2199

[155] Gaj T, Gersbach CA, Barbas CFIII. ZFN, TALEN, and CRISPR/Cas-based methods for genome engineering. Trends in Biotechnology. 2013;31:397-405. DOI: 10.1016/j.tibtech.2013.04.004